智能系统与技术丛书

Practical Machine Learning
for Data Analysis Using Python

机器学习实践
基于Python进行数据分析

［沙］阿卜杜勒哈密特·苏巴西（Abdulhamit Subasi）著
陆小鹿 何楚 蒲薇榄 译

机械工业出版社
China Machine Press

图书在版编目(CIP)数据

机器学习实践：基于 Python 进行数据分析 /（沙特）阿卜杜勒哈密特·苏巴西（Abdulhamit Subasi）著；陆小鹿，何楚，蒲薇榄译. -- 北京：机械工业出版社，2021.12
（智能系统与技术丛书）

书名原文：Practical Machine Learning for Data Analysis Using Python
ISBN 978-7-111-69818-0

I. ① 机⋯ II. ① 阿⋯ ② 陆⋯ ③ 何⋯ ④ 蒲⋯ III. ① 软件工具 - 程序设计 ② 机器学习
IV. ① TP311.56 ② TP181

中国版本图书馆 CIP 数据核字（2021）第 263464 号

本书版权登记号：图字 01-2021-2293

Practical Machine Learning for Data Analysis Using Python
Abdulhamit Subasi
ISBN: 9780128213797
Copyright © 2020 Elsevier Inc. All rights reserved.
Authorized Chinese translation published by China Machine Press.

《机器学习实践：基于 Python 进行数据分析》（陆小鹿 何 楚 蒲薇榄 译）
ISBN: 9787111698180
Copyright © Elsevier Inc. and China Machine Press. All rights reserved.

No part of this publication may be reproduced or transmitted in any form or by any means, electronic or mechanical, including photocopying, recording, or any information storage and retrieval system, without permission in writing from Elsevier (Singapore) Pte Ltd. Details on how to seek permission, further information about the Elsevier's permissions policies and arrangements with organizations such as the Copyright Clearance Center and the Copyright Licensing Agency, can be found at our website: www.elsevier.com/permissions.

This book and the individual contributions contained in it are protected under copyright by Elsevier Inc. and China Machine Press (other than as may be noted herein).

This edition of *Practical Machine Learning for Data Analysis Using Python* is published by China Machine Press under arrangement with ELSEVIER INC.

This edition is authorized for sale in Chinese mainland (excluding Hong Kong, Macau and Taiwan). Unauthorized export of this edition is a violation of the Copyright Act. Violation of this Law is subject to Civil and Criminal Penalties.

本版由 ELSEVIER INC. 授权机械工业出版社在中国大陆地区（不包括香港、澳门特别行政区及台湾地区）出版发行。

本版仅限在中国大陆地区（不包括香港、澳门特别行政区及台湾地区）出版及标价销售。未经许可之出口，视为违反著作权法，将受民事及刑事法律之制裁。

本书封底贴有 Elsevier 防伪标签，无标签者不得销售。

注意

本书涉及领域的知识和实践标准在不断变化。新的研究和经验拓展我们的理解，因此须对研究方法、专业实践或医疗方法作出调整。从业者和研究人员必须始终依靠自身经验和知识来评估和使用本书中提到的所有信息、方法、化合物或本书中描述的实验。在使用这些信息或方法时，他们应注意自身和他人的安全，包括注意他们负有专业责任的当事人的安全。在法律允许的最大范围内，爱思唯尔、译文的原文作者、原文编辑及原文内容提供者均不对因产品责任、疏忽或其他人身或财产伤害及 / 或损失承担责任，亦不对由于使用或操作文中提到的方法、产品、说明或思想而导致的人身或财产伤害及 / 或损失承担责任。

出版发行：机械工业出版社（北京市西城区百万庄大街 22 号 邮政编码：100037）
责任编辑：王春华 刘 锋   责任校对：殷 虹
印 刷：三河市宏达印刷有限公司   版 次：2022 年 1 月第 1 版第 1 次印刷
开 本：186mm×240mm 1/16   印 张：28.5
书 号：ISBN 978-7-111-69818-0   定 价：139.00 元

客服电话：（010）88361066 88379833 68326294   投稿热线：（010）88379604
华章网站：www.hzbook.com   读者信箱：hzjsj@hzbook.com

版权所有·侵权必究
封底无防伪标均为盗版
本书法律顾问：北京大成律师事务所 韩光 / 邹晓东

DEDICATION
题　　献

首先非常感谢我的父母，他们总是期望我做到最好，让我知道我可以完成任何事情。

致我的妻子 Rahime，感谢她给予的耐心和支持。

致我的孩子 Seyma Nur、Tuba Nur 和 Muhammed Enes，他们永远在我心里，成为我生活的乐趣。

致所有阅读本书的读者，感谢你们对本书的投入。如果有任何意见或建议，请告知我。

Abdulhamit Subasi

THE TRANSLATOR'S WORDS

译 者 序

作为一名 IT 行业的从业者,最近几年越来越感受到机器学习技术在产业上的应用不断深入;而作为一名 IT 产品的用户,也越来越感受到机器学习技术在各种 App 中的应用不断普及,比如内容推荐、语音识别等。可以预见人工智能在未来的应用将更加广泛。

随着算力的提升以及数据规模的增加,机器学习逐渐被应用于许多行业领域。如何将机器学习应用于实践中是本书所关注的话题。有别于其他理论性较强的书籍,本书从实际用例出发,介绍诸如聚类、分类等基本算法,并且每章都有相关的代码示例。同时,本书更强调机器学习应用的完整流程,因此,除机器学习算法之外,还介绍了许多实用的预处理、特征工程以及结果评估的方法。

对于大众来说,机器学习既触手可及又充满神秘;对于专业人士来说,机器学习是热门的领域也是基础的技能。而本书可以很好地兼顾各类读者的需求,不仅非常详尽且全面地介绍机器学习的基础知识,而且通过丰富的 Python 示例代码演示在不同场景中的使用。刚入门的读者可以从基础入手,了解常规的理论知识。具有一定经验的读者或者对某些领域非常感兴趣的读者,可以在相应的章节深挖下去,比如,某些章节对分类算法进行了非常全面的讲解,并提供不同场景的算法选型和实战介绍。此外,本书中的代码大多源自 scikit-learn、Keras 等 Python 机器学习库中的经典示例,采用的数据集也都非常权威。

第 1 章概述机器学习的应用场景及一些关键技术,包括机器学习框架流程(数据收集、预处理、特征提取及缩放等)和基本的模型评估技术。同时,本章还介绍主流的基于 Python 语言的机器学习环境。

第 2 章主要介绍数据的预处理方法,包括基本的特征类型和基础的特征转换方法,例如阈值化、离散化和归一化等,同时介绍通过特征的构造、选择和消除等方式对特征进行降维。

第 3 章介绍机器学习的基本概念以及常用的算法模型,较为全面地囊括了监督学习和无监督学习所涉及的算法。同时，使用 scikit-learn、TensorFlow、Keras 等 Python 机器学习库对算法的实践进行举例说明。通过本章的学习，读者能够对机器学习方法的全貌有所了解。

第 4 章主要介绍分类算法在医疗保健领域的应用，尤其是如何使用分类算法预测各类病症。例如，如何使用分类算法针对 EEG 信号识别是否患有癫痫。

第 5 章主要介绍分类算法在其他领域的应用，例如网络安全和银行安全领域的网络入侵检测、钓鱼网站检测、信用卡欺诈检测等，在图像分类和文本分类领域有手写数字识别、文本挖掘等。

第 6 章介绍回归算法在众多领域的预测应用，包括股价预测、通货膨胀预测、电力负荷预测、风速预测、旅游需求预测、房价预测、单车使用情况预测等。对于每个应用场景，都包含完整的代码示例供读者参考。

第 7 章介绍聚类算法。聚类算法是无监督学习中的一类重要算法，本章从简单的 k 均值算法开始，由浅入深，不仅从理论角度介绍不同的聚类方法，并且辅以示例，使读者能够更好地理解并应用这类算法。另外，本章还介绍聚类算法的评估方法以及如何将聚类算法应用于特征选取中。

本书较为适合机器学习实践者阅读。在本书中，算法模型主要通过语言文字和代码示例进行介绍，而不是通过复杂烦琐的公式。通过各章的示例，我们可以更好地了解特定算法的实际应用场景。同时，由于本书所涉及的算法都较为经典，因此初学者也可以通过阅读本书来更好地了解理论知识在实际场景中的应用。

最后，衷心祝愿各位读者愉快地完成本书的学习，并且能够将机器学习适当地应用于现实生活和日常工作中。

陆小鹿　何楚　蒲薇榄

2021 年 5 月 18 日

PREFACE

前　言

　　机器学习解决方案的飞速发展及其在工业界的广泛应用极大地推动了其从不同领域中观察（学习）数据，从而学习到复杂模型解决实际问题的能力。通常，创造出高效的学习模型并且得到可靠的结果需要付出大量的时间和成本。掌握项目的主要概念一般可以通过以下方式完成：构建可靠的数据流程管道，通过特征提取及选择进行数据分析和可视化，建模。因此，对于一个可靠的机器学习解决方案而言，不仅包括机器学习模型开发框架本身，还包括成功的预处理模块、可视化、系统集成以及健全的运行时部署和维护设定。Python是一种具有多种功能的创新编程语言，其简单的实现和集成、活跃的开发者社区以及不断成长的机器学习生态系统，对于机器学习的发展及广泛应用做出了极大的贡献。

　　技术的不断进步使得智能组织以及数据驱动的企业成为现实。如今，当数据成了重中之重，市场对于机器学习以及数据科学从业人员的需求十分庞大。实际上，我们正面临着数据科学家以及机器学习专家的短缺。从事21世纪最热门的职业毫无疑问要在这个领域有专家级的经验。

　　机器学习技术是一系列计算机算法，包括人工神经网络、k近邻算法、支持向量机、决策树算法以及深度学习。机器学习当前应用于许多领域，尤其是经济学、安全、医疗保健、生物医学以及生物医学工程。本书介绍如何使用机器学习技术来分析这些领域的数据。

　　本书作者有很多使用Python及其机器学习生态系统来解决实际问题的经验。本书旨在提升读者的技能水平，使大家能够创建实际的机器学习解决方案。同时，本书亦是一本构建实际智能系统的问题解决指南，它提供了一个包括原则、过程、实际案例以及代码的系统框架。同时，本书也包括读者在理解和解决不同的机器学习问题时所需的重要技能。

　　对于正在进行机器学习开发的读者而言，本书是一本绝佳的参考，因为书中包含了众多使用Python机器学习环境进行开发的实际案例。本书旨在为读者使用机器学习知识解决不同领域的实际问题打下坚实基础，这些实际问题涉及生物医学信号分析、医疗保健、安全、经济以及金融领域。此外，本书还介绍了一系列机器学习模型，包括回归、分类、聚类以及预测等方向。

本书共七章。第 1 章主要介绍基于机器学习的数据分析。第 2 章概述一些数据预处理技术，例如特征提取、转换、特征选择以及降维。第 3 章概述一些常见的用于预报、预测和分类的机器学习技术，例如朴素贝叶斯、k 近邻、人工神经网络、支持向量机、决策树、随机森林、装袋、提升、堆叠、投票、深度神经网络、循环神经网络和卷积神经网络。第 4 章主要呈现一些医疗保健领域中的分类案例，包括常用于生物医学信号分析和识别的技术，例如心电图（ECG）、脑电图（EEG）和肌电图（EMG）信号处理。此外，第 4 章还会介绍一些医疗数据分类案例，例如人体行为识别，基于微阵列基因表达的癌症、乳腺癌、糖尿病和心脏病检测等。第 5 章主要介绍一些实际应用，包括入侵检测、钓鱼网站检测、垃圾邮件检测、信用评分、信用卡欺诈检测、手写数字识别、图像分类和文本分类。第 6 章主要介绍一些回归技术的案例，例如股市分析、经济变量预测、电力负荷预测、风速预测、旅游需求预测以及房价预测。第 7 章包括一些无监督学习技术的案例（聚类）。

本书主要目的是帮助包括 IT 专业人员、分析师、开发人员、数据科学家和工程师在内的广大读者掌握解决实际问题的能力。此外，本书也可作为数据科学和机器学习领域的研究生教材。同时，本书还能帮助研究人员建立起使用机器学习技术进行数据分析的基础。另外，本书还将帮助包括研究人员、专业人士、学者和一系列学科的研究生在内的广大读者，尤其是那些刚开始寻求在生物医学信号分析、医疗数据分析、金融和经济数据预测以及计算机安全等领域应用机器学习技术的读者。

执行本书所提供的代码示例需要在 macOS、Linux 或 Microsoft Windows 上安装 Python 3.x 或更高版本。本书中的代码示例经常使用 Python 的基本库，例如 SciPy、NumPy、scikit-learn、matplotlib、pandas、OpenCV、TensorFlow 和 Keras。

ACKNOWLEDGEMENTS

致　　谢

首先，我想要感谢出版商 Elsevier 及其团队的专业人员，他们使得本书的写作之旅非常轻松简单，同时也感谢所有幕后工作者为本书取得成功所付出的努力。

感谢 Sara Pianavilla 和 Rafael Trombaco 的大力支持。

感谢 Paul Prasad Chandramohan 耐心地完成了本书出版所需要的一切工作。

Abdulhamit Subasi

目 录

译者序
前言
致谢

第1章　简介 ·· 1
 1.1　什么是机器学习 ···························· 1
  1.1.1　为什么需要使用机器学习 ········ 2
  1.1.2　做出数据驱动决策 ················· 3
  1.1.3　定义以及关键术语 ················· 4
  1.1.4　机器学习的关键任务 ·············· 6
  1.1.5　机器学习技术 ······················· 6
 1.2　机器学习框架 ································ 6
  1.2.1　数据收集 ······························ 7
  1.2.2　数据描述 ······························ 7
  1.2.3　探索性数据分析 ···················· 7
  1.2.4　数据质量分析 ······················· 8
  1.2.5　数据准备 ······························ 8
  1.2.6　数据集成 ······························ 8
  1.2.7　数据整理 ······························ 8
  1.2.8　特征缩放和特征提取 ·············· 9
  1.2.9　特征选择及降维 ···················· 9
  1.2.10　建模 ··································· 9
  1.2.11　选择建模技术 ······················ 9
  1.2.12　构建模型 ···························· 10
  1.2.13　模型评估及调优 ·················· 10
  1.2.14　实现以及检验已经创建的模型 ··································· 10
  1.2.15　监督学习框架 ······················ 11
  1.2.16　无监督学习框架 ·················· 11
 1.3　性能评估 ······································ 12
  1.3.1　混淆矩阵 ····························· 13
  1.3.2　F值分析 ······························ 14
  1.3.3　ROC分析 ···························· 15
  1.3.4　Kappa统计量 ······················· 15
  1.3.5　度量了什么 ·························· 16
  1.3.6　如何度量 ····························· 17
  1.3.7　如何解释估计 ······················· 17
  1.3.8　scikit-learn中的k折交叉验证 ··································· 18
  1.3.9　如何选择正确的算法 ············· 18
 1.4　Python机器学习环境 ······················ 18

			1.4.1	缺陷 ········ 20
			1.4.2	缺点 ········ 20
			1.4.3	NumPy 库 ······ 20
			1.4.4	Pandas ······· 20
	1.5	本章小结 ············ 21		
	1.6	参考文献 ············ 22		

第 2 章 数据预处理 ········ 23

	2.1	简介 ··············· 23		
	2.2	特征提取和转换 ········ 24		
			2.2.1	特征类型 ······ 24
			2.2.2	统计特征 ······ 25
			2.2.3	结构化特征 ····· 27
			2.2.4	特征转换 ······ 28
			2.2.5	阈值化和离散化 ··· 28
			2.2.6	数据操作 ······ 28
			2.2.7	标准化 ········ 29
			2.2.8	归一化和校准 ···· 33
			2.2.9	不完整的特征 ···· 34
			2.2.10	特征提取的方法 ·· 36
			2.2.11	使用小波变换进行特征提取 ······· 38
	2.3	降维 ··············· 45		
			2.3.1	特征构造和选择 ··· 47
			2.3.2	单变量特征选择 ··· 48
			2.3.3	递归式特征消除 ··· 51
			2.3.4	从模型选择特征 ··· 52
			2.3.5	主成分分析 ····· 53
			2.3.6	增量 PCA ······ 57
			2.3.7	核 PCA ······· 58
			2.3.8	邻近成分分析 ···· 59

			2.3.9	独立成分分析 ···· 61
			2.3.10	线性判别分析 ··· 65
			2.3.11	熵 ·········· 67
	2.4	基于聚类的特征提取和降维 ···· 68		
	2.5	参考文献 ············ 75		

第 3 章 机器学习技术 ······ 77

	3.1	简介 ··············· 77		
	3.2	什么是机器学习 ········ 78		
			3.2.1	理解机器学习 ···· 78
			3.2.2	如何让机器学习 ··· 78
			3.2.3	多学科领域 ····· 79
			3.2.4	机器学习问题 ···· 80
			3.2.5	机器学习的目标 ··· 80
			3.2.6	机器学习的挑战 ··· 81
	3.3	Python 库 ··········· 81		
			3.3.1	scikit-learn ···· 81
			3.3.2	TensorFlow ···· 83
			3.3.3	Keras ········ 84
			3.3.4	使用 Keras 构建模型 · 84
			3.3.5	自然语言工具包 ··· 85
	3.4	学习场景 ············ 87		
	3.5	监督学习算法 ········· 88		
			3.5.1	分类 ········· 89
			3.5.2	预报、预测和回归 ·· 90
			3.5.3	线性模型 ······ 90
			3.5.4	感知机 ········ 98
			3.5.5	逻辑回归 ······ 100
			3.5.6	线性判别分析 ··· 102
			3.5.7	人工神经网络 ··· 105
			3.5.8	k 近邻 ······· 109

3.5.9 支持向量机 ………………………… 113
3.5.10 决策树分类器 ……………………… 118
3.5.11 朴素贝叶斯 ………………………… 123
3.5.12 集成学习 …………………………… 126
3.5.13 bagging 算法 ……………………… 127
3.5.14 随机森林 …………………………… 131
3.5.15 boosting 算法 ……………………… 136
3.5.16 其他集成方法 ……………………… 146
3.5.17 深度学习 …………………………… 151
3.5.18 深度神经网络 ……………………… 152
3.5.19 循环神经网络 ……………………… 155
3.5.20 自编码器 …………………………… 157
3.5.21 长短期记忆网络 …………………… 157
3.5.22 卷积神经网络 ……………………… 160
3.6 无监督学习 ………………………………… 162
3.6.1 k 均值算法 ………………………… 163
3.6.2 轮廓系数 …………………………… 165
3.6.3 异常检测 …………………………… 167
3.6.4 关联规则挖掘 ……………………… 170
3.7 强化学习 …………………………………… 170
3.8 基于实例的学习 …………………………… 171
3.9 本章小结 …………………………………… 171
3.10 参考文献 ………………………………… 172

第 4 章 医疗保健分类示例 …………… 174
4.1 简介 ………………………………………… 174
4.2 脑电图信号分析 …………………………… 175
 4.2.1 癫痫症的预测和检测 ……………… 176
 4.2.2 情绪识别 …………………………… 194
 4.2.3 局灶性和非局灶性癫痫 EEG
   信号的分类 ………………………… 201
 4.2.4 偏头痛检测 ………………………… 212
4.3 EMG 信号分析 …………………………… 217
 4.3.1 神经肌肉疾病的诊断 ……………… 218
 4.3.2 假体控制中的 EMG 信号 ………… 225
 4.3.3 康复机器人中的 EMG
   信号 ………………………………… 232
4.4 心电图信号分析 …………………………… 238
4.5 人类活动识别 ……………………………… 247
 4.5.1 基于传感器的人类活动
   识别 ………………………………… 248
 4.5.2 基于智能手机的人类活动
   识别 ………………………………… 250
4.6 用于癌症检测的微阵列基因表达数据
 分类 ………………………………………… 256
4.7 乳腺癌检测 ………………………………… 257
4.8 预测胎儿风险的心电图数据分类 …… 260
4.9 糖尿病检测 ………………………………… 263
4.10 心脏病检测 ……………………………… 267
4.11 慢性肾脏病的诊断 ……………………… 270
4.12 本章小结 ………………………………… 273
4.13 参考文献 ………………………………… 273

第 5 章 其他分类示例 …………………… 277
5.1 入侵检测 …………………………………… 277
5.2 钓鱼网站检测 ……………………………… 280
5.3 垃圾邮件检测 ……………………………… 283
5.4 信用评分 …………………………………… 287
5.5 信用卡欺诈检测 …………………………… 290
5.6 使用 CNN 进行手写数字识别 ………… 297

5.7 使用 CNN 进行 Fashion-MNIST 图像分类 ………… 306
5.8 使用 CNN 进行 CIFAR 图像分类 ……… 313
5.9 文本分类 …………………… 321
5.10 本章小结 …………………… 334
5.11 参考文献 …………………… 334

第 6 章 回归示例 ……………… 337
6.1 简介 ………………………… 337
6.2 股票市场价格指数收益预测 ………… 338
6.3 通货膨胀预测 ……………… 356
6.4 电力负荷预测 ……………… 358
6.5 风速预测 …………………… 365
6.6 旅游需求预测 ……………… 370
6.7 房价预测 …………………… 380
6.8 单车使用情况预测 ………… 395
6.9 本章小结 …………………… 399
6.10 参考文献 …………………… 400

第 7 章 聚类示例 ……………… 402
7.1 简介 ………………………… 402
7.2 聚类 ………………………… 403
  7.2.1 评估聚类输出 …………… 404

7.2.2 聚类分析的应用 …………… 404
7.2.3 可能的聚类数 …………… 405
7.2.4 聚类算法种类 …………… 405
7.3 k 均值聚类算法 …………… 406
7.4 k 中心点聚类算法 ………… 408
7.5 层次聚类 …………………… 409
  7.5.1 聚集聚类算法 …………… 409
  7.5.2 分裂聚类算法 …………… 412
7.6 模糊 c 均值聚类算法 ……… 416
7.7 基于密度的聚类算法 ……… 418
  7.7.1 DBSCAN 算法 …………… 418
  7.7.2 OPTICS 聚类算法 ………… 420
7.8 基于期望最大化的混合高斯模型聚类算法 …………………… 423
7.9 贝叶斯聚类 ………………… 426
7.10 轮廓分析 …………………… 428
7.11 基于聚类的图像分割 ……… 430
7.12 基于聚类的特征提取 ……… 433
7.13 基于聚类的分类 …………… 439
7.14 本章小结 …………………… 442
7.15 参考文献 …………………… 442

CHAPTER 1

第 1 章

简　　介

1.1　什么是机器学习

随着计算机的计算能力和存储性能的提高，我们的时代已成为"信息时代"或"数据时代"。此外，我们必须分析大数据并利用人工智能、数据科学、数据挖掘和机器学习的概念和技术创建智能系统。当然，大多数人已经学会了这些术语并意识到"数据就是新的石油"。在过去的十年中，组织和企业最重要的任务就是理解和分析数据，并且利用这些信息做出更好的决策。实际上，技术的飞速发展为诸如机器学习、人工智能和深度学习等领域创造了良好的环境。研究人员、工程师和数据科学家创建了框架、工具、技术、算法以及方法论以实现可以用于自动化任务、检测异常、实现复杂分析以及预测事件的智能系统和模型（Sarkar，Bali，&Sharma，2018）。

机器学习被定义为利用经验提高预测性能或准确率的计算技术。其中，经验表示以前的信息，学习器可以从以往的电子数据记录和可供调查使用的数据中获得这些信息。这些数据可能是数字化的由人标注的训练集，或是在与生态系统交互中所获得的其他形式的数据。无论在何种情况下，数据大小以及质量都是预测器能否完成预测任务的关键。机器学习包括创建能够准确进行预测的算法。正如计算机科学的其他领域一样，衡量这些方法好坏的主要标准是它们的时间和空间复杂度。尽管如此，在机器学习中，需要衡量样本复杂度，因为它可以评估算法学习一组概念需要的样本大小。通常，机器学习的理论保证了模型的复杂度以及训练需要的时间在可接受范围内。由于机器学习技术的性能基于所使用的数据及特征，因此其特点是与统计和数据分析有着不可分割的联系。通常情况下，机器学习算法是数据驱动的，它兼具计算机科学、概率论、统计学以及优化算法的重要概念。此外，这些类型的应用涉及更广泛的机器学习问题的种类。机器学习问题的主要类别有分类（classification）、回归（regression）、排序（ranking）、聚类（clustering）以及降维（dimension reduction)(Mohri, Rostamizadeh, & Talwalkar, 2018)。

在分类问题中,每个对象都会对应一个类别。类别的数量可多可少,取决于具体的问题。在回归中,每个对象都会有一个实数值对应。常见的回归问题包括股票价格预测以及经济变量预测。在回归问题中,我们使用实际值与预测值的差值来衡量预测的好坏,然而在分类问题中,我们无法使用这种方法进行衡量,因为不同类别之间没有差值的概念。在排序问题中,一组对象需要根据某种衡量方式进行排序。在聚类问题中,则需要将相似的对象放到同一个分区内。聚类方法通常用于分析大数据集。例如,在社交网络数据分析中,我们通常使用聚类算法来进行社区识别。流形学习(manifold learning)或降维是用来将对象的初始表示转换到一个低维空间表示的方法,这种转换通常保留了原有数据的重要特征。机器学习的主要目的是对未知数据进行准确预测,设计稳定、有效的预测算法,这些算法甚至可以应用于大型问题(Mohri et al., 2018)。

机器学习需要使用正确的特征来创建能够完成当前任务的准确模型。实际上,特征定义为领域中相关的对象。任务是对于关于这些领域对象的实际问题抽象。这里的一般形式是将其分类为两个或更多个类别。大多数这些任务可以定义为将数据点映射到输出空间的问题。这一映射(或模型)本身就是通过使用训练数据集所构建的机器学习方法的输出(Flach, 2012)。我们将会在本书中讨论一些具体的可以使用机器学习解决的问题。无论是何种机器学习模型,它们都是基于某些特征被设计出来,用于解决部分任务的。

大多数情况下,我们期望从数据中得到的知识和洞察无法直接从原始数据中观察得到。机器学习将数据转变成信息。机器学习属于统计学、工程以及计算机科学的交叉领域,并且也常见于其他领域。它可以被广泛应用于诸多领域,例如金融、经济、政治、地理科学以及医学。同时,它也是解决不同问题的工具。任何需要与数据打交道的领域都可以从使用机器学习中获益。实际上,很多问题的解决方法都是不确定的。因此,对于这些问题,我们需要统计学(Harrington, 2012)。

本书使用基于案例的方法来介绍机器学习的不同实践、概念以及相关问题。主要思想是向读者传授"如何解决机器学习问题"以及"如何在数据分析中使用机器学习的主要构建块"的知识。我们希望读者在读完本书后能够了解如何使用机器学习来进行数据分析。

1.1.1　为什么需要使用机器学习

人类是世界上最聪明的生物。他们可以定义、创造、评估以及解决复杂问题。人类的大脑迄今尚未被完全理解,这也是为什么人工智能在很多方面还未能超越人类的原因。基于你目前所知,尽管传统的编程模型已经很好,并且领域知识以及人类智能在进行数据驱动的决策方面至关重要,然而若想能够更加快速准确地做出决策,机器学习必不可少。机器学习技术考虑数据以及预期的结果,并且通过计算机来创建程序,这个程序可以被认为是一个模型。模型可以应用于未来新的数据中产业预期的输出,做出需要的决策。机器将会尝试利用输入数据以及预期的结果来从数据中学习特定的模式。最终创建一个类似于计算机程序的模型，在将来新的数据上做出数据驱动的决策（分类或预测）。我们可以通过一个实际例子来帮助理解这一概念，例如为一家决策公司处理基础设施。若想使用机器学习来解决问题，我们需要实现以下步骤：

- 使用设备数据和日志来从数据仓库中获取足够的历史数据。
- 确定关键的数据属性，这些属性将会对建模至关重要。
- 长时间监控和记录设备属性及其行为，其中包括正常和异常的设备行为或离群值。
- 使用这些收集到的数据以及某种机器学习方法，来构建一个模型，学习其特征设计模式，并且能够有持续的输出。
- 将模型应用于新的数据上，看模型是否能够判断一个设备是否能够正常运行或是否能产生预期输出。基于设备数据，我们可以创建一个框架来包括新开发的机器学习模型。基于开发的框架，我们不仅可以持续监控这些设备，甚至可以预测设备异常，预先对设备进行维修。

实际上，开发一个机器学习模型的步骤远远比上述复杂。我们简化描述的主要目的是让你关注于理论而非具体的技术细节。同时，我们希望你能够从传统的思维方式转向更加数据驱动的思维方式。机器学习的神奇之处就在于其不会被某个领域所局限，而是可以广泛应用于各种不同的领域、行业以及业务。同样，数据标签也并非构建模型所必须。某些时候，我们可以使用无监督学习来构建模型（Sarkar et al., 2018）。

1.1.2 做出数据驱动决策

从数据中提取关键的洞察和信息是公司和企业深度投资人工智能和机器学习的主要目的。数据驱动决策并不是一个创新的概念，它早已被用于统计学、信息管理系统以及运筹学领域中数十年，用来提高决策效果。显然，讨论远比实现简单，因为我们可以清楚地使用数据来做出任何明智的决定。另外的问题在于，我们通常利用直觉或推理的力量基于过去的经验进行决策。我们的大脑十分强大，它可以帮助我们识别图像中的人物、理解朋友或同事所说的意思以及决定是否要接受一个商业交易等。大脑承担着大多数思考的职责。这正是机器难以学习和解决计算退税额和贷款利息等问题的原因。针对这些问题，我们通常使用数据驱动的机器学习技术等不同的方法来提高决策准度。尽管数据驱动的决策十分重要，它需要大规模、高效率地实施。使用人工智能或者机器学习的主要目的是自动化一些能够从数据中学习到固定模式的任务（Sarkar et al., 2018）。

如今，发达国家的劳动力已经由手工型转为知识型。尽管现在还有很大的不确定性，我们经常听到诸如"风险最小化"、"利润最大化"和"找到最佳的市场策略"等词汇。互联网使得知识更易获取，导致了我们对于员工的知识水平要求更高。从数据中获取智慧成了一个更加重要的才能。由于如此多的经济活动依赖于数据，我们无法承受数据损失带来的后果。机器学习能够帮助分析数据并从中提取有价值的信息（Harrington, 2012）。

1.1.3 定义以及关键术语

通常，我们习惯于先计算然后再找出重要的部分。需要评估的项目通常称为构成实例的特征或属性（Harrington，2012）。机器学习中的一个重要步骤就是特征提取。因此，由多个点组成的待处理数据，其特点和有效特征可以通过采用不同的特征提取技术来获得。这些信息以及特征参数描述了数据行为，具有十分重要的作用。信息度高且有特点的特征可以更好地描述数据。这些特征可以通过使用多元化的特征提取算法进行抽取。通常，特征提取是数据分析中的一个步骤，并且可以为后续分类算法的成功打下基础（Graimann，Allison，& Pfurtscheller，2009）。处理少量但能较好表现数据特征的样本在提高算法的性能上具有重要作用。特征提取通常是将数据转化成特征向量。数据分类结构通常根据具有独特特点的特征向量来决定数据的类别（Subasi，2019a）。

大多数情况下，抽取的特征不足以完全解释数据，尤其是采用次优或冗余的特征集时。相较于寻找更好的特征，我们更应该假设给定系统的输入和输出之间存在非线性关系。例如，疾病诊断系统通常使用处理后的生物信号波，目的是学习疾病与系统输入的数据之间的关系，从而使其能够通过输入的信息寻找相匹配的疾病。这种类型的任务可以轻松被机器学习技术完成（Begg，Lai，& Palaniswami，2008；Subasi，2019b）。

常见的智能计算技术有监督学习、无监督学习、强化学习以及深度学习。在这些技术中，研究最多的是基于函数估计的监督学习。外部监督者会将标记过的数据样本作为输入提供给监督学习公式，从而让系统识别样本以及预期的输出之间的潜在联系。经过训练的模型将能够轻松地预测未知样本的输出。强化学习、随机学习以及风险最小化是一些常见的监督学习方法（Begg et al.，2008；Subasi，2019b）。

分类是机器学习的常见任务之一。例如，假设我们想要区分正常EEG信号和癫痫症EEG信号，必须使用脑电图设备然后雇佣神经科医生（EEG专家）来分析从受试者身上获取的EEG信号。这样的做法十分昂贵并且笨拙，更重要的是，神经科医生每次只能出现在一个地方。我们可以通过使用计算机识别癫痫症脑电图来自动化这一过程。分辨是否是癫痫症EEG是一个分类问题。解决分类问题的机器学习算法有很多。这一问题中包含两个类别，即癫痫和正常。当我们决定好需要使用的分类算法后，下一步就是进行模型训练。模型训练需要高质量的训练集。训练集中包含许多模型训练所需的样本。每个训练实例都具有许多特征以及一个目标变量（类别）。目标变量就是机器学习技术所需要预测的输出值。在分类问题中，目标变量是一个标称值；而在回归任务中，目标变量可以是一个连续值。目标变量（类别）在训练集中是已知的。机器通过发现输入特征以及目标变量之间的关系进行学习。由于目标变量是类别，所以可以简化为标称值。我们通常假设分类问题中类别的总数是有限的。测试机器学习算法需要使用部分训练集（隔开的数据集），通常称为测试集。首先，在训练模型的过程中将训练集的样本作为程序的输入，然后将测试集的样本作为输入。注意，测试集的目标变量不会输入程序中，并且程序需要预测每个样本所属分类。程序预测的输出和实际的目标变量将会进行比较，从而评估模型的性能（Harrington，2012）。

我们将使用癫痫发作检测的例子来介绍一些基本的定义以及演示如何将机器学习应用于实际中。癫痫发作检测涉及自动化分类癫痫症和正常 EEG 信号。

样例：用于学习或评估的数据项或数据实例。在癫痫发作检测问题中，这些样例是用于训练和测试的 EEG 信号。

特征：与样例有关的属性集，通常使用向量形式表示。在 EEG 信号的例子中，相关的特征可以是平均值、标准差、平均功率、偏度和峰度等。

标签：样例的值或者类别。在分类问题中，每个样例都有一个特定的类别。例如，二元分类中的"正常"和"癫痫"类别。在回归问题中，标签通常是实值。

训练样本：用于训练机器学习算法的样例。在 EEG 问题中，训练样本包含 EEG 信号集合以及其对应的标签。不同的机器学习场景有不同的训练样本。

验证样本：在处理有标签的数据时，用于机器学习算法调参的样例。机器学习算法通常会有一个或多个可调参数，验证样本用于为这些模型参数寻找合适的值。

测试样本：用于评估算法性能的样本。测试样本是训练集和验证集的一部分，但是不可用于模型的训练阶段。在癫痫预测的例子中，测试样本包含 EEG 信号记录，而算法需要根据这些 EEG 信号的特征做出预测。随后，我们将会对这些预测与测试样本的实际标签进行对比，从而对模型性能做出评估。

损失函数：计算预测的标签以及实际标签之间的差或是损失的函数。

现在，我们已经对"判断癫痫"这个问题的训练过程有了一些了解。我们可以从一些有标签的数据实例开始着手。首先，这些数据应该被随机分为训练集、验证集和测试集。每个部分数据的多少由很多条件决定。例如，验证集的大小取决于算法有多少可调参数。此外，由于有标签的数据通常比较少并且模型的性能直接受到训练集的影响，训练集一般远远大于测试集。接着，需要从数据中提取相关特征——这是机器学习中至关重要的一步。有效且信息量大的特征能够成功指引机器学习模型，而信息不全或信息匮乏的特征将会对模型的性能造成不好的影响。特征选取十分重要，这个任务是开发者需要注意的。特征的选取通常能够反映开发者对于当前任务的先验知识，通常在实际开发中，这些知识会对学习器的性能造成很大的影响。现在，选定的特征将会通过参与模型调参的过程用于模型训练。对于每一个可调参数，学习算法将会选择一个值，而我们将会使用验证集上效果最好的一组参数值作为最终模型的参数。最终，我们将会使用选中的模型在测试集上进行预测。分类器性能可以根据问题相关的损失函数，通过比较预测结果与实际标签值差得出。可见，我们是在测试集上评估算法的性能。机器学习算法可能是一致性的，即，算法在训练集上表现良好，但在测试集上表现很差。这种情况通常出现在一个一致性的学习器上，算法专注于优化小规模的训练集，然而由于问题的复杂度较高，使得算法泛化性能较差。我们再次强调记忆和泛化的关键区别，它们是算法是否精确的基本属性（Mohri et al.，2018）。

1.1.4 机器学习的关键任务

癫痫发作检测是一个二元分类问题，可以算是机器学习中最常见的问题之一。显然，我们可以推广到两个类别以上的分类问题。例如，分离不同种类的 EEG 信号。这个问题可以看成两个二元分类问题的组合：第一个问题是区分正常和异常的 EEG，第二个问题是在异常的 EEG 中区别癫痫或其他类型疾病的信号。然而，在这种转化方式下，我们可能会损失一部分有价值的信息。因为一些异常的 EEG 信号可能会被视作癫痫信号而非其他疾病信号。因此，多类分类问题常被单独视作一个任务。

分类问题的主要任务是预测一个数据样本属于哪一类。另一类机器学习任务，回归，则是预测数值。回归和分类都属于监督学习。这类问题被分为监督学习是因为我们会告诉算法预测什么。与其对应的无监督学习则是让算法在没有目标预测值或是标签的数据上学习。将相似的数据项分为一组的任务称为聚类。另外，在无监督学习中，我们需要用统计数据来描述当前数据，这一过程定义为密度估计。无监督学习的另一项任务是将数据的特征减少，使其能够在二维或是三维中进行可视化（Harrington，2012）。关联规则是市场应用中十分流行的一种模式，其结果常应用于在线购物网站（Flach，2012）。

1.1.5 机器学习技术

我们可以在数据上应用许多机器学习的算法、技术和方法创建模型来解决现实生活中的实际问题。当然，相同的机器学习方法可以根据不同的条件分在不同的类别之下（Sarkar et al.，2018）。机器学习方法的一些主要领域如下。

1）根据训练过程中所需要的人为监督的多少：
- 监督学习
- 无监督学习
- 半监督学习
- 强化学习

2）根据从增量数据样本中学习的能力：
- 批量学习
- 在线学习

3）根据从数据样本中泛化的方式：
- 基于实例的学习
- 基于模型的学习

1.2 机器学习框架

解决实际生活中的机器学习或数据分析问题的最好方法就是使用一个机器学习框架，从数据收集以及转化为有价值的信息开始，然后应用机器学习技术。机器学习框架主要由数据检索与抽取、数据准备、建模、评估以及部署这几个部分组成（Harrington，2012；

Sarkar et al.，2018）。

图 1.1 展示了一个典型的机器学习框架及其主要步骤。我们将会在以下几节详细介绍使用机器学习框架的重要过程。

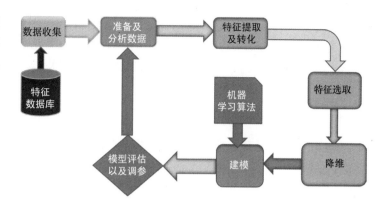

图 1.1 一个典型的机器学习框架

1.2.1 数据收集

这个任务主要是提取、记录以及收集分析所需要的数据。通常，其中涉及使用存有历史数据的数据仓库、数据集市、数据湖等。同时需要根据组织内部现有的数据进行评估，以确定是否需要额外的数据。如果需要，我们可以通过网站、RSS 订阅或者 API 进行数据获取。这些数据可以来自网络（例如开源数据），或者是其他渠道所得，例如调查表、购买记录、试验记录以及仿真。甚至，你可以从设备中获取风速信息，或者是血糖水平，或者是任何可以衡量的东西。获取数据的方式和渠道有很多种。为了节省时间和成本，你可以优先考虑公开的数据集。这个阶段包括了从不同数据源中收集、提取以及获取数据（Harrington，2012；Sarkar et al.，2018）。

1.2.2 数据描述

数据描述包括对数据进行基本分析来更多地了解数据、数据源、属性、数据量以及关系。在进行数据描述时，以下几个方面至关重要：

- 数据源（SQL、NoSQL、大数据）、原始记录（ROO）、参考记录（ROR）
- 数据量（大小、数据记录数量、数据库总数、表总数）
- 数据属性以及它们的描述（变量、数据类型）
- 关系以及映射方案（理解属性表示）
- 基本的描述性统计（均值、中位数、方差）
- 重点关注与实际应用相关的重要属性

1.2.3 探索性数据分析

探索性数据分析是生命周期中第一个主要步骤。该步骤的主要目的是对数据细节有较好的了解。通常用描述性统计数据、图表、图形以及可视化技术来帮助我们寻找数据属性中的关系以及关联性。收集数据之后，我们需要保证数据的格式是可用的。一些算法对于输入的特征有特殊的格式需求，一些算法对于目标变量和特征的数据类型有要求，如字符串、整数等。在这个阶段，数据已经经历了预处理、清洗以及更改。数据预处理、数据清洗、数据整理以及初步的探索性数据分析也是在这个阶段完成的。下列是本阶段所包含的一些主要任务（Harrington，2012；Sarkar et al.，2018）：
- 探索、描述以及可视化数据属性
- 根据对问题的重要性，从所有数据中选择合适的子集和属性
- 广泛地对数据进行评估，以发现数据属性的潜在关系和关联，用来检验假设
- 如果有缺失的数据，需要进行记录

1.2.4 数据质量分析

数据质量分析是理解数据的最后一步，在这个阶段中，对数据集的数据质量进行分析，以发现数据的潜在不足、错误以及问题。这些问题都需要在进一步分析数据或正式建模之前得到解决。数据分析可以是简单地将前一步的输出粘贴到编辑器中去确认前一步操作是否正确。可以通过这样的检查来确定是否有明显的模式，或是某些数据是否与其他数据有极大不同。将数据绘制在不同的维度也会有所帮助。数据质量分析的重点包括以下几个方面（Harrington，2012；Sarkar et al.，2018）：
- 缺失值
- 不一致的值
- 由于数据错误（人为或自动）导致的信息错误
- 错误的元数据信息

1.2.5 数据准备

数据准备是对于问题和有关数据集有所了解之后的另一个步骤。数据准备包含一系列任务，包括数据清洗、整理以及在进行建模和训练模型前准备数据。记住，数据准备通常是数据挖掘中花费时间最多的阶段。然而，这个阶段的步骤需要十分仔细地执行，因为数据质量的好坏将会直接影响模型的最终性能——使用质量差的数据将会得到性能差的模型（Sarkar et al.，2018）。

1.2.6 数据集成

当有很多数据源或者数据集的时候，我们需要进行数据集成。数据集成可以由两种方法实现。当数据集有相同属性时，可以考虑直接将新的数据集加在之前的数据集之后。对于属性或列不同的数据集之间的集成，通常采用通过相似的键等常用字段来进行合并的方法（Sarkar et al.，2018）。

1.2.7 数据整理

数据整理的过程包括了数据处理、归一化、清洗以及数据格式转化。通常，原始数据很难直接为机器学习算法所利用。因此，我们需要根据数据的形式进行处理，修正潜在的错误及不一致性，并且最终转化成为机器学习算法接受的格式。数据整理中主要的工作如下（Sarkar et al.，2018）：

- 管理缺失值（删除整行，补充缺失值）
- 处理数据不一致（删除整行、删除属性、修正不一致）
- 纠正不适当的元数据和注释
- 管理不明确的属性值
- 将数据存储成适当的格式（CSV、JSON、关系型）

1.2.8 特征缩放和特征提取

在这一阶段，从原始数据中提取重要的特征或属性，或是从现有的特征中创造新的特征。数据特征通常需要进行缩放或归一化处理，以防止在训练过程中模型产生偏差。另外，我们通常从所有的特征中根据特征的重要性以及质量选取子集作为学习算法的输入。这一步骤被称为特征提取。特征提取是基于现有的属性以及一定的逻辑、规则或假设创造新的属性或变量的过程（Sarkar et al., 2018）。

1.2.9 特征选择及降维

特征选择就是基于属性质量、重要度、约定、显著性以及限制条件等一系列参数从特征或属性中选取子集的过程。有时甚至需要使用机器学习算法来帮助我们进行特征选择（Sarkar et al., 2018）。降维是将原有的特征向量维度降低但依然保持特征高分辨度，去掉冗余无关信息的过程。降维的主要目的是能够降低诸如分类器等模型的计算复杂度（Phinyomark et al., 2013）。大多数的特征抽取过程中都会产生冗余的特征。实际上，若我们想要提高分类器的分类准确率，减少分类错误，需要选择一部分的特征进行使用。有很多特征选取以及降维的算法都可以达到增强分类器准确率的目的（Wołczowski and Zdunek, 2017）。

1.2.10 建模

在建模阶段，我们通常使用之前从数据中提取到的特征作为算法输入。机器学习算法必须使用之前阶段所获得的干净数据或特征作为输入，这样才能提取有价值的知识和信息。理论上，机器学习的算法应以减少错误和增强数据泛化能力表达作为训练目标。这些获取到的知识将会被用于下一阶段。模型在此阶段形成无监督学习是没有训练阶段的，因为不存在目标值。建模是整个分析过程中最主要的步骤之一，它需要使用干净并且符合格式要求的数据以及数据的属性创建能够解决问题的模型。这是一个迭代的过程，如图 1.1 所示，模型评估以及建模之前的阶段都与建模的步骤息息相关。简而言之，这个过程就是创建许多不同的模型并且根据评估条件选择最好的模型来解决问题（Sarkar et al., 2018）。

1.2.11 选择建模技术

在这个阶段，我们会选择一系列合适的数据挖掘以及机器学习技术、框架、工具以及算法。我们通常根据数据中所表现的见解、输入数据的特征以及是否健壮和适合解决这个问题而选择相应的技术。通常由现有数据、数据挖掘的目标、实际目标、算法限制以及优缺点而决定的。

1.2.12 构建模型

构建模型的过程也是通过使用现有数据集中的数据以及提取的特征进行模型训练的过程。结合使用数据（特征）和机器学习的技术可以让我们得到期望的见解以及（或）预测。通常情况下，我们会在同样的数据集上使用多种的建模技术来解决问题，以求达到预期的性能评估标准。关键问题是如何追踪所产生的模型、所使用的参数值以及每个模型的输出（Sarkar et al., 2018）。

1.2.13 模型评估及调优

本阶段需要使用到前一阶段所学到的信息。不同的模型使用的评估指标都可能是不一样的，例如模型的准确率、精确度、召回率、F值以及平均绝对误差。模型的参数也需要根据不同的技术进行调优，从而达到最好的性能，常见的技术有网格搜索和交叉验证等。调优的模型也需要和数据挖掘的目标相比较，从而确定是否已经达到了预期的模型性能以及结果。在机器学习领域中，模型调优又称为超参数优化。最终模型的性能通过不同的评估指标进行评估，从而判断模型的效果如何。如果模型有很多参数，那么需要通过超参数优化从而找到参数的最优值，进而最大化模型性能。如果是监督学习的模型，通常使用一些约定俗成的评价指标。针对无监督学习的模型，可能需要使用其他评估指标（Harrington，2012；Sarkar et al., 2018）。

建模阶段所得的模型需要满足数据挖掘的目标，并且达到期望的评估条件，例如模型准确率的要求。模型评估的过程包括详细的模型评估以及评估模型的输出。当模型达到预期的评价指标并且能够输出理想的预测值，需要考虑下述条件，对模型性能进行全面的评估（Sarkar et al., 2018）：

- 根据成功标准所定下的模型性能
- 模型的结果是否是可靠且可复现的
- 健壮性、可扩展性和是否易实现
- 模型的潜在可扩展性
- 对模型输出的满意度
- 根据实际目标和模型输出质量以及它们的重要性对最终模型进行排序
- 任何模型过优化的假设或局限
- 实现整个机器学习框架的开销，从数据提取和处理到建模以及预测

1.2.14 实现以及检验已经创建的模型

我们需要经常对已实现模型的输出进行定期检验。对所选模型进行持续部署并且对于

模型从训练到产品化有信心。根据现有的硬件条件、软件以及服务器情况制定合适的实现方案。模型已经经过验证、保存并且已经在服务器和系统中安装之后，还需要制定计划，对模型进行观察，包括模型的性能、模型的合理性以及是否需要对模型进行更新或替换（Sarkar et al.，2018）。

1.2.15 监督学习框架

监督学习技术需要有标签的数据来训练模型并且在未见过的测试集数据上预测结果。在测试集上，当使用模型对未知数据进行预测时，特征缩放、特征提取以及特征选择的过程必须与训练集保持一致。图 1.2 中表示了一个经典的监督学习框架。图中突出了训练和预测两个阶段。此外，如前文所述，在模型预测阶段，需要保证数据处理、特征缩放、提取、特征选择以及降维的操作与模型训练阶段一致。所有监督学习的方法都需要保证这点。

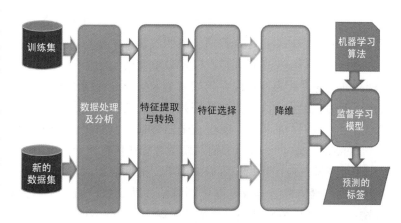

图 1.2　监督学习框架

同时，如图所示，在模型训练阶段，需要输入数据的特征以及标签，而在测试阶段，我们只需要输入特征，模型需要预测标签值（Sarkar et al.，2018）。

1.2.16 无监督学习框架

通常使用无监督学习来提取数据存在的模式、关联、关系以及聚类。同监督学习一样，无监督学习也需要特征缩放、提取、选择以及降维的过程，但是，无监督学习并没有标签数据。因此，无监督学习框架与监督学习框架有点不同。图 1.3 表示了

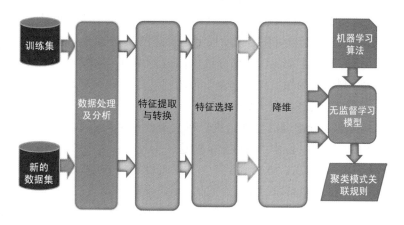

图 1.3　无监督学习框架

一个经典的无监督学习的框架。如图1.3所示，标签信息并不包括在模型训练的过程中。在无监督学习中，无标签的训练数据也需要经历与监督学习一样的转化过程。我们使用无监督学习算法，根据输入的特征数据进行模型的训练。在预测阶段，需要对新数据进行与训练阶段一致的转化操作，根据不同的预测任务以及算法如聚类、模式识别、关联规则或降维，给出相应的预测输出。（Sarkar et al., 2018）。

1.3 性能评估

模型评估是根据模型的类别、特征以及参数在多个模型中选择最佳模型的过程。这是通过对模型预测值进行评估，从而选择最佳模型。通常，使用不同的评估指标以及根据模型是否能够可靠预测目标来评估模型的质量。根据具体应用对于模型的要求，可以使用不同的评估指标对模型进行评估。模型最重要的质量指标是泛化性。评估模型是否具有良好的泛化性，通常将模型应用在领域一个未采样的数据子集上。然而，这种评估方式并没有被广泛采用，因为训练数据的排序并不是分类任务的最终目的。模型实际的性能与需要体现在对全域数据上的应用。通过在未知数据集上的评估有助于评估模型在全域数据上的性能（Cichosz, 2014）。学习如何对模型性能评估有助于对模型的预测输出做出可靠的评估。根据模型不同的应用场景，可以考虑不同的性能指标。模型的泛化性在模型评估的过程中极为重要，因为模型本身是基于训练集，即全域中的一小部分数据创建的。因此，根据Cichosz（2014），在说明模型性能的时候，同样要表明该性能是在何数据集（训练集、验证集、测试集或是全集）上所取得的。

另外，我们需要记住，相较于所熟悉的其他计算机科学问题，机器学习问题大多数时候没有一个"正确"答案。在某些情况下，所使用的特征对于数据所属的类别仅仅是一个"标志"作用，并不能当作是正确预测类别的"信号"。因为这些原因，评估机器学习算法的性能是一个重要的问题：我们必须对于算法在新数据集上的表现有所了解。假设我们想知道某个训练的新模型在EEG数据集上的表现，可以简单地计算包括癫痫和正常在内的分类正确的EEG信号的数量，除以数据总数，得到一个比例值。这个所得值就被称为分类器的准确率。然而这并不能说明模型是否过拟合（Flach, 2012）。

更好的方法是使用一部分数据作为训练集，保留一部分数据作为测试集。当发生过拟合时，测试集的性能将会远差于训练集。即使测试集数据实例是从数据中随机选取并且有很多噪声，大多数测试实例也应该与训练集相似。在实际应用中，我们通常使用交叉验证的方法重复分割训练–测试集。在交叉验证中，将数据随机平分成10份，其中9份作为训练集，剩下一份作为测试集。将这一个过程重复10次，每一份都作为测试集使用过。最后，计算模型在这10个测试集上的平均性能。交叉验证通常可以应用在所有的监督学习模型中，对于无监督学习模型，我们考虑不同的评估方法（Flach, 2012）。

当评估回归（预测）这类机器学习模型时，我们需要使用统计质量指标，例如均方根误差（RMSE）、平均绝对误差（MAE）、相对绝对误差（RAE）、相对平方根误差（RRSE）和相关系数（R）（见表1.1），而非准确率、F值、受试者操作特征（ROC，receiver operating characteristic 下称ROC）区域和Kappa统计量。在回归中，相较于关心预测值"正确与否"，我们应该考虑预测值与实际值相差多少。假设有 n 个测试用例，第 i 个测试用例的实际值（观察值）为 a_i，预测值（估计值）为 P_i，则评估指标计算方式如下：

$$\bar{a} = \frac{1}{n}\sum_{i=1}^{n}a_i, \bar{p} = \frac{1}{n}\sum_{i=1}^{n}p_i, S_A = \frac{1}{n-1}\sum_{i=1}^{n}(a_i-\bar{a})^2, S_p = \frac{1}{n-1}\sum_{i=1}^{n}(p_i-\bar{p})^2,$$
$$S_{PA} = \frac{1}{n-1}\sum_{i=1}^{n}(p_i-\bar{p})(a_i-\bar{a})$$
（1.1）

模型在数据集上的性能可以根据真实标签与预测值，使用某个评估指标计算。模型在数据集上的性能表现了与数据集目标概念的相似程序。另一方面，可以通过在训练集上对模型进行评估从而得知模型性能。模型在训练集上的性能能够更好地帮助我们理解这个模型，然而，这并非我们的兴趣所在，因为最终目的并不是在训练集上使用模型进行预测。最后，模型的预期性能代表了其在全部数据域上的真实性能。模型的真实性能表示它可以在某些条件下对于数据域中的任意新实例做出正确的预测。虽然我们不知道模型对未知数据的预测

表1.1 回归模型评估指标

平均绝对误差（MAE）	MAE= $\frac{|p_1-a_1|+\cdots+|p_n-a_n|}{n}$
均方根误差（RMSE）	RMSE= $\sqrt{\frac{(p_1-a_1)^2+\cdots+(p_n-a_n)^2}{n}}$
相对绝对误差（RAE）	RAE= $\frac{\sum_{i=1}^{n}|p_i-a_i|}{\sum_{i=1}^{n}|a_i-\bar{a}|}$
相对平方根误差（RRSE）	RRSE= $\sqrt{\frac{\sum_{i=1}^{n}(p_i-a_i)^2}{\sum_{i=1}^{n}(a_i-\bar{a})^2}}$
相关系数（R）	R= $\frac{S_{PA}}{S_p S_A}$

性能，但是通过真实的性能或是预期性能，可以近似估计模型的性能。若想得知模型的真实性能，需要采用合适的步骤对模型进行预测，通常需要使用模型之前并未见过的数据（即使用模型未见过的数据进行模型性能评估，并且将其作为模型在全域上的性能）(Cichosz，2014)。

"没有免费的午餐"定理指出，一旦评估，任何学习算法都无法在所有分类问题上优于另一个算法，而对于所有可能的学习问题，随机猜测所得到的结论可能会有与之相当的性能。避免"没有免费的午餐"定理诅咒的唯一方法是更多地探索数据分布并且根据这个知识选择相应的学习算法（Flach，2012）。

1.3.1 混淆矩阵

分类器的性能可以通过混淆矩阵进行说明。该矩阵对模型在特定的类上预测性能及其泛化属性有重要见解。但是混淆矩阵并不总是能够像预期的那样，根据模型的功能以及作用，对于各个模型进行排序并选择出最佳模型。虽然混淆矩阵可以对模型性能提供不同的评价指标，但是它通常是用来评估二元分类器的。我们可以用 $C=\{0,1\}$ 来表示这类模型，

其中 0 表示阴性类，1 表示阳性类。对于某一个数据实例 x，我们使用 $c(x)=1$ 表示它属于阳性类，使用 $c(x)=0$ 表示其属于阴性类。混淆矩阵仅在两类不对称时使用。当我们需要对多类分类器进行评估时，需要基于混淆矩阵考虑额外的方法。在混淆矩阵中，我们使用阳性和阴性来表示分类器的输出正确与否（Cichosz，2014）。图 1.4 展示了一个 2×2 的混淆矩阵。

		预测的类别	
		类=Yes	类=No
实际类别	类=Yes	(TP)	(FN)
	类=No	(FP)	(TN)

图 1.4 混淆矩阵的表示

在这个矩阵中，每一行表示测试集中的实际标签数据，每一列表示分类器所预测的输出。基于混淆矩阵，我们可以计算很多评估指标。其中，最简单的就是准确率，即预测准确的比例。

- 误分类的错误表示为错误的分类比上所有的测试实例

$$误分类错误率 = \frac{FP+FN}{TP+TN+FP+FN} \tag{1.2}$$

- 模型的准确率是所有正确分类的数目除以所有的测试实例

$$准确率 = \frac{TP+TN}{TP+TN+FP+FN} \tag{1.3}$$

- 真阳性率（TPR）或敏感度是根据阳性实例正确分类比例计算得出的，如下：

$$TPR = \frac{TP}{TP+FN} \tag{1.4}$$

- 假阳性率（FPR）或特异性是通过计算错误分类为阳性类的阴性类实例所得，如下：

$$FPR = \frac{TN}{TN+FP} \tag{1.5}$$

- 精确度是指在所有被分为阳性类的实例中，实际为阳性类的实例的占比，如下：

$$Precision = \frac{TP}{TP+FP} \tag{1.6}$$

- 召回率是指正确分类为阳性类的实例在所有的阳性实例的占比，如下：

$$Recall = \frac{TP}{TP+FN} \tag{1.7}$$

1.3.2 F 值分析

在信息检索文献中，将精确度与召回率的调和平均值称为 F 值（F-measure）（Flach，2012）。这个指标实际上借由精确度和召回率的加权调和平均值对模型的性能做出统计分析。根据 Sokolova、Japkowicz 和 Szpakowicz（Sokolova, Japkowicz, & Szpakowicz, 2006），

F 值是使用具有更高灵敏度的辅助算法的复合度量，并且遇到了更高特异性的模拟。理想的状态下，即完美的精确度与召回率，F 值为 1。若精确度和召回率都很差时，就意味着一个很低的 F 值，即 0。更加实际的 F 值可以通过在测试集上计算精确度和召回率得到。通常在信息检索或是文档分类与分类查询性能中，我们会使用 F 值进行模型评估。然而，F 值并没有考虑真实的负向实例。在评估二元分类模型时，我们更倾向于使用如 Cohen's kappa 系数和 Matthews 相关系数。F 值在诸如分词与实体识别评价等自然语言处理中较常使用。

$$F 值 = \frac{Precision \times Recall}{Precision+Recall} \tag{1.8}$$

1.3.3 ROC 分析

ROC 曲线是覆盖曲线归一化至 [0，1] 区间的结果。一般而言，覆盖曲线可以是矩形但是 ROC 曲线永远只占单元格。此外，在覆盖图中，覆盖曲线下的面积给出了正确排序对的绝对数量。在 ROC 图中，ROC 曲线下面积（AUC）是排名准确率（Flach，2012）。ROC 分析是在操作、比较和选择的多个点上增强分类器性能评价/评估的工具。ROC 原本是二战时期用来检测雷达信号的，如今被应用于分类器评估中。它使用笛卡儿坐标系，y 轴表示 TPR（真阳性率），x 轴表示 RPR（假阳性率）。这两个轴使 ROC 的平面成为平面上的单个点，可以可视化真阳性和假阳性之间的基本权衡，代表了离散分类器的性能。类似地，分类器集中的单个操作点也被建立为 ROC 平面上的一个点。TPR 为 1 且 FPR 为 0（即（0，1））显示了所有实例被正确分类的理想操作点。TPR 为 0 并且 FPR 为 1（即（1，0））的点，是所有的实例都被分类错误的最差操作点。点相关因子（0，0）代表一个分类器，该分类器通常会预测一个非正非负的 0 类。平面上的点（1，1）代表了永远会预测分类为 1 的分类器。将所有分类器输出的操作点在平面上相连，我们就可以进行 ROC 分析。因此，分类器性能的可视化与截断值不相关。该图显示了各种不同的操作点，因此在同一张图上会呈现不同的 FPR 和 TPR 的权衡。仅依赖于其评分函数元素的分类器性能可以使用 ROC 曲线以图形方式表示。分类器中的评分表达式包含了有关属性和类值之间关系的知识。根据为整个数据集域生成的评分，找到分类器评分的所有可能操作点以生成 ROC 曲线非常重要（Cichosz，2014）。

1.3.4 Kappa 统计量

Kappa 统计量通过将期望模型从预测器中剥离的方式进行模型评估。它将当前模型表示为理想模型的一部分。相较于简单计算模型之间的一致程度，它能够更加可靠地反映模型的性能，因为该方法将偶然发生的模型间的一致考虑在内。Kappa 统计量不仅能够表示模型预测与实际数据的一致程度，也能包括了偶然发生的一致事件。然而，与简单的成功率相似，它并不考虑花销（Hall，Witten，& Frank，2011）。

$$k = \frac{p_o - p_e}{1 - p_e} = 1 - \frac{1 - p_o}{1 - p_e} \quad (1.9)$$

其中 P_0 是观察到的一致程度，定义如下：

$$P_0 = \frac{TP+TN}{TP+TN+FP+FN} \quad (1.10)$$

P_e 是随机发生的一致事件的概率（Yang & Zhou，2015）。随机一致的概率就是模型与实际数据在 Yes 和 No 上一致的概率，如下：

$$P_e = P_{YES} + P_{NO} \quad (1.11)$$

即

$$P_{YES} = \frac{TP+FP}{TP+TN+FP+FN} * \frac{TP+FN}{TP+TN+FP+FN} \quad (1.12)$$

$$P_{NO} = \frac{FN+TN}{TP+TN+FP+FN} * \frac{FP+TN}{TP+TN+FP+FN} \quad (1.13)$$

当模型与原数据高度统一时，$k=1$。若仅存在随机一致性而不存在强一致性，那么 P_e，$k=1$。Kappa 值可以为负，此时意味着两者之间没有一致关系，或者存在一种比随机一致性更糟糕的关系。

1.3.5 度量了什么

性能指标有时也被称为评估指标。但是，评估指标不一定是标量，因此 ROC 曲线也可以作为评估指标。任何这些评估指标的适用性取决于基于实验目标所定义的性能。我们尤其要注意的是不能够混淆评估指标和实验目标，因为这两者之间通常有很大的区别。在机器学习中这种情况更加明显，例如，实验目标（准确率）是可以在未见过的数据上衡量的。然而，可能需要考虑一些未知的因素。例如，模型可能会部署在多元化的操作环境中，有着不同的类分布。尽管在未来的情境下评估模型性能是我们的实验目标，但是实际上可能遇见各种类的分布，使得准确率不再是模型评估的最佳选择，可能需要考虑平均召回率作为评估指标。这个例子表明了如果将准确率作为评估指标，实际上无形中假设了模型部署的环境中的数据分布与测试数据集一致。因此，更好的做法是记录足够的信息，以便能够复制列联表。我们可以使用一系列的评估指标，例如真阳率、真阴率、假阳率、假阴率、类分布以及测试集大小。精确度与召回率常见于评估信息检索任务。而精确度与召回率的结合，即 F 值，对于实际阴性总数不敏感。这不是 F 值的缺点，恰恰相反，在阴性实例较多的领域，这正是 F 值的优势。当阴性实例较多时，模型很容易通过将所有的实例预测为阴性而取得较高准确率，而这并不是我们所期待的模型的行为。因此，当你选择 F 值作为评估指标时，就表示你默认了阴性实例对于你当前的操作环境没有影响。另一个评估指标是阳性预测率，即预测为阳性的实例在总数据实例中的比例。尽管阳性预测率不能够告诉我们很多分类器性能的好坏，但是它告诉了分类器所预测的类的分布。在某些情况下，AUC 也是一个好的评估指标，因为其与期望准确率线性相关（Flach，2012）。

1.3.6 如何度量

到这里，我们只讨论了从混淆矩阵中可以计算的评估指标。问题在于：应该选择什么数据进行模型评估，和如何考虑与评价指标相关的不确定性。我们如何得到 k 个准确率的独立估计？在实际应用中，有标签的数据通常很少，不足以单独设置一个验证集，否则就会导致训练集数据不足。另一个广泛使用的方法叫作 k 折交叉验证，它常使用于模型选择以及训练。当我们有足够多的数据，可以从数据中随机采样 k 个大小为 n 的数据集，并且在每一个数据集上计算模型的准确率。如果我们要评估一个学习算法，而不是一个给定的模型，则需要将训练集和测试集分开。当数据集不够多时，通常使用交叉验证（CV）的方式。使用交叉验证的时候，需要将数据随机分成 k 份，保留一份作为测试集，将其余 k-1 份作为训练集，在保留的测试集上进行模型评估。这个过程需要重复 k 次，直到 k 份数据都被作为测试集使用过。尽管我们可以随机选取 k 值进行交叉验证，通常选择 k=10。另外，也可以使用 k=n，留一份作为测试集，其余作为训练集，并且将这个过程重复 n 次。这被称为留一法交叉验证。如果算法对于类的分布十分敏感，可以使用分层交叉验证的方法评估模型，其中每一份数据集中的类分布应该保证大致相同（Flach，2012）。

k 折交叉验证的特例，即 k=m 时，被称为留一法交叉验证，因为在每一次迭代中，只有一个数据实例从训练集被保留下来作为测试用例。将所有的测试数据实例的评价指标值平均之后，我们就大概得到算法性能的无偏估计，并且对算法的性能有大致了解。通常，留一法交叉验证有十分昂贵的计算开销，因为它需要我们将模型在大小为 m-1 的数据集上训练 k 次，然而，对于某些算法，它是一种有效的估计方法。另外，k 折交叉验证通常被用来评估模型性能，并且应用于模型选择中。我们将有标签的数据集随机分成 k 份，针对每个特定的参数，使用同样的训练和测试集对模型进行评估和测试。我们不仅考虑模型在全部的数据集上的平均表现作为模型的性能，同时也需要考虑评估值的标准差（Mohri et al.，2018）。

1.3.7 如何解释估计

一旦计算了相关的评估指标，我们就可以根据评估指标值选择最好的模型和学习算法。然而，其中最重要的问题就是如何对待这些估计中内在的不确定性。置信区间和显著性检验是两个需要讨论的重要指标。值得注意的是，置信区间是关于估计值，而非评价指标的实际值。p 值通常在显著性检验中使用。显著性检验可以在算法的交叉验证中使用。对于一对算法，我们可以用显著性检验估计这两个算法在每一份上准确率的差别。p 值是根据正态分布估计所得，如果 p 值低于显著性水平，就拒绝零假设。有时候，过程中的不确定性会产生钟形抽样分布，虽然其与正态分布相似，但是比正态分布尾部更长。这个分布被称为 t 分布。t 分布比正态分布更重尾的程度由自由度的数量决定。整个过程称为配对 t 检验（Flach，2012）。

1.3.8 scikit-learn 中的 k 折交叉验证

在大多数实际应用中，我们通常把有标签的数据分成三份，第一份作为训练集，第二份作为模型选择的验证集。在选择最佳模型之后，需要在第三份数据上测试模型性能（测试集）(Shalev-Shwartz & Ben-David，2014)。

在这一过程中定义的验证集通常需要很多数据，我们从其中采样一部分作为新的验证集。但是在某些数据有限的情况下，我们不想要在验证集上"浪费"数据。k 折交叉验证通常能够在不浪费很多数据的情况下较为准确地估计模型的性能。在 k 折交叉验证中，将原始训练集分为 k 个子集（折），每折的大小为 m/k。在每一次迭代中，我们保留一份用作测试，计算模型的错误率，在其他的 $k-1$ 份上进行训练。将这个过程重复 k 次，将错误率平均 k 次后作为模型性能的估计。在 $k=m$，即 m 为所有数据数量的特殊情况下，将这一过程称为留一法（LOO）。k 折交叉检验通常应用于模型选择中（或者调参）。当最佳的参数值被选择后，算法需要在整个训练数据集上重新训练 (Shalev-Shwartz & Ben-David，2014)。下面是 k 折交叉验证的示例伪代码：

```
from sklearn.model_selection import cross_val_score
from sklearn.datasets import load_iris
from sklearn.discriminant_analysis import LinearDiscriminantAnalysis
iris = load_iris()
clf = LinearDiscriminantAnalysis(solver = "svd", store_covariance = True)
scores = cross_val_score(clf, iris.data, iris.target, cv = 5)
scores.mean()
```

1.3.9 如何选择正确的算法

根据我们的目标来在众多算法中选择正确的算法。如果目标是预测一个目标值，那么需要考虑监督学习算法。如果不是，那么无监督学习可能是最佳选择。如果监督学习是所选方法，那么目标值是什么？如果目标是离散值，那么可以选择分类算法；如果目标值是连续值，那么可以考虑回归算法。在第 3 章中，我们会介绍监督学习算法，例如分类、回归以及聚类。在算法选择的过程中，第二个需要考虑的点是数据。数据的特征是标量还是连续值？特征中是否有缺失值？数据中是否有异常值？数据中所有的特征都可以帮助你缩小算法选择的范围。另外，对于最佳算法的选择并不只有唯一的答案。你需要做的是尝试不同的算法并且选择最佳的一个。还有很多其他的机器学习工具可以帮助提高机器学习技术的性能（Harrington，2012）。

1.4 Python 机器学习环境

现在我们可以讨论一下实现机器学习算法的编程语言了。这个编程语言必须能够被很多人理解并且有很多支持不同任务的库。我们需要一个有活跃开发者社区的编程语言。由于这些原因，Python 是最佳选择，并且是实现机器学习算法的最好语言。首先，Python 具有清晰的语法。Python 清晰的语法使得其被称为可执行的伪代码。Python 默认的安装包总已经包括了许多高级数据类型，例如列表、元组、字典、集合和队列。你可以使用熟悉的任意编程方式编写代码，例如面向对象、过程式以及函数式。Python 能够方便地操作文本类型数据，这使得它能够方便地处理非数字型数据。由于很多用户以及组织使用 Python，使其有丰富的开发文档。由于 Python 十分流行，并且有丰富的案例，可以帮助我们快速学习。另外，Python 的流行意味着存在许多适用于很多应用的模块。Python 在科学界和金融界中也十分流行。诸如 SciPy 和 NumPy 的科学计算库可以帮助你快速帮助你进行向量和矩阵的操作。另外，由于 SciPy 和 NumPy 是由低级语言如 C 编译的库，使得计算速度很快。Python 中的科学计算工具可以和绘图库 Matplotlib 配合使用，能够绘制 2D 和 3D 图，并且可以处理大部分科学计算中使用到的类型。Python 中的一个新的模块 Pylab，旨在将 NumPy、SciPy 以及 Matplotlib 合并成一个 Python 环境进行安装（Harrington，2012）。

  Python 是解释性编程语言，意味着其源代码可以转化成为字节码，然后通过 Python 虚拟机进行执行。与 C 和 C++ 等编译型语言不同，Python 不需要执行构建和连接步骤。Python 与其他语言有两个主要区别：

- Python 代码开发较快。因为 Python 代码不需要编译和构建，其可以快速被改变并执行。这可以缩短开发周期。
- Python 代码执行起来比其他语言慢。因为 Python 代码不需要编译执行并且需要增加额外的 Python 虚拟环境层进行执行，因此相较于 C 和 C++ 等传统语言而言，Python 代码执行起来较慢。

  Python 在广泛使用的编程语言排行榜中稳步上升，根据多项调查和研究，它是世界上第五重要的编程语言。最近有几项调查显示 Python 是机器学习和数据科学中最受欢迎的语言！我们将汇总一份 Python 提供优势的简要清单，这些优势可能可以解释其受欢迎程度。

  易于学习：Python 是一种相对易于学习的语言。它的语法对于初学者来说很简单，易于学习和理解。与 C 或 Java 之类的语言相比，执行 Python 程序所需的样板代码最少。

  支持多种编程范式：Python 是一种多范式、多用途的编程语言。它支持面向对象的程序设计、结构化程序设计、函数式编程，甚至面向方面的编程。这种多功能性使它可供众多程序员使用。

  可扩展性：Python 的可扩展性是其最重要的特征之一。Python 有大量易于安装和使用的模块。这些模块涵盖了编程的各个方面，从数据访问到流行算法的实现。易于扩展的特性可确保 Python 开发人员的生产率更高，因为可用库可以解决大量问题。

  活跃的开源社区：Python 是开源的，并得到了大型开发人员社区的支持。这使其具有健壮性和自适应性。Python 社区很容易修复遇到的错误。由于开放源代码，开发人员可以根据需要修改 Python 源代码。

1.4.1 缺陷

尽管 Python 是一种非常流行的编程语言，但它也有自己的缺陷。它的最重要限制之一是执行速度。作为一种解释语言，与编译语言相比，它速度较慢。在需要极高性能代码的情况下，此限制可能会对 Python 有所局限。对于将来的 Python 实现，这是主要的改进领域，每个后续的 Python 版本都改进该问题。尽管必须承认它永远无法像编译型语言一样快，但是我们坚信它可以通过其他部分的高效和有效地弥补这一不足。

可以使用必要的库管理工具（即 Anaconda）随附的预打包的 Python 发行版分别安装 Python 和必要的库。Anaconda 是 Python 的打包编译以及一整套库，其中包括在数据科学中广泛使用的核心库。这种发行版的主要优点是不需要复杂的设置，可以在所有类型的操作系统和平台（尤其是 Windows）上正常运行。Anaconda 发行版附带一个出色的 IDE，Spyder（科学 Python 开发环境）以及其他有用的工具，如 Jupyter Notebooks、IPython 控制台和出色的程序包管理工具 Conda（Sarkar et al., 2018）。

1.4.2 缺点

Python 的唯一真正缺点是它不如 Java 或 C 快。但是，你可以从 Python 调用 C 编译的程序。这提供了两全其美的优势，并允许增量开发程序。如果该程序是以模块化方式构建的，则可以首先在 Python 中启动它并运行，然后为了提高速度，开始在 C 语言中构建部分代码。Boost C++ 库使此操作变得容易。其他工具（例如 Cython 和 PyPy）允许编写 Python 的类型化版本，并且性能比常规 Python 高。如果某个程序或应用程序的构想存在缺陷，那么它在低速和高速下都会存在缺陷。如果一个主意是一个坏主意，编写代码可以使其快速发展或扩展（Harrington，2012）。

1.4.3 NumPy 库

NumPy 在 Python 中是支持机器学习的重要库。这是 Python 科学计算中最重要的库之一。它增加了对核心 Python 的多维数组和矩阵的支持，以及对这些数组的快速矢量化操作。它被几乎所有的机器学习和科学计算库所采用。NumPy 的流行程度已得到以下事实的验证：主要的操作系统发行版（如 Linux 和 MacOS）将 NumPy 捆绑为默认软件包，而不是将其视为附加软件包（Sarkar et al., 2018）。

1.4.4 Pandas

Pandas 是一个至关重要的 Python 库，用于数据操作、处理和分析。它是一组直观、易于使用的工具，可对任何类型的数据执行操作。Pandas 允许使用截面数据（cross-sectional data）和基于时间序列（time-series）的数据。DataFrame 是最重要和最有用的数据结构，可用于 Pandas 中几乎所有类型的数据表示和操作。与 NumPy 数组不同，DataFrame 可以包含异构数据。自然地，表格数据是通过 DataFrame 来表征的，该数据框类似于 Excel 工作表或 SQL 表。这对于在机器学习和数据科学中表示原始数据集和已处理的特征集非常有帮助。可以沿着 DataFrame 中的轴、行和列执行所有操作。

1.5 本章小结

模型构成了机器学习的基本概念，因为它们包含了从数据中学到的东西，以解决给定的任务。现在有大量的模型可供选择。选择模型的目的之一是根据具体的任务，例如，分类、回归、聚类以及关联发现。几乎在科学和工程的每个分支中都可以找到这些任务。数学家、工程师、心理学家、计算机科学家和许多其他人已经发现（有时是重新发现）解决这些任务的方法。它们都有自己特定的背景，因此创建这些模型的原理也有所不同。这种多样性是一件好事，因为它支持机器学习作为一门强大而令人兴奋的学科（Flach，2012）。

机器学习是一种利用有价值的信息特征来创建正确模型以产生正确任务的技术。这些任务包括二元和多类分类、回归、聚类和描述性建模。前几个任务的模型是在受监督的方式下学习的，需要标记训练数据集。如果想知道模型的质量，还需要从训练数据集中分离出带有标签的测试数据集，以在训练数据集上评估模型。需要测试数据集以暴露发生的任何过拟合。另一方面，无监督学习适用于未标记的数据集，因此，没有这样的数据集专供无监督学习评估。例如，要评估将数据集划分为聚类的特定方式，可以估算距聚类中心的平均距离。其他形式的无监督学习包括学习关联和对诸如电影类型的隐藏变量进行分类。可以区分输出包括目标变量的预测模型和描述数据集中有趣结构的描述性模型。通常，预测模型是以监督方式学习的，而描述性模型是通过无监督学习技术生成的（Flach，2012）。

因为这是一本实用的机器学习书，所以在后续章节中，我们将重点介绍特定的用例、问题和实际案例研究。实现与学习算法、数据管理、模型构建、评估和安排相关的形式化描述、概念和基础至关重要。因此，我们尝试涵盖所有这些知识点，包括与数据挖掘和机器学习工作流相关的问题，以便你获得可以使用的基础框架，并且可以解决后续章节中介绍的任何现实问题。此外，我们涵盖了与机器学习相关的几个跨学科领域。本书主要研究用于数据分析的实用或应用机器学习，因此本章的重点是应用机器学习算法来解决实际问题。因此，具有 Python 和机器学习的基本水平将会对阅读本书有所帮助。但是，由于本书考虑了众多读者的不同专业水平，第 2 章和第 3 章介绍了机器学习的关键方面和构建机器学习管道。我们还将讨论 Python 在构建机器学习系统中的用法以及用于解决机器学习问题的主要工具和框架。本书提供了很多代码片段、示例和多个案例研究。我们利用 Python 3 并为所有示例提供相关的代码文件，以提供更交互式的体验。在阅读本书时，可以自己尝试所有示例，并采用它们来解决自己的实际问题。

本书讨论了特征在机器学习中的作用。模型中必须利用特征，在少数情况下，只需要一个特征就足以创建模型。数据并不总是具有现成的特征，通常我们必须构造或转换特征。

由于这个原因，机器学习也被称为迭代过程。在此过程中，我们在创建模型后便已提取或选择了正确的特征，如果模型无法获得良好的性能，则需要分析其性能以确定如何通过改变特征来提升模型性能。因此，在第2章中，我们会讨论预处理，包括降维、特征提取、特征选择和特征转换。第3章介绍了一些机器学习技术。第4章以用医疗保健分类为例，介绍了分类问题，并且在第5章中介绍了其他类型的分类问题。第6章中提到了回归（预测）。第7章主要介绍了无监督学习（聚类）的例子。

1.6 参考文献

Begg, R., Lai, D. T., & Palaniswami, M. (2008). *Computational intelligence in biomedical engineering.* Boca Raton, FL: CRC Press.

Cichosz, P. (2014). *Data mining algorithms: Explained using R.* West Sussex, UK: John Wiley & Sons.

Flach, P. (2012). *Machine learning: The art and science of algorithms that make sense of data.* Cambridge, UK: Cambridge University Press.

Graimann, B., Allison, B., & Pfurtscheller, G. (2009). Brain–computer interfaces: A gentle introduction. In *Brain-Computer Interfaces* (pp. 1–27). Berlin, Heidelberg: Springer.

Hall, M., Witten, I., & Frank, E. (2011). *Data mining: Practical machine learning tools and techniques.* Burlington, MA: Morgan Kaufmann.

Harrington, P. (2012). *Machine learning in action.* Shelter Island, NY: Manning Publications Co.

Mohri, M., Rostamizadeh, A., & Talwalkar, A. (2018). *Foundations of machine learning.* Cambridge, MA: MIT press.

Phinyomark, A., Quaine, F., Charbonnier, S., Serviere, C., Tarpin-Bernard, F., & Laurillau, Y. (2013). EMG feature evaluation for improving myoelectric pattern recognition robustness. *Expert Systems with Applications*, 40(12), 4832–4840.

Sarkar, D., Bali, R., & Sharma, T. (2018). Practical Machine Learning with Python. A Problem-Solvers Guide To Building Real-World Intelligent Systems, Springer Science+Business Media, New York.

Shalev-Shwartz, S., & Ben-David, S. (2014). *Understanding machine learning: From theory to algorithms.* New York, NY: Cambridge University press.

Sokolova, M., Japkowicz, N., & Szpakowicz, S. (2006). Beyond accuracy, F-score and ROC: a family of discriminant measures for performance evaluation. 1015-1021. Springer.

Subasi, A. (2019a). Electroencephalogram-controlled assistive devices. In K. Pal, H. -B. Kraatz, A. Khasnobish, S. Bag, I. Banerjee, & U. Kuruganti (Eds.), *Bioelectronics and medical devices* (pp. 261–284). https://doi.org/10.1016/B978-0-08-102420-1.00016-9.

Subasi, A. (2019b). Practical Guide for Biomedical Signals Analysis Using Machine Learning Techniques, A MATLAB Based Approach (1st Edition). United Kingdom: Academic Press.

Wołczowski, A., & Zdunek, R. (2017). Electromyography and mechanomyographysignal recognition: experimental analysis using multi-wayarray decomposition methods. *Biocybernetics and Biomedical Engineering*, 37(1), 103–113 https://doi.org/10.1016/j.bbe.2016.09.004.

Yang, Z., & Zhou, M. (2015). Kappa statistic for clustered physician–patients polytomous data. *Computational Statistics & Data Analysis*, 87, 1–17.

CHAPTER 2

第 2 章

数据预处理

2.1 简介

在机器学习中,无论是分类算法还是回归算法,数据都被作为输入提供给学习器用于决策。理想情况下,数据无须单独的特征提取或筛选处理,除了不相关的特征,每个特征都必须被分类器(或回归器)使用。但是基于许多因素需要考虑将降维作为单独的预处理步骤:

- 在大多数的学习算法中,复杂度依赖于输入数据的维度以及数据样本的大小,为了降低内存和计算量,我们企图降低问题的维度。降维也会降低测试时学习算法的复杂度。
- 如果输入数据的信息量很少,我们可以通过提取它来节省计算成本。
- 简单的模型在小数据集上更加稳定。简单的模型方差更小,也就是说它们不依赖特定的样本,包括离群值、噪声等。
- 如果数据能用更少的特征表示,那么我们可以更好地了解从数据中获得灵感的过程,也就是知识提取。
- 如果可以在不丢失信息的情况下使用更少的维度表示数据,就能够通过绘制来可视化地分析其结构和离群值。

任何学习器的复杂度都依赖于输入数据的数量。这将同时影响空间、时间复杂度和训练这种学习器所需要的训练样本的数量(Alpaydin, 2014)。本章将讨论特征选择方法,它们从一批带有信息的特征中筛选出一个子集并剔除其他特征,以及特征提取方法,它们从原始的输入中提炼出一些新特征,还会讨论降维技术。

当数据存在大量特征时,我们有必要对数据进行降维或者找到数据某些属性的低维描述。关于降维(或者流形学习)技术的主要讨论包括:

计算:压缩原始数据,这作为一个预处理步骤可以加快对数据的后续操作。

可视化：将输入数据映射到两维或者三维空间，可视化数据以进行初步分析。

特征提取：可靠地生成一组更小、更有效、更有信息量或者更有价值的特征集合。

降维的优势通常通过模拟数据（比如 Swiss roll 数据集）表示，例如：

假设有样本 S=（x_1, ..., x_m），特征映射 $\Phi:X \to R_N$，数据矩阵 $X \in R_N \times m$ 定义为（$\Phi(x_1)$, ..., $\Phi(x_m)$）。第 i 个数据点表示为 $x_i=\Phi(x_i)$，它是一个 N 维向量，也是 X 的第 i 列。降维方法通常是为了找到数据的 k 维表示 $Y \in Rk \times m$，它在某种程度上可以如实反映原始表示 X。本章将讨论几个解决该问题的方法。首先介绍最普遍使用的降维技术——主成分分析（PCA）。然后介绍 PCA 的核版本（KPCA），以及 KPCA 和流形学习算法的关系（Mohri, Rostamizadeh, & Talwalkar, 2018；Subasi, 2019）。

本章将通过各式各样的 Python 案例介绍数据分析中的各种特征提取方法和降维技术。此外本章还会介绍包括主成分分析（PCA）、独立成分分析（ICA）、线性判别分析（LDA）、熵和统计值在内的降维技术。本章的目的在于帮助研究者在数据分析中选择合适的预处理技术，因此本章也将讨论数据分类中基础的预处理方法。在相应章节的结尾我们也将通过示例来演示 Python 函数的不同应用。这些例子大多数来自 Python-scikit-learn 库（https://scikit-learn.org/stable/），然后进行了改编。

2.2 特征提取和转换

机器学习中可能用到特征的地方很多。通常来说，在信号处理和文本分类中，生物医学信号或文档都没有内置的特征，因此机器学习算法的开发者必须构造特征。特征构造（或者提取）过程对机器学习算法的实现来说至关重要。在生物医学信号分析中，特征提取设法放大与分类任务相关的"信号"并减弱"噪声"，从而对有用且有价值的信息进行精准的描述。不过，很容易发现在某些情况下这样做并不合适，比如在我们试图训练一个分类器去区分正常和异常的生物医学信号时。最自然的做法是根据已知的特征构建一个模型，但是可以用我们认为合适的方式修改特征，或者甚至生成新的特征。例如，真实的特征值通常包含可以通过离散化消除的冗余特征。特征的迷人之处之一及其多维特性在于可以以多种方式协作。这种相互作用有时会被加以利用，有时会被忽略，有时会带来挑战。特征还有被关联的其他方式（Flach, 2012）。

2.2.1 特征类型

考虑两个特征，一个特征描述人的年龄，另一个描述他们的门牌号。虽然这两个特征都由整数表示，但是我们使用这两个特征的方式可能完全不同。计算一群人的平均年龄是有意义的，但是计算门牌号的平均值却毫无意义。这又取决于特征值是否在有意义的尺度上表示。虽然门牌号表面上看起来是整数，但实际上是序数。它们可以用于判断 10 号的邻居是 8 号和 12 号，但是我们无法推测 8 号和 10 号的距离与 10 号和 12 号的距离一样。由于缺乏线性尺度，对门牌号进行加减是没有意义的，也就没有必要对门牌号做类似求平均值的操作（Flach，2012）。

2.2.2 统计特征

许多统计特征可以从各个子样本数据点中提取出来，因为它们是描述数据分布的主要区别值。这些特征包括数据向量的最小值、最大值、平均值、中位数、众数、标准差、方差、第一四分位数、第三四分位数和四分位距（IQR）（Subasi，2019）。对特征的各种计算通常称为统计或者总计，主要有形状统计、离散度统计和集中趋势统计三种类型。其中的每一项都可以表示为给定样本（样本统计）的有形属性或未知总体的假设属性。统计值——即平均值、标准差、偏度和峰度——通常用于数据的降维。一阶和二阶统计量在数据分析中至关重要。另一方面，二阶统计量对于许多时间序列数据而言是不够的。因此，为了更好地描述数据，还可以使用高阶统计量。一阶和二阶统计量指的是均值和方差，而高阶统计量指的是高阶矩（Mendel，1991）。如果 $X(n)$ 是一个随机过程，$X(n)$ 的矩可以表示为矩生成函数的泰勒展开的系数（Kutlu & Kuntalp，2012；Subasi，2019）。

$$m_2(i) = E[X(n), X(n+i)] \tag{2.1}$$

$$\phi_x(w) = E[\exp(jwx)] \tag{2.2}$$

如果离散的时间序列均值为零，那么矩定义为：

$$m_3(i,j) = E[X(n), X(n+i) \cdot X(n+j)] \tag{2.3}$$

$$m_4(i,j,k) = E[X(n), X(n+i) \cdot X(n+j) \cdot X(n+k)] \tag{2.4}$$

这里 $E[\cdot]$ 是随机过程 $X(\cdot)$ 的期望值（Kutlu & Kuntalp，2012）。

高阶统计量（HOS）表示三阶和更高阶计算数的累积量，它们是低阶矩和低阶累积量的线性组合。基于时间序列样本是取自未知分布 $P(x)$ 的随机变量这一假设，矩和累积量被计算并用作特征。连续分布 $P(x)$ 的零点矩 $\mu_n(0)$（原点矩）的数学公式表示如下：

$$\mu_n(0) = \int x^n P(x) \tag{2.5}$$

对于时间序列，离散分布 $P(x)$ 的零点矩用公式 2.6 表示：

$$\mu_n(0) = \sum x^n P(x) \tag{2.6}$$

矩可以基于某个点 a，那么公式 2.6 可以计算为公式 2.7：

$$\mu_n(a) = \sum (x-a)^n P(x) \tag{2.7}$$

一阶矩 μ_1 表示平均值；二阶矩 μ_2 表示方差；三阶矩 μ_3 表示计算非对称度分布的偏度。四阶是峰度，用于测量分布的峰态。中心矩是取自分布均值的矩。分布的累积量 k_n 源自概率密度函数 $P(x)$。高阶累积量表示为矩的线性组合（Begg, Lai, & Palaniswami, 2008；Subasi, 2019）。假设用原点矩，前四阶累积量可以用公式 2.8 明确表示：

$$\begin{aligned} k_1 &= \mu_1 \\ k_2 &= \mu_2 - \mu_1^2 \\ k_3 &= 2\mu_1^3 - 3\mu_1\mu_2 + \mu_3 \\ k_4 &= -6\mu_1^4 + 12\mu_1^2\mu_2 - 3\mu_2^2 - 4\mu_1\mu_3 + \mu_4 \end{aligned} \qquad (2.8)$$

集中趋势中最显著的统计量为平均值

$$\mu = \frac{1}{M}\sum_{j=1}^{M}|y_j| \qquad (2.9)$$

以及中位数和众数（Flach，2012）。

基于特征的第二类计算是离散度（分散度）统计，即与平均值之差的平方的平均值（方差）和它的平方根（标准差）。

$$\sigma = \sqrt{\frac{1}{M}\sum_{j=1}^{M}(y_j - \mu)^2} \qquad (2.10)$$

方差和标准差大体上测量的是同样的东西，但是后者的好处是它和特征表示的是同一尺度。还有一种更简单的离散度统计是极差，即最大值和最小值之间的差值。另一种离散度统计是百分位数。第 p 百分位数是令 $p\%$ 的实例低于该值的一个数。假设有 100 个实例，第 80 百分位数表示一组从小到大排列的值中的第 81 个实例。如果 p 是 25 的倍数，百分位数也叫四分位数。百分位数和四分位数是分位数的特殊情况。一旦知道分位数，我们就可以用分位数间的距离测量出离散度。例如，四分位距是第三和第一四分位数（即第 75 百分位数和第 25 百分位数）之差。

一个分布的偏斜情况和峰态是用偏度和峰度来衡量的，它们分别是样本的三阶和四阶中心矩。显然，一阶中心矩是与均值的平均差，二阶中心矩是与均值的平均平方偏差，也称作方差。三阶中心矩 m_3 可为正可为负。偏度的定义为：

$$\varphi = \sqrt{\frac{1}{M}\sum_{j=1}^{M}\frac{(y_j - \mu)^3}{\sigma^3}}, \qquad (2.11)$$

其中 σ 是样本的标准差。偏度为正数表示分布为右偏分布，偏度为负数则为左偏分布。峰度的定义为：

$$\phi = \sqrt{\frac{1}{M}\sum_{j=1}^{M}\frac{(y_j - \mu)^4}{\sigma^4}} \qquad (2.12)$$

峰度值为正表示该分布的峰部比正态分布更尖锐（Flach，2012）。

同时，一些特征提取方法得到的特征向量太大而无法用作分类器的输入。通过使用一阶、二阶、三阶和四阶累积量，就可以计算出缩小后的特征集。在任何学习任务中，将一组系数转化为更小的特征集是关键的一步,因为缩小过的特征集能够更好地反映时间序列数据的行为(Subasi,2019)。

例 2.1 以下 Python 代码用于计算统计值,例如计算向量的平均值、中位数、标准差、偏度和峰度。请注意本例改编自 Python-scikit-learn。

```python
# descriptive statistics
import scipy as sp
import scipy.stats as stats
import numpy as np
from matplotlib import pyplot as plt
from sklearn.datasets.samples_generator import make_blobs
#Create a dataset
X, y = make_blobs(n_samples = 300, centers = 4, cluster_std = 0.60, random_state = 0)
plt.scatter(X[:,0], X[:,1])

X_mean = sp.mean(X[:,1])
print('Mean = ',X_mean)

X_mean = np.mean(X[:,1])
print('Mean = ',X_mean)

X_SD = sp.std(X[:,1])
print('SD = ',X_SD)
X_SD = np.std(X[:,1])
print('SD = ',X_SD)

X_median = sp.median(X[:,1])
print('Median = ',X_median)
X_median = np.median(X[:,1])
print('Median = ',X_median)

X_skewness = stats.skew(X[:,1])
print('Skewness = ',X_skewness)

X_kurtosis = stats.kurtosis(X[:,1])
print('Kurtosis = ',X_kurtosis)
```

2.2.3 结构化特征

人们通常潜在地认为实例就是由特征值组成的向量。这说明除了从实例的特征值中提取的信息以外不存在其他信息。用特征值向量定义实例称为抽象,即过滤掉冗余信息的结果。抽象的一个例子是用词频向量定义邮件。但偶尔也要避免这种抽象,需要保留更多关于实例的信息,这些信息可以在有限的特征值向量获取到。例如,一封邮件可以表示为一个长字符串,或者一个单词和标点符号组成的序列,或者一棵由 HTML 标签组成的树等。

描述这种结构化实例空间的特征称为结构化特征。结构化特征的显著特点是，它们涉及表示对象而不是实例本身的局部变量。虽然如此，还可以对局部变量采用其他形式的聚合。结构化特征可以在学习模型之前构建，也可以在学习时构建。第一种场景称为命题化，因为特征可以看作从一阶逻辑到没有局部变量的命题逻辑的转换。命题化方法的主要挑战是如何处理潜在特征数量的组合爆炸问题。

2.2.4 特征转换

特征转换的目标是通过排除、修改或者添加信息的方式提高特征的有效性。最常见的特征转换是将一类特征转换成序列中的下一种类型的特征。不过，也有转换会修改定量特征的尺度或者为顺序特征、分类特征和布尔特征增加范围（或者条目）。最简单的特征转换是完全演绎，在这种情况下，它们得出明确的结果而不需要做任何选择。二值化是将一个分类特征转换成一组布尔特征，分类特征中的每个值对应一个布尔特征。这么做将丢失信息，因为单个分类特征的值是互斥的，但是这适用于模型不能处理两个以上的特征值的情况。无序化简单地通过去掉特征值的顺序将顺序特征变成分类特征。由于大多数学习模型无法直接处理有序的特征，因此通常需要无序化。另一个有趣的选择是通过校准方法为特征添加一个尺度（Flach，2012）。

sklearn.preprocessing 包提供几个定制的工具函数和转换器类，可用于将原始特征向量转换成更适合下游估计器的表示。通常，数据集的标准化对学习算法有益。对存在一些离群值的集合，比较适合进行定标、稳定转换和归一化处理。

2.2.5 阈值化和离散化

阈值化通过特征值划分数据，将有序特征或定量特征转换为布尔特征。这些阈值可以通过有监督或无监督的方式筛选出来。典型的无监督阈值化包括对数据进行一些统计计算，而有监督阈值化需要对特征值数据进行排序，并通过该排序来优化特定的目标函数，比如信息增益。如果阈值化泛化出多个阈值，则得到了最常用的非演绎特征转换之一。在离散化中，一个定量特征被转换成一个有序特征。每个有序值被表示成一个二进制数并关联一个原始定量特征的区间（Flach，2012）。

2.2.6 数据操作

将不同的特征转换为同一尺度主要有两种方法：标准化和归一化。这些术语往往相当宽泛地用于各种领域，并且它们的含义可能根据上下文会有所不同。通常，归一化表示将特征的范围重新调整为 [0，1] 的区间，它是最小最大缩放的特殊情况。归一化数据时可以简单地将最小最大缩放应用到每个特征列上（Raschka，2015）。

> **例 2.2** 以下 Python 代码用 scikit-learn 实现了最小最大缩放处理。该例子改编自 Python-scikit-learn。另一种标准化方式是将特征按比例缩小至给定的最小值和最大值之

间，通常在 0 和 1 之间，或者使每个特征的最大绝对值按比例缩小到单位大小。这分别可通过 MinMaxScaler 和 MaxAbsScaler 实现。调整比例可以确保特征有极小标准差并保留稀疏数据中的零条目项。这里是一个将数据矩阵缩小到 [0，1] 范围的例子。

```
from sklearn import preprocessing
import numpy as np
X_train = np.array([[ 1., -1., 2.],
            [ 2., 0., 0.],
            [ 0., 1., -1.]])
print('Original Matrix:\n',X_train)
min_max_scaler = preprocessing.MinMaxScaler()
X_train_minmax = min_max_scaler.fit_transform(X_train)
print('Scaled Matrix:\n',X_train_minmax)
```

2.2.7 标准化

众所周知，scikit-learn 中实现的多种机器学习算法都需要对数据集进行标准化。如果各个特征没有或多或少看上去像标准正态分布的数据，即零均值、单位方差的高斯分布数据，那么这些算法将不会达到很好的效果。实际上，分布的形态通常被忽略掉，而仅仅通过去掉每个特征的平均值将数据转换至中心点，然后通过将非常量特征除以其标准差来进行缩放。该预处理模块额外提供了一个工具类 StandardScaler，该工具类利用 Transformer API 对训练集计算平均值和标准差，之后也可以用这种方式在测试集上做同样的转换（scikit-learn, n.d.）。

例 2.3 以下 Python 代码用 scikit-learn 实现了简单的缩放处理。请注意本例改编自 Python-scikit-learn。该缩放函数提供对单个类似数组的数据集实现简单快捷的操作。

```
from sklearn import preprocessing
import numpy as np
X_train = np.array([[ 1., -1., 2.],
           [ 2., 0., 0.],
           [ 0., 1., -1.]])
X_scaled = preprocessing.scale(X_train)
X_scaled
#%%
#Scaled data has zero mean and unit variance:

X_scaled.mean(axis = 0)
#%%
X_scaled.std(axis = 0)
```

最小最大缩放归一化比较合适需要数据在一个有界区间的情况。标准化可能对某些机器学习技术更加实用，因为不同的线性模型初始化权重为 0 或者接近 0 的小随机值。通过标准化,特征列被中心化为以平均值0、标准差为1的方式呈正态分布。此外,相较于最小最大缩放将数据缩放至有限的范围,标准化保留了关于离群值的有效信息并且使算法对这些信息不那么敏感。像 MinMaxScaler 一样,在 scikit-learn 中标准化也作为一个 StandardScaler 类被实现。还必须强调的是 StandardScaler 类只适用于训练数据以及利用这些参数转换测试集或任何新的数据点(Raschka,2015)。

例 2.4 以下 Python 代码用 scikit-learn 实现缩放处理。该预处理模块还提供一个实现 Transfromer API 的工具类 StandarScaler 来对训练集计算平均值和标准差,后面还可以再次用该类对测试集进行同样的转换。因此该类适合用于 sklearn.pipeline.Pipeline 中的早期步骤。请注意本例改编自 Python-scikit-learn。StandardScaler 可以如下使用:

```python
from sklearn import preprocessing
import numpy as np
X_train = np.array([[ 2., -2., 5.],
          [ 4., 0., 0.],
          [ 0., 2., -4.]])
print('Original Matrix:\n',X_train)
scaler = preprocessing.StandardScaler().fit(X_train)
scaler
print('Scalar Mean:',scaler.mean_)
print('Scalar:',scaler.scale_)
X_train_scaled = scaler.transform(X_train)
print('Scaled Matrix:\n',X_train_scaled)
```

我们利用前面的代码从 preprocessing 模块中加载 StandardScaler 类,并初始化一个新的 StandardScaler 对象。StandardScaler 使用 fit 方法为来自训练数据的每个特征维度估算参数 μ(样本均值)和 σ(标准差)。然后我们利用得出的参数 μ 和 σ 通过调用 transform 方法标准化训练数据。请注意,我们在标准化测试集时也采用了同样的缩放参数,以便训练集和测试集中的数据可以相互比较(Raschka,2015)。scikit-learn 还实现了大量不同的性能指标,这些指标可以通过 metrics 模块获得。

例 2.5 以下 Python 代码用 scikit-learn 实现缩放处理。StandardScaler 适用于 sklearn.pipeline.Pipeline 的早期步骤。在这个例子中,我们使用 sklearn.datasets.samples_generator 创建数据集。请注意,本例改编自 Python-scikit-learn。StandardScaler 可以如下使用:

```python
from matplotlib import pyplot as plt
from sklearn.datasets.samples_generator import make_blobs
from sklearn import preprocessing
import numpy as np
#Create a dataset
```

```
X, y = make_blobs(n_samples = 100, centers = 4, cluster_std = 0.60, random_
state = 0)
plt.scatter(X[:,0], X[:,1])

print('Original Matrix:\n',X)
scaler = preprocessing.StandardScaler().fit(X)
scaler

print('Scalar Mean:',scaler.mean_)
print('Scalar:',scaler.scale_)
X_scaled = scaler.transform(X)
print('Scaled Matrix:\n',X_scaled)
```

另一种转换是非线性转换。该转换分为两种：分位数转换和幂转换。两种都是基于特征的单调转换，从而保留每个特征值的秩。QuantileTransformer 和 quantile_transform 提供非参数转换，将数据映射到 0 ~ 1 的均匀分布上。

例 2.6 以下 Python 代码用 scikit-learn 实现 QuantileTransformer 处理。在该例子中，我们使用 sklearn.datasets 中的 Iris 数据集以及二维散点图。请注意，本例改编自 Python-scikit-learn。QuantileTransformer 可以如下使用：

```
import numpy as np
import matplotlib.pyplot as plt
from sklearn.datasets import load_iris
from sklearn import preprocessing

iris = load_iris()
X, y = iris.data, iris.target

X = iris.data[:, :2] # we only take the first two features.
y = iris.target

plt.figure(2, figsize = (8, 6))
plt.clf()

#%%
#Transform data
quantile_transformer = preprocessing.QuantileTransformer(random_state = 0)
X_trans = quantile_transformer.fit_transform(X)
np.percentile(X[:, 0], [0, 25, 50, 75, 100])

np.percentile(X_trans[:, 0], [0, 25, 50, 75, 100])
#%%
# Plot the original data points
plt.scatter(X[:, 0], X[:, 1], c = y, cmap = plt.cm.Set1,
            edgecolor = 'k')
```

```
plt.xlabel('Sepal length')
plt.ylabel('Sepal width')
plt.title('Original data points')
plt.xticks(())
plt.yticks(())

#%%
# Plot the transformed data points
plt.scatter(X_trans[:, 0], X[:, 1], c = y, cmap = plt.cm.Set1,
            edgecolor = 'k')
plt.xlabel('Sepal length')
plt.ylabel('Sepal width')
plt.title('Transformed data points')
plt.xticks(())
plt.yticks(())
```

例 2.7 本例是例子 2.6 的一个修改版本,我们同样使用 QuantileTransformer,但绘制的是三维散点图:

```
import numpy as np
import matplotlib.pyplot as plt
from sklearn.datasets import load_iris
from mpl_toolkits.mplot3d import Axes3D
from sklearn.model_selection import train_test_split
from sklearn import preprocessing

iris = load_iris()
X, y = iris.data, iris.target

X = iris.data[:, :3] # we only take the first two features.
y = iris.target

plt.figure(3, figsize = (8, 6))
plt.clf()

#%%
#Transform data
quantile_transformer = preprocessing.QuantileTransformer(random_state = 0)
X_trans = quantile_transformer.fit_transform(X)

np.percentile(X[:, 0], [0, 25, 50, 75, 100])
np.percentile(X_trans[:, 0], [0, 25, 50, 75, 100])

#%%
# Plot the original data points
fig = plt.figure(1, figsize = (8, 6))
ax = Axes3D(fig, elev = -150, azim = 110)
ax.scatter(X[:, 0], X[:, 1], X[:, 2], c = y,
           cmap = plt.cm.Set1, edgecolor = 'k', s = 40)
ax.set_title('Original data points')
ax.set_xlabel('Sepal length')
ax.w_xaxis.set_ticklabels([])
```

```
ax.set_ylabel('Sepal width')
ax.w_yaxis.set_ticklabels([])
ax.set_zlabel('Sepal height')
ax.w_zaxis.set_ticklabels([])
plt.show()
#%%
# Plot the transformed data points
fig = plt.figure(1, figsize = (8, 6))
ax = Axes3D(fig, elev = -150, azim = 110)
ax.scatter(X_trans[:, 0], X_trans[:, 1], X_trans[:, 2], c = y,
           cmap = plt.cm.Set1, edgecolor = 'k', s = 40)
ax.set_title('Transformed data points')
ax.set_xlabel('Sepal length')
ax.w_xaxis.set_ticklabels([])
ax.set_ylabel('Sepal width')
ax.w_yaxis.set_ticklabels([])
ax.set_zlabel('Sepal height')
ax.w_zaxis.set_ticklabels([])

plt.show()
```

2.2.8 归一化和校准

特征转换中的阈值化和离散化可以去掉定量特征的度量。用无监督的方式对定量特征的范围进行调整一般称为归一化，而特征校准则使用有监督的方式。特征归一化主要是为了消除在不同尺度上测量的一些定量特征的影响。如果特征是正态分布的，以平均值为中心并除以标准差则可以将其变换到 z 分数。特征校准被认为是有监督的特征变换，其为任意特征添加一个有意义的度量，并且该度量携带分类信息（Flach，2012）。

例 2.8 以下 Python 代码用 scikit-learn 实现归一化处理。该例子使用 sklearn.datasets 中的 Iris 数据集，并且分别绘制使用 preprocessing.normalize 进行归一化前后的三维散点图。请注意，本例改编自 Python-scikit-learn。归一化是缩放单个样本以具有单位范数的过程。如果你计划使用二次形式（如点积或任何其他核函数）来量化任何样本对间的相似度，那么此过程将非常有用。normalize 函数使用 l1 范数或 l2 范数对几个数据集进行快速且简单的归一化处理。

```
import numpy as np
import matplotlib.pyplot as plt
from sklearn.datasets import load_iris
from mpl_toolkits.mplot3d import Axes3D
from sklearn.model_selection import train_test_split
from sklearn import preprocessing
```

```
iris = load_iris()
X, y = iris.data, iris.target
X = iris.data[:, :3] # we only take the first two features.
y = iris.target
plt.figure(3, figsize = (8, 6))
plt.clf()
#%%
# Plot the original data points
# Plot the original data points
fig = plt.figure(1, figsize = (8, 6))
ax = Axes3D(fig, elev = -150, azim = 110)
ax.scatter(X[:, 0], X[:, 1], X[:, 2], c = y,
           cmap = plt.cm.Set1, edgecolor = 'k', s = 40)
ax.set_title('Original data points')
ax.set_xlabel('Sepal length')
ax.w_xaxis.set_ticklabels([])
ax.set_ylabel('Sepal width')
ax.w_yaxis.set_ticklabels([])
ax.set_zlabel('Sepal height')
ax.w_zaxis.set_ticklabels([])

plt.show()
#%%
X_normalized = preprocessing.normalize(X, norm = 'l2')
# Plot the normalized data points
fig = plt.figure(1, figsize = (8, 6))
ax = Axes3D(fig, elev = -150, azim = 110)
ax.scatter(X_normalized[:, 0], X_normalized[:, 1], X_normalized[:, 2],
c = y,
           cmap = plt.cm.Set1, edgecolor = 'k', s = 40)
ax.set_title('Transformed data points')
ax.set_xlabel('Sepal length')
ax.w_xaxis.set_ticklabels([])
ax.set_ylabel('Sepal width')
ax.w_yaxis.set_ticklabels([])
ax.set_zlabel('Sepal height')
ax.w_zaxis.set_ticklabels([])
plt.show()
```

2.2.9 不完整的特征

在训练时缺失特征值时会产生问题。首先，缺失的特征值可能与目标变量有关。这些特征必须指定为"缺失"值，因为树模型可能会在其上分叉，但是线性模型可能不会如此。插补是填充缺失值的过程，该过程可以通过在特征的观察值上取每个类的平均值、中位数或者众数实现。一种更复杂的策略是找到特征的相关性，即通过为每个不完整的特征创建一个预测模型并使用该模型"预测"缺失值（Flach, 2012）。

由于各种原因，真实世界的许多数据集都包含缺失值，它们通常被编码成空格、非数值或者其他的占位符。但是这样的数据集并不能与scikit-learn的估计器兼容，因为这些估计器会假设数组中的元素都是数值的且都有自己的含义。想要使用不完整的数据集，一种简单的方式是丢弃掉包含缺失数据的整行和（或）整列。不过，带来的代价是丢失可能比较重要（即使不完整）的数据。更好的策略是计算出缺失值，即从数据的已知部分推导出它们。MissingIndicator转换器可用于将数据集转换为相应的二进制矩阵来证明数据集中缺失值的存在。这种转换与插补相结合是非常有用的。当使用插补时，知道哪些值是缺失的可能会有用。非数值通常被用作缺失值的占位符。但这会强制数据为浮点数类型。missing_values参数允许标识其他类型的占位符，比如整型。在下面的例子中，-1将被作为缺失值（scikit-learn, n.d.）。

例2.9 以下Python代码用scikit-learn实现填充缺失值的插补过程。该例子使用sklearn.datasets中的Iris数据集，并且分别绘制使用SimpleImputer和MissingIndicator进行插补前后的三维散点图。请注意，本例改编自Python-scikit-learn。

```
import numpy as np
from sklearn.datasets import load_iris
from sklearn.impute import SimpleImputer, MissingIndicator
from sklearn.model_selection import train_test_split
from sklearn.pipeline import FeatureUnion, make_pipeline
from sklearn.tree import DecisionTreeClassifier
X, y = load_iris(return_X_y = True)
mask = np.random.randint(0, 2, size = X.shape).astype(np.bool)
X[mask] = np.nan
X_train, X_test, y_train, _ = train_test_split(X, y, test_size = 100,ran-
dom_state = 0)

#%%
"""Now a FeatureUnion is created. All features will be imputed utilizing
SimpleImputer,
in order to enable classifiers to work with this data. Moreover, it adds
the
indicator variables from MissingIndicator."""
transformer = FeatureUnion(
                transformer_list = [
                ('features', SimpleImputer(strategy = 'mean')),
                ('indicators', MissingIndicator())])
transformer = transformer.fit(X_train, y_train)
results = transformer.transform(X_test)
results.shape

#%%%
"""Of course, the transformer cannot be utilized to make any predictions.
this should be wrapped in a Pipeline with a classifier (e.g., a
DecisionTreeClassifier)
to be able to make predictions."""
clf = make_pipeline(transformer, DecisionTreeClassifier())
```

```
clf = clf.fit(X_train, y_train)
results = clf.predict(X_test)
results.shape
```

2.2.10 特征提取的方法

特征提取的目的是寻找信息量最大、最鲜明和最简化的特征集，从而提高数据存储和处理的成功率。对于分类问题，重要的特征向量依然是最常见、最合适的信号表示方式。各领域对数据建模和分类感兴趣的许多科学家正团结力量去改善特征提取面临的问题。如今在数据分析和机器学习领域的进步令创造识别系统成为可能，这使得过去无法完成的任务得以完成。在数据分析中，特征提取是应用进步的核心（Guyon，Gunn，Nikravesh，& Zadeh，2006；Subasi，2019）。

在特征提取中，我们关心的是如何找到一个新的 k 维向量的集合，它们是原有 d 维向量的组合。广为人知且最常用的特征提取方法是主成分分析（principal component analysis）和线性判别分析（linear discriminant analysis），无监督和有监督学习技术。主成分分析被认为类似于另外两种无监督的线性方法，即因子分析（factor analysis）和多维缩放（multidimensional scaling）。当观测变量不是一组而是两组时，可以利用典型相关分析（canonical correlation analysis）来寻找联合特征，从而解释两组变量之间的依赖关系（Alpaydin，2014）。

传统的分类器不包含处理类边界的过程。因此，如果输入变量（特征数量）与训练数据的数量相比很大，那么类边界可能不会重叠。在这种情况下，分类器的泛化能力可能不够。因此，为了提高泛化能力，通常从原始输入变量中通过特征提取、降维或特征选择等方法形成一小组特征。在构建一个具有高泛化能力的模型时，最有效且最典型的做法是采用信息量大且特点鲜明的特征集。不过，由于没有有效的方法可以为某个分类问题找到一组原始特征，因此有必要通过试错来找到原始特征集。如果特征的数量很大且每个特征对分类的影响微不足道，更合适的方案是将特征集转换成一个更小的特征集。在数据分析中，通过线性变换的方法将原始数据转换成特征集。如果原始特征集中的每个特征对分类都有影响，则通过特征提取、特征选择或者降维去缩小特征集。通过特征选择或降维去掉无效或冗余的特征后，可以获得更高的泛化性能，而且比通过初始特征集进行分类更快（Abe，2010；Subasi，2019）。

对于有效的分类来说，需要精确的特征提取技术来从原始数据集中提取信息量大且特点鲜明的特征集。本质上，如果提取的特征无法准确地表示使用的数据且没有意义，那么使用这些特征的分类算法在描述特征的类别时可能会存在问题（Siuly et al.Siuly，Li，& Zhang，2016）。相应地，分类的性能可能较差。生物医学信号分类的关键步骤之一就是特征提取。因此，由多个数据点组成的生物医学信号可以通过不同的特征提取方式提取那些信息量大且特点鲜明的特征。这些信息量大且特点鲜明的参数可以描述生物医学信号的行为，这些行为可能指向特定的动作或活动。生物医学信号分析中采用的信号模式可以用频率和振幅来描述。这些特征可以利用各种特征提取算法提取出来，这是信号处理中简化分类的后续阶段的另一个关键步骤（Graimann，Allison，& Pfurtscheller，2009）。由于所有波形的持续时间和频率都是有限的，因此需要对生物医学信号进行有效的分解，以实现时间、频率和空间维度的整合。生物医学信号可以使用能检测时间和频率变化的时频（TF）方法来分解（Kevric & Subasi，2017；Sanei，2013；Subasi，2019）。

用更少的数据来描述正确的信号特征，这对分类来说很重要且可以达到更好效果。通常通过将信号转换成相关的特征收集到一个特征向量中，这称为特征提取。信号分类框架分析出信号的显著特征，并根据这些显著特征决定信号的类别（Siuly et al.，2016）。特征提取技术可以分为四种：参数方法、非参数方法、特征向量方法和时频方法。基于模型的方法或者参数方法使用已知的函数形式产生信号模型，然后估计产生的模型中的参数。自回归（AR）模型、滑动平均（MA）模型、自回归滑动平均（ARMA）模型和Lyapunov指数是常用的参数方法。AR模型适用于表示窄峰的光谱（Kay，1993；Kay & Marple，1981；Proakis & Manolakiss；Stoica & Moses，1997）。非参数方法基于对功率谱密度（PSD）的描述来进行谱估计。两种著名的非参数方法是周期图和相关图，这些方法的确为足够长的数据长度提供了实际的高分辨率，但谱估计能力很差，因为它们的方差很高而且不会随着数据长度的增加而减少。特征向量方法用于从带噪声的测量中估计信号的频率和功率。这些方法是由带噪声的信号的相关矩阵的特征分解产生的。对于由若干个正弦波组成且噪声较小的信号来说，特征向量方法，如Pisarenko法、最小范数法和多信号分类法（MUSIC）对信号的拟合效果最好（Proakis & Manolakis，2007；Stoica & Moses，1997）。时频方法被广泛用于生物医学信号的处理和分析。时频方法，如短时傅里叶变换（STFT）、小波变换（WT）、离散小波变换（DWT）、平稳小波变换（SWT）和小波包分解（WPD）对信号进行时域和频域分解（Siuly et al.，2016）。短时傅里叶变换、Wigner-Ville分布、Cohen类核函数、小波变换、离散小波变换、平稳小波变换、小波包分解、双树复小波变换（DT-CWT）、可调Q小波变换（TQWT）、经验小波变换（EWT）、经验模态分解（EMD）、集成经验模态分解（EEMD）、变分模态分解（VMD）和完备集成经验模态分解（CEEMD）都是广为人知的时频方法（Subasi，2019）。

特征提取的目的是从原始信号中提取特征，从而实现可靠的分类。对生物医学信号的分类来说，特征提取是最关键的一步，因为错误的提取方式将导致分类效果大打折扣。特征提取步骤需要对原始数据降维，降维的结果需包含原始向量中多数的有价值信息。因此，根据数据集的本质找出能描述整个数据集的关键特征至关重要。统计特征是定义生物医学信号分布的最典型的值，可以从每个子样本数据点中提取多样化的统计特征。这些特征可以是生物医学信号的最小值、最大值、平均值、中位数、众数、标准差、方差、第一四分位数、第三四分位数和四分位距（IQR）(Siuly et al.，2016；Subasi，2019）。

最近几十年，一些特征提取算法已经被广泛地用于数据分析。分类器的性能取决于待分类数据的基本特征。没有一个分类器可以在所有问题上表现优秀。一些真实的测试已经用于比较分类器的性能并证明数据的特性会影响分类器的性能。广为人知的评估分类系统性能的方法有总体分类精确度和混淆矩阵。最近，受试者操作特征（ROC）曲线被用来权衡给定分类算法的真阳性和假阳性率（Siuly et al., 2016；Subasi, 2019）。

2.2.11 使用小波变换进行特征提取

每种变换都会提供额外的信息帮助我们重新了解原始波形。一些时频方法无法完全解决时频问题。小波变换是一种描述波形的时频特性的方法。但波形是被分割成尺度段而不是时间段（Semmlow, 2004）。小波由时间上的伸缩和滑动两个参数组成。小波变换是时间上能量集中的振荡函数，它用于增强对瞬态信号的表示。带通滤波器特性只是小波函数必须具备的几个数学特性之一。小波分析试图在时间和频率上都达到令人满意的局部化。滑动和伸缩是两个新的自由度，它们可以分析信号中的细微结构和全局波形。多分辨率分析的重要思想是以越来越高的分辨率水平分析不同尺度的信号（Sörnmo & Laguna, 2005）。一些好书中有全面讲述如何运用数学方法进行小波分析，其中也包括滤波库的话题（Basseville & Nikiforov, 1993；Gustafsson, 2000）。

1. 连续小波变换

连续小波变换（CWT）是小波家族的成员，用 $\psi_{s,\tau}(t)$ 表示，它是根据连续值参数 τ 和 s 对母小波 $\psi(t)$ 的滑动和伸缩：

$$\psi_{s,\tau}(t) = \frac{1}{\sqrt{s}} \psi\left(\frac{t-\tau}{s}\right) \qquad (2.13)$$

其中因子 $1/\sqrt{s}$ 确保所有的伸缩函数包含相同的能量。小波在 $s>1$ 时膨胀，$0<s<1$ 时收缩。

探测函数 $\psi_{s,\tau}(t)$ 总是呈振荡形式。当 $s=1$，$\tau=0$ 时表现为其天然形态，称为母小波 $\psi_{1,0}(t) \equiv \psi(t)$，连同它的几个家族成员一起由收缩和膨胀产生。如果一个小波被收缩到更小的时间尺度，那就会使它在时间上更局部，而在频率上更不局部，因为由此产生的带通频率响应的带宽增大、频率更高。连续时间信号 $x(t)$ 的连续小波变换 $\omega(s,\tau)$ 是通过比较信号 $x(t)$ 与探测函数 $\psi_{s,\tau}(t)$ 来识别的：

$$\omega(s,\tau) = \int_{-\infty}^{+\infty} x(t) \psi_{s,\tau}(t) \mathrm{d}t \qquad (2.14)$$

该变换创建时间尺度域上的二维映射。由于以上方程表示信号 $x(t)$ 与具有脉冲响应 $\psi(-t/s)/\sqrt{s}$ 的滤波器之间的卷积，因此 CWT 可以看作一个线性滤波器。由于 CWT 将波形分解为关于变量 s 和 τ 的系数，因此我们需要对小波系数进行两重积分来重建原始波形。

$$x(t) = \frac{1}{C_\psi} \int_{-\infty}^{\infty} \int_{0}^{\infty} \omega(s,\tau) \psi_{s,\tau}(t) \frac{\mathrm{d}\tau \mathrm{d}s}{s^2}, \qquad (2.15)$$

其中 $C_\psi = \int_0^\infty \dfrac{|\psi(f)|^2}{|f|} df < \infty$, （2.16）

$\psi(f)$ 表示 $\psi(t)$ 的傅里叶变换。最简单的小波是 Haar 小波，它是 Walsh 基础函数的成员。另一个常见的小波是墨西哥帽小波，它的定义方程为：

$$\psi(t) = (1 - 2t^2)e^{-t^2}$$ （2.17）

Morlet 小波是以一个小波分析的先驱者命名，它的方程为：

$$\psi(t) = e^{-t^2} \cos\left(\pi\sqrt{\dfrac{2}{\ln 2}}t\right)$$ （2.18）

目前已经有各种各样的小波被提出，每种小波都具有一些适合特定应用的特征。小波提供了时间和频率局部化之间的平衡。然而，它们并不是以精确的时间或频率出现的。更准确地说，它们并不完全包含在时间或频率中，而是很好地包含在两者中。这些范围也与 CWT 的时间和频率分辨率有关。较短的小波时间范围提供了更好的隔离局部时间事件的能力，但以频率分辨率为代价，因为小波只响应高频分量。另一方面，CWT 为较长的小波长度提供更高的频率分辨率。这种时间和频率分辨率之间的统一折中使得 CWT 适用于分析包含被缓慢变化（低频）分量覆盖的快速变化（高频）分量的信号（Semmlow, 2004；Subasi, 2019）。

例 2.10 以下 Python 代码使用连续小波变换提取 ECG 信号特征。

```
#===================================================================
# Continuous wavelet transform
#===================================================================
import numpy as np
import matplotlib.pyplot as plt
import pywt
import pywt.data

ecg = pywt.data.ecg()
plt.plot(ecg)
plt.xlabel("Samples")
plt.ylabel("ECG in mV")
plt.show()

#Continuous wavelet transform.
from scipy import signal
import matplotlib.pyplot as plt
widths = np.arange(1, 31)
cwtmatr = signal.cwt(ecg, signal.ricker, widths)
plt.imshow(cwtmatr, extent = [-1, 1, 31, 1], cmap = 'PRGn', aspect = 'auto',
           vmax = abs(cwtmatr).max(), vmin = -abs(cwtmatr).max())
plt.show()
```

2. 离散小波变换

CWT 的唯一主要问题是它的无限冗余，因为产生了无数的系数，超出了精确描述原始信号的实际需要。只有需要重建原始信号时，这些冗余才会变得昂贵，因为所有的系数都会被使用，这将产生许多不必要的计算量。离散小波变换（DWT）一般是通过限制伸缩和滑动的变化为 2 的幂来产生系数，因此有时也称为二进小波变换，其缩写相同。然而，我们仍然可以从二进小波变换的离散系数精确地形成原始信号（Bodenstein et al., 1985）。如果选中的小波属于正交家族，DWT 甚至表示了一个非冗余的双边变换（Semmlow, 2004）。

两个小波参数的二元采样定义为：

$$s = 2^{-j}, \tau = k2^{-j} \tag{2.19}$$

其中 j 和 k 都是整数。因此离散探测函数为：

$$\psi_{j,k}(t) = 2^{j/2}\psi(2jt - k) \tag{2.20}$$

插入方程 2.20 到 2.14 后，我们得到离散小波变换（DWT）：

$$\omega_{j,k} = \int_{-\infty}^{\infty} x(t)\psi_{j,k}(t)dt \tag{2.21}$$

通过 DWT 的逆变换或小波序列展开来恢复原信号：

$$x(t)\sum_{j=-\infty}^{\infty}\sum_{k=-\infty}^{\infty}\omega_{j,k}\psi_{j,k}(t) \tag{2.22}$$

其中 $\psi_{j,k}(t)$ 是一组正交基函数。小波序列展开表示为在两个指标 j 和 k 上面的总和，这与基函数 $\psi_{j,k}(t)$ 的伸缩和滑动有关。

这里我们提出了一个称为尺度函数的新概念，它将简化 DWT 的应用和计算。首先计算最佳分辨率，接着通过原始波形的平滑形式而不是原始波形本身来计算粗糙分辨率。平滑形式采用尺度函数获得，有时候也称为平滑函数（Semmlow, 2004; Subasi, 2019）。

例 2.11 下面的 Python 代码使用 6 级分解的离散小波变换（DWT）提取 ECG 信号特征。它将打印然后绘制近似系数和细节系数。

```
# ================================================================
# ================================================================
# Discrete wavelet transform
# ================================================================
#
# Created on Sat Sep 14 23:20:26 2019
#
# @author: asubasi
# ================================================================
print(__doc__)
```

```python
import numpy as np
import matplotlib.pyplot as plt
from scipy import signal
from scipy.misc import electrocardiogram
ecg = electrocardiogram()
fs = 360
time = np.arange(ecg.size) / fs
plt.plot(time, ecg)
plt.xlabel("time in s")
plt.ylabel("ECG in mV")
plt.xlim(9, 10.2)
plt.ylim(-1, 1.5)
plt.show()
#%%
import numpy as np
import matplotlib.pyplot as plt
import pywt
import pywt.data
ecg = pywt.data.ecg()
mode = pywt.Modes.sp1DWT = 1
#db1 = pywt.Wavelet('db1')
waveletname = 'db1'
#waveletname = 'sym5'
coeff = pywt.wavedec(ecg, waveletname, level = 6)
cA6,cD6,cD5,cD4, cD3, cD2, cD1 = coeff

print('cA1\n',cD1)
print('cA2\n',cD2)
print('cA3\n',cD3)
print('cA4\n',cD4)
print('cA5\n',cD5)
print('cD5\n',cD6)
print('cD5\n',cA6)
fig, ax = plt.subplots(figsize = (6,1))
ax.set_title("Original ECG Signal: ")
ax.plot(ecg)
plt.show()
#%%
fig, axarr = plt.subplots(nrows = 7, ncols = 1, figsize = (9,9))
axarr[0].plot(cD1, 'r')
axarr[0].set_ylabel("cD1", fontsize = 14, rotation = 90)

axarr[1].plot(cD2, 'r')
axarr[1].set_ylabel("cD2", fontsize = 14, rotation = 90)
axarr[2].plot(cD3, 'r')
axarr[2].set_ylabel("cD3", fontsize = 14, rotation = 90)
```

```
axarr[3].plot(cD4, 'r')
axarr[3].set_ylabel("cD4", fontsize = 14, rotation = 90)
axarr[4].plot(cD5, 'r')
axarr[4].set_ylabel("cD5", fontsize = 14, rotation = 90)
axarr[5].plot(cD6, 'r')
axarr[5].set_ylabel("cD6", fontsize = 14, rotation = 90)
axarr[6].plot(cA6, 'r')
axarr[6].set_ylabel("cA6", fontsize = 14, rotation = 90)
axarr[1].set_yticklabels([])
axarr[0].set_title("Coefficients", fontsize = 14)
plt.tight_layout()
plt.show()
```

3. 平稳小波变换

平稳小波变换（SWT）一次性计算出给定信号的所有离散小波变换（DWT）。更具体地说，对于第1级，SWT是通过用合适的滤波器对信号进行卷积来完成的，但没有使用下采样。那么第1级的细节系数和近似系数将与信号长度相同。通过利用不进行下采样的合适滤波器，一般的步骤 j 对 $j-1$ 级的近似系数进行卷积得到 j 级的细节系数和近似系数（Zhang et al.，2015）。

DWT是对信号 $x(t)$ 的分解，其被认为是连续的带通滤波和下采样。$x(t) = x_0(t)$ 被分解为两个部分：$y_1(t)$ 代表 $x_0(t)$ 的高频部分，而 $x_1(t)$ 代表低频部分。利用连续的滤波器组可以使DWT的计算更快。遗憾的是，一旦使用离散时间序列 $x(t)$，则DWT不是平移不变的。如果输入的时间序列 $x(t)$ 被平移，那么结果系数可能变得完全不同。平稳小波变换没有此类问题（Nason & Silverman，1995），主要是因为虽然SWT是DWT的一种，但是下采样被替换成了上采样（Sudakov et al.，2017；Subasi，2019）。

例2.12 以下Python代码利用5级分解的平稳小波变换（SWT）提取ECG信号特征。它将打印然后绘制近似系数和细节系数。

```
# ================================================================
# Stationary wavelet transform
# ================================================================
"""
Created on Sat Sep 14 23:20:26 2019
@author: asubasi
"""
print(__doc__)

import numpy as np
import matplotlib.pyplot as plt
```

```python
from scipy import signal
from scipy.misc import electrocardiogram
ecg = electrocardiogram()
fs = 360
time = np.arange(ecg.size) / fs
plt.plot(time, ecg)
plt.xlabel("time in s")
plt.ylabel("ECG in mV")
plt.xlim(9, 10.2)
plt.ylim(-1, 1.5)
plt.show()
#%%
import pywt
import matplotlib.pyplot as plt
import numpy as np

#db1 = pywt.Wavelet('db1')
waveletname = 'db1'
#waveletname = 'sym5'
coeffs = pywt.swt(ecg, waveletname, level = 5)
print('cA1\n',coeffs[0][0])
print('cA2\n',coeffs[1][0])
print('cA3\n',coeffs[2][0])
print('cA4\n',coeffs[3][0])
print('cA5\n',coeffs[4][0])

print('cD5\n',coeffs[4][1])
print('cD4\n',coeffs[3][1])
print('cD3\n',coeffs[2][1])
print('cD2\n',coeffs[1][1])
print('cD1\n',coeffs[0][1])

fig, ax = plt.subplots(figsize = (6,1))
ax.set_title("Original ECG Signal: ")
ax.plot(ecg)
plt.show()
#%%
fig, axarr = plt.subplots(nrows = 5, ncols = 2, figsize = (6,6))

axarr[0, 0].plot(coeffs[0][0], 'r')
axarr[0, 1].plot(coeffs[0][1], 'g')
axarr[0, 0].set_ylabel("Level {}".format(1), fontsize = 14, rotation = 90)
axarr[1, 0].plot(coeffs[1][0], 'r')
axarr[1, 1].plot(coeffs[1][1], 'g')
axarr[1, 0].set_ylabel("Level {}".format(2), fontsize = 14, rotation = 90)
axarr[2, 0].plot(coeffs[2][0], 'r')
axarr[2, 1].plot(coeffs[2][1], 'g')
axarr[2, 0].set_ylabel("Level {}".format(3), fontsize = 14, rotation = 90)
axarr[3, 0].plot(coeffs[3][0], 'r')
axarr[3, 1].plot(coeffs[3][1], 'g')
```

```
axarr[3, 0].set_ylabel("Level {}".format(4), fontsize = 14, rotation = 90)
axarr[4, 0].plot(coeffs[4][0], 'r')
axarr[4, 1].plot(coeffs[4][1], 'g')
axarr[4, 0].set_ylabel("Level {}".format(5), fontsize = 14, rotation = 90)
axarr[1, 0].set_yticklabels([])
axarr[0, 0].set_title("Approximation coefficients", fontsize = 14)
axarr[0, 1].set_title("Detail coefficients", fontsize = 14)
axarr[0, 1].set_yticklabels([])
plt.tight_layout()
plt.show()
```

4. 小波包分解

小波变换通过将信号分解成一组基函数以获得更好的时间分辨率。小波包分解（WPD）被看作 DWT 的扩展，其中低频分量（即近似值）被分解。另一方面，WPD 同时使用近似值（低频分量）和细节（高频分量）（Daubechies，1990；Learned & Willsky，1995；Unser & Aldroubi，1996）。DWT 和 WPD 互不相同，因为 WPD 同时将低频与高频分量划分成它们的分段。所以，WPD 对分解后的信号产生增强的频率分辨率。WPD 被认为是一种连续时间的小波变换，其适用不同尺度或层级的不同频率。小波包分解有助于结合不同层级的分解来建立原始信号（Kutlu & Kuntalp，2012）。WPD 的分解分两步实现。第一步，修改滤波器 / 下采样级联。在 WPD 结构的每一级级联中，两个分支（近似系数和细节系数）都被进一步过滤和下采样。第二步，采用基于熵的标准，对树进行修改从而为给定信号选择最适合的分解。这个过程称为分解树的裁剪（Blinowska & Zygierewicz，2011；Subasi，2019）。

例 2.13 以下 Python 代码利用 5 级分解的小波包分解（WPD）提取 ECG 信号特征。然后绘制近似系数和细节系数。

```
# ==================================================================
# Wavelet packed decomposition (WPD)
# ==================================================================
"""
Created on Sat Sep 14 23:20:26 2019
@author: asubasi
"""

print(__doc__)

import numpy as np
import matplotlib.pyplot as plt

from scipy import signal
from scipy.misc import import electrocardiogram

ecg = electrocardiogram()
```

```
fs = 360
time = np.arange(ecg.size) / fs
plt.plot(time, ecg)
plt.xlabel("time in s")
plt.ylabel("ECG in mV")
plt.xlim(9, 10.2)
plt.ylim(-1, 1.5)
plt.show()
#%%
import pywt
import matplotlib.pyplot as plt
import numpy as np
#waveletname = 'db1'
waveletname = pywt.Wavelet('db1')
fig, ax = plt.subplots(figsize = (6,1))
ax.set_title("Original ECG Signal: ")
ax.plot(ecg)
plt.show()

fig, axarr = plt.subplots(nrows = 5, ncols = 2, figsize = (6,6))
wp = pywt.WaveletPacket(ecg, waveletname, mode = 'symmetric', maxlevel = 6)
axarr[0, 0].plot(wp['a'].data, 'r')
axarr[0, 1].plot(wp['d'].data, 'g')
axarr[0, 0].set_ylabel("Level {}".format(1), fontsize = 14, rotation = 90)

axarr[1, 0].plot(wp['aa'].data, 'r')
axarr[1, 1].plot(wp['dd'].data, 'g')
axarr[1, 0].set_ylabel("Level {}".format(2), fontsize = 14, rotation = 90)

axarr[2, 0].plot(wp['aaa'].data, 'r')
axarr[2, 1].plot(wp['ddd'].data, 'g')
axarr[2, 0].set_ylabel("Level {}".format(3), fontsize = 14, rotation = 90)

axarr[3, 0].plot(wp['aaaa'].data, 'r')
axarr[3, 1].plot(wp['dddd'].data, 'g')
axarr[3, 0].set_ylabel("Level {}".format(4), fontsize = 14, rotation = 90)

axarr[4, 0].plot(wp['aaaaa'].data, 'r')
axarr[4, 1].plot(wp['ddddd'].data, 'g')
axarr[4, 0].set_ylabel("Level {}".format(5), fontsize = 14, rotation = 90)

axarr[1, 0].set_yticklabels([])
axarr[0, 0].set_title("Approximation coefficients", fontsize = 14)
axarr[0, 1].set_title("Detail coefficients", fontsize = 14)
axarr[0, 1].set_yticklabels([])
plt.tight_layout()
plt.show()
```

2.3 降维

降维是一个降低原始特征向量维度的过程，保留最特别的信息并去掉无关的信息，从而缩短分类器的计算时间（Phinyomark et al.，2013）。大多数特征提取方法都会产生冗余特征。事实上，通过使用某类特征选择/裁剪方法产生一个新的特征集可以提高分类器的性能并实现最小分类误差。为了达到更高的分类准确率，许多技术已经被用于降维和特征选择（Wołczowski & Zdunek，2017）。在特征选择和降维中，需要从原始特征集中选择能够实现最大泛化能力的最小特征子集。为此，特征子集的泛化能力必须在特征选择的过程中评估出来（Abe，2010；Subasi，2019）。

数据的维度也许是数据分析中获得更准确结果所需要的。少量的参数以不同的方式被用于减少数据的维度。此外，为了实现更好的分类准确率，必须将特征或维度最小化。例如，基于小波的时频方法产生小波系数来描述信号能量在时域和频域的分布，它们用一组小波系数来描述生物医学信号。由于基于小波的特征提取工具产生的特征向量规模过大而无法作为分类器的输入，因此应该利用降维技术从小波系数中提取更少量的特征。近来，各种降维方法被用于降维，如 Lyapunov 指数、低阶或高阶统计量、熵等。近似熵作为复杂度的衡量标准，可以应用于有噪声的数据集，并且优于谱熵、Kolmogorov-Sinai 熵和分形维数。样本熵对数据长度的依赖性较小。模糊熵起源于模糊集理论，它是另一种衡量复杂性的方法。降维的另一种方法是利用子带（sub-bands）的一阶、二阶、三阶和四阶统计量。缩小后的特征集是从时频分解的子带中计算出来的（Subasi，2019）。

降维是特征选择的另一种方法。比如，主成分分析（PCA）已经被有效地用于各种研究中。它与特征选择相似的是会产生低维的表示，这有助于创造更低容量的预测器从而提高泛化能力。但是与特征选择不同的是，它可能会保留来自所有原始输入变量的信息。本质上，这些方法是完全无监督的，它们可能会去除对目标标签有高度预测性的低方差变化，或者保留一些与当前的分类任务无关的高方差变化。还可以在降维中结合特征提取算法选择出更适合的特征，反之亦然（Bengio et al.，2006；Subasi，2019）。

通常来说，相比使用所有特征，利用原始特征集合的一个子集能实现更好的分类准确率。存在的冗余特征会导致辨析度降低，还可能会使分类器产生混淆。因此，特征选择对于产生更准确的模型具有终极意义。有很多特征选择方法可以选择出最佳的特征集去描述问题（Begg et al.，2008；Subasi，2019）。

特征提取是将数据从高维空间变换到低维空间。数据变换可以是线性的，如主成分分析（PCA），然而还有许多非线性的降维技术。降维可以通过线性判别分析（LDA）、主成分分析、独立成分分析（ICA）等方式实现。这些方法的定义使我们对降维的理解更加清晰。对于二维非高斯数据集，PCA 提取方差最大的分量，ICA 提取最独立的分量。ICA 认为信号元素就像具有高斯分布和最小二阶统计量的随机变量。显然，对于任何非高斯分布，最大方差不符合 PCA 基向量。ICA 最小化输入数据中的二阶和高阶依赖，并试图找到数据（投影到数据上时）在统计上独立的基准。LDA 在潜在空间中找到最能区分类的向量（Delac et al.，2005；Subasi，2019）。

降维是将数据从高维空间映射到低维空间的方法。该过程与压缩的理念直接相关。对数据降维的理由有很多。首先，高维数据带来计算的挑战。此外，在某些场景中高维会导致学习算法的泛化能力变差。最后，降维可以用于解释数据和找到数据中有价值的部分并用于说明目的。本章介绍广泛使用的降维技术。这些技术通过对原始数据进行线性变换实现降维（Shalev-Shwartz & Ben-David，2014）。

大数据集，包括具有许多特征的数据，在预测模型的训练中会产生计算问题。对数据集进行降维的方法，比如k均值聚类，将减少数据集的特征维度或特征数量。传统降维方式主成分分析（PCA）常用于数据分析，对于预测模型数据中的特征降维来说是相对欠佳的工具。但是，PCA所展示的基础数学原型——矩阵分解，提供了一种有价值的方法组织我们对于许多重要学习模型的思考（Watt, Borhani, & Katsaggelos, 2016）。

当我们开始从原始数据样本中提取特征或属性时，有些时候会发现特征空间中的特征数量极大。由此我们将面临各种挑战，分析和可视化拥有成千上万特征的数据导致特征空间极其复杂，还会引起与训练模型、内存和空间限制有关的问题。这被称为"维度灾难"。在这些场景中，无监督技术也可以用来减少每个数据样本中特征或属性的数量。这些技术通过提取或者筛选出一组主要的或者具有代表性的特征从而减少特征变量的数量。目前有许多受欢迎的降维算法，比如主成分分析（PCA）、最近邻分类和判别分析（Sarkar, Bali, & Sharma, 2018）。

2.3.1 特征构造和选择

新的特征可以从不同的原始特征中构造出来。一个新的特征可以通过生成两个布尔或者分类特征的笛卡儿乘积构建。当新的特征生成以后，在学习之前从它们中选出合适的子集是很容易的。这将加快学习并避免过拟合。特征选择的方法很多。过滤式方法根据特定的指标对特征进行评分，然后选出得分靠前的特征。可以采用多种指标为特征评分，如信息增益、χ^2统计量和相关系数。简单的过滤式方法的主要缺点是没有考虑到特征之间的冗余性。此外，特征过滤器不区分特征之间的依赖性，因为它们只依赖边际分布（Flach, 2012）。

特征选择关注的是从d维特征中找出最有信息量的k维特征，并丢弃另外$(d-k)$维特征。在子集选择中，我们关心找出特征集的最佳子集，其利用最少数量的维度贡献最高的准确率。剩余的且不重要的维度将被丢弃。在分类和回归问题中可以使用合适的误差函数。有两种主要的方法。前进法（forward selection）从0个变量开始依次加入变量，每次添加那个使误差减少最多的变量，直到无法降低误差为止。后退法（backward selection）从所有变量开始依次移除变量，每次移除减少误差最多的变量，直到任何要移除的变量都会明显地增加误差为止。在这两种情况下，检查误差必须在训练集之外的验证集上进行，因为需要测试普遍的准确率。通常来说特征越多训练误差越低，但验证误差不一定越低。如果删除某个特征只会小幅地增加误差，我们可以决定删除该特征以降低复杂度。子集选择是有监督的，因为它的输出被分类器或回归器用来评估误差，但它可以用于任何分类或回归方法（Alpaydin, 2014）。

2.3.2 单变量特征选择

单变量特征选择基于单变量统计测试选择出最佳特征。它是估计器的一个预处理步骤。scikit-learn 将特征选择程序作为实现转换方法的对象来实现。

- **SelectKBest** 去除评分最高的 k 个特征以外的所有特征。
- **SelectPercentile** 去除用户指定的评分最高百分比以外的所有特征。
- 对特征的常见单变量统计检验：假阳性率 **SelectFpr**、错误发现率 **SelectFdr**、或总体错误率 **SelectFwe**。
- **GenericUnivariateSelect** 允许用可配置的策略进行单变量特征选择。它允许用超参搜索估计器选择最佳单变量选择策略。

这些对象以一个评分函数作为输入，返回单变量分数和 p 值。(或者只对 **SelectKBest** 和 **SelectPercntile** 评分)：

- 回归：f_regersion、mutual_info_regression
- 分类：chi2、f_classif、mutual_info_classif

基于 F 检验的方法评估两个随机变量之间的线性依赖程度。另一方面，互信息法可以捕捉到任何形式的统计依赖，但由于是非参数化的，它们需要更多的样本来进行准确的估计 (scikit-learn, n.d.)。

例 2.14 以下 Python 代码使用 scikit-learn 实现特征选择的过程。该例子使用 sklearn.datasets 中的 Iris 数据集，并且使用 sklearn.feature_selection.SelectKBest 进行特征选择。特征的数量由 k 决定，该例子中 $k=2$。请注意，本例改编自 Python-scikit-learn。

```
# ====================================================================
# Univariate feature selection
# ====================================================================
"""
Created on Sat Sep 14 23:20:26 2019
@author: asubasi
"""

from sklearn.datasets import load_iris
from sklearn.feature_selection import SelectKBest
from sklearn.feature_selection import chi2
iris = load_iris()
X, y = iris.data, iris.target
print(X)

X.shape

#%%
X_new = SelectKBest(chi2, k = 2).fit_transform(X, y)
print(X_new)
X_new.shape
```

例 2.15 以下 Python 代码使用 scikit-learn 实现单变量特征选择的处理。该例子中，我们使用 sklearn.datasets 中的 Breast Cancer 数据集，并且利用 sklearn.feature_selection.SelectPercentile 和 sklearn.feature_selection.f_classif 进行特征选择。请注意，本例改编自 Python-scikit-learn。在 Breast Cancer 数据中加入（无信息量的）噪声特征并且采用单变量特征选择。对于每个特征，单变量特征选择的 p 值和对应的 SVM 权重都被绘制出来。单变量特征选择筛选出信息量较大的特征，它们的 SVM 权重较大。在整个特征集合中，只有前四个特征是重要且有信息量的，它们在单变量特征选择中评分最高。SVM 对其中的一个特征赋予较大的权重，但同时也选择了许多非信息特征。在使用 SVM 之前先使用单变量特征选择提高那些重要且有信息量的特征的 SVM 权重，由此提高性能。

```
# ==================================================================
# Univariate feature selection
# ==================================================================
"""
Created on Fri Oct 4 00:16:28 2019
@author: asubasi
"""

print(__doc__)

import numpy as np
import matplotlib.pyplot as plt

from sklearn import datasets, svm
from sklearn.feature_selection import SelectPercentile, f_classif
# #################################################################
#######
# Import some data to play with

# The Breast Cancer dataset
Breast_Cancer = datasets.load_breast_cancer()
# Some noisy data not correlated
E = np.random.uniform(0, 0.1, size = (len(Breast_Cancer.data), 20))

# Add the noisy data to the informative features
X = np.hstack((Breast_Cancer.data, E))
y = Breast_Cancer.target
plt.figure(1)
plt.clf()
X_indices = np.arange(X.shape[-1])
# #################################################################
#######
# Univariate feature selection with F-test for feature scoring
# We use the default selection function: the 10% most significant features
selector = SelectPercentile(f_classif, percentile = 10)
selector.fit(X, y)
scores = -np.log10(selector.pvalues_)
```

```python
scores /= scores.max()
plt.bar(X_indices - .45, scores, width = .2,
    label = r'Univariate score ($-Log(p_{value})$)', color = 'darkorange',
    edgecolor = 'black')
# ###################################################################
#######
# Compare to the weights of an SVM
clf = svm.SVC(kernel = 'linear')
clf.fit(X, y)

svm_weights = (clf.coef_ ** 2).sum(axis = 0)
svm_weights /= svm_weights.max()

plt.bar(X_indices - .25, svm_weights, width = .2, label = 'SVM weight',
            color = 'navy', edgecolor = 'black')

clf_selected = svm.SVC(kernel = 'linear')
clf_selected.fit(selector.transform(X), y)

svm_weights_selected = (clf_selected.coef_ ** 2).sum(axis = 0)
svm_weights_selected /= svm_weights_selected.max()

plt.bar(X_indices[selector.get_support()] - .05, svm_weights_selected,
    width = .2, label = 'SVM weights after selection', color = 'c',
    edgecolor = 'black')

plt.title("Comparing feature selection")
plt.xlabel('Feature number')
plt.yticks(())
plt.axis('tight')
plt.legend(loc = 'upper right')
plt.show()
```

例2.16 以下 Python 代码利用 scikit-learn 实现单变量特征选择的过程。该例子使用 sklearn.datasets 中的 Breast Cancer 数据集。我们将展示如何在运行 SVC（支持向量分类器）之前执行单变量特征选择从而提高模型的性能。从图中看出，当所有特征都被用到时，模型达到最好的性能。请注意本例改编自 Python-scikit-learn。

```python
# =====================================================================
# SVM-Anova: SVM with univariate feature selection
# =====================================================================
"""
Created on Fri Oct 4 14:24:39 2019
@author: asubasi
"""
print(__doc__)
import numpy as np
import matplotlib.pyplot as plt
from sklearn.feature_selection import SelectPercentile, chi2
from sklearn.model_selection import cross_val_score
from sklearn.pipeline import Pipeline
```

```python
from sklearn.preprocessing import StandardScaler
from sklearn.svm import SVC
from sklearn import datasets
# #################################################################
#######
# Import some data to play with
Breast_Cancer = datasets.load_breast_cancer()
X = Breast_Cancer.data
y = Breast_Cancer.target
# #################################################################
#######
# Create a feature-selection transform, a scaler and an instance of SVM that we
# combine together to have an full-blown estimator
clf = Pipeline([('anova', SelectPercentile(chi2)),
                ('scaler', StandardScaler()),
                ('svc', SVC(gamma = "auto"))])
# #################################################################
#######
# Plot the cross-validation score as a function of percentile of features
score_means = list()
score_stds = list()
percentiles = (1, 3, 6, 10, 15, 20, 30, 40, 60, 80, 100)

for percentile in percentiles:
    clf.set_params(anova__percentile = percentile)
    this_scores = cross_val_score(clf, X, y, cv = 5)
    score_means.append(this_scores.mean())
    score_stds.append(this_scores.std())

plt.errorbar(percentiles, score_means, np.array(score_stds))
plt.title(
    'Performance of the SVM-Anova varying the percentile of features selected')
plt.xticks(np.linspace(0, 100, 11, endpoint = True))
plt.xlabel('Percentile')
plt.ylabel('Accuracy Score')
plt.axis('tight')
plt.show()
```

2.3.3 递归式特征消除

对于指定特征权重（如线性模型的系数）的机器学习模型来说，递归式特征消除（RFE）通过递归的方式筛选越来越小的特征集来选择特征。首先，在初始特征集上训练机器学习模型，每个特征的重要性是通过 coef_attribute 或 feature_importances_attribute 来确定的。然后消除当前特征集中最没有信息量的特征。该过程在被删减的集合上递归地重复执行直到选出最佳数量的特征（scikit-learn，n.d.）。

例 2.17 以下 Python 代码使用 scikit-learn 实现包含交叉验证的递归式特征消除过程。该例子使用 sklearn.datasets 中的 Breast Cancer 数据集,其利用 sklearn.feature_selection.RFECV 选择特征,还利用递归特征消除与交叉验证自动调整选择特征的数量。RFECV 在交叉验证的循环中执行 RFE 寻找最佳特征数量。请注意,本例改编自 Python-scikit-learn。

```
# =============================================================
# Recursive feature elimination with cross-validation
# =============================================================
"""
Created on Thu Oct 3 23:42:44 2019
@author: asubasi
"""
print(__doc__)
import matplotlib.pyplot as plt
from sklearn.svm import SVC
from sklearn.model_selection import StratifiedKFold
from sklearn.feature_selection import RFECV
from sklearn.datasets import load_breast_cancer
Breast_Cancer = load_breast_cancer()
X = Breast_Cancer.data
y = Breast_Cancer.target
# Create the RFE object and compute a cross-validated score.
svc = SVC(kernel = "linear")
# The "accuracy" scoring is proportional to the number of correct
# classifications
rfecv = RFECV(estimator = svc, step = 1, cv = StratifiedKFold(2),
        scoring = 'accuracy')
rfecv.fit(X, y)

print("Optimal number of features : %d" % rfecv.n_features_)
# Plot number of features VS. cross-validation scores
plt.figure()
plt.xlabel("Number of features selected")
plt.ylabel("Cross validation score (nb of correct classifications)")
plt.plot(range(1, len(rfecv.grid_scores_) + 1), rfecv.grid_scores_)
plt.show()
```

2.3.4 从模型选择特征

SelectFromModel 是一个元转换器,它可以与拟合之后带有 coef_ 或者 feature_importances_ 属性的任意模型一起使用。如果与特征相关的 coef_ 或者 feature_importances_ 值低于预先设置的阈值,则可以认为该特征无意义并移除它。除了指定数值上的阈值以外,还可以通过指定字符串参数调用内置的启发式方法得到阈值。可以使用的启发式方法有"平均数""中位数"以及与浮点数相乘,例如"0.1*平均数"(scikit-learn,n.d.)。

例 2.18 以下 Python 代码使用 scikit-learn 实现从模型选择特征的过程。该例子使用 sklearn.datasets 中的 Breast Cancer 数据集,并且利用 sklearn.feature_selection.SelectFromModel 进行特征选择。请注意,本例改编自 Python-scikit-learn。

```
# ================================================================
# Feature selection from a model
# ================================================================
from sklearn.ensemble import ExtraTreesClassifier
from sklearn.feature_selection import SelectFromModel
from sklearn import datasets
Breast_Cancer = datasets.load_breast_cancer()
X = Breast_Cancer.data
y = Breast_Cancer.target
X.shape
#%%
clf = ExtraTreesClassifier(n_estimators = 50)
clf = clf.fit(X, y)
clf.feature_importances_
model = SelectFromModel(clf, prefit = True)
X_new = model.transform(X)
X_new.shape
```

2.3.5 主成分分析

主成分分析是一种代数特征构建方法,它通过对给定特征进行线性组合构造新的特征。第一个主成分由数据中最大方差的方向指定,第二主成分是与第一个成分正交的最大方差方向,以此类推(Flach, 2012)。PCA 是一种无监督技术,因为它不使用输出信息,原则是使方差最大化。样本投影到 ω_1 上后主分量 ω_1 才计算样本,这使样本点间的差异变得最明显(Alpaydin, 2014)。PCA 是统计学的一个分支,称为多元分析。顾名思义,多元分析解决多个元素或者度量的分析问题。多元数据可以用 M 维空间表示,每个空间维度包含一个信号。一般来说,多元分析旨在得出考虑多个变量之间以及变量内部关系的结果,并使用对所有数据都有效的工具。多元分析的关键问题是找到多元数据的转换,即产生一个更小的数据集。例如,它可能包括多元变量中的相关信息,可以通过使用更少的维度(即变量)来表征,减少后的变量集可能比原始数据集更有意义。转换将原始变量集转换成新的变量集,同时产生新变量,新变量与原始数据相比更少从而最终减少多元数据集的维度。由于这些变量的值相对较小,它们对整个数据集的有益信息可能不多,因此可以删除。通过恰当的转换可以去掉大量对整个信息贡献极少的变量。因为线性转换的计算更简单,因此获取一组新变量的数据转换是线性函数。线性变换的过程可以表示为原始数据集在 M 维空间的旋转以及可能的缩放。PCA 的目标是转换数据集形成一组不相关的新变量(主成分)。它的目的是降低数据的维度,但不一定产生更多有意义的变量。PCA 可以在不丢失信息的情况下减少数据集中变量的数量,并且找出更有意义的新变量。它将一组相关变量转换成一组新的不相关变量。如果数据集中的变量已经是不相关的，那么 PCA 将不实用。此外，主成分间相互正交并且它们描述的可变性组织有序（Semmlow，2004；Subasi，2019）。

在拥有许多特征的大数据集中，使用特征转换找到更少且更精炼的特征表示是更合适的。PCA 是其中一种技术，它对特征进行投影从而产生一个缩小的特征表示。假设特征集包含 n 个训练样本，即 $X=\{x_1, x_2, ..., x_j\}$，然后该算法生成主成分 P_k，即原始特征 X（Begg et al.，2008）的线性组合。P_k 可以写成：

$$Pk = a_{k1}x_1 + a_{k2}x_2 + ... + a_n x_{kn}, \text{其中} \sum_i a_{ki}^2 = 1 \quad (2.23)$$

由于主成分向量是通过相互正交形成的，因此方差（主成分）最大。通常在使用 PCA 算法之前首先需要将训练数据进行标准化为零均值和单位方差（Begg et al.，2008；Subasi，2019）。

降维通常是在保证丢失最少信息的基础上将高维空间转换成低维空间。该过程称为特征提取。PCA 是一种著名的特征提取方法，它可以消除给定随机过程间的二阶相关性。PCA 通过计算输入信号的协方差矩阵的特征向量将输入的高维向量线性转换为分量不相关的低维向量。PCA 通常采用最值衡量信息标量，如最大的投影信号方差或最小的重建误差。PCA 的目标是从输入空间提取出 m 个可以描述信号最小方差的正交方向 $w_i \in R^n, i=1,2,...,M$。接着，输入向量 $x \in R^n$ 在不丢失重要内在信息的情况下被转换到更低的 M 维空间。向量 x 可以重新表示为通过内积 $x^T w_i$ 将其投影到 w_i 所在的 M 维子空间。这就产生了降维（Du & Swamy，2006；Subasi，2019）。

PCA 是一种将多通道信号分解为线性独立分量的方法，即在时间和空间上不相关。在给定的时间间隔内，来自所有通道的样本被视为空间中的一个点，该空间的维度等于通道数。通过忽略最小的方差，原始的时间序列可以进一步表示为这些分量的线性组合以降低数据的维度。PCA 可以通过奇异值分解（SVD）算法来实现（Blinowska & Zygierewicz，2011）。PCA 被用来降低数据集的维度并产生更少的不相关变量作为特征以便更好地对数据进行分类。通常来说，记录的多通道信号数量巨大且高度相关，同时也包含很多冗余信息。PCA 可以将大量的相关变量转换为少量的不相关信息，称作主成分。主成分描述并传递出原始信号中最有信息量的数据和差异信息。因此，PCA 特征在不同的信号分类中效果较好（Siuly et al.，2016；Subasi，2019）。处理预测建模问题时，特征选择是一种实用且通用的特征空间降维方案。主成分分析（PCA）是一种常见的降维技术用于将数据转换到适当的低维特征子空间（Watt et al.，2016）。

PCA 是 scikit-learn 的转换类之一，其在使用相同模型参数对训练数据和测试数据进行变换之前先对使用训练数据的模型进行拟合。

例 2.19 以下的 Python 代码使用 scikit-learn 实现用 PCA 进行降维处理。该例子使用 sklearn.datasets 的 Iris 数据集，降维采用 sklearn.decomposition.PCA。PCA 用于将多变量数据集分解为一组连续的正交分量，这些分量代表了方差的最大值。在 scikit-learn 中，PCA 是以 decomposition 对象的形式实现的，即 sklearn.decomposition.PCA，可以用在将新数据投影到分量上。PCA 在使用 SVD 之前对每个特征的输入数据进行中心化处理但不进行缩放。可通过设置参数 whiten = True 使数据在投影到奇异空间的同时将每个分量缩放为单位方差。这通常在下游模型对信号的各向同性做了强假设时很有帮助。下面的例子将由四个特征组成的 Iris 数据集投影到方差最大的两个维度上。请注意，本例改编自 Python-scikit-learn。

```
# ====================================================================
# Principal component analysis (PCA)
# ====================================================================
import matplotlib.pyplot as plt
from sklearn import datasets
from sklearn.decomposition import PCA

iris = datasets.load_iris()
X = iris.data
y = iris.target
target_names = iris.target_names

pca = PCA(n_components = 2)
X_r = pca.fit(X).transform(X)
# Percentage of variance explained for each components
print('explained variance ratio (first two components): %s'
                % str(pca.explained_variance_ratio_))
#%%
# ====================================================================
# 2D presentation
# ====================================================================
# Plot the original data points
plt.figure(2, figsize = (6, 5))
plt.clf()
plt.scatter(X[:, 0], X[:, 1], c = y, cmap = plt.cm.Set1,
    edgecolor = 'k')
plt.xlabel('Sepal length')
plt.ylabel('Sepal width')
plt.title('Original data points')
plt.xticks(())
plt.yticks(())
#%%
plt.figure()
colors = ['navy', 'turquoise', 'darkorange']
lw = 2

for color, i, target_name in zip(colors, [0, 1, 2], target_names):
    plt.scatter(X_r[y == i, 0], X_r[y == i, 1], color = color, alpha = .8,
```

```python
           lw = lw,
                   label = target_name)
plt.legend(loc = 'best', shadow = False, scatterpoints = 1)
plt.title('PCA of IRIS dataset')
Below is an example of the iris dataset, which is projected on the 3
dimensions that explain most variance.
#%%
# ================================================================
# #3D presentation
# ================================================================
import numpy as np
import matplotlib.pyplot as plt
from sklearn.decomposition import PCA
from sklearn.datasets import load_iris
from mpl_toolkits.mplot3d import Axes3D
from sklearn import preprocessing
iris = load_iris()
X, y = iris.data, iris.target


X = iris.data
y = iris.target


plt.figure(3, figsize = (8, 6))
plt.clf()


#%%
# Plot the original data points
fig = plt.figure(1, figsize = (6, 5))
ax = Axes3D(fig, elev = -150, azim = 110)
ax.scatter(X[:, 0], X[:, 1], X[:, 2], c = y,
     cmap = plt.cm.Set1, edgecolor = 'k', s = 40)
ax.set_title('Original data points')
ax.set_xlabel('Sepal length')
ax.w_xaxis.set_ticklabels([])
ax.set_ylabel('Sepal width')
ax.w_yaxis.set_ticklabels([])
ax.set_zlabel('Sepal height')
ax.w_zaxis.set_ticklabels([])
plt.show()
#%%
#Transform data
pca = PCA(n_components = 3)
X_PCA = pca.fit(X).transform(X)
# Plot the PCA transformed data points
fig = plt.figure(1, figsize = (6, 5))
ax = Axes3D(fig, elev = -150, azim = 110)
ax.scatter(X_PCA[:, 0], X_PCA[:, 1], X_PCA[:, 2], c = y,
```

```
    cmap = plt.cm.Set1, edgecolor = 'k', s = 40)
ax.set_title('PCA of IRIS dataset')
ax.set_xlabel('PCA1')
ax.w_xaxis.set_ticklabels([])
ax.set_ylabel('PCA2')
ax.w_yaxis.set_ticklabels([])
ax.set_zlabel('PCA3')
ax.w_zaxis.set_ticklabels([])
plt.show()
```

2.3.6 增量 PCA

当需要分解的数据超出内存容量时,增量主成分分析(IPCA)天然地用来替代 PCA。IPCA 利用独立于样本数的内存空间对输入数据构造低秩近似。IPCA 仍然依赖于输入数据的特征,但可以通过改变批处理的大小来控制内存使用量。下面给出的例子有助于直观地看到 IPCA 在每次只处理少量样本的情况下产生的数据投影与采用 PCA 时类似(sckit-learn, n.d.)。

例 2.20 以下的 Python 代码使用 scikit-learn 实现用增量 PCA 进行降维处理。该例子使用 sklearn.datasets 的 Breast Cancer 数据集,降维采用 sklearn.decomposition. IncrementalPCA。增量 PCA 用于将多变量数据集分解为一组连续的正交分量,这些分量代表了方差的最大值。在 scikit-learn 中,PCA 是以 decomposition 对象的形式实现的,sklearn.decomposition.IncrementalPCA 可以用于将新数据投影到分量上。请注意,本例改编自 Python-scikit-learn。

```
# =====================================================================
# Incremental PCA with breast cancer dataset
# =====================================================================
import numpy as np
from sklearn.model_selection import train_test_split
from sklearn.decomposition import PCA, IncrementalPCA
from sklearn import datasets
# #####################################################################
#######
# Import some data to play with

Breast_Cancer = datasets.load_breast_cancer()
X = Breast_Cancer.data
y = Breast_Cancer.target

n_components = 2
ipca = IncrementalPCA(n_components = n_components, batch_size = 10)
X_ipca = ipca.fit_transform(X)

pca = PCA(n_components = n_components)
X_pca = pca.fit_transform(X)
```

```
colors = ['navy', 'turquoise', 'darkorange']
for X_transformed, title in [(X_ipca, "Incremental PCA"), (X_pca, "PCA")]:
    plt.figure(figsize = (8, 8))
            for color, i, target_name in zip(colors, [0, 1, 2], Breast_Cancer.target_names):
            plt.scatter(X_transformed[y == i, 0], X_transformed[y == i, 1],
                color = color, lw = 2, label = target_name)
       if "Incremental" in title:
       err = np.abs(np.abs(X_pca) - np.abs(X_ipca)).mean()
       plt.title(title + " of iris dataset\nMean absolute unsigned error "
           "%.6f" % err)
       else:
       plt.title(title + " of iris dataset")
       plt.legend(loc = "best", shadow = False, scatterpoints = 1)
          plt.axis([-1000, 2000, -500, 500])
plt.show()
```

2.3.7 核 PCA

大多数机器学习算法可以对输入数据的线性可分离性做出假设。然而，如果我们处理的是实际应用中经常遇到的非线性问题，线性变换的降维技术，如 PCA 和 LDA 可能不是最佳选择。在这种情况下，内核版本的 PCA 或者说核 PCA 可能更适合。利用核 PCA 将不可线性分离的数据转换到一个新的低维子空间上，这适合于线性分类器（Raschka, 2015）。scikit-learn 已经在 sklearn.decomposition 子模块中实现了核 PCA 类。它的使用方式与标准的 PCA 类相似并且可以通过内核参数指定内核。

例 2.21 以下的 Python 代码使用 scikit-learn 实现用核 PCA 进行降维处理。该例子使用 sklearn.datasets 的 Iris 数据集，降维采用 sklearn.decomposition.KernelPCA。核 PCA 在数据不是线性可分离的情况下使用。线性不可分的数据可以转换成一个新的低维子空间，适合于线性分类器。在 scikit-learn 中，PCA 是以 decomposition 对象的形式实现的，sklearn.decomposition.KernelPCA 可以用于将新数据投影到分量上。请注意，本例改编自 Python-scikit-learn。

```
# ================================================================
# Kernel PCA example
# ================================================================
import matplotlib.pyplot as plt
from sklearn import datasets
from sklearn.decomposition import PCA, KernelPCA
```

```
iris = datasets.load_iris()
X = iris.data
y = iris.target
target_names = iris.target_names

pca = PCA(n_components = 2)
X_pca = pca.fit(X).transform(X)

kpca = KernelPCA(kernel = "rbf", fit_inverse_transform = True, gamma = 10)
X_kpca = kpca.fit_transform(X)
# Percentage of variance explained for each components
print('explained variance ratio (first two components): %s'
    % str(pca.explained_variance_ratio_))
#%%
# Plot the original data points
plt.figure(2, figsize = (8, 6))
plt.clf()

plt.scatter(X[:, 0], X[:, 1], c = y, cmap = plt.cm.Set1,
    edgecolor = 'k')
plt.xlabel('Sepal length')
plt.ylabel('Sepal width')
plt.title('Original data points')
plt.xticks(())
plt.yticks(())

#%%
#Plot PCA applied Iris Data
plt.figure()
colors = ['navy', 'turquoise', 'darkorange']
lw = 2
for color, i, target_name in zip(colors, [0, 1, 2], target_names):
    plt.scatter(X_pca[y == i, 0], X_pca[y == i, 1], color = color, alpha = .8,
    lw = lw,
        label = target_name)
plt.legend(loc = 'best', shadow = False, scatterpoints = 1)
plt.title('PCA of IRIS dataset')
#%%
#Plot Kernel PCA applied Iris Data
plt.figure()
for color, i, target_name in zip(colors, [0, 1, 2], target_names):
    plt.scatter(X_kpca[y == i, 0], X_kpca[y == i, 1], alpha = .8, color = color,
        label = target_name)
plt.legend(loc = 'best', shadow = False, scatterpoints = 1)
plt.title('KPCA of IRIS dataset')
plt.show()
```

2.3.8 邻近成分分析

主成分分析（PCA）用于识别主成分的组合，它描述了数据中最大的方差。邻近成分分析（NCA）试图找到一个特征空间使得随机最近邻算法准确率最高。和 LDA 一样，它是一种有监督的方法。NCA 使用数据的聚类虽然降低了过多维度，但这在感官上是有意义的。NCA 可以用来实现有监督的降维。输入的数据被投影到一个线性子空间上，该子空间由最小化 NCA 目标的方向组成。期望的维数可以通过参数 n_components 来设置（scikit-learn，n.d.）。

例 2.22 该例子将不同的（线性）降维方法应用于 sklearn.datasets 中的 Iris 数据集并进行比较。该数据集的每个类别包含 50 个样本。此外，这个例子比较了 PCA（sklearn.decomposition.PCA）、稀疏 PCA（sklearn.decomposi-tion.SparsePCA）和 NCA（NeighborhoodComponentsAnalysis）应用在 Iris 数据集上的降维效果。PCA 用于识别构成数据中最大方差的属性（主成分或特征空间中的方向）组合。其中不同的样本被绘制在前两个主成分上。稀疏 PCA 是另一种方法，它采用套索（弹性网）来产生具有稀疏载荷的修改后的主成分。NCA 试图找到一个特征空间使得随机最近邻算法准确率最高。它是一种有监督的技术且需要对数据进行聚类，这虽然降低了过多维度但在感官上是有意义的。请注意，本例改编自 Python-scikit-learn。

```
# ====================================================================
# Dimensionality reduction with PCA, sparse PCA, and NCA
# ====================================================================
import numpy as np
import matplotlib.pyplot as plt
from sklearn import datasets
from sklearn.model_selection import train_test_split
from sklearn.decomposition import PCA
from sklearn.discriminant_analysis import LinearDiscriminantAnalysis
from sklearn.neighbors import (KNeighborsClassifier,
                NeighborhoodComponentsAnalysis)
from sklearn.pipeline import make_pipeline
from sklearn.preprocessing import StandardScaler
from sklearn.decomposition import SparsePCA

n_neighbors = 3
random_state = 0
# Load Iris dataset
iris = datasets.load_iris()
X = iris.data
y = iris.target
target_names = iris.target_names
# Split into train/test
X_train, X_test, y_train, y_test = \
    train_test_split(X, y, test_size = 0.5, stratify = y,
            random_state = random_state)

dim = len(X[0])
n_classes = len(np.unique(y))

# Reduce dimension to 2 with PCA
```

```
pca = make_pipeline(StandardScaler(),
            PCA(n_components = 2, random_state = random_state))
# Reduce dimension to 2 with Sparce PCA
spca = make_pipeline(StandardScaler(),SparsePCA(n_components = 2,
        normalize_components = True, random_state = 0))
# Reduce dimension to 2 with NeighborhoodComponentAnalysis
nca = make_pipeline(StandardScaler(),
            NeighborhoodComponentsAnalysis(n_components = 2,
                random_state = random_state))
# Use a nearest neighbor classifier to evaluate the methods
knn = KNeighborsClassifier(n_neighbors = n_neighbors)
# Make a list of the methods to be compared
dim_reduction_methods = [('PCA', pca), ('Sparce PCA', spca), ('NCA', nca)]
# plt.figure()
for i, (name, model) in enumerate(dim_reduction_methods):
    plt.figure()
  # plt.subplot(1, 3, i + 1, aspect = 1)

    # Fit the method's model
    model.fit(X_train, y_train)

    # Fit a nearest neighbor classifier on the embedded training set
    knn.fit(model.transform(X_train), y_train)

    # Compute the nearest neighbor accuracy on the embedded test set
    acc_knn = knn.score(model.transform(X_test), y_test)

    # Embed the data set in 2 dimensions using the fitted model
    X_embedded = model.transform(X)

    # Plot the projected points and show the evaluation score
    plt.scatter(X_embedded[:, 0], X_embedded[:, 1], c = y, s = 30, cmap = 'Set1')
        plt.title("{}, KNN (k = {})\nTest accuracy = {:.2f}".format(name,
  n_neighbors,
acc_knn))
plt.show()
```

2.3.9 独立成分分析

独立成分分析（ICA）是一个统计建模技术，也是 PCA 的一种延伸。ICA 最初为盲源分离（BSS）而引入然后经过修改用于降维和特征提取。ICA 可以用于 BSS、降维、特征提取和信号探测。不完备 ICA（undercomplete ICA）可以用于特征提取，而过完备 ICA（overcomplete ICA）可以用于多尺度且冗余的基础集合的降维（Du & Swamy, 2006）。ICA 的思想是将信号分解成基础的独立分量。假定连接的源信号是相互独立的，那么该思想在信号的去噪和分离中起着至关重要的作用（Sanei & Chambers, 2013; Subasi, 2019）。

ICA 是一种统计信号处理方法，它将多通道信号分解为统计上独立的成分。ICA 的目标表示为

$$y = W^T X \qquad (2.24)$$

其中 y 的成分必须是统计上独立的。估算信号需要原始信号的高阶统计量而不是 PCA 中使用的二阶矩或样本的协方差。基于所有独立分量都具有有限方差这一假设，Cramer-Rao 边界被用来计算 ICA 中的源信号（Du & Swamy，2006）。要想根据中心极限定理找到独立分量，应该找到最小高斯分布的分量。为此，必须遵循启发式的方法，即假设所需的独立分量应具有相同的分布（Hyvärinen & Oja，2000；Subasi，2019）。在 scikit-learn 中，独立成分分析使用 Fast ICA 算法实现。由于 ICA 模型不包含噪声项，可以通过白化处理确保模型准确。这可以利用内部的 whiten 参数或手动使用 PCA 某个变量来实现。它天然地适用于分离混合信号（盲源分离）(scikit-learn, n.d.)。

例 2.23　下面的 Python 代码直观地列举并比较了在特征空间中利用两种不同的分量分析方法（即 PCA 和 ICA）进行分析的结果。在特征空间中表示 ICA 提供了"几何 ICA"的视图。ICA 是一种在特征空间中寻找方向的算法，这些方向与强非高斯投影有关。这些方向在原始特征空间中不需要是正交的，但在去噪后的特征空间中是正交的，所有的方向都关联相同的方差。另一方面，PCA 在原始特征空间中找到与寻找最大方差方向相关的正交方向。该例子利用具有低自由度数（程序运行后的左上图，下同）的一个强非高斯过程模拟独立源。它们被结合起来产生观测值（右上图）。在这个原始观测空间中，PCA 描述方向显示成绿色向量。与 PCA 向量相关的方差经过白化处理后，信号显示在 PCA 空间中（左下图）。ICA 负责在这个空间中寻找一个旋转以确定最大的非高斯方向（右下）。请注意，本例改编自 Python-scikit-learn。

```
# =================================================================
# FastICA on 2D point clouds
# =================================================================
# Authors: Alexandre Gramfort, Gael Varoquaux
# License: BSD 3 clause
import numpy as np
import matplotlib.pyplot as plt
from sklearn.decomposition import PCA, FastICA
# ################################################################
#######
# Generate sample data
rng = np.random.RandomState(42)
S = rng.standard_t(1.5, size=(20000, 2))
S[:, 0] *= 2.

# Mix data
A = np.array([[1, 1], [0, 2]]) # Mixing matrix

X = np.dot(S, A.T) # Generate observations

pca = PCA()
```

```python
S_pca_ = pca.fit(X).transform(X)
ica = FastICA(random_state = rng)
S_ica_ = ica.fit(X).transform(X)  # Estimate the sources
S_ica_ /= S_ica_.std(axis = 0)
# ######################################################################
#######
# Plot results
def plot_samples(S, axis_list = None):
    plt.scatter(S[:, 0], S[:, 1], s = 2, marker = 'o', zorder = 10,
        color = 'steelblue', alpha = 0.5)
        if axis_list is not None:
        colors = ['green', 'red']
        for color, axis in zip(colors, axis_list):
            axis /= axis.std()
            x_axis, y_axis = axis
            # Trick to get legend to work
            plt.plot(0.1 * x_axis, 0.1 * y_axis, linewidth = 2, color = color)
            plt.quiver(0, 0, x_axis, y_axis, zorder = 11, width = 0.01, scale = 6,
                color = color)
            plt.hlines(0, -3, 3)
            plt.vlines(0, -3, 3)
            plt.xlim(-3, 3)
            plt.ylim(-3, 3)
            plt.xlabel('x')
            plt.ylabel('y')
plt.figure()
plt.subplot(2, 2, 1)
plot_samples(S / S.std())
plt.title('True Independent Sources')

axis_list = [pca.components_.T, ica.mixing_]
plt.subplot(2, 2, 2)
plot_samples(X / np.std(X), axis_list = axis_list)
legend = plt.legend(['PCA', 'ICA'], loc = 'upper right')
legend.set_zorder(100)

plt.title('Observations')

plt.subplot(2, 2, 3)
plot_samples(S_pca_ / np.std(S_pca_, axis = 0))
plt.title('PCA recovered signals')

plt.subplot(2, 2, 4)
plot_samples(S_ica_ / np.std(S_ica_))
plt.title('ICA recovered signals')

plt.subplots_adjust(0.09, 0.04, 0.94, 0.94, 0.26, 0.36)
plt.show()
```

ICA 被用来评估给定噪声测量数据的源头。想象一下同时演奏三个乐器并且用三个麦克风记录混合信号。ICA 用于恢复源头，即各个乐器的演奏信号。值得注意的是，PCA 无法恢复乐器的演奏信号，因为相关信号反映的是非高斯过程（scikit-learn, n.d.）。

> **例 2.24** 使用 FastICA 进行盲源分离：以下 Python 代码使用 scikit-learn 实现用 ICA 进行降维处理。ICA 用于评估给定噪声测量数据的源头。想象一下同时演奏三个乐器并且用三个麦克风记录混合信号。ICA 用于恢复源头，即各个乐器的演奏信号。值得注意的是，PCA 无法恢复乐器的演奏信号，因为相关信号反映的是非高斯过程。在这个例子中，我们采用带有加性噪声分量的三种不同信号的组合，并利用 sklearn.decomposition.FastICA 分离信号源。在 scikit-learn 中，fastICA 实现成一个 decompostion 对象，即 sklearn.decomposition.FastICA，可以用于分离信号源。请注意，本例改编自 Python-scikit-learn。
>
> ```
> # ================================================================
> # Blind source separation using FastICA
> # ================================================================
> import numpy as np
> import matplotlib.pyplot as plt
> from scipy import signal
> from sklearn.decomposition import FastICA, PCA
> # ####################################################################
> #######
> # Generate sample data
> np.random.seed(0)
> n_samples = 2000
> time = np.linspace(0, 8, n_samples)
>
> s1 = np.sin(2 * time) # Signal 1 : sinusoidal signal
> s2 = np.sign(np.sin(3 * time)) # Signal 2 : square signal
> s3 = signal.sawtooth(2 * np.pi * time) # Signal 3: saw tooth signal
>
> S = np.c_[s1, s2, s3]
> S += 0.2 * np.random.normal(size = S.shape) # Add noise
>
> S /= S.std(axis = 0) # Standardize data
> # Mix data
> A = np.array([[1, 1, 1], [0.5, 2, 1.0], [1.5, 1.0, 2.0]]) # Mixing matrix
> X = np.dot(S, A.T) # Generate observations
>
> # Compute ICA
> ica = FastICA(n_components = 3)
> S_ = ica.fit_transform(X) # Reconstruct signals
> A_ = ica.mixing_ # Get estimated mixing matrix
> # We can 'prove' that the ICA model applies by reverting the unmixing.
> assert np.allclose(X, np.dot(S_, A_.T) + ica.mean_)
>
> # For comparison, compute PCA
> pca = PCA(n_components = 3)
> ```

```
H = pca.fit_transform(X)  # Reconstruct signals based on orthogonal compo-
nents

# #########################################################################
#######
# Plot results
plt.figure(2, figsize = (8, 6))
models = [X, S, S_, H]
names = ['Observations (mixed signal)',
    'True Sources',
    'ICA recovered signals',
    'PCA recovered signals']
colors = ['red', 'steelblue', 'orange']
#colors = ['navy', 'turquoise', 'darkorange']
for ii, (model, name) in enumerate(zip(models, names), 1):
    plt.subplot(4, 1, ii)
    plt.title(name)
    for sig, color in zip(model.T, colors):
        plt.plot(sig, color = color)
plt.subplots_adjust(0.09, 0.04, 0.94, 0.94, 0.26, 0.46)
plt.show()
```

2.3.10 线性判别分析

线性判断分析（LDA）是用于解决分类问题的有监督的降维技术。

PCA被广泛用于特征提取。作为PCA的一种改进，核PCA已被广泛接受。由于PCA不使用类别信息，第一主成分实际上对分类并不重要。另一方面，LDA所寻找的轴可以最大限度地将投射在该轴上的训练数据分成两个类。LDA的应用仅限于每个类由一个聚类组成的情况，并且它们不紧密重叠。通过适当地选择核及其参数值，核判别分析（KDA）消除了LDA的局限性。它也可以扩展到多类问题。KDA已被用作特征选择、核选择以及特征提取的标准（Abe, 2010; Subasi, 2019）。

LDA是众多降维方法中著名且被广泛使用的方法。LDA旨在令类间散度和总数据散度在投影空间中的比值最大，并且需要对每个数据进行标注。然而在实际应用中，有标签的数据很少而无标签的数据数量很大，所以LDA在这种情况下很难应用（Subasi, 2019; Wang, Lu, Gu, Du, & Yang, 2016）。

例2.25 下面的Python代码使用LDA进行降维。以下例子将包含四种特征的Iris数据集投影到方差最大的二维空间。该数据集包括三种类型的鸢尾花（Setosa、Versicolour和Virginica）和鸢尾花的四个属性：萼片长度、萼片宽度、花瓣长度和花瓣宽度。LDA发现构成"类间"差异最大的属性。相较于PCA，LDA利用已知的类别标签，是一种典型的监督学习方法。本例中我们使用sklearn.discriminant_ analysis.LinearDiscriminantAnalysis 进行降维。在 scikit-learn 中，LDA 可以用来降维并且是以 discriminant_analysis 对象的形式实现，即 sklearn.discriminant_analysis.LinearDiscriminantAnalysis。请注意，本例改编自 Python-scikit-learn。

```python
# ================================================================
# LDA 2D projection of Iris dataset
# ================================================================
import matplotlib.pyplot as plt
from sklearn import datasets
from sklearn.discriminant_analysis import LinearDiscriminantAnalysis

iris = datasets.load_iris()
X = iris.data
y = iris.target
target_names = iris.target_names

lda = LinearDiscriminantAnalysis(n_components = 2)
X_r2 = lda.fit(X, y).transform(X)

# Percentage of variance explained for each components
print('explained variance ratio (first two components): %s'
      % str(lda.explained_variance_ratio_))
#%%
# ================================================================
# 2D presentation
# ================================================================
# Plot the original Iris data points
plt.figure(2, figsize = (6, 5))
colors = ['turquoise', 'blue', 'red']
for color, i, target_name in zip(colors, [0, 1, 2], target_names):
    plt.scatter(X[y == i, 0], X[y == i, 1], alpha = .8, color = color,
        label = target_name)
plt.legend(loc = 'best', shadow = False, scatterpoints = 1)
plt.title('Original IRIS dataset')
plt.xlabel('Sepal length')
plt.ylabel('Sepal width')

plt.show()
#%%
#LDA applied Iris Data
plt.figure(2, figsize = (6, 5))
colors = ['turquoise', 'blue', 'red']
for color, i, target_name in zip(colors, [0, 1, 2], target_names):
    plt.scatter(X_r2[y == i, 0], X_r2[y == i, 1], alpha = .8, color = color,
        label = target_name)
plt.legend(loc = 'best', shadow = False, scatterpoints = 1)
plt.title('LDA applied IRIS dataset')
plt.show()
```

2.3.11 熵

熵是一种不确定性的程度。数据的混乱程度可以用系统的熵来计算。熵越高表示不确定性越高,系统越混乱。熵定义为:

$$x(n) \int_{\min(x)}^{\max(x)} P_x \log\left(\frac{1}{p_x}\right) dx \qquad (2.25)$$

其中 p_x 是信号 $x(n)$ 的概率密度函数(PDF)。一般来说,一旦数据通道被合并处理,其分布就可以是一个合并的 PDF。尽管这种测量方法是用于单通道信号,但它可以很容易地扩展到多通道甚至多维信号。另一方面,在一个事件的发生受到另一个事件限制时,PDF 可以被条件 PDF 所取代。在这种情况下,熵被称为条件熵。熵非常容易受到噪声的影响。即使原始信号是有序的,噪声也会增加不确定性,而且噪声信号的熵更高。熵还被用来计算许多其他有用的参数,如互信息、负熵、非高斯性和 Kulback-Leibler 发散。这些变量被全面用于估计系统的非线性程度和信号的对应关系。

例 2.26 下面的 Python 代码使用熵和相对熵进行降维。我们将使用一个随机产生的样本数据集,它有四个不同的中心。在这个例子中,我们利用 scipy.special.entr 找到给定向量的熵。在 scipy 中,entr 被实现成一个特殊的对象,即 scipy.special.entr,我们可以利用它来计算熵。同样,elementwise 函数 scipy.special.rel_entr 也被用来计算相对熵。

```
# ====================================================================
# Entropy dimension reduction
# ====================================================================
from matplotlib import pyplot as plt
from sklearn.datasets.samples_generator import make_blobs
from scipy import special

#Create a dataset
X, y = make_blobs(n_samples = 300, centers = 4, cluster_std = 0.60, random_
state = 0)

#%%
# Plot the original data points
plt.figure(2, figsize = (6, 5))
plt.clf()
plt.scatter(X[:, 0], X[:, 1], c = y, cmap = plt.cm.Set1,
            edgecolor = 'k')
plt.xlabel('X')
plt.ylabel('Y')
plt.title('Original data points')
plt.xticks(())
plt.yticks(())

#%%
```

```
#Calculate the Entropy of a vector
X_entropy = special.entr(X[:,1])
print('Entropy = ',X_entropy)

#Elementwise function for computing relative entropy
#special.rel_entr(x, y)
X_rel_entr = special.rel_entr(X[:,0], X[:,1])
print('Relative Entropy = ',X_rel_entr)
```

2.4 基于聚类的特征提取和降维

k 均值聚类通过从数据点的聚类或者组群中寻找合适的代表点或者中心点实现降维。然后每个聚类中的所有元素根据对应的中心点来描述特征。因此，聚类问题是将数据分割成具有相似特征的聚类，对 k 均值来说这个特征就是在特征空间中的几何接近程度。当清楚这一点后，就可以放弃不切实际的概念，然后用它来创建一个学习问题，即准确地复原聚类的中心点。用 c_k 表示第 k 个聚类的中心点，用 S_k 表示属于这个聚类的 P 个数据点子集的索引集合，P 个数据点分别表示为 $x_1...x_P$，属于第 k 个聚类的点必须靠近它的中心点，且所有中心点数学上可以用 $k=1...k$ 表示。首先将中心点排列至中心点矩阵中可以恰当地表示所需要的这种关系。

$$C = c_1 c_2 \cdots c_k \qquad (2.26)$$

然后用 e_k 表示第 k 个标准基向量（即 $k \times 1$ 向量，第 k 点为 1，其他地方为 0），我们可以表示 $C_{e_k} = c_k$，因此对于每个 k 方程（2.26）中的关系可以表示为

$$C_{e_k} \approx x_p \ (对于所有 p \in S_k) \qquad (2.27)$$

接下来，为了更恰当地写出这些方程，我们将数据逐列叠加到数据矩阵 $X = x_1\ x_2 \cdots x_P$ 中，并生成一个 $K \times P$ 的分配矩阵 W，该矩阵的第 p 列表示为 w_p，它是第 p 个点所属聚类对应的标准基向量，即是说如果 $p \in S_k$，则 $w_p = e_k$。利用 w_p 我们可以将式（2.27）表示为 $Cw_p \approx x_p$ 对所有 $p \in S_k$，或者用矩阵同时表示所有的 K 个关系

$$CW \approx X \qquad (2.28)$$

我们可以忘掉这样的假设：已知聚类中心点的位置以及哪些点被分配给它们，也就是对中心点矩阵 C 和分配矩阵 W 的准确描述。我们希望学习出这两个矩阵的正确值。特别是知道理想的矩阵 C 和矩阵 W 满足式（2.28）所描述的简化关系，即 $CW \approx X$，或者换句话说，$CW-X$2F 非常小，同时 W 由标准基向量组成，这些向量与数据点到它们各自的中心点有关。需要注意目标非凸，既然不能同时对 C 和 W 进行最小化，那么通过交替最小化来解决问题，也就是说交替地对其中一个变量（C 或 W）进行目标函数最小化，同时保持另一个变量固定。

例 2.27 下面的 Python 代码利用 k 均值进行聚类。我们将使用 Iris 数据集，它包含三种类型的鸢尾花数据（Setosa、Versicolour 和 Virginica），具有四个属性：萼片长度、萼片宽度、花瓣长度和花瓣宽度。在这个例子中，我们利用 sklearn.cluster.KMeans 对 Iris 数据集进行聚类。在 scikit-learn 中，k 均值被实现为聚类对象并用它来寻找聚类，即 sklearn.cluster.KMeans。请注意，本例改编自 Python-scikit-learn。

```
# ================================================================
# K-means clustering
# ================================================================
from sklearn.datasets import load_iris
from sklearn.cluster import KMeans
from sklearn.svm import SVC
import numpy as np
iris = load_iris()
X = iris['data']
y = iris['target']
""" 
sklearn.cluster.KMeans(n_clusters = 8, init = 'k-means + +', n_init = 10,
max_iter = 300, tol = 0.0001,
precompute_distances = 'auto', verbose = 0, random_state = None, copy_x = True,
n_jobs = None,
algorithm = 'auto')
"""
kmeans = KMeans(n_clusters = 3)
kmeans.fit(X)
y_kmeans = kmeans.predict(X)

plt.scatter(X[:, 0], X[:, 1], c = y_kmeans, s = 50, cmap = 'viridis')
centers = kmeans.cluster_centers_
plt.scatter(centers[:, 0], centers[:, 1], c = 'black', s = 200, alpha = 0.5);
#%%
from sklearn.metrics import pairwise_distances_argmin
def find_clusters(X, n_clusters, rseed = 2):
    # 1. Randomly choose clusters
    rng = np.random.RandomState(rseed)
    i = rng.permutation(X.shape[0])[:n_clusters]
    centers = X[i]

    while True:
        # 2a. Assign labels based on closest center
        labels = pairwise_distances_argmin(X, centers)

    # 2b. Find new centers from means of points
        new_centers = np.array([X[labels == i].mean(0)
                                for i in range(n_clusters)])

    # 2c. Check for convergence
        if np.all(centers == new_centers):
```

```
            break
        centers = new_centers
                return centers, labels
centers, labels = find_clusters(X, 3)
plt.scatter(X[:, 0], X[:, 1], c = labels,
        s = 50, cmap = 'viridis');
#%%
centers, labels = find_clusters(X, 3, rseed = 0)
plt.scatter(X[:, 0], X[:, 1], c = labels,
        s = 50, cmap = 'viridis');
#%%
labels = KMeans(3, random_state = 0).fit_predict(X)
plt.scatter(X[:, 0], X[:, 1], c = labels,
        s = 50, cmap = 'viridis');
```

例2.28 下面的Python代码利用高斯混合模型进行聚类。我们将使用Breast Cancer数据集，它包含拥有30种属性的两类数据（Malign和Benign）。我们在这个例子中使用sklearn.mixture.GaussianMixture对Breast Cancer数据集进行聚类。在scikit-learn中，高斯混合模型被实现成一个mixture对象，即sklearn.mixture.GaussianMixture。请注意，本例改编自Python-scikit-learn。

```
# ====================================================================
# Gaussian mixture model for clustering
# ====================================================================
import numpy as np
from matplotlib import pyplot as plt
from sklearn import datasets
# ####################################################################
######
# Import some data to play with
Breast_Cancer = datasets.load_breast_cancer()
X = Breast_Cancer.data
y = Breast_Cancer.target

from sklearn.mixture import GaussianMixture
gmm = GaussianMixture(n_components = 2).fit(X)
proba = gmm.predict_proba(X)
y_gmm = gmm.predict(X)

plt.scatter(X[:, 0], X[:, 1], c = y_gmm, s = 50, cmap = 'viridis')
centers = gmm.cluster_centers_
plt.scatter(centers[:, 0], centers[:, 1], c = 'black', s = 200, alpha = 0.5);
#%%
from sklearn.metrics import pairwise_distances_argmin

def find_clusters(X, n_clusters, rseed = 2):
    # 1. Randomly choose clusters
    rng = np.random.RandomState(rseed)
```

```
        i = rng.permutation(X.shape[0])[:n_clusters]
        centers = X[i]

    while True:
        # 2a. Assign labels based on closest center
        labels = pairwise_distances_argmin(X, centers)
        # 2b. Find new centers from means of points
        new_centers = np.array([X[labels == i].mean(0)
                                for i in range(n_clusters)])

        # 2c. Check for convergence
        if np.all(centers == new_centers):
            break
        centers = new_centers

    return centers, labels
centers, labels = find_clusters(X, 2)
plt.scatter(X[:, 0], X[:, 1], c = labels,
            s = 50, cmap = 'viridis');
#%%
centers, labels = find_clusters(X, 2, rseed = 0)
plt.scatter(X[:, 0], X[:, 1], c = labels,
            s = 50, cmap = 'viridis');
```

例 2.29 下面的 Python 代码使用 k 均值聚类作为特征提取器。我们将使用 Iris 数据集，它包含三种类型的鸢尾花数据（Setosa、Versicolour 和 Virginica），具有四个属性：萼片长度、萼片宽度、花瓣长度和花瓣宽度。在这个例子中，我们利用 sklearn.cluster.KMeans 提取 Iris 数据集中的特征。在 scikit-learn 中，k 均值被实现为一个聚类对象并用它来提取特征，即 sklearn.cluster.KMeans。请注意，本例改编自 Python-scikit-learn。

```
# =====================================================================
# K-means as a feature extractor
# =====================================================================
from sklearn.model_selection import train_test_split
from sklearn.datasets import load_iris
from sklearn.cluster import KMeans
from sklearn.svm import SVC
import numpy as np
iris = load_iris()
X = iris['data']
y = iris['target']
#%%
#Classify the original Iris data
Xtrain, Xtest, ytrain, ytest = train_test_split(X, y, test_size = 0.3, random_state = 0)
svm = SVC().fit(Xtrain,ytrain)
ypred = svm.predict(Xtest)
```

```python
#%%
from sklearn import metrics
print('Accuracy:', np.round(metrics.accuracy_score(ytest,ypred),4))
print('Precision:', np.round(metrics.precision_score(ytest,
            ypred,average = 'weighted'),4))
print('Recall:', np.round(metrics.recall_score(ytest,ypred,
            average = 'weighted'),4))
print('F1 Score:', np.round(metrics.f1_score(ytest,ypred,
            average = 'weighted'),4))
#print('AUC:', np.round(metrics.roc_auc_score(ytest, ypred)))
print('Cohen Kappa Score:', np.round(metrics.cohen_kappa_score(ytest,
ypred)))
print('Matthews Corrcoef:', np.round(metrics.matthews_corrcoef(ytest,
ypred)))
print('\t\tClassification Report:\n', metrics.classification_report(ypred,
ytest))
#%%
from sklearn.metrics import confusion_matrix
from io import BytesIO #neded for ploting
import seaborn as sns; sns.set()
import matplotlib.pyplot as plt

mat = confusion_matrix(ytest, ypred)
sns.heatmap(mat.T, square = True, annot = True, fmt = 'd', cbar = False)
plt.xlabel('true label')
plt.ylabel('predicted label');

plt.savefig("Confusion.jpg")
# Save SVG in a fake file object.
f = BytesIO()
plt.savefig(f, format = "svg")
#%%
#Extract Features using K-Means and Then Classify
kmeans = KMeans(n_clusters = 10).fit(X)
distances = np.column_stack([np.sum((X - center)**2, axis = 1)**0.5 for
center in kmeans.cluster_centers_])

Xtrain, Xtest, ytrain, ytest = train_test_split(distances, y, test_
size = 0.3, random_state = 0)
svm = SVC().fit(Xtrain,ytrain)
ypred = svm.predict(Xtest)

#%%
from sklearn import metrics
print('Accuracy:', np.round(metrics.accuracy_score(ytest,ypred),4))
print('Precision:', np.round(metrics.precision_score(ytest,
            ypred,average = 'weighted'),4))
print('Recall:', np.round(metrics.recall_score(ytest,ypred,
            average = 'weighted'),4))
print('F1 Score:', np.round(metrics.f1_score(ytest,ypred,
            average = 'weighted'),4))
#print('AUC:', np.round(metrics.roc_auc_score(ytest, ypred)))
```

```
print('Cohen Kappa Score:', np.round(metrics.cohen_kappa_score(ytest, 
ypred)))
print('Matthews Corrcoef:', np.round(metrics.matthews_corrcoef(ytest, 
ypred)))
print('\t\tClassification Report:\n', metrics.classification_report(ypred, 
ytest))
#%%
from sklearn.metrics import confusion_matrix
from io import BytesIO #neded for plot
import seaborn as sns; sns.set()
import matplotlib.pyplot as plt

mat = confusion_matrix(ytest, ypred)
sns.heatmap(mat.T, square = True, annot = True, fmt = 'd', cbar = False)
plt.xlabel('true label')
plt.ylabel('predicted label');

plt.savefig("Confusion.jpg")
# Save SVG in a fake file object.
f = BytesIO()
plt.savefig(f, format = "svg")
```

例 2.30 下面的 Python 代码使用高斯混合模型作为特征提取器。我们将使用 Breast Cancer 数据集, 它包含拥有 30 种属性的两类数据 (Malign 和 Benign)。我们在这个例子中使用 sklearn.mixture.GaussianMixture 提取 Breast Cancer 数据集中的特征。在 scikit-learn 中, 高斯混合模型被实现成一个 mixture 对象并用它来提取特征, 即 sklearn.mixture.GaussianMixture。请注意, 本例改编自 Python-scikit-learn。

```
# ======================================================================
# Gaussian mixture model as a feature extractor
# ======================================================================
from sklearn.model_selection import train_test_split
from sklearn.cluster import KMeans
from sklearn.svm import SVC
import numpy as np
from sklearn import datasets

# #####################################################################
#######
# Import Breast Cancer data to play with
Breast_Cancer = datasets.load_breast_cancer()
X = Breast_Cancer.data
y = Breast_Cancer.target

#%%
# Classify the original Breast Cancer Dataset
Xtrain, Xtest, ytrain, ytest = train_test_split(X, y, test_size = 0.3, random_state = 0)
svm = SVC().fit(Xtrain,ytrain)
```

```python
ypred = svm.predict(Xtest)
#%%
from sklearn import metrics
print('Accuracy:', np.round(metrics.accuracy_score(ytest,ypred),4))
print('Precision:', np.round(metrics.precision_score(ytest,
            ypred,average = 'weighted'),4))
print('Recall:', np.round(metrics.recall_score(ytest,ypred,
            average = 'weighted'),4))
print('F1 Score:', np.round(metrics.f1_score(ytest,ypred,
            average = 'weighted'),4))
#print('AUC:', np.round(metrics.roc_auc_score(ytest, ypred)))
print('Cohen Kappa Score:', np.round(metrics.cohen_kappa_score(ytest,
ypred)))
print('Matthews Corrcoef:', np.round(metrics.matthews_corrcoef(ytest,
ypred)))
print('\t\tClassification Report:\n', metrics.classification_report(ypred,
ytest))
#%%
from sklearn.metrics import confusion_matrix
from io import BytesIO #neded for plot
import seaborn as sns; sns.set()
import matplotlib.pyplot as plt

mat = confusion_matrix(ytest, ypred)
sns.heatmap(mat.T, square = True, annot = True, fmt = 'd', cbar = False)
plt.xlabel('true label')
plt.ylabel('predicted label');
plt.savefig("Confusion.jpg")
# Save SVG in a fake file object.
f = BytesIO()
plt.savefig(f, format = "svg")

#%%
# Extract Features Using Gaussian Mixture Model and then Classify
from sklearn.mixture import GaussianMixture
gmm = GaussianMixture(n_components = 15).fit(X)
proba = gmm.predict_proba(X)
Xtrain, Xtest, ytrain, ytest = train_test_split(proba, y, test_size = 0.3,
random_state = 0)
svm = SVC().fit(Xtrain,ytrain)
ypred = svm.predict(Xtest)

#%%
from sklearn import metrics
print('Accuracy:', np.round(metrics.accuracy_score(ytest,ypred),4))
print('Precision:', np.round(metrics.precision_score(ytest,
            ypred,average = 'weighted'),4))
print('Recall:', np.round(metrics.recall_score(ytest,ypred,
            average = 'weighted'),4))
print('F1 Score:', np.round(metrics.f1_score(ytest,ypred,
            average = 'weighted'),4))
```

```
#print('AUC:', np.round(metrics.roc_auc_score(ytest, ypred)))
print('Cohen Kappa Score:', np.round(metrics.cohen_kappa_score(ytest,
ypred)))
print('Matthews Corrcoef:', np.round(metrics.matthews_corrcoef(ytest,
ypred)))
print('\t\tClassification Report:\n', metrics.classification_report(ypred,
ytest))
#%%
from sklearn.metrics import confusion_matrix
from io import BytesIO #neded for plot
import seaborn as sns; sns.set()
import matplotlib.pyplot as plt

mat = confusion_matrix(ytest, ypred)
sns.heatmap(mat.T, square = True, annot = True, fmt = 'd', cbar = False)
plt.xlabel('true label')
plt.ylabel('predicted label');
plt.savefig("Confusion.jpg")
# Save SVG in a fake file object.
f = BytesIO()
plt.savefig(f, format = "svg")
```

2.5 参考文献

Abe, S. (2010). Feature selection and extraction. In Support Vector Machines for Pattern Classification (pp. 331-341). Springer, London, UK.

Alpaydin, E. (2014). *Introduction to machine learning*. Cambridge, MA, London, England: MIT press.

Basseville, M., & Nikiforov, I.V. (1993). Detection of abrupt changes: Theory and application (Vol. 104). Prentice Hall Englewood Cliffs.

Begg, R., Lai, D. T., & Palaniswami, M. (2008). *Computational intelligence in biomedical engineering*. Boca Raton, FL: CRC Press.

Bengio, Y., Delalleau, O., Roux, N., Paiement, J.-F., Vincent, P., & Ouimet, M. (2006). Feature Extraction: Foundations and Applications, chapter Spectral Dimensionality Reduction (pp. 519–550). Springer, Berlin, Heidelberg.

Blinowska, K. J., & Zygierewicz, J. (2011). *Practical biomedical signal analysis using MATLAB®*. Boca Raton, FL: CRC Press.

Bodenstein, G., Schneider, W., & Malsburg, C. (1985). Computerized EEG pattern classification by adaptive segmentation and probability-density-function classification. Description of the method. *Computers in Biology and Medicine*, 15(5), 297–313.

Daubechies, I. (1990). The wavelet transform, time-frequency localization and signal analysis. *IEEE Transactions on Information Theory*, 36(5), 961–1005.

Delac, K., Grgic, M., & Grgic, S. (2005). A comparative study of PCA, ICA, and LDA. 99-106.

Du, K. -L., & Swamy, M. (2006). Principal component analysis networks. In *Neural Networks in a Softcomputing Framework* (pp. 295–351). London, UK: Springer.

Flach, P. (2012). *Machine learning: The art and science of algorithms that make sense of data*. Cambridge, UK: Cambridge University Press.

Graimann, B., Allison, B., & Pfurtscheller, G. (2009). Brain–computer interfaces: A gentle introduction. In *Brain-Computer Interfaces* (pp. 1–27). Berlin, Heidelberg: Springer.

Gustafsson, F., & Gustafsson, F. (2000). Adaptive filtering and change detection (Vol. 1). Wiley, New York.

Guyon, I., Gunn, S., Nikravesh, M., & Zadeh, L. (2006). Feature Extraction: Foundations and Applications (Studies in Fuzziness and Soft Computing) Springer. Secaucus, NJ, USA.

Hyvärinen, A., & Oja, E. (2000). Independent component analysis: Algorithms and applications. *Neural Networks, 13*(4–5), 411–430.

Kay, S. M. (1993). *Fundamentals of statistical signal processing, vol. I: estimation theory.* NJ: Prentice Hall Upper Saddle River.

Kay, S. M., & Marple, S. L. (1981). Spectrum analysis—A modern perspective. *Proceedings of the IEEE, 69*(11), 1380–1419.

Kevric, J., & Subasi, A. (2017). Comparison of signal decomposition methods in classification of EEG signals for motor-imagery BCI system. *Biomedical Signal Processing and Control, 31,* 398–406.

Kutlu, Y., & Kuntalp, D. (2012). Feature extraction for ECG heartbeats using higher order statistics of WPD coefficients. *Computer Methods and Programs in Biomedicine, 105*(3), 257–267.

Learned, R. E., & Willsky, A. S. (1995). A wavelet packet approach to transient signal classification. *Applied and Computational Harmonic Analysis, 2*(3), 265–278.

Mendel, J. M. (1991). Tutorial on higher-order statistics (spectra) in signal processing and system theory: Theoretical results and some applications. *Proceedings of the IEEE, 79*(3), 278–305.

Mohri, M., Rostamizadeh, A., & Talwalkar, A. (2018). *Foundations of machine learning.* Cambridge, MA: MIT press.

Nason, G.P., & Silverman, B.W. (1995). The stationary wavelet transform and some statistical applications. In Wavelets and statistics (pp. 281-299). Springer, New York, NY.

Phinyomark, A., Quaine, F., Charbonnier, S., Serviere, C., Tarpin-Bernard, F., & Laurillau, Y. (2013). EMG feature evaluation for improving myoelectric pattern recognition robustness. *Expert Systems with Applications, 40*(12), 4832–4840.

Proakis, John G., & Manolakis, Dimitris G. (2007). *Digital signal processing: Principles, algorithms and applications.* Englewood Cliffs: Pearson Prentice Hall.

Raschka, S. (2015). *Python machine learning.* Birmingham, UK: Packt Publishing Ltd.

Sanei, S. (2013). *Adaptive processing of brain signals.* West Sussex, UK: John Wiley & Sons.

Sanei, S., & Chambers, J. A. (2013). *EEG signal processing.* West Sussex, UK: John Wiley & Sons.

Sarkar, D., Bali, R., & Sharma, T. (2018). Practical machine learning with Python. A problem-solvers guide to building real-world intelligent systems. Apress, Berkeley, CA.

Scikit-learn: Machine learning in Python—Scikit-learn 0.21.3 documentation. (n.d.). Retrieved September 25, 2019, from https://scikit-learn.org/stable/index.html.

Semmlow, J. (2004). Biosignal and biomedical image processing: MATLAB-based applications. 2004. Marcel Decker. Inc., New York, NY.

Shalev-Shwartz, S., & Ben-David, S. (2014). *Understanding machine learning: From theory to algorithms.* New York, NY: Cambridge University Press.

Siuly, S., Li, Y., & Zhang, Y. (2016). *EEG signal analysis and classification.* Cham, Switzerland: Springer International Publishing.

Sörnmo, L., & Laguna, P. (2005). *Bioelectrical signal processing in cardiac and neurological applications (Vol. 8).* London, UK: Academic Press.

Stoica, P., & Moses, R. L. (1997). *Introduction to spectral analysis (Vol. 1).* NJ: Prentice Hall Upper Saddle River.

Subasi, A. (2019). In *Practical guide for biomedical signals analysis using machine learning techniques, a MATLAB based approach* (1st ed.). London, UK: Academic Press, Elsevier Inc.

Sudakov, O., Kriukova, G., Natarov, R., Gaidar, V., Maximyuk, O., Radchenko, S., & Isaev, D. (2017). Distributed system for sampling and analysis of electroencephalograms. *IEEE, 1,* 306–310.

Unser, M., & Aldroubi, A. (1996). A review of wavelets in biomedical applications. *Proceedings of the IEEE, 84*(4), 626–638.

Wang, S., Lu, J., Gu, X., Du, H., & Yang, J. (2016). Semi-supervised linear discriminant analysis for dimension reduction and classification. *Pattern Recognition, 57,* 179–189.

Watt, J., Borhani, R., & Katsaggelos, A. K. (2016). *Machine learning refined: Foundations, algorithms and applications.* Cambridge, UK: Cambridge University Press.

Wołczowski, A., & Zdunek, R. (2017). Electromyography and mechanomyography signal recognition: Experimental analysis using multi-way array decomposition methods. *Biocybernetics and Biomedical Engineering, 37*(1), 103–113 https://doi.org/10.1016/j.bbe.2016.09.004.

Zhang, Y., Dong, Z., Liu, A., Wang, S., Ji, G., Zhang, Z., & Yang, J. (2015). Magnetic resonance brain image classification via stationary wavelet transform and generalized eigenvalue proximal support vector machine. *Journal of Medical Imaging and Health Informatics, 5*(7), 1395–1403.

CHAPTER 3

第 3 章

机器学习技术

3.1 简介

机器学习模型被定义为由一些参数生成的计算机程序,这些参数通过使用训练数据或过去的经验来优化。机器学习通常使用统计分析方法来生成模型,主要目标在于根据训练样本进行推理预测。在某些情况下,训练算法的效率与其分类的准确性同等重要。现如今,机器学习技术已经作为决策支持系统应用于各个领域(Alpaydin, 2014)。

学习属于一种跨学科活动,涉及数学、统计学、计算机科学、经济学、物理学、心理学和神经科学等众多学科。值得注意的是,并非所有的人类活动都与智力有关,因此在某些方面机器可以表现或响应得更好,甚至还包括一些人类不能真正执行的智力任务。尽管如此,结构化的行为和决策使得理解系统响应和系统组件对于高效决策变得非常重要,而复杂系统中的传统机器学习模型可能无法实现所需的智能响应。因为模型对每个行为、行动或决策都要有系统的理解。相反,从系统的角度来看,每个活动都可能由其他事件或一系列事件引起,这些关系非常复杂而且难以理解。学习具有两大特征,即对环境中可预测行为的学习和对系统中不可预测行为的学习。因此,有必要从新的视角去看待学习的概念和模型,因为即使是在不可预测的场景中,系统和机器也被期望能够表现得智能。这些期望使得持续学习和从多种信息源中学习变得至关重要。在学习过程中,一个重要的部分是为这些系统及其有效应用进行数据的解释和适配(Kulkarni, 2012)。

不同类型的学习技术被用于不同目的。聚类方法可用于将样本或对象划分为多个类,用户只需要识别聚类数量并将其输入到聚类算法中。分类与聚类相似,但是需要将一些标签数据用于训练分类器。它们通过使用测试数据与标签数据类别的相似性对其进行分类,目标是根据这些数据的特征找到并确定两个或多个类的边界。在聚类和分类中,关于使用哪些特征以及如何提取和增强特征上始终存在争议。在过去的几十年中,已经提出了许多分类算法。其中最受欢迎的机器学习算法包括：线性判别分析（LDA）、朴素贝叶斯（NB）、k 近邻（k-NN）、人工神经网络（ANN）、支持向量机（SVM）、决策树算法。此外，聚类算法也有很多，如 k 均值、模糊 c 均值、基于密度的 DBSCAN 算法和 OPTICS 算法（Sanei, 2013；Subasi, 2019）。

3.2 什么是机器学习

机器学习是对计算机算法的研究，用于帮助在某些情况下形成准确的预测和反应，或智能地行动。通常，机器学习是指根据过去的经验学习如何在将来创造更好的情况。机器从现有的信息、知识和经验中学习。因此，机器学习是开发这样的一个程序，它允许我们从不同数据源中选择相关数据，并利用这些数据进行分析，以预测系统在相似或不同的场景下的行为。机器学习还可以对对象和活动进行分类，以支持针对新场景的决策。机器学习的动机是需要额外的智能和学习策略来应对不确定的场景（Kulkarni, 2012）。

3.2.1 理解机器学习

在 20 世纪 90 年代，自研究人员开始将机器学习作为人工智能（AI）的重要子领域，机器学习就变得流行起来。作为借用了概率、统计和 AI 思想的算法，它比其他需要人工干预的、固定的、基于规则的模型更为成功。此外，机器学习是一个多学科领域，随着时间的推移逐渐发展，并且仍在不断发展。显然，自 20 世纪 90 年代以来，随着支持向量机、随机森林、长短期记忆（LSTM）网络等算法的发现，以及 scikit-learn、PyTorch、TensorFlow 和 Theano 等机器学习和深度学习框架的进步，机器学习发展迅速。近年来，智能系统（包括 IBM Watson、DeepFace 和 AlphaGo）也开始兴起（Sarkar, Bali, & Sharma, 2018）。

3.2.2 如何让机器学习

个人、公司和组织都在不断努力解决许多现实的问题以获取利益。在许多情况下，对机器进行训练是大有裨益的，其中一些如下：
- 在某个领域里缺乏足够的人类专业知识。
- 场景和行为在不断变化。
- 人们在某个领域具有足够的专业知识，但是很难正确地解释该技能或将其转化为计算任务，例如语音识别、翻译、场景识别、认知任务等。
- 大规模解决特定领域的问题，其中包括具有过多动态需求和限制的数据。

上述情况是一些简单的例子，对于这些情况，学习机器比试图建立智能框架能更有效地节省能源、时间和资源，因为智能框架在效率、覆盖面、范围和智慧上都会受到限制。由于存在大量的历史数据，机器学习模型可以用来完成一些任务。它通过分析一段时间内的数据模式获取足够的经验，然后利用这些经验，通过有限的、劳动密集型干预手段解决未来遇到的问题。主要思想仍然是解决机器学习方法在概念上易于定义的问题（Sarkar et al., 2018)。

常见的机器学习任务可以描述如下：

- 分类或归类：通常包括一系列任务或问题，机器应针对这些任务或问题使用数据点或样本，并为每个样本分配一个特定的类别或类。
- 回归：这类任务通常以输出实际数值的方式进行预测，而不是针对输入的数据点输出一个类别或类。
- 异常检测：此项活动包括在传输的事件日志、交易报告以及其他以这种模式出的数据点中，检测不同于正常行为的异常和反常的模式或事件。
- 结构化注释：通常包括对输入数据点进行分析，并添加结构化元数据作为对原始数据的注释，用以描述附加信息以及数据元素之间的关系。
- 翻译：自动化的机器翻译功能当然是机器学习的标志，它能够将某种特定语言的输入示例翻译成另一种语言。
- 聚类或分组：通常由输入数据样本组成，利用机器学习检查输入数据相互之间的特征潜在模式、相似性和输入数据点本身之间的关系。一般来说，这类任务缺少预标记或预注释的数据，因此它们构成了无监督机器学习的一部分（Sarkar et al., 2018）。

3.2.3 多学科领域

机器学习通常被认为是人工智能的子领域，在某些方面，甚至被认为是计算机科学的子领域。机器学习包含自身已经传承了一段时间以及从多个学科中吸纳而来的思想，由此机器学习成了一个真正的多学科和跨学科领域。机器学习相关的主要领域如下，需要记住的关键点是，这不是一个完整的列表，但反映了机器学习所涉及的主要领域。

- 计算机科学
- 数学
- 统计学
- 人工智能
- 数据挖掘
- 深度学习
- 数据科学
- 自然语言处理

数据科学是一个广泛的跨学科领域，横跨前面所有其他子领域。数据科学背后的思想是使用一些方法、算法和技术从数据和领域知识中提取信息。关系数据库出现后，发展出了数据挖掘和模式识别技术的概念，例如数据库知识发现（KDD, knowledge discovery of databases）。这些领域更多地关注从大数据集中提取信息的能力和方法，而机器学习衍生出的概念与数据分析更相关。人工智能（AI）是一个超集，机器学习是其重点领域之一。AI的基本思想是，基于机器对其环境和输入参数/属性的感知，以及对预期任务的执行，发展出机器智能。机器学习通常所涉及的算法和技术可用于识别数据、构建表征并完成任务（如预测）。另一个与机器学习相关的人工智能重要子领域是自然语言处理（NLP），它主要衍生自计算机科学和计算语言学。目前，文本分析是数据科学家们用于处理、提取和理解人类自然语言的突出领域。深度学习是机器学习的一个子领域，它涉及与表征学习相关的方法，通过获取经验来改进数据。它采用了层次化的方式，利用嵌套的分层结构来表示给定的输入属性及其当前环境。因此，机器学习可以用来解决现实世界的问题。这为我们提供了一个不错的关于机器学习多学科领域广阔前景的概览（Sarkar et al., 2018）。

3.2.4 机器学习问题

需要智能解决的问题都归属于机器学习问题的范畴典型的有人脸识别、字符识别、垃圾邮件过滤、语音识别、文档分类、欺诈检测、异常检测、股市预测、天气预报、入住率预测。值得注意的是，许多涉及决策的复杂问题也可以视为可以用机器学习解决的问题。这些问题包括从经验和观察中学习以及在已知和未知搜索空间中寻找解决方案等。它们包括对对象进行分类，并将其映射到决策和解决方案。任何类型的对象或事件的分类都属于机器学习问题（Kulkarni，2012）。机器学习在一些传统方法无法解决的真实世界问题中被广泛使用（Sarkar et al., 2018）。其真实世界应用可以分为如下几类：

- 医疗保健数据分析
- 异常检测
- 欺诈检测和预防
- 在线购物平台的产品推荐
- 情绪与情感分析
- 天气预报
- 股市预测
- 内容推荐
- 购物篮分析
- 客户细分
- 语音识别
- 客户流失预测
- 点击率预测
- 故障/缺陷检测和预防
- 图像和视频中物体和场景的识别
- 垃圾邮件过滤

3.2.5 机器学习的目标

机器学习的关键任务是产生具有实用价值的学习算法，其目标被定义为开发和改进计算机算法和模型，以满足现实世界中的决策需求。值得注意的是，从洗衣机和微波炉到可以自动降落的飞机，机器学习在几乎所有现代应用程序和设备中都扮演着至关重要的角色。机器学习时代引入了从简单的数据分析和模式识别到模糊逻辑和推理的技术。由于机器学习是数据驱动的，数据源通常很少，并且在大多数情况下很难识别有用的数据，但这些源提供了大量的信息，包括数据之间的重要关系和相关性。机器学习可以提取这些关系，这是数据挖掘应用的一个领域。机器学习的目的在于构建可用于解决现实生活中问题的智能系统。算法的复杂度、信息的数量和质量、计算机的计算能力以及系统架构的有效性和可靠性决定了智能化的程度，而智能化的程度取决于算法的发展、学习和演化（Kulkarni, 2012）。

3.2.6 机器学习的挑战

机器学习是一个快速发展且令人兴奋的科学领域，具有很多机会和前景。但是由于算法的复杂性、对数据的依赖性以及它们并不属于更为经典的计算模型，机器学习也面临着一系列挑战。以下是机器学习的一些主要挑战。

- 数据质量问题导致的数据处理和特征提取方面的问题。
- 数据获取、处理、检索是非常烦琐且耗时的过程。
- 在许多情况下缺乏高质量和充足的训练数据。
- 特征提取，特别是手工加工特征，是机器学习中最困难的任务之一。近来，深度学习似乎在该领域取得一些进展。
- 用定义良好的目标和目的明确表达业务障碍，可能是有问题的。
- 过拟合和欠拟合可能会导致模型从训练数据中学习到的配置和关系质量较差，从而导致性能下降。
- 维度灾难可能是一个真正的挑战，即具有太多的特征。
- 在现实世界中实现复杂的模型并不容易。

这并不是机器学习当前所面临挑战的完整列表，但的确是数据科学家和分析师们面临的热门问题，尤其是在机器学习项目或任务中。

3.3 Python 库

Python 是一种在学术界和企业界广泛用于创建和使用机器学习算法的编程语言。我们将介绍不同的机器学习框架，例如 TensorFlow，可用于在数据集上创建基于神经网络的模型。再则，我们将说明如何使用 Keras，它是一个高级接口，可轻松创建深度学习模型并且具有精简的 API，能够在 TensorFlow 上运行。此外，还有许多出色的深度学习框架，例如 Theano、PyTorch、MXNet、Caffe 和 Lasagne。

3.3.1 scikit-learn

scikit-learn（https://scikit-learn.org/stable/）是 Python 最流行、最重要的机器学习和数

据科学应用之一（Pedregosa et al.，2011）。它适用于广泛的机器学习算法，包括机器学习的关键领域，如分类、回归和聚类。该库包含了所有标准机器学习算法，例如支持向量机、逻辑回归、随机森林、k均值聚类以及层次聚类，可以说是实用机器学习的基石。scikit-learn 主要用 Python 实现，但为了达到更好的性能，部分核心代码是用 Cython 实现的。它还对常用的学习算法实现进行了封装。scikit-learn 将在后续章节中广泛使用，所以这里的目的是熟悉该库的结构及其核心组件（Sarkar et al.，2018）。scikit-learn 库是由一系列相当小而适度的核心 API 思想以及设计模式构成。这些核心 API 如下：

数据集表示：大多数机器学习任务的数据表示方式都比较相似。一般来说，数据点的集合将由数据点向量堆积来表示，数据中的每一行都表征了数据点的某一观测向量，称为数据集。数据点向量包括多个自变量（或特征）以及一个或多个因变量。

估计器：估计器接口是 scikit-learn 库最重要的元素之一，包中的所有机器学习算法都使用估计器接口。学习阶段分两步进行：第一步是估计器对象的初始化，包括选择适合算法的类对象并提供超参数或参数；第二步是对数据进行拟合。拟合函数可以学习算法的性能参数，并将其描述为对象的一般属性，用于简单的最终模型检验。拟合函数数据通常是以输入输出对的形式给出。除了机器学习算法外，许多数据转换方法也是使用估计器 API 引入，例如特征缩放和主成分分析（PCA）。这使得以适当的方式查看简单数据转换和简单机制的转换过程成为可能。

预测器：预测器接口将训练得到的估计器用于未知数据，进行预测、预报等。例如，在监督学习问题中，预测器接口将提供未知测试集的预测类。预测器接口还包括提供测量的性能值的支持。其设计要求是创建一个打分函数，为样本输出提供一个标量值以衡量模型的性能。在将来，这些值将被用来调整机器学习模型。

转换器：在机器学习中，创建模型之前将输入数据进行转换是非常常见的。许多数据转换都是适度的，比如使用对数转换将缺失数据替换成常数，而有些数据转换则类似于 PCA 等学习算法。许多估计器对象可以帮助转换器接口简化转换过程。转换器接口帮助我们对输入数据构建非平凡转换，并将转换结果提供给学习算法。由于转换器对象保留了用于数据转换的估计器，因此利用转换函数，可以非常直接地对未知测试数据进行相同的转换（Sarkar et al.，2018）。

例 3.1 以下 Python 代码通过使用 scikit-learn 库 API 来实现回归。在本例中，我们使用了 sklearn.datasets 中的波士顿房价数据集。请注意，本例改编自 Python-scikit-learn，它使用波士顿房价数据集以二维图的方式说明了这种回归技术。该数据集包含 506 个实例，每个实例有 13 个数值/类别属性和 1 个目标值。

该数据集是加利福尼亚大学（UCI）机器学习房价数据集的一个副本，可以通过该网站访问：https://archive.ics.uci.edu/ml/machine-learning-databases/housing/。

```
# Code source: Jaques Grobler
# License: BSD 3 clause

# ==================================================================
# Linear regression example using scikit-learn API
# ==================================================================
import matplotlib.pyplot as plt
import numpy as np
from sklearn import datasets, linear_model
from sklearn.metrics import mean_squared_error, r2_score
from sklearn.metrics import mean_absolute_error
from sklearn.model_selection import train_test_split

# Load the Boston House prices dataset
boston = datasets.load_boston()
X = boston.data
y = boston.target

Xtrain, Xtest, ytrain, ytest = train_test_split(X, y, test_size = 0.5, random_state = 0)
# Create linear regression object
regr = linear_model.LinearRegression()
# Train the model using the training sets
regr.fit(Xtrain, ytrain)
# Make predictions using the testing set
ypred = regr.predict(Xtest)

# The coefficients
print('Coefficients: \n', regr.coef_)
#Intercept
print('Intercept: \n', regr.intercept_ )
# The mean absolute error
print("MAE = %5.3f" % mean_absolute_error(ytest, ypred))
# Explained variance score: 1 is perfect prediction
print("R^2 = %0.5f" % r2_score(ytest, ypred))

# The mean squared error
print("MSE = %5.3f" % mean_squared_error(ytest, ypred))

# Plot outputs
fig, ax = plt.subplots()
ax.scatter(ytest, ypred, edgecolors = (0, 0, 0))
ax.plot([y.min(), y.max()], [y.min(), y.max()], 'k--', lw = 4)
ax.set_xlabel('Measured')
ax.set_ylabel('Predicted')
plt.show()
```

3.3.2　TensorFlow

TensorFlow是由Google实现的开源机器学习库，它构建于Google的内部系统之上，以推动其在机器学习领域的发展和研究。TensorFlow 可以看作 Google 为更新 Theano 提供的新方案，它为深度学习、神经网络和机器学习提供了简单易用的框架，强调快速原型化和建模。这就引入了符号化的数学表达式可以转化为概念图的定义。然后，这些图被编译并有效地实现为底层代码中。TensorFlow 还允许使用中央处理器（CPU）和图形处理单元（GPU）。此外，它比由 Google 设计的张量处理单元（TPU）工作得更好。TensorFlow 除了提供 Python API，对其他语言（如 C++、Haskell、Java 和 Go）也提供了 API。TensorFlow 支持更高级的服务，它基于模型的创建，以及模型的开发和服务在各个流程中的实现，使机器学习的过程变得简单。同样地，TensorFlow 旨在使应用程序易于理解并提供详尽的文档。可以使用 pip 或 conda 安装功能安装 TensorFlow 库。请记住，你需要在框架中更新 dask 和 pandas 库以有效地使用 TensorFlow（Sarkar et al., 2018）。在网站 https://www.tensorflow.org/ 上，你可以找到 TensorFlow 的相关文档，其中提供了几个例子可以了解更多细节。我们将在这里介绍一些将 TensorFlow 用于深度学习的基础实例。

3.3.3　Keras

Keras 是 Python 的高级深度学习平台，可以在 TensorFlow 之上运行。Keras 由 Francois Chollet 创建，它最重要的优势在于可以节省时间，其简单易用且高效的高级 API 能够将概念快速模型化。Keras 帮助我们以更直接、更用户友好的方式使用 TensorFlow，而无须编写不必要的模板软件来创建深度学习模型。Keras 取得成功的主要原因在于其可伸缩性和灵活性。除了提供对专用库的简单访问，Keras 还确保我们仍然可以使用 TensorFlow 包提供的优势。使用常见的 pip 或 conda 安装命令，可以很容易地安装 Keras。我们必须假定已经安装了 TensorFlow，因为它需要用作创建 Keras 模型的后端（Sarkar et al., 2018）。

3.3.4　使用 Keras 构建模型

Keras 模型的构建过程有三个步骤。第一步是定义模型的结构，通过选择想要使用的基础模型来实现，可以是顺序模型或者函数式模型。当定义了问题的基础模型后，可以通过增加模型的层数进一步扩展模型。我们将从输入层开始，将输入数据的特征向量添加到该层。具体要添加到模型中的层取决于模型的需求。Keras 提供了一些可以添加到系统中的层（隐藏层、全连接层、卷积神经网络（CNN）、长短期记忆（LSTM）、循环神经网络等），其中的大部分将在我们讨论深度学习模型时介绍。我们需要将这些层以复杂的方式堆叠在一起，以实现最终的原型模型，并插入最后的输出层。机器学习周期的下一阶段是编译在第一阶段定义的框架模型。除了模型架构，学习过程需要定义以下三个重要参数（Sarkar et al., 2018）：

优化器：对于训练过程最简单的解释就是优化损失函数。一旦我们创建了模型和损失函数，就需要确定优化器来定义优化的具体方法或算法，用于训练模型，降低损失或误差。这可以是已经实现的优化器的字符串标识符，或者是可以实现的优化器类的函数或对象。

损失函数：损失函数，也被定义为目标函数，需要定义最小化损失/误差的目标，使模型在数个迭代内实现最佳输出。对于一些预先实现的损失函数，例如用于分类的交叉熵损失，或用于回归的均方误差，可能会是一个字符串标识符，也可能是一个自定义的损失函数。

性能度量：度量是学习过程的一种表示，可以被量化。我们可以在生成模型的同时，确定一个想要监控的性能度量（如分类方法的准确率），这将告诉我们训练过程是否成功。它有助于评估模型的性能。

构建模型的最后一步是实现编译的方法，开始训练。这将执行底层编译代码以在训练过程中找出模型所需的参数和权重（Sarkar et al., 2018）。

3.3.5 自然语言工具包

人类语言与计算机编程语言有所不同，它们并不像编程语言那样被设计成一组有限的数学运算。自然语言是人们用来交流的，而编程语言编写的计算机程序，是用来精确地要求机器做什么。但是我们没有自然语言编译器或解释器。自然语言处理是计算机科学和人工智能（AI）中有关对自然语言的处理的研究领域。具体来说，这个过程包括将自然语言翻译成可以被机器用来理解世界的信息（数字），而且有时候会用来产生代表这种理解的自然语言文本。如果你创建的计算机程序能够理解自然语言，那么它可以对这些语句进行操作，甚至做出响应。但这些行动和反应并没有精确定义，这给自然语言管道的设计者提供了更多的灵活性（Hapke, Lane, & Howard, 2019）。

自然语言处理系统通常称为管道，因为它通常包括几个处理阶段，其中自然语言从一端流入，而处理后的数据从另一端流出。最终，你可以通过编写代码完成一些令人兴奋且不可预知的事情，比如发起一个对话，让计算机看起来更人性化一些。这可能看起来像魔法，所有先进的技术一开始都是这样。但我们只是拉开了帷幕让你可以在里面玩，你很快就会发现，所有的工具和设备都需要自己的魔法技巧。让机器具备处理任何自然事物的能力是不自然的。就像是用建筑图纸去建造一个可以做有用的事的建筑，想要让计算机去解释那些不能被编程为机器所理解的语言，几乎是不可能的（Hapke et al., 2019）。

Python 从一开始就被设计成一门实用的语言，并且拥有许多自己的语言制作工具。这两个特点使其成为学习自然语言处理的合理选择。在企业环境中，Python 是一门很好的语言，它拥有众多贡献者对公共代码库做出贡献，可以为 NLP 算法创建稳定的输出管道。他们甚至尽可能地使用 Python 来代替作为"通用语言"的算术和计算符号（Hapke et al., 2019）。

自然语言不能直接转化为精确的数学运算符集合，但它们包含可以推导的信息和指引。这些信息或指令可以被即时处理、索引、检查和响应。其中对一个语句的响应可以是产生一连串的文字，这就是"你要建立的对话引擎或聊天机器人角色"。自然语言的另一大挑战是解码，这比其他问题更难解决。自然语言的发送方（说和写的人）会设想接收方（听和读的人）是人类，而不是计算机（Hapke et al.，2019）。

人类语言解释者，这个心智假设成了一个强有力的设想。如果我们假定处理器能够获取一段时间内对世界的认知，将使我们能用少量的文字表达大量含义。然而，这种程度的信息压缩对计算机来说仍然是失控的。在 NLP 管道中，没有简单的"心智理论"可以作为参考。尽管如此，帮助机器创建常识本体或知识库的策略，还是有助于它们基于这些信息理解语句（Hapke et al.，2019）。

搜索引擎可以通过结合自然语言文本上下文的方式对网站和文档进行索引，从而产生更有意义的搜索结果。自动补全使用 NLP 实现其补全功能，并广泛应用于手机键盘和搜索引擎。大多数具有拼写校验、语法检查、用语索引、指导格式等功能的文字处理器、浏览器插件或文本编辑器都使用了 NLP。许多对话引擎（聊天机器人）通过使用自然语言搜索来找到答案，对聊天信息进行回复。除了聊天机器人或虚拟助手中的简单答复之外，能够产生或组织文本的 NLP 管道还可以用于编写更长的段落。美联社使用 NLP 作为"机器人记者"撰写整篇财经新闻或体育赛事的报道。机器人可以写出看起来像你家乡天气的天气预报，因为人类气象学家使用 NLP 功能的文字处理机起草了这些脚本。在垃圾邮件过滤器和垃圾邮件制造者之间的猫鼠游戏中，垃圾邮件过滤器一直保持着自己的优势，但在社交网络等其他上下文中可能会失去优势（Hapke et al.，2019）。

聊天机器人产生了大约 20% 的关于 2016 年美国总统大选的信息，而这些机器人反映了其所有者和程序员的观点。NLP 系统不仅可以创建较短的社交网络帖子。在亚马逊等其他公司，NLP 可以用来撰写长电影或产品的评论。大多数评论者都在使用自主的 NLP 管道，他们从未踏入电影院或购买被评论的产品。在 Slack、IRC 甚至一些客服网站中，聊天机器人需要对不明确的指令和查询进行交互。而配备了语音识别和生产系统的聊天机器人甚至可以管理具有不确定目的或"目标功能"的冗长对话，如预订当地餐厅。对于那些无法满足于电话树，但又不愿意雇佣人工服务的公司，可以使用 NLP 程序答复客户电话。由于计算资源的限制，早期的 NLP 框架不得不利用人脑的计算能力，对复杂的逻辑规则进行建模和实现，从自然语言中获取信息。这是一种基于模式的 NLP 方法。模式，就像正则表达式一样，不一定是简单的字符序列变化。此外，NLP 通常还包括单词序列或语言部分的模式，或其他"更高级的"模式（Hapke et al.，2019）。

Python 最强大的处理文本内容的库应该是自然语言工具包（NLTK）。本节将介绍 NLTK 及其重要模块，并对其主要模块进行简要说明。NLTK 类似于其他标准库，但也有一点很大不同。在其他库中，我们不需要访问额外的辅助数据。但要想充分发挥 NLTK 的潜力，需要一些辅助数据，主要是各种语料库，因为库中的函数和模块需要使用这些信息（Sarkar et al.，2018）。

任何的文本分析都以在某个数据集中收集感兴趣的文档作为起点，该数据集对于下一步的处理和评估至关重要。这样的文档集合通常称为语料，不同版本的语料称为语料库。NLTK 的 nltk.corpus 模块提供了必要的函数，能够以多种不同的形式读取语料库数据。同时，NLTK 捆绑的数据集也促进了业务训练（Sarkar et al.，2018）。

分词是语言预处理和标准化的关键步骤之一。每个文本文件都由一系列元素构成，如句子、短语和术语。分词方法可以用来将文件分解成小的组件，句子、术语、分句等都可以使用。句子分词和单词分词是标记文本最常见的方法。NLTK 的 nltk.tokenize 模块提供了这样的特性，它能够使任何文本数据易于分词处理（Sarkar et al.，2018）。

文本文件是根据不同的语法规则和结构来构建的，而语法以文本文档的语言为基础。标记方法包括获取文本语料库、标记文本、以及为语料库元数据中的每个单词分配标签等详细信息。nltk.tag 模块包含了可以用来标记以及其他相关操作的各种算法（Sarkar et al.，2018）。

一个单词可以有多种形式，这取决于它描述语言的哪个部分。词干提取用于将一个词的不同形式全部转换为基本形式，即所谓的词根步骤。词形还原类似于词干提取，但是其基本形式是已知的词根，并且是一个在语义和词法上都正确的单词。这种转换是至关重要的，因为很多时候核心词包含了更多关于文档的信息，而这些不同的形式会稀释这些信息。NLTK 的 nltk.stem 框架包含了可用于对语料库进行词干提取和词形还原的各种方法（Sarkar et al.，2018）。

分块是类似语法解析和分词的另一个处理过程，但关键的区别在于，分块是针对文本中的短语进行处理，而不是逐一解析每个单词。我们可能会使用分块处理为短语标记其他的话语信息，这对理解文档的结构至关重要。NLTK 的 nltk.chunk 模块包含了所需技术，可以应用于语料库实现分块过程（Sarkar et al.，2018）。

情感分析是文本数据处理中最流行的技术之一。情感处理是指获取文本文档进行分析，试图确定文档所呈现的观点和极性的过程。文本文件引用中的极性可以表明该信息反映的情绪，例如积极的、消极的或是中立的。利用不同的算法和不同层次的文本分割，可以完成对文本数据的情感分析。nltk.sentiment 模块可用于对文本文档进行各种各样的情感分析（Sarkar et al.，2018）。

文本文档的分类包括从多个文本文档（语料库）中学习情感、话题、主题、类别等，然后使用训练模型给将来未知的文档打标签。与普通结构化数据的主要区别在于，我们将在特征表示中使用非结构化文本。聚类是指根据某种相似度将不同的文档聚集到一起，如语义相似度。通常，当完成了需求工程和功能提取之后，会使用 nltk.classify 和 nltk.cluster 模块来执行这些操作。此外，还有许多其他的文本分析框架，如 pattern、genism、textblob 和 spacy 等（Sarkar et al.，2018）。

3.4 学习场景

机器学习的常见场景在学习器可获得的训练数据类别、处理训练数据的顺序和算法、评估学习算法所利用的测试数据等方面都有所不同。

监督学习：学习器使用一系列已标记的样本作为训练数据，预测所有的未知实例。这是分类回归和排名问题最常见的场景。监督学习的例子包括异常检测、人脸识别、信号和图像分类、天气预报以及股市预测。

无监督学习：学习器只接收未标记的训练数据，并预测所有未知的实例。由于在无监督学习的环境中没有已标记的实例，学习器性能的定量评估具有挑战性。聚类和降维属于无监督学习问题。

半监督学习：学习器接收的训练样本包括已标记和未标记的实例，并预测所有未知的实例。半监督学习的典型场景是很容易获得未标记的数据，但标签的获取成本很高。各种类型的应用相关问题，如分类回归和排名过程，都可以作为半监督学习的例子。其基本思想是，利用未标记实例的分布，帮助学习器获得比在监督环境下更好的表现。现代经验主义和实用机器学习研究的重点是分析在什么情况下可以真正实现这些方法。

转导推理：在半监督学习中，连同一组未标记的测试点，学习者会得到一个标记的学习样本。然而，转导推理的目的只是预测这些特定测试点的标签。转导推理似乎是一个比较容易的任务，也符合一些现代应用的场景。尽管如此，和半监督系统一样，在这种环境下什么样的假设下可以获得更好的性能，这些研究问题还没有得到妥善解决。

在线学习：与以往的场景不同，该交互式的方法包含了多个回合，并且将训练和测试步骤交织在一起。学习器在每个阶段都会得到一个无标签的学习点，进行预测，得到真实标签，并造成损失。在线学习的目标是减轻所有回合中的累积损失。不同于前面提到的方法，在线学习不允许有任何分布性的假设。此外，在这种情况下，实例及其标签可能被对抗性地选择。

强化学习：学习器不断地与环境接触，收集信息，在某些情况下影响环境，并在每次活动中获得即时的奖励。学习器的目标是在一系列行为和迭代中优化环境带来的奖励。尽管如此，环境并不会提供长期的激励反馈，学习器面临着发现和利用的问题，因为它必须在探索未知的活动以发现信息或直接利用已经收集的信息之间做出选择。

主动学习：学习器自适应地、交互式地接收训练样本，通常是通过查询oracle数据库以获取新实例上的标签。主动学习的目标是通过使用较少的带标签样例，达到与标准监督学习案例相当的性能。主动学习通常应用于标签获取昂贵的系统中，如计算生物学的应用。此外，还有许多其他的方法和要求更高一些的学习场景（Mohri, Rostamizadeh, &Talwalkar, 2018）。

3.5 监督学习算法

监督学习算法在模型训练过程中对每个数据样本使用训练数据和相关标签，目的是从输入数据样本对应的输出映射或关系中学习。然后可以利用这个训练得到的模型，对任何一组新的输入数据实例进行输出预测，而这些数据实例是之前在模型训练过程中没有出现过的。这些方法被定义为有监督的，因为模型从数据样本中学习，而训练阶段所需的性能标签/响应是预先知道的。监督学习试图对训练数据中的输入与其对应的输出之间的关系进行建模，这样就可以根据之前学习到的输入与其目标输出之间的关系和映射知识来预测新输入数据的输出响应。根据所要解决的机器学习问题的类型，监督训练方法包括分类和回归两大类（Sarkar et al.，2018）。

3.5.1 分类

基于分类的任务是监督学习的子领域，其关键目标是预测输入数据类别性质的输出标签或响应，这与模型在训练阶段所学内容相关。因此，每个输出响应都属于一个特定的离散类别或类。常用的分类算法包括逻辑回归（LR）、线性判别分析（LDA）、人工神经网络（ANN）、支持向量机（SVM）、k近邻（k-NN）、朴素贝叶斯（NB）、决策树（DT）、随机森林、梯度提升以及深度学习技术（Sarkar et al.，2018）。

机器学习技术可以解决各种各样的问题。在分类任务中，需要从训练数据中学习一个适合的分类器。分类只是学习模型任务的其中一类，其他还包括回归和聚类。对于这些任务，我们会解释如何使用机器学习技术来解决问题，需要哪些信息，如何评估模型性能，以及怎样选择最佳模型。机器学习中的对象一般称为实例，所有可能实例的集合称为实例空间。为了完成所考虑的任务，需要建立一个模型，该模型使用包含标签实例的训练集（也称为样本）进行创建。通常会预留一部分标签数据——称为测试集，用于测试或评估分类器。一旦为实例指定了特定数量的特征（属性），输入空间的基本类型就产生了（Flach，2012）。

贷款是指银行等金融机构借出的一笔连本带息的资金，通常是分期偿还。能够提前预测贷款的风险（即客户有可能违约，无法全额还款）对银行来说是至关重要的。这既是为了保证银行的利润，也是为了不给客户带来超出其经济能力的贷款负担。银行在给定信贷额度和客户信息的情况下，对信贷评分的风险进行衡量。从具体的应用中获取这些信息的目的是推断出一个通用规则，对客户的属性和相关风险之间的关系进行编码。也就是说，为了计算贷款申请的风险，机器学习程序会将过去的数据应用于计算模型，然后据此选择接受或拒绝（Alpaydin，2014）该申请。

在数据科学领域有一些机器学习的应用，其中之一是光学字符识别，它涉及从图像中识别字符代码。人们的笔迹风格各异，字符的书写通常有大有小，有铅笔或钢笔，还可能几个图像关联同一个字符。有一些机器学习算法可以学习输入序列，并建立不同的变化模型。在人脸识别中，输入是一个图像，类别是被识别的人，学习算法需要通过学习将身份和人脸图像关联起来。该问题比光学字符识别更困难，因为这里有多个类，输入图像更大，而且人脸是三维的，不同的姿势和灯光会造成显著的差异。此外，某些输入可能会有遮挡，例如，眼镜可能会遮挡眼睛和眉毛，胡子可能会遮挡下巴。在医学诊断中，输入是患者的相关信息，类别是疾病。输入的内容包括患者的年龄、性别、既往病史、当前症状、血液检查、脑电图、肌电图、心电信号、核磁共振或PET图像等。有些测试不适用于所有患者，因此这些输入属性可能会缺失。在医学诊断中，错误的决策可能会导致治疗不当或缺乏治疗，在有疑问的情况下，分类器最好拒绝并将决定权交给人类专家。在语音识别中，输入数据是声音信号，类是口语单词。在这种情况下，要学习的是由声音信号到语言术语的交互。由于性别、年龄、口音的差异，对于同一个词语，不同的人有不同的发音，这使得语音识别更加困难。创建语言模型最好的方法是从大量的数据中学习。机器学习在自然语言处理方面的应用不断扩大，垃圾邮件过滤是其中之一。在垃圾邮件过滤中，一方的垃圾邮件生成器和另一方的过滤器不断产生更多更巧妙的方法来战胜对方。文本分析是另一个不错的例子，例如分析社交网站中的帖子和博客，以提取热点话题或决定投放什么广告。生物统计学是利用人的生理或行为特征对人进行认证或识别，需要整合各种模式的输入。生理特征可能是指纹、人脸、手掌和虹膜的图像，行为特征可能是签名、步法、声音和击键。可以利用机器学习来识别不同的模式，以获得一个整体的接受/拒绝决策（Alpaydin，2014）。

3.5.2 预报、预测和回归

回归是一种机器学习任务，其主要目的是对值进行预估。不同于分类具有明确的类或类别，回归是基于输入数据和输出（这些输出是连续的数值）响应的集合。回归模型利用输入数据的特征或属性及其伴随的数值输出来学习输入与其相应输出之间的基本关系和关联。最常见的实际例子之一是股市预测。根据以往股票市场价值的相关数据，可以建立一个简单的回归模型来预测股票价格。简单的线性回归模型试图用一个特征或描述性变量 x 和一个单一的响应变量 y 来模拟数据关系，其目标是预测 y。Lasso 回归是一种特殊的回归类型，它可以进行正态回归，通过对特征或变量的正则化和选择，能够很好地广义化模型。Lasso 又称最小绝对值收敛和选择算子。岭回归是另一种特殊的回归类型，它进行正态回归，通过正则化对模型进行广义化，以防止模型过拟合。广义线性模型是基本的实现方式，可以用来对数据进行建模，以预测各种类型的输出响应。

3.5.3 线性模型

在机器学习中，线性模型具有特殊的意义，因为它们很简单。线性模型是参数化的，这意味着它们具有固定的形式，有少量的数值参数可以通过从数据中学习得到。线性模型是稳定的，所以训练数据的微小变化对学习模型的影响有限。相比于其他一些模型，线性模型不太容易过拟合训练数据，因为它们的参数相对较少。反之，有时会造成欠拟合的结果。此外，线性模型的方差较小，但偏差较大。因此，该模型通常在数据有限且需要避免过拟合的情况下值得考虑。线性模型存在于所有的预测任务中，包括分类、概率估计和回归（Flach，2012）。

一个简单的线性回归模型假设回归函数 $E(Y|X)$ 对于输入 $X_1, \cdots, X_p$ 呈线性分布。线性模型主要是在前计算机时代的统计学中发展起来的，但即使是在当今的计算机时代，仍然有充分的理由去研究和使用线性模型。线性模型都很简单，且大多包含了对输入输出之

间影响因子的概括。在预测方面，它们甚至有时候会胜过复杂的非线性模型，特别是在训练实例少、信噪比低或数据稀疏的情况下。最后，线性方法还可以用于输入转换，这大大扩展了其使用范围。这种一般化的方法有时候被称为基本函数形式（Hastie, Tibshirani, Friedman, & Franklin, 2005）。

例3.2 以下 Python 代码通过使用 scikit-learn 库中的 API 来表示线性回归。在本例中，我们使用了 sklearn.datasets 中的糖尿病数据集。该数据集包含取自 442 名糖尿病患者的 10 个基线变量——年龄、性别、体重指数、平均血压和 6 个血清测量值，以及感兴趣的响应——1 年后疾病进展的定量测量。因此，该数据集总共有 442 个具有 10 个属性的实例。

其中，第 11 列是测量基线数据 1 年后疾病进展的定量测量。每个属性都以均值为中心，并以标准差乘以 n_samples 进行缩放（也就是说，每一列的平方和为 1）。原始数据可从以下网站下载：https://www4.stat.ncsu.edu/~boos/var.select/diabetes.html。

本例使用糖尿病数据集的所有特征，以二维图的形式描述了回归方法。图中的直线展示了线性回归如何绘制出一条直线，它使数据集中观察到的值与线性近似预测值之间的残差平方和最小化。同时计算了相关系数 R^2、平均绝对误差（MAE）和均方误差（MSE）。请注意，本例改编自 Python-scikit-learn。

```
# Code source: Jaques Grobler
# License: BSD 3 clause
# ========================================================================
# Linear regression example
# ========================================================================
import matplotlib.pyplot as plt
import numpy as np
from sklearn import datasets, linear_model
from sklearn.metrics import mean_squared_error, r2_score
from sklearn.metrics import mean_absolute_error
from sklearn.model_selection import train_test_split

# Load the diabetes dataset
diabetes = datasets.load_diabetes()
diabetes_X = diabetes.data
y = diabetes.target
Xtrain, Xtest, ytrain, ytest = train_test_split(X, y, test_size = 0.5, random_state = 0)

# Create linear regression object
regr = linear_model.LinearRegression()
# Train the model using the training sets
regr.fit(Xtrain, ytrain)
# Make predictions using the testing set
ypred = regr.predict(Xtest)
```

```python
# The coefficients
print('Coefficients: \n', regr.coef_)
#Intercept
print('Intercept: \n', regr.intercept_ )
# Explained variance score: 1 is perfect prediction
print("R^2 = %0.5f" % r2_score(ytest, ypred))
# The mean absolute error
print("MAE = %5.3f" % mean_absolute_error(ytest, ypred))
# The mean squared error
print("MSE = %5.3f" % mean_squared_error(ytest, ypred))

# Plot outputs
fig, ax = plt.subplots()
ax.scatter(ytest, ypred, edgecolors = (0, 0, 0))
ax.plot([y.min(), y.max()], [y.min(), y.max()], 'k--', lw = 4)
ax.set_xlabel('Measured')
ax.set_ylabel('Predicted')
plt.show()
```

例 3.3 以下 Python 代码通过使用 scikit-learn 库中的 API 来表示岭回归。在本例中，我们使用了 sklearn.datasets 中的糖尿病数据集（详见前述）。本例使用糖尿病数据集的所有特征，以二维图的形式描述了回归方法。图中的直线展示了线性回归如何绘制出一条直线，它使数据集中观察到的值与线性近似预测值之间的残差平方和最小化。同时计算了相关系数 R^2、MAE 和 MSE。请注意，本例改编自 Python-scikit-learn。

```
# ================================================================
# Ridge regression example
# ================================================================
import matplotlib.pyplot as plt
from sklearn import datasets, linear_model
from sklearn.metrics import mean_squared_error, r2_score
from sklearn.metrics import mean_absolute_error
from sklearn.model_selection import train_test_split

# Load the diabetes dataset
diabetes = datasets.load_diabetes()
diabetes_X = diabetes.data
y = diabetes.target
Xtrain, Xtest, ytrain, ytest = train_test_split(X, y, test_size = 0.5, random_state = 0)

# Create linear regression object
regr = linear_model.Ridge(alpha = .5)
# Train the model using the training sets
regr.fit(Xtrain, ytrain)
# Make predictions using the testing set
ypred = regr.predict(Xtest)
```

```python
# The coefficients
print('Coefficients: \n', regr.coef_)
#Intercept
print('Intercept: \n', regr.intercept_ )
# Explained variance score: 1 is perfect prediction
print("R^2 = %0.5f" % r2_score(ytest, ypred))
# The mean absolute error
print("MAE = %5.3f" % mean_absolute_error(ytest, ypred))
# The mean squared error
print("MSE = %5.3f" % mean_squared_error(ytest, ypred))
# Plot outputs
fig, ax = plt.subplots()
ax.scatter(ytest, ypred, edgecolors = (0, 0, 0))
ax.plot([y.min(), y.max()], [y.min(), y.max()], 'k--', lw = 4)
ax.set_xlabel('Measured')
ax.set_ylabel('Predicted')
plt.show()
```

例 3.4 以下 Python 代码通过使用 scikit-learn 库中的 API 来表示 Lasso 回归。在本例中，我们使用了 sklearn.datasets 中的波士顿房价数据集。本例使用波士顿房价数据集的所有特征，以二维图的形式描述了回归方法。图中的直线展示了线性回归如何绘制出一条直线，它使数据集中观察到的值与线性近似预测值之间的残差平方和最小化。同时计算了相关系数 R^2、MAE 和 MSE。请注意，本例改编自 Python-scikit-learn。

```python
# ========================================================================
# Lasso regression example
# ========================================================================
import matplotlib.pyplot as plt
import numpy as np
from sklearn import datasets, linear_model
from sklearn.metrics import mean_squared_error, r2_score
from sklearn.metrics import mean_absolute_error
from sklearn.model_selection import train_test_split
# Load the Boston House prices dataset
boston = datasets.load_boston()
X = boston.data
y = boston.target
Xtrain, Xtest, ytrain, ytest = train_test_split(X, y, test_size = 0.5, random_state = 0)

# Create linear regression object
regr = linear_model.Lasso(alpha = 0.1)
# Train the model using the training sets
regr.fit(Xtrain, ytrain)

# Make predictions using the testing set
ypred = regr.predict(Xtest)
```

```
# The coefficients
print('Coefficients: \n', regr.coef_)
#Intercept
print('Intercept: \n', regr.intercept_ )
# Explained variance score: 1 is perfect prediction
print("R^2 = %0.5f" % r2_score(ytest, ypred))
# The mean absolute error
print("MAE = %5.3f" % mean_absolute_error(ytest, ypred))
# The mean squared error
print("MSE = %5.3f" % mean_squared_error(ytest, ypred))
# Plot outputs
fig, ax = plt.subplots()
ax.scatter(ytest, ypred, edgecolors = (0, 0, 0))
ax.plot([y.min(), y.max()], [y.min(), y.max()], 'k--', lw = 4)
ax.set_xlabel('Measured')
ax.set_ylabel('Predicted')
plt.show()
```

例 3.5 以下 Python 代码通过使用 scikit-learn 库中的 API 来表示回归中的目标转换。在本例中，我们使用了 sklearn.datasets 中的波士顿房价数据集。本例使用 TransformedTargetRegressor，利用波士顿房价数据集转换后的特征。TransformedTargetRegressor 在拟合回归模型之前对目标 y 进行转换。通过逆向转换，预测结果被映射回原始空间。用于预测的回归器和应用于目标变量的转换器可以作为参数传入 TransformedTargetRegressor。图中的直线展示了线性回归如何绘制出一条直线，它使数据集中观察到的值与线性近似预测值之间的残差平方和最小化。同时计算了相关系数 R^2、MAE 和 MSE。从本例中可以看出，目标转换改进了回归模型的性能。请注意，本例改编自 Python-scikit-learn。

```
# ====================================================================
#Transforming target in regression
# ====================================================================
from sklearn.compose import TransformedTargetRegressor
from sklearn.preprocessing import QuantileTransformer
from sklearn.linear_model import LinearRegression
from sklearn.model_selection import train_test_split
from sklearn.metrics import mean_squared_error, r2_score
from sklearn.metrics import mean_absolute_error
from sklearn.datasets import load_boston

# Load the Boston house prices dataset
boston = load_boston()
X = boston.data
y = boston.target
X_train, X_test, y_train, y_test = train_test_split(X, y, random_state = 0)
```

```python
# ================================================================
# Prediction without transformation
# ================================================================

# Create linear regression object
regr = LinearRegression()
# Train the model using the training sets
regr.fit(X_train, y_train)
# Make predictions using the testing set
y_pred = regr.predict(X_test)

print('R^2 score without Transformation: {0:.2f}'.format(regr.score(X_test, y_test)))
# Explained variance score: 1 is perfect prediction
print("R^2 = %0.5f" % r2_score(y_test, y_pred))
# The mean absolute error
print("MAE = %5.3f" % mean_absolute_error(y_test, y_pred))
# The mean squared error
print("MSE = %5.3f" % mean_squared_error(y_test, y_pred))

# Plot outputs
fig, ax = plt.subplots()
ax.scatter(y_test, y_pred, edgecolors = (0, 0, 0))
ax.plot([y.min(), y.max()], [y.min(), y.max()], 'k--', lw = 4)
ax.set_xlabel('Measured')
ax.set_ylabel('Predicted')
plt.title('Without Transformation')
plt.show()

# ================================================================
# Prediction with transformation
# ================================================================
transformer = QuantileTransformer(output_distribution = 'normal')
regressor = LinearRegression()
regr = TransformedTargetRegressor(regressor = regressor,
            transformer = transformer)
X_train, X_test, y_train, y_test = train_test_split(X, y, random_state = 0)
regr.fit(X_train, y_train)
y_pred = regr.predict(X_test)

print('R^2 score with Transformation: {0:.2f}'.format(regr.score(X_test, y_test)))
# Explained variance score: 1 is perfect prediction
print("R^2 = %0.5f" % r2_score(y_test, y_pred))
# The mean absolute error
print("MAE = %5.3f" % mean_absolute_error(y_test, y_pred))
# The mean squared error
print("MSE = %5.3f" % mean_squared_error(y_test, y_pred))

# Plot outputs
```

```
fig, ax = plt.subplots()
ax.scatter(y_test, y_pred, edgecolors = (0, 0, 0))
ax.plot([y.min(), y.max()], [y.min(), y.max()], 'k--', lw = 4)
ax.set_xlabel('Measured')
ax.set_ylabel('Predicted')
plt.title('With Transformation')
plt.show()
```

例 3.6 以下 Python 代码通过使用 scikit-learn 库中的 API，对贝叶斯岭回归和普通最小二乘法（OLS）进行比较。在本例中，我们使用了 sklearn.datasets 中的波士顿房价数据集。图中展示了真实房价和通过使用贝叶斯岭回归和普通最小二乘法的评估房价。同时计算了相关系数 R^2、MAE 和 MSE。请注意，本例改编自 Python-scikit-learn。

```
# Bayesian ridge regression

# ====================================================================
print(__doc__)
import matplotlib.pyplot as plt
from scipy import stats
from sklearn.model_selection import train_test_split
from sklearn.metrics import mean_squared_error, r2_score
from sklearn.metrics import mean_absolute_error
from sklearn.datasets import load_boston
from sklearn.linear_model import BayesianRidge, LinearRegression

# Load the Boston house prices dataset
boston = load_boston()

X = boston.data
y = boston.target
X_train, X_test, y_train, y_test = train_test_split(X, y, random_state = 0)

# ####################################################################
# Fit the Bayesian ridge regression and an OLS for comparison
regr = BayesianRidge(compute_score = True)
regr.fit(X_train, y_train)
y_pred_bayesian = regr.predict(X_test)
ols = LinearRegression()
ols.fit(X_train, y_train)
y_pred_linear = ols.predict(X_test)

# ####################################################################
# Plot true house prices, estimated house prices
lw = 2
plt.figure(figsize = (6, 5))
plt.title("Real and Predicted House Prices")
plt.plot(y_pred_bayesian, color = 'lightgreen', linewidth = lw,
```

```
label = "Bayesian Ridge estimate")
plt.plot(y_test, color = 'gold', linewidth = lw, label = "Ground truth")
plt.plot(y_pred_linear, color = 'navy', linestyle = '--', label = "OLS estimate")
plt.xlabel("Houses")
plt.ylabel("House Prices")
plt.legend(loc = "best", prop = dict(size = 12))
```

例 3.7 以下 Python 代码通过使用 scikit-learn 库中的 API 来表示岭线性分类器。在本例中,我们使用了 sklearn.datasets 中的 Iris 数据集(鸢尾属植物数据集)。Iris 数据集包含 150 个实例(三类,每类 50 个),每个实例包含 4 个数值预测属性和分类信息。这些属性分别是花萼长度、花萼宽度、花瓣长度、花瓣宽度,单位都是厘米(cm)。三类鸢尾属植物分别是 Iris-Setosa、Iris-Versicolour 和 Iris-Virginica。Iris 数据集使用岭线性分类器进行分类。这里计算了分类的准确率、精确度、召回率、F1 得分、Cohen kappa 得分和 Matthews 相关系数,同时给出了分类报告和混淆矩阵。请注意,本例改编自 Python-scikit-learn。

```
# ===================================================================
# Ridge classifier example
# ===================================================================
import numpy as np
from sklearn.datasets import load_iris
from sklearn.model_selection import train_test_split
from sklearn.linear_model import RidgeClassifier

iris = load_iris()
X, y = iris.data, iris.target
Xtrain, Xtest, ytrain, ytest = train_test_split(X, y, test_size = 0.3, random_state = 0)

clf = RidgeClassifier(tol = 1e-2, solver = "sag")
clf.fit(Xtrain,ytrain)
ypred = clf.predict(Xtest)

from sklearn import metrics
print('Accuracy:', np.round(metrics.accuracy_score(ytest,ypred),4))
print('Precision:', np.round(metrics.precision_score(ytest,
        ypred,average = 'weighted'),4))
print('Recall:', np.round(metrics.recall_score(ytest,ypred,
        average = 'weighted'),4))
print('F1 Score:', np.round(metrics.f1_score(ytest,ypred,
        average = 'weighted'),4))
print('Cohen Kappa Score:', np.round(metrics.cohen_kappa_score(ytest,
ypred),4))
print('Matthews Corrcoef:', np.round(metrics.matthews_corrcoef(ytest,
```

```
ypred),4))
print('\t\tClassification Report:\n', metrics.classification_report(ypred,
ytest))

from sklearn.metrics import confusion_matrix
from io import BytesIO #neded for plot
import seaborn as sns; sns.set()
import matplotlib.pyplot as plt

mat = confusion_matrix(ytest, ypred)
sns.heatmap(mat.T, square = True, annot = True, fmt = 'd', cbar = False)
plt.xlabel('true label')
plt.ylabel('predicted label');

plt.savefig("Confusion.jpg")
# Save SVG in a fake file object.
f = BytesIO()
plt.savefig(f, format = "svg")
```

3.5.4 感知机

如果有一个线性判定边界用于划分类，则称这些标签数据是线性可分的。如果该边界存在，最小二乘法分类器可以找到该划分判定边界，但不能完全确保。能够在线性可分数据上完成精确分离的线性分类器称为感知机，最初是作为简单的神经网络提出的。感知机在训练集上重复学习，每次遇到错误分类实例时更新权重向量（Flach，2012）。

感知机的概念最初由 F. Rosenblatt 提出（Rosenblatt，1958），它可以用于对模式进行分类。感知机是一种前馈神经元，其中数据流从输入到输出是单向的。在输入层，每个输入数据项都乘以一个常量权重因子 ω_{ij}，然后传递到下一层神经元。输入层的每个神经元可以将其输出传播给中间层的一个或多个神经元。中间层的每个元素将其输入相加到一个净值，并乘以一个可变的权重，然后传播到输出层的神经元。输出层的每个神经元与中间层的所有元素相连接，根据它们的加权输入网络，最后得到的值为 0 或 1。在感知机的学习过程中，输出层中每个神经元预期的输出 z_i 是已知的，因此，可以不断地与计算输出 o_i 进行比较（Veit，2012）。

例 3.8 以下 Python 代码通过调用 scikit-learn 库中的 API 来对感知机分类器进行举例说明。在本例中，我们使用了 sklearn.datasets 中的手写体数字光学识别数据集。手写体数字光学识别数据集包括 5620 个实例，每个实例包括 64 个数值的预测属性（8×8 的整数像素图像，数值范围为 0 到 16）和类别。从以下网站可以访问 UCI 机器了解手写数字数据集副本：（https://archive.ics.uci.edu/ml/datasets/Optical+Recognition+of+Handwritten+Digits）。

该数据集由手写数字图像组成，分为 10 类，每类代表一个数字。将手写数字数据集划分为训练集和测试集，然后使用感知机分类器进行分类。这里计算了分类的准确率、精确度、召回率、F1 得分、Cohen kappa 得分和 Matthews 相关系数，同时给出了分类报告和混淆矩阵。请注意，本例改编自 Python-scikit-learn。

```
# =====================================================================
# Perceptron example
# =====================================================================
import numpy as np
from sklearn.datasets import load_digits
from sklearn.linear_model import Perceptron
from sklearn.model_selection import train_test_split

X, y = load_digits(return_X_y = True)
Xtrain, Xtest, ytrain, ytest = train_test_split(X, y, test_size = 0.3, random_state = 0)

#Create the Model
clf = Perceptron(tol = 1e-3, random_state = 0)
#Train the Model with Training dataset
clf.fit(Xtrain,ytrain)
#Predict the Model with Test dataset
ypred = clf.predict(Xtest)
print("Score:",clf.score(Xtest, ytest))

#Evaluate the Model and Print Performance Metrics
from sklearn import metrics
print('Accuracy:', np.round(metrics.accuracy_score(ytest,ypred),4))
print('Precision:', np.round(metrics.precision_score(ytest,
        ypred,average = 'weighted'),4))
print('Recall:', np.round(metrics.recall_score(ytest,ypred,
        average = 'weighted'),4))
print('F1 Score:', np.round(metrics.f1_score(ytest,ypred,
        average = 'weighted'),4))
print('Cohen  Kappa  Score:', np.round(metrics.cohen_kappa_score(ytest,
ypred),4))
print('Matthews  Corrcoef:', np.round(metrics.matthews_corrcoef(ytest,
ypred),4))
print('\t\tClassification Report:\n', metrics.classification_report(ypred,
ytest))
from sklearn.metrics import confusion_matrix
print("Confusion Matrix:\n",confusion_matrix(ytest, ypred))

#Plot Confusion Matrix
from sklearn.metrics import confusion_matrix
from io import BytesIO #neded for plot
import seaborn as sns; sns.set()
import matplotlib.pyplot as plt
```

```
mat = confusion_matrix(ytest, ypred)
sns.heatmap(mat.T, square = True, annot = True, fmt = 'd', cbar = False)
plt.xlabel('true label')
plt.ylabel('predicted label');
plt.savefig("Confusion.jpg")
# Save SVG in a fake file object.
f = BytesIO()
plt.savefig(f, format = "svg")
```

3.5.5 逻辑回归

逻辑回归，虽然名字里包含了回归二字，但它是一种分类模型而不是回归模型。在二元分类和线性分类问题中，逻辑回归是一种简单且更加有效的解决方法。作为分类模型，逻辑回归非常容易实现，并且能够在线性可分类中获得非常好的性能。同时，它也是工业上广泛采用的一种分类算法。与自适应线性元件（Adaline）和感知机一样，逻辑回归是一种二元分类的统计方法，它可以被泛化为多类分类。scikit-learn 具有高度优化的逻辑回归实现版本，支持多类分类任务（Raschka，2015）。

例 3.9 以下 Python 代码通过使用 scikit-learn 库中的 API 来对逻辑回归分类器进行举例说明。在本例中，我们使用了 sklearn.datasets 中的手写体数字光学识别数据集。将手写数字数据集划分为训练集和测试集，然后使用逻辑回归分类器进行分类。这里计算了分类的准确率、精确度、召回率、F1 得分、Cohen kappa 得分和 Matthews 相关系数，同时给出了分类报告和混淆矩阵。请注意，本例改编自 Python-scikit-learn。

```
# =====================================================================
# Classification with logistic regression
# =====================================================================
print(__doc__)

# Authors: Alexandre Gramfort <alexandre.gramfort@inria.fr>
#          Mathieu Blondel <mathieu@mblondel.org>
#          Andreas Mueller <amueller@ais.uni-bonn.de>
# License: BSD 3 clause

import numpy as np
import matplotlib.pyplot as plt
from sklearn.model_selection import train_test_split
from sklearn.linear_model import LogisticRegression
from sklearn import datasets
from sklearn.preprocessing import StandardScaler

digits = datasets.load_digits()
X, y = digits.data, digits.target
```

```python
Xtrain, Xtest, ytrain, ytest = train_test_split(X, y, test_size = 0.3, ran-
dom_state = 0)
clf = LogisticRegression(C = 50, multi_class = 'multinomial',
penalty = 'l1', solver = 'saga', tol = 0.1)
clf.fit(Xtrain, ytrain)
ypred = clf.predict(Xtest)

#Print performance metrics
from sklearn import metrics
print('Accuracy:', np.round(metrics.accuracy_score(ytest,ypred),4))
print('Precision:', np.round(metrics.precision_score(ytest,
        ypred,average = 'weighted'),4))
print('Recall:', np.round(metrics.recall_score(ytest,ypred,
        average = 'weighted'),4))
print('F1 Score:', np.round(metrics.f1_score(ytest,ypred,
        average = 'weighted'),4))
print('Cohen   Kappa   Score:',   np.round(metrics.cohen_kappa_score(ytest,
ypred),4))
print('Matthews   Corrcoef:',   np.round(metrics.matthews_corrcoef(ytest,
ypred),4))
print('\t\tClassification Report:\n', metrics.classification_report(ypred,
ytest))

from sklearn.metrics import confusion_matrix
from io import BytesIO #neded for plot
import seaborn as sns; sns.set()
import matplotlib.pyplot as plt

mat = confusion_matrix(ytest, ypred)
sns.heatmap(mat.T, square = True, annot = True, fmt = 'd', cbar = False)
plt.xlabel('true label')
plt.ylabel('predicted label');

plt.savefig("Confusion.jpg")
# Save SVG in a fake file object.
f = BytesIO()
plt.savefig(f, format = "svg")
```

例3.10 以下 Python 代码通过使用 scikit-learn 库中的 API 来对逻辑回归分类器进行举例说明。在本例中，我们使用了 sklearn.datasets 中的 Iris 数据集（鸢尾属植物数据集）。Iris 数据集采用 10 折交叉验证，使用逻辑回归分类器进行分类。这里计算了分类的准确率、精确度、召回率、F1 得分、Cohen kappa 得分和 Matthews 相关系数。请注意，本例改编自 Python-scikit-learn。

```
# ======================================================================
# Logistic regression example with cross-validation
# ======================================================================
```

```
from sklearn.linear_model import LogisticRegressionCV
from sklearn.model_selection import cross_val_score
from sklearn.datasets import load_iris
import warnings

#To prevent warnings
warnings.filterwarnings("ignore")

iris = load_iris()
X, y = iris.data, iris.target

CV = 10 #10-fold cross validation
model = LogisticRegressionCV(cv = CV, random_state = 0, multi_
class = 'multinomial').fit(X, y)

#Evaluate Model Using 10-Fold Cross Validation and Print Performance
Metrics
Acc_scores = cross_val_score(model, X, y, cv = CV)
print("Accuracy: %0.3f (+/- %0.3f)" % (Acc_scores.mean(), Acc_scores.
std() * 2))
f1_scores = cross_val_score(model, X, y, cv = CV,scoring = 'f1_macro')
print("F1 score: %0.3f (+/- %0.3f)" % (f1_scores.mean(), f1_scores.std()
* 2))
Precision_scores = cross_val_score(model, X, y,
cv = CV,scoring = 'precision_macro')
print("Precision score: %0.3f (+/- %0.3f)" % (Precision_scores.mean(),
Precision_scores.std() * 2))
Recall_scores = cross_val_score(model, X, y, cv = CV,scoring = 'recall_
macro')
print("Recall score: %0.3f (+/- %0.3f)" % (Recall_scores.mean(), Recall_
scores.std() * 2))
from sklearn.metrics import cohen_kappa_score, make_scorer
kappa_scorer = make_scorer(cohen_kappa_score)
Kappa_scores = cross_val_score(model, X, y, cv = CV,scoring = kappa_scorer)
print("Kappa score: %0.3f (+/- %0.3f)" % (Kappa_scores.mean(), Kappa_
scores.std() * 2))
```

3.5.6 线性判别分析

线性判别分析（LDA）作为一种分类器，用于寻找特征的线性组合，将两类或多类数据分离开。进一步的组合可以作为线性分类器使用。在 LDA 中，类别预期是呈正态分布的。与 PCA 类似，LDA 可同时被用于降维和分类。在二分类数据集中，类别 1 和类别 2 的先验概率为 p_1 和 p_2；类别平均值分别为 μ_1、μ_2，而整体平均值为 μ；协方差分别为 cov_1 和 cov_2。

$$\mu = p_1 \times \mu_1 + p_2 \times \mu_2 \tag{3.1}$$

然后，使用类内和类间的散度来表示所需的类分离标准。类内散度为：

$$S_w = \sum_{j=1}^{C} p_j x cov_j \qquad (3.2)$$

其中 C 表示类别的数量，cov_j 表示如下：

$$cov_j = (x_j - \mu_j)(x_j - \mu_j)^T \qquad (3.3)$$

类间散度为：

$$S_b = \frac{1}{C} \sum_{j=1}^{C} (\mu_j - \mu)(\mu_j - \mu)^T \qquad (3.4)$$

然后，目的是找到一个判别平面，最大可能地提高类间散度和类内散度（方差）的比率，即类内方差最小，类间方差最大：

$$J_{LDA} = \frac{wS_b w^T}{wS_w w^T} \qquad (3.5)$$

在实际案例中，类的协方差和均值是未知的，但它们可以通过训练集计算得到。同时，可以用最大似然估计或最大后验估计代替上述方程中的精确值（Sanei，2013；Subasi，2019）。

例 3.11 以下 Python 代码通过使用 scikit-learn 库中的 API 来对 LDA 分类器进行举例说明。在本例中，我们使用了 MNIST 手写数字数据集，可以通过以下网站访问该数据集：http://yann.lecun.com/exdb/mnist/。

MNIST 的训练集包含 60 000 个实例，测试集包含 10 000 个实例。该数据集是 NIST 提供的更大数据集的子集，这些数字的大小被归一化，并在固定尺寸的图像中进行了居中处理。来自 NIST 的原始黑白图像被归一化成大小为 20×20 像素的图像，同时保留其纵横比。由此产生的图像保持其灰度级别，这是归一化算法利用抗锯齿技术处理的结果。通过计算像素的质心，并对图像进行平移使其位于图片区域的中心，从而使数字图像居中到一个 28×28 的图片中。MNIST 手写数字数据集划分为训练集和测试集，然后使用 LDA 分类器进行分类。这里计算了分类的准确率、精确度、召回率、F1 得分、Cohen kappa 得分和 Matthews 相关系数，同时给出了分类报告和混淆矩阵。请注意，本例改编自 Python-scikit-learn。

```
# ==================================================
# LDA example with training and test set
# ==================================================
import numpy as np
import time
from sklearn.discriminant_analysis import LinearDiscriminantAnalysis
from sklearn.model_selection import train_test_split
from sklearn.datasets import fetch_openml
```

```python
# Turn down for faster convergence
train_samples = 5000

# Load data from https://www.openml.org/d/554
X, y = fetch_openml('mnist_784', version = 1, return_X_y = True)
Xtrain, Xtest, ytrain, ytest = train_test_split(X, y, test_size = 0.3, random_state = 0)

#lda = LinearDiscriminantAnalysis(solver = "svd", store_covariance = True)
#clf = LinearDiscriminantAnalysis(solver = 'lsqr', shrinkage = 'auto')
clf = LinearDiscriminantAnalysis(solver = 'lsqr', shrinkage = None)
clf.fit(Xtrain,ytrain)
ypred = clf.predict(Xtest)

from sklearn import metrics
print('Accuracy:', np.round(metrics.accuracy_score(ytest,ypred),4))
print('Precision:', np.round(metrics.precision_score(ytest,
        ypred,average = 'weighted'),4))
print('Recall:', np.round(metrics.recall_score(ytest,ypred,
        average = 'weighted'),4))
print('F1 Score:', np.round(metrics.f1_score(ytest,ypred,
        average = 'weighted'),4))
print('Cohen Kappa Score:', np.round(metrics.cohen_kappa_score(ytest,
ypred),4))
print('Matthews Corrcoef:', np.round(metrics.matthews_corrcoef(ytest,
ypred),4))
print('\t\tClassification Report:\n', metrics.classification_report(ypred,
ytest))
from sklearn.metrics import confusion_matrix
print("Confusion Matrix:\n", confusion_matrix(ytest, ypred))
```

例 3.12 以下 Python 代码通过使用 scikit-learn 库中的 API 来对 LDA 分类器进行举例说明。在本例中，我们使用了 sklearn.datasets 中的 Iris 数据集（鸢尾属植物数据集）。Iris 数据集采用 10 折交叉验证，使用 LDA 分类器进行分类。这里计算了分类的准确率、精确度、召回率、F1 得分、Cohen kappa 得分和 Matthews 相关系数。请注意，本例改编自 Python-scikit-learn。

```python
# ==================================================================
# LDA example with cross-validation
# ==================================================================
from sklearn.discriminant_analysis import LinearDiscriminantAnalysis
from sklearn.model_selection import cross_val_score
from sklearn.datasets import load_iris
from sklearn.metrics import cohen_kappa_score, make_scorer
iris = load_iris()
X, y = iris.data, iris.target
```

```
# fit model no training data
#lda = LinearDiscriminantAnalysis(solver = "svd", store_covariance = True)
#model = LinearDiscriminantAnalysis(solver = 'lsqr', shrinkage = 'auto')
model = LinearDiscriminantAnalysis(solver = 'lsqr', shrinkage = None)

CV = 10 #10-Fold Cross Validation
#Evaluate Model Using 10-Fold Cross Validation and Print Performance Metrics
Acc_scores = cross_val_score(model, X, y, cv = CV)
print("Accuracy: %0.3f (+/- %0.3f)" % (Acc_scores.mean(), Acc_scores.std() * 2))
f1_scores = cross_val_score(model, X, y, cv = CV,scoring = 'f1_macro')
print("F1 score: %0.3f (+/- %0.3f)" % (f1_scores.mean(), f1_scores.std() * 2))
Precision_scores = cross_val_score(model, X, y, cv = CV,scoring = 'precision_macro')
print("Precision score: %0.3f (+/- %0.3f)" % (Precision_scores.mean(), Precision_scores.std() * 2))
Recall_scores = cross_val_score(model, X, y, cv = CV,scoring = 'recall_macro')
print("Recall score: %0.3f (+/- %0.3f)" % (Recall_scores.mean(), Recall_scores.std() * 2))
kappa_scorer = make_scorer(cohen_kappa_score)
Kappa_scores = cross_val_score(model, X, y, cv = CV,scoring = kappa_scorer)
print("Kappa score: %0.3f (+/- %0.3f)" % (Kappa_scores.mean(), Kappa_scores.std() * 2))
```

3.5.7 人工神经网络

人工神经网络（ANN）模型受到了人脑及其工作方式的启发，其中的节点以及节点之间的连接与我们大脑中的神经元相似。但 ANN 和人脑之间也存在着很大的区别：大脑有许多神经元作为处理单元并行工作，而计算机只有有限数量的处理器。此外，与计算机处理器相比，神经元更简单，速度更慢；计算机处理器具有更大规模的计算能力。神经元由并行方式工作的网络或突触组成。计算机系统中的处理器是主动的，存储器是被动的。然而，大脑的存储器和处理单元是分布在一起的，因为大脑通过神经元进行处理，通过突触进行存储（Alpaydin，2014）。标准的 ANN 包括一个输入层和一个输出层，在输入层和输出层之间，至少有一个隐藏层。ANN 还包括几层的节点、确定的链接模式和层连接、连接权重，以及将加权输入映射到输出的节点（神经元）激活函数。在整个训练过程中，权重会发生改变。反向传播算法用于训练 ANN，它包括以下两个关键步骤：传播和更新权重。

传播

1）为了生成输出层的输出值，输入数据样本向量通过神经网络向前传播。

2）将产生的输出向量与该输入数据向量的实际（期望）输出向量进行比较。

3）计算输出单元的误差。

4）将误差值反向传播到每个节点（神经元）。

更新权重

1）将输出误差和输入激活相乘，计算当前权重梯度。

2）用学习率乘以梯度，得到当前位置下降的距离，然后更新节点权重。通过多次迭代重复这两个步骤，直到得到可靠的结果。反向传播通常与优化算法（如随机梯度下降）一起使用。多层感知器（MLP）是一种全连接前馈人工神经网络，至少有三层（输入、输出和至少一个隐藏层）。可以采用反向传播来训练 MLP，以及深度神经网络（多层 MLP）（Sarkar et al., 2018）。

例 3.13 以下 Python 代码通过使用 scikit-learn 库中的 API 来对 MLP 分类器进行举例说明。在本例中，我们使用了 sklearn.datasets 中的 Iris 数据集（鸢尾属植物数据集）。Iris 数据集划分为训练集和测试集，使用 MLP 分类器进行分类。这里计算了分类的准确率、精确度、召回率、F1 得分、Cohen kappa 得分和 Matthews 相关系数，同时给出了分类报告和混淆矩阵。请注意，本例改编自 Python-scikit-learn。

```
# ===================================================================
# MLP example with training and test set
# ===================================================================
import numpy as np
from sklearn.datasets import load_iris
from sklearn.model_selection import train_test_split
from sklearn.neural_network import MLPClassifier

#Load Iris Dataset
iris = load_iris()
X, y = iris.data, iris.target
#Create Train and Test set
Xtrain, Xtest, ytrain, ytest = train_test_split(X, y, test_size = 0.3, random_state = 0)

#Create the Model
mlp = MLPClassifier(hidden_layer_sizes = (100, ), learning_rate_init = 0.001,
       alpha = 1, momentum = 0.9, max_iter = 1000)
#Train the Model with Training dataset
mlp.fit(Xtrain, ytrain)
#Test the Model with Testing dataset
ypred = mlp.predict(Xtest)

from sklearn import metrics
print('Accuracy:', np.round(metrics.accuracy_score(ytest, ypred), 4))
print('Precision:', np.round(metrics.precision_score(ytest,
       ypred, average = 'weighted'), 4))
print('Recall:', np.round(metrics.recall_score(ytest, ypred,
       average = 'weighted'), 4))
print('F1 Score:', np.round(metrics.f1_score(ytest, ypred,
       average = 'weighted'), 4))
```

```
print('Cohen    Kappa    Score:',    np.round(metrics.cohen_kappa_score(ytest,
ypred)))
print('Matthews      Corrcoef:',     np.round(metrics.matthews_corrcoef(ytest,
ypred)))
print('\t\tClassification   Report:\n',   metrics.classification_report(ypred,
ytest))

from sklearn.metrics import confusion_matrix
from io import BytesIO #neded for plot
import seaborn as sns; sns.set()
import matplotlib.pyplot as plt

mat = confusion_matrix(ytest, ypred)
sns.heatmap(mat.T, square = True, annot = True, fmt = 'd', cbar = False)
plt.xlabel('true label')
plt.ylabel('predicted label');

plt.savefig("Confusion.jpg")
# Save SVG in a fake file object.
f = BytesIO()
plt.savefig(f, format = "svg")
```

例 3.14 以下 Python 代码通过使用 scikit-learn 库中的 API 来对 MLP 分类器进行举例说明。在本例中，我们使用了 sklearn.datasets 中的 Iris 数据集（鸢尾属植物数据集）。Iris 数据集采用 10 折交叉验证，使用 MLP 分类器进行分类。这里计算了分类的准确率、精确度、召回率、F1 得分、Cohen kappa 得分和 Matthews 相关系数。请注意，本例改编自 Python-scikit-learn。

```
# ======================================================================
# MLP example with cross-validation
# ======================================================================
import numpy as np
from sklearn.datasets import load_iris
from sklearn.neural_network import MLPClassifier
from sklearn.model_selection import cross_val_score
from sklearn.metrics import cohen_kappa_score, make_scorer

#Load Iris Dataset
iris = load_iris()
X, y = iris.data, iris.target
#Create the Model
model = MLPClassifier(hidden_layer_sizes = (100, ), learning_rate_
init = 0.001,
        alpha = 1, momentum = 0.9,max_iter = 1000)

CV = 10 #10-Fold Cross Validation
#Evaluate Model Using 10-Fold Cross Validation and Print Performance
```

```
Metrics
Acc_scores = cross_val_score(model, X, y, cv = CV)
print("Accuracy: %0.3f (+/- %0.3f)" % (Acc_scores.mean(), Acc_scores.
std() * 2))
f1_scores = cross_val_score(model, X, y, cv = CV,scoring = 'f1_macro')
print("F1 score: %0.3f (+/- %0.3f)" % (f1_scores.mean(), f1_scores.std()
* 2))
Precision_scores = cross_val_score(model, X, y,
cv = CV,scoring = 'precision_macro')
print("Precision score: %0.3f (+/- %0.3f)" % (Precision_scores.mean(),
Precision_scores.std() * 2))
Recall_scores = cross_val_score(model, X, y, cv = CV,scoring = 'recall_
macro')
print("Recall score: %0.3f (+/- %0.3f)" % (Recall_scores.mean(), Recall_
scores.std() * 2))
kappa_scorer = make_scorer(cohen_kappa_score)
Kappa_scores = cross_val_score(model, X, y, cv = CV,scoring = kappa_scorer)
print("Kappa score: %0.3f (+/- %0.3f)" % (Kappa_scores.mean(), Kappa_
scores.std() * 2))
```

例 3.15 以下 Python 代码通过使用 scikit-learn 库中的 API 来对 MLP 回归进行举例说明。在本例中，我们使用了 sklearn.datasets 中的加州住房数据集，其中包括从 1990 年美国人口普查中提取的数据。所有自变量和因变量为零项的块组被排除在外，最终得到的数据集包括 20 640 个观测实例，每个实例包括 9 个属性。该数据集也可以从 StatLib 镜像下载。本例使用了加州住房数据集的所有特征，图中的直线展示了 MLP 回归如何绘制出一条直线，它使数据集中观察到的值与 MLP 预测值之间的残差平方和最小化。同时计算了相关系数 R^2、平均绝对误差（MAE）和均方误差（MSE）。请注意，本例改编自 Python-scikit-learn。

```
# =================================================================
# MLP regression example with California housing dataset
# =================================================================
import matplotlib.pyplot as plt
import numpy as np
from sklearn.neural_network import MLPRegressor
from sklearn.datasets.california_housing import fetch_california_housing
from sklearn.metrics import mean_squared_error, r2_score
from sklearn.metrics import mean_absolute_error
from sklearn.model_selection import train_test_split

# Load the California Housing dataset
cal_housing = fetch_california_housing()
X, y = cal_housing.data, cal_housing.target
names = cal_housing.feature_names
Xtrain, Xtest, ytrain, ytest = train_test_split(X, y, test_size = 0.1,
random_state = 0)
```

```
# Create MLP regression object
print("Training MLPRegressor...")
regr = MLPRegressor(activation = 'logistic')
# Train the model using the training sets
regr.fit(Xtrain, ytrain)
# Make predictions using the testing set
ypred = regr.predict(Xtest)
# Explained variance score: 1 is perfect prediction
print("R^2 = %0.5f" % r2_score(ytest, ypred))
# The mean absolute error
print("MAE = %5.3f" % mean_absolute_error(ytest, ypred))
# The mean squared error
print("MSE = %5.3f" % mean_squared_error(ytest, ypred))
# Plot outputs
fig, ax = plt.subplots()
ax.scatter(ytest, ypred, edgecolors = (0, 0, 0))
ax.plot([y.min(), y.max()], [y.min(), y.max()], 'k--', lw = 4)
ax.set_xlabel('Measured')
ax.set_ylabel('Predicted')
plt.show()
```

3.5.8 k近邻

在具有大量训练数据集的情况下，k近邻法似乎是劳动密集型的，因为它主要用于模式识别。类比学习是最近邻算法的基石，详细来说，其实就是用类比学习来进行最近邻的分类，通过对密切相关的训练元组和给定的测试元组进行比较来完成。因此，n个属性用于识别训练元组，在n维空间中，每个元组都关联到一个独立的点。对于未识别的元组，k近邻分类器会探索模式空间中所有位置接近的k个训练元组，它们被称为k近邻。距离度量用于定义元组之间的相似程度，比如欧几里得距离。对于任何k个最近邻，总有最基本的类别与未识别的元组相关。当k等于1时，意味着在模式空间中，未识别的元组与其距离最近的训练元组所属类别相关。通常情况下，当训练元组的数量较大时，k的值也较大（Han，Pei，& Kamber，2011）。该方法具有三个要素：已标签的对象或记录集合，用于确定对象之间距离的距离度量方法，以及用于识别最近邻的k值。在对未标记的对象进行分类时，需要计算标记对象与对象本身之间的距离，从而识别得到k个最近邻。因此，最近邻的类标签被用于确定该对象的类标签（Wu et al.，2008）。

训练集记录了所有的训练实例，除非训练集包括不同类的相同实例，否则在训练集上可以完美地将类分开。此外，通过选择合适的实例，任何决策边界都可以或多或少地得到表示。因此，k近邻分类器的偏差较小，但方差较大。如果训练数据有限、有噪声或不具代表性，这将导致过拟合的风险。事实上，对于维度灾难来说，高维实例空间也是一个挑战。高维空间往往是极其稀疏的，点与点之间几乎都相距甚远，因此距离通常无法提供有用的信息。但即使消除了维度灾难，也不是简单的统计特征数的问题，因为有很多原因导致实例空间的有效维度可能远小于特征数。例如，在距离计算中有些特征可能是不相关的，反而淹没了相关特征的信号。在这种情况下，最好是在创建基于距离的模型之前，使用特征选择来降低维度。另一方面，数据可能能够存在于比实例空间维度更低的流形中，因此可以使用其他的降维技术，如主成分分析。显然，样本数据库中只能保留输出目标（或样本），但如果能找到一种方法将这些与降维技术结合起来，就可以通过应用 k 近邻方法消除这一局限性。k 近邻分类器在待分类实例的 $k \geqslant 1$ 个最近的实例之间进行投票，将多数邻居属于的类别作为预测结果。如果将 k 近邻用于回归问题，则是把 k 个邻居的平均值（也可以是距离加权）作为预测结果（Flach，2012）。

例 3.16 以下 Python 代码通过使用 scikit-learn 库中的 API 来对 k-NN 分类器进行举例说明。在本例中，我们使用了 MNIST 手写数字数据集（详见前述）。MNIST 手写数字数据集划分为训练集和测试集，然后使用 k-NN 分类器进行分类。这里计算了分类的准确率、精确度、召回率、F1 得分、Cohen kappa 得分和 Matthews 相关系数，同时给出了分类报告和混淆矩阵。请注意，本例改编自 Python-scikit-learn。

```
# ==================================================================
# k-NN example with training and test set
# ==================================================================
import time
import numpy as np
from sklearn.datasets import fetch_openml
from sklearn.model_selection import train_test_split
from sklearn.preprocessing import StandardScaler
from sklearn.utils import check_random_state
from sklearn.neighbors import KNeighborsClassifier

print(__doc__)

# Turn down for faster convergence
train_samples = 5000
# Load data from https://www.openml.org/d/554
X, y = fetch_openml('mnist_784', version = 1, return_X_y = True)

random_state = check_random_state(0)
permutation = random_state.permutation(X.shape[0])
X = X[permutation]
y = y[permutation]
X = X.reshape((X.shape[0], -1))
Xtrain, Xtest, ytrain, ytest = train_test_split(X, y, train_size = train_samples, test_size = 10000)
#Transform data using standart scaler
scaler = StandardScaler()
Xtrain = scaler.fit_transform(Xtrain)
Xtest = scaler.transform(Xtest)
```

```python
#Create the Model
clf = KNeighborsClassifier(n_neighbors = 5)
#Train the model with Training Dataset
clf.fit(Xtrain,ytrain)
#Test the model with Testset
ypred = clf.predict(Xtest)

#Evaluate the Model and Print Performance Metrics
from sklearn import metrics
print('Accuracy:', np.round(metrics.accuracy_score(ytest,ypred),4))
print('Precision:', np.round(metrics.precision_score(ytest,
        ypred,average = 'weighted'),4))
print('Recall:', np.round(metrics.recall_score(ytest,ypred,
        average = 'weighted'),4))
print('F1 Score:', np.round(metrics.f1_score(ytest,ypred,
        average = 'weighted'),4))
print('Cohen Kappa Score:', np.round(metrics.cohen_kappa_score(ytest,
ypred),4))
print('Matthews Corrcoef:', np.round(metrics.matthews_corrcoef(ytest,
ypred),4))
print('\t\tClassification Report:\n', metrics.classification_report(ypred,
ytest))
from sklearn.metrics import confusion_matrix
print("Confusion Matrix:\n",confusion_matrix(ytest, ypred))

#Plot Confusion Matrix
from sklearn.metrics import confusion_matrix
from io import BytesIO #neded for plot
import seaborn as sns; sns.set()
import matplotlib.pyplot as plt

mat = confusion_matrix(ytest, ypred)
sns.heatmap(mat.T, square = True, annot = True, fmt = 'd', cbar = False)
plt.xlabel('true label')
plt.ylabel('predicted label');

plt.savefig("Confusion.jpg")
# Save SVG in a fake file object.
f = BytesIO()
plt.savefig(f, format = "svg")
```

例 3.17 以下 Python 代码通过使用 scikit-learn 库中的 API 来对 k-NN 分类器进行举例说明。在本例中,我们使用了 sklearn.datasets 中的 Iris 数据集(鸢尾属植物数据集)。Iris 数据集采用 10 折交叉验证,使用 k-NN 分类器进行分类。这里计算了分类的准确率、精确度、召回率、F1 得分、Cohen kappa 得分和 Matthews 相关系数。请注意,本例改编自 Python-scikit-learn。

```
# ==================================================================
# k-NN example with cross-validation
# ==================================================================
from sklearn.neighbors import KNeighborsClassifier
from sklearn.model_selection import cross_val_score
from sklearn.datasets import load_iris
iris = load_iris()
X, y = iris.data, iris.target

#Create a Model
model = KNeighborsClassifier(n_neighbors = 5)

CV = 10 #10-Fold Cross Validation
#Evaluate Model Using 10-Fold Cross Validation and Print Performance Metrics
Acc_scores = cross_val_score(model, X, y, cv = CV)
print("Accuracy: %0.3f (+/- %0.3f)" % (Acc_scores.mean(), Acc_scores.std() * 2))
f1_scores = cross_val_score(model, X, y, cv = CV,scoring = 'f1_macro')
print("F1 score: %0.3f (+/- %0.3f)" % (f1_scores.mean(), f1_scores.std() * 2))
Precision_scores = cross_val_score(model, X, y, cv = CV,scoring = 'precision_macro')
print("Precision score: %0.3f (+/- %0.3f)" % (Precision_scores.mean(), Precision_scores.std() * 2))
Recall_scores = cross_val_score(model, X, y, cv = CV,scoring = 'recall_macro')
print("Recall score: %0.3f (+/- %0.3f)" % (Recall_scores.mean(), Recall_scores.std() * 2))
from sklearn.metrics import cohen_kappa_score, make_scorer
kappa_scorer = make_scorer(cohen_kappa_score)
Kappa_scores = cross_val_score(model, X, y, cv = CV,scoring = kappa_scorer)
print("Kappa score: %0.3f (+/- %0.3f)" % (Kappa_scores.mean(), Kappa_scores.std() * 2))
```

例 3.18 以下 Python 代码通过使用 scikit-learn 库中的 API 来对 k-NN 回归进行举例说明。在本例中，我们使用了 sklearn.datasets 中的加州住房数据集（详见前述）。图中的直线展示了 k-NN 回归如何绘制出一条直线，它使数据集中观察到的值与 k-NN 预测值之间的残差平方和最小化。同时计算了相关系数 R^2、平均绝对误差（MAE）和均方误差（MSE）。请注意，本例改编自 Python-scikit-learn。

```
# ==================================================================
# k-NN regression example with California housing dataset
# ==================================================================
import matplotlib.pyplot as plt
import numpy as np
from sklearn.neighbors import KNeighborsRegressor
```

```python
from sklearn.datasets.california_housing import fetch_california_housing
from sklearn.metrics import mean_squared_error, r2_score
from sklearn.metrics import mean_absolute_error
from sklearn.model_selection import train_test_split

# Load the California Housing dataset
cal_housing = fetch_california_housing()
X, y = cal_housing.data, cal_housing.target
names = cal_housing.feature_names
Xtrain, Xtest, ytrain, ytest = train_test_split(X, y, test_size = 0.1,
random_state = 0)

# Create a Regression object
print("Training Regressor...")
regr = KNeighborsRegressor(n_neighbors = 2)
# Train the model using the training sets
regr.fit(Xtrain, ytrain)
# Make predictions using the testing set
ypred = regr.predict(Xtest)

# Explained variance score: 1 is perfect prediction
print("R^2 = %0.5f" % r2_score(ytest, ypred))
# The mean absolute error
print("MAE = %5.3f" % mean_absolute_error(ytest, ypred))
# The mean squared error
print("MSE = %5.3f" % mean_squared_error(ytest, ypred))

# Plot outputs
fig, ax = plt.subplots()
ax.scatter(ytest, ypred, edgecolors = (0, 0, 0))
ax.plot([y.min(), y.max()], [y.min(), y.max()], 'k--', lw = 4)
ax.set_xlabel('Measured')
ax.set_ylabel('Predicted')
plt.show()
```

3.5.9 支持向量机

支持向量机（SVM）是主要的机器学习算法之一，不仅精确度高，而且稳定性强。在二分类学习任务中，SVM 的目标是找到最适合的分类函数来分离训练数据中的类。对于线性可分的数据集，该线性分类函数的作用是求解经过两类中心的分离超平面，将两类分离开。因为有很多线性超平面，SVM 需要找到最合适的分类函数，以确保类之间的间隔最大。顾名思义，间隔是指类之间的空间。用数学术语来说，间隔是离超平面最近的样本点到超平面的最短距离。这样的几何定义对于增大间隔是非常重要的，尽管有大量的超平面，但SVM 可以利用的不超过两个。SVM 的目的在于确定最大间隔的超平面，用以对模型进行最大限度的泛化。它能够在训练数据上具有最适宜的分类性能，同时对未知数据进行完美的分类（Wu et al., 2008）。SVM 还可以用于泛化高度分布的数据。相比于其他的分类算法，如贝叶斯和 ANN，SVM 表现出较高的准确性。其训练过程使用了序列极小化技术，能够提高泛化能力，从而具有较高的分类准确率（Ray, Mohanty, & Panigrahi, 2019）。

例 3.19 以下 Python 代码通过调用 scikit-learn 库中的 API 来对 SVM 分类器进行举例说明。在本例中，我们使用了 sklearn.datasets 中的手写体数字光学识别数据集。该数据集被划分为训练集和测试集，然后使用 SVM 分类器进行分类。这里计算了分类的准确率、精确度、召回率、F1 得分、Cohen kappa 得分和 Matthews 相关系数，同时给出了分类报告和混淆矩阵。请注意，本例改编自 Python-scikit-learn。

```python
# ======================================================================
# SVM example with training and test set
# ======================================================================
import numpy as np
from sklearn.model_selection import train_test_split
from sklearn.preprocessing import StandardScaler
from sklearn.utils import check_random_state
from sklearn import svm
from sklearn.datasets import load_digits

#Load dataset and split to training and testing set
X, y = load_digits(return_X_y = True)
Xtrain, Xtest, ytrain, ytest = train_test_split(X, y, test_size = 0.3, random_state = 0)

#Transfrom the data using Standard Scaler
scaler = StandardScaler()
Xtrain = scaler.fit_transform(Xtrain)
Xtest = scaler.transform(Xtest)

C = 10.0 # SVM regularization parameter
#Create the Model
clf = svm.SVC(kernel = 'linear', C = C)
#Train the model with Training Dataset
clf.fit(Xtrain,ytrain)
#Test the model with Testset
ypred = clf.predict(Xtest)

#Evaluate the Model and Print Performance Metrics
from sklearn import metrics
print('Accuracy:', np.round(metrics.accuracy_score(ytest,ypred),4))
print('Precision:', np.round(metrics.precision_score(ytest,
        ypred,average = 'weighted'),4))
print('Recall:', np.round(metrics.recall_score(ytest,ypred,
        average = 'weighted'),4))
print('F1 Score:', np.round(metrics.f1_score(ytest,ypred,
        average = 'weighted'),4))
```

```python
print('Cohen Kappa Score:', np.round(metrics.cohen_kappa_score(ytest,
ypred),4))
print('Matthews Corrcoef:', np.round(metrics.matthews_corrcoef(ytest,
ypred),4))
print('\t\tClassification Report:\n', metrics.classification_report(ypred,
ytest))
from sklearn.metrics import confusion_matrix
print("Confusion Matrix:\n",confusion_matrix(ytest, ypred))

#Plot Confusion Matrix
from sklearn.metrics import confusion_matrix
from io import BytesIO #neded for plot
import seaborn as sns; sns.set()
import matplotlib.pyplot as plt

mat = confusion_matrix(ytest, ypred)
sns.heatmap(mat.T, square = True, annot = True, fmt = 'd', cbar = False)
plt.xlabel('true label')
plt.ylabel('predicted label');

plt.savefig("Confusion.jpg")
# Save SVG in a fake file object.
f = BytesIO()
plt.savefig(f, format = "svg")
```

例 3.20 以下 Python 代码通过使用 scikit-learn 库中的 API 来对 SVM 分类器进行举例说明。在本例中，我们使用了 sklearn.datasets 中的 Iris 数据集（鸢尾属植物数据集）。Iris 数据集采用 10 折交叉验证，使用 SVM 分类器进行分类。这里计算了分类的准确率、精确度、召回率、F1 得分、Cohen kappa 得分和 Matthews 相关系数。请注意，本例改编自 Python-scikit-learn。

```python
# ====================================================================
# SVM example with cross-validation
# ====================================================================
from sklearn import svm
from sklearn.model_selection import cross_val_score
from sklearn.datasets import load_iris
iris = load_iris()
X, y = iris.data, iris.target
C = 50.0 # SVM regularization parameter
# fit model no training data
model = svm.SVC(kernel = 'linear', C = C)

CV = 10 #10-Fold Cross Validation
#Evaluate Model Using 10-Fold Cross Validation and Print Performance
Metrics
Acc_scores = cross_val_score(model, X, y, cv = CV)
```

```
print("Accuracy: %0.3f (+/- %0.3f)" % (Acc_scores.mean(), Acc_scores.
std() * 2))
f1_scores = cross_val_score(model, X, y, cv = CV,scoring = 'f1_macro')
print("F1 score: %0.3f (+/- %0.3f)" % (f1_scores.mean(), f1_scores.std()
 * 2))
Precision_scores = cross_val_score(model, X, y,
cv = CV,scoring = 'precision_macro')
print("Precision score: %0.3f (+/- %0.3f)" % (Precision_scores.mean(),
Precision_scores.std() * 2))
Recall_scores = cross_val_score(model, X, y, cv = CV,scoring = 'recall_
macro')
print("Recall score: %0.3f (+/- %0.3f)" % (Recall_scores.mean(),
Recall_scores.std() * 2))
from sklearn.metrics import cohen_kappa_score, make_scorer
kappa_scorer = make_scorer(cohen_kappa_score)
Kappa_scores = cross_val_score(model, X, y, cv = CV,scoring = kappa_scorer)
print("Kappa score: %0.3f (+/- %0.3f)" % (Kappa_scores.mean(), Kappa_
scores.std() * 2))
```

例 3.21 以下 Python 代码通过使用 scikit-learn 库中的 API 来表示使用 SVR 进行回归中的目标转换。在本例中，我们使用了 sklearn.datasets 中的波士顿房价数据集。本例使用 TransformedTargetRegressor，利用波士顿房价数据集转换后的特征。TransformedTargetRegressor 在拟合回归模型之前对目标 y 进行转换。通过逆向转换，预测结果被映射回原始空间。用于预测的回归器和应用于目标变量的转换器可以作为参数传入 TransformedTargetRegressor。图中的直线展示了 SVR 回归如何绘制出一条直线，它使数据集中观察到的值与 SVR 近似预测值之间的残差平方和最小化。同时计算了相关系数 R^2、MAE 和 MSE。从本例中可以看出，目标转换改进了 SVR 回归的性能。请注意，本例改编自 Python-scikit-learn。

```
# =====================================================================
# SVR regression example with Boston house prices dataset
# =====================================================================
# =====================================================================
#Transforming target in regression
# =====================================================================
from sklearn.compose import TransformedTargetRegressor
from sklearn.preprocessing import QuantileTransformer
from sklearn.svm import SVR
from sklearn.model_selection import train_test_split
from sklearn.metrics import mean_squared_error, r2_score
from sklearn.metrics import mean_absolute_error
from sklearn.datasets import load_boston

# Load the Boston House prices dataset
boston = load_boston()
```

```python
X = boston.data
y = boston.target
X_train, X_test, y_train, y_test = train_test_split(X, y, random_state = 0)

# ====================================================================
# Prediction without Transformation
# ====================================================================

#Create Regression object
regr = SVR(gamma = 'scale', C = 200.0, epsilon = 0.2)
# Train the model using the training sets
regr.fit(X_train, y_train)
# Make predictions using the testing set
y_pred = regr.predict(X_test)

print('R^2 score without Transformation: {0:.2f}'.format(regr.score(X_test, y_test)))
# Explained variance score: 1 is perfect prediction
print("R^2 = %0.5f" % r2_score(y_test, y_pred))
# The mean absolute error
print("MAE = %5.3f" % mean_absolute_error(y_test, y_pred))
# The mean squared error
print("MSE = %5.3f" % mean_squared_error(y_test, y_pred))

# Plot outputs
fig, ax = plt.subplots()
ax.scatter(y_test, y_pred, edgecolors = (0, 0, 0))
ax.plot([y.min(), y.max()], [y.min(), y.max()], 'k--', lw = 4)
ax.set_xlabel('Measured')
ax.set_ylabel('Predicted')

plt.title('Without Transformation')
plt.show()

# ====================================================================
# Prediction with Transformation
# ====================================================================
transformer = QuantileTransformer(output_distribution = 'normal')
regressor = SVR(gamma = 'scale', C = 200.0, epsilon = 0.2)
regr = TransformedTargetRegressor(regressor = regressor,
            transformer = transformer)
X_train, X_test, y_train, y_test = train_test_split(X, y, random_state = 0)
regr.fit(X_train, y_train)
y_pred = regr.predict(X_test)

print('R^2 score with Transformation: {0:.2f}'.format(regr.score(X_test, y_test)))
# Explained variance score: 1 is perfect prediction
print("R^2 = %0.5f" % r2_score(y_test, y_pred))
# The mean absolute error
```

```
print("MAE = %5.3f" % mean_absolute_error(y_test, y_pred))
# The mean squared error
print("MSE = %5.3f" % mean_squared_error(y_test, y_pred))

# Plot outputs
fig, ax = plt.subplots()
ax.scatter(y_test, y_pred, edgecolors = (0, 0, 0))
ax.plot([y.min(), y.max()], [y.min(), y.max()], 'k--', lw = 4)
ax.set_xlabel('Measured')
ax.set_ylabel('Predicted')
plt.title('With Transformation')
plt.show()
```

3.5.10 决策树分类器

决策树是一种分而治之的递归数据结构。它是一种强大的非参数方法，可用于分类和回归。在参数估计中，模型被用于表示整个输入空间，然后通过完整的训练数据进行训练，估计其未知参数。我们很容易将相同的参数模板和集合用于任何测试输入。在非参数估计中，输入空间被划分为局部区域，由欧几里得法则等距离度量进行描述，然后通过该区域的训练数据创建相关的局部模型。决策树是一种分层监督学习模型，其中少部分阶段以一系列顺序分割的方式定义局部区域。决策树由内部决策节点和终端叶子节点组成，每个决策节点会使用一个测试函数，以离散的分数值标记各个分支。每个节点会对输入进行测试，然后根据测试结果选择一个分支。该过程从根节点开始，递归地进行，直到到达叶子节点。决策树是一种非参数模型，即不假定类别密度的参数结构，且树的产生也不是静态的先验过程，而是在学习过程中根据数据中问题的性质，通过插入树叶和分支来扩展树。在分类问题中，每个叶子节点的输出标签是一个类别代码，而在回归问题中，输出是一个数值。叶子节点标识了输入空间中的一个局部区域，该区域中的实例具有相同的分类标签或是与回归中非常相似的数值输出。区域边界由判定测试序列明确指定，这些判别测试序列在从根节点到叶子节点路径上的内部节点中编码得到。在包含输入的区域中，很容易得到分层分配的决策结果。此外，决策树可以转换成易于理解的IF-THEN规则，因此被广泛地接受，并优于更准确但更难解释的方法（Alpaydin, 2014; Subasi, 2019）。

决策树学习模型是一棵壮大的树，由根节点开始使用完整的训练信息进行学习，然后在每个阶段进行最佳分割。决策树将训练数据一分为二或者更多，这取决于所选择的特征是离散型还是数值型。然后继续递归地对相关集合进行分裂，直到不再需要分割，即叶子节点被生成并标记。在将决策树用于分类，即分类树的情况下，利用杂质测量来衡量分裂的好坏。如果对所有分支进行划分后，分支选择的实例属于同一类别，那么该分裂是有效的。熵是度量杂质的一种潜在方式（Quinlan and Ross, 1986），但并不是唯一可行的方式。对于所有属性（二元和数值）都应该进行杂质计算，选择使熵最小的属性。然后对所有不纯的分支进行递归平行的扩展，直到所有的分支都是纯的。这是CART（分类和回归树）算法

（Breiman，Friedman，Olshen，& Stone，1984）ID3 算法（Quinlan 和 Ross，1986），及其扩展算法 C4.5（Alpaydin，2014；Quinlan，1993；Subasi，2019）的基础。

通常情况下，如果到达节点的训练实例数量小于训练集的一定比例，就不会再划分节点。这种提前停止构建树的行为称为树的预剪枝。另一种简化树的方法是后剪枝，事实上后剪枝比预剪枝效果更好。在后剪枝的整个过程中，树会一直生长，直到所有叶子都是纯的，没有任何训练错误。随后，需要识别和修剪导致过拟合的子树。用于剪枝的数据集需要与初始标签数据集分离开，不能在训练过程中使用。每棵被修剪的子树由其训练实例所标记的叶子节点来代替。如果该叶子节点的表现不比剪枝前的子树差，那么将删除子树，并保留该叶子节点，因为由子树带来的额外的复杂度是不够的，否则，应当保留子树。如果将预剪枝和后剪枝相比较，预剪枝更快，但后剪枝通常能得到更加准确的树（Alpaydin，2014；Subasi，2019）。

树模型是最常见的机器学习模型之一。由于是递归的"分而治之"的存在，所以树是具有描述性的且容易理解，计算机科学家特别感兴趣。树模型的使用不仅限于分类任务，还可以解决几乎任何机器学习问题，包括排序、概率估计、回归以及聚类。可以为所有的这些模型定义共同的树结构（Flach，2012）。

例 3.22 以下 Python 代码通过使用 scikit-learn 库中的 API 来对 DT 分类器进行举例说明。在本例中，我们使用了 MNIST 手写数字数据集（详见前述）。MNIST 手写数字数据集被划分为训练集和测试集，然后使用 DT 分类器进行分类。这里计算了分类的准确率、精确度、召回率、F1 得分、Cohen kappa 得分和 Matthews 相关系数，同时给出了分类报告和混淆矩阵。请注意，本例改编自 Python-scikit-learn。

```
# ========================================================================
# Decision tree example with training and testset
# ========================================================================
import time
import numpy as np
from sklearn.datasets import fetch_openml
from sklearn.model_selection import train_test_split
from sklearn.preprocessing import StandardScaler
from sklearn.utils import check_random_state
from sklearn import tree
print(__doc__)

# Turn down for faster convergence
train_samples = 5000

# Load data from https://www.openml.org/d/554
X, y = fetch_openml('mnist_784', version = 1, return_X_y = True)

random_state = check_random_state(0)
```

```python
permutation = random_state.permutation(X.shape[0])
X = X[permutation]
y = y[permutation]
X = X.reshape((X.shape[0], -1))

Xtrain, Xtest, ytrain, ytest = train_test_split(
X, y, train_size = train_samples, test_size = 10000)
scaler = StandardScaler()
Xtrain = scaler.fit_transform(Xtrain)
Xtest = scaler.transform(Xtest)

#Create the Model
clf = tree.DecisionTreeClassifier()
#Train the Model with Training dataset
clf.fit(Xtrain,ytrain)
#Test the Model with Testing dataset
ypred = clf.predict(Xtest)

#Evaluate the Model and Print Performance Metrics
from sklearn import metrics
print('Accuracy:', np.round(metrics.accuracy_score(ytest,ypred),4))
print('Precision:', np.round(metrics.precision_score(ytest,
          ypred,average = 'weighted'),4))
print('Recall:', np.round(metrics.recall_score(ytest,ypred,
          average = 'weighted'),4))
print('F1 Score:', np.round(metrics.f1_score(ytest,ypred,
          average = 'weighted'),4))
print('Cohen Kappa Score:', np.round(metrics.cohen_kappa_score(ytest,
ypred),4))
print('Matthews Corrcoef:', np.round(metrics.matthews_corrcoef(ytest,
ypred),4))
print('\t\tClassification Report:\n', metrics.classification_report(ypred,
ytest))

from sklearn.metrics import confusion_matrix
print("Confusion Matrix:\n",confusion_matrix(ytest, ypred))
from sklearn.metrics import confusion_matrix
from io import BytesIO #neded for plot
import seaborn as sns; sns.set()
import matplotlib.pyplot as plt

mat = confusion_matrix(ytest, ypred)
sns.heatmap(mat.T, square = True, annot = True, fmt = 'd', cbar = False)
plt.xlabel('true label')
plt.ylabel('predicted label');

plt.savefig("Confusion.jpg")
# Save SVG in a fake file object.
f = BytesIO()
plt.savefig(f, format = "svg")
```

例 3.23　以下 Python 代码通过使用 scikit-learn 库中的 API 来对 DT 分类器进行举例说明。在本例中，我们使用了 sklearn.datasets 中的 Iris 数据集（鸢尾属植物数据集）。Iris 数据集采用 10 折交叉验证，使用 DT 分类器进行分类。这里计算了分类的准确率、精确度、召回率、F1 得分、Cohen kappa 得分和 Matthews 相关系数。请注意，本例改编自 Python-scikit-learn。

```
# ================================================================
# Decision tree example with cross-validation
# ================================================================

from sklearn import tree
from sklearn.model_selection import cross_val_score
from sklearn.datasets import load_iris
iris = load_iris()
X, y = iris.data, iris.target


# fit model no training data
model = tree.DecisionTreeClassifier()

CV = 10  #10-Fold Cross Validation
#Evaluate Model Using 10-Fold Cross-Validation and Print Performance Metrics
Acc_scores = cross_val_score(model, X, y, cv = CV)
print("Accuracy: %0.3f (+/- %0.3f)" % (Acc_scores.mean(), Acc_scores.std() * 2))
f1_scores = cross_val_score(model, X, y, cv = CV, scoring = 'f1_macro')
print("F1 score: %0.3f (+/- %0.3f)" % (f1_scores.mean(), f1_scores.std() * 2))
Precision_scores = cross_val_score(model, X, y, cv = CV, scoring = 'precision_macro')
print("Precision score: %0.3f (+/- %0.3f)" % (Precision_scores.mean(), Precision_scores.std() * 2))
Recall_scores = cross_val_score(model, X, y, cv = CV, scoring = 'recall_macro')
print("Recall score: %0.3f (+/- %0.3f)" % (Recall_scores.mean(), Recall_scores.std() * 2))
from sklearn.metrics import cohen_kappa_score, make_scorer
kappa_scorer = make_scorer(cohen_kappa_score)
Kappa_scores = cross_val_score(model, X, y, cv = CV, scoring = kappa_scorer)
print("Kappa score: %0.3f (+/- %0.3f)" % (Kappa_scores.mean(), Kappa_scores.std() * 2))
```

例 3.24　以下 Python 代码通过使用 scikit-learn 库中的 API 来表示使用 DT 回归器进行回归中的目标转换。在本例中，我们使用了 sklearn.datasets 中的波士顿房价数据集。本例使用 TransformedTargetRegressor，利用波士顿房价数据集转换后的特征。

TransformedTargetRegressor 在拟合回归模型之前对目标 y 进行转换。通过逆向转换，预测结果被映射回原始空间。用于预测的回归器和应用于目标变量的转换器可以作为参数传入 TransformedTargetRegressor。图中的直线展示了 DT 回归如何绘制出一条直线，它使数据集中观察到的值与 DT 回归预测值之间的残差平方和最小化。同时计算了相关系数 R^2、MAE 和 MSE。从本例中可以看出，目标转换改进了 DT 回归的性能。请注意，本例改编自 Python-scikit-learn。

```
# ================================================================
# Decision tree regression example with Boston house prices dataset
# ================================================================
# ================================================================
#Transforming target in regression
# ================================================================
from sklearn.compose import TransformedTargetRegressor
from sklearn.preprocessing import QuantileTransformer
from sklearn.tree import DecisionTreeRegressor
from sklearn.model_selection import train_test_split
from sklearn.metrics import mean_squared_error, r2_score
from sklearn.metrics import mean_absolute_error
from sklearn.datasets import load_boston

# Load the Boston house prices dataset
boston = load_boston()
X = boston.data
y = boston.target
X_train, X_test, y_train, y_test = train_test_split(X, y, random_
state = 0)

# ================================================================
# Prediction without Transformation
# ================================================================

#Create Regression object
regr = DecisionTreeRegressor(random_state = 0)
# Train the model using the training sets

regr.fit(X_train, y_train)
# Make predictions using the testing set
y_pred = regr.predict(X_test)

print('R^2 score without Transformation: {0:.2f}'.format(regr.score(X_
test, y_test)))
# Explained variance score: 1 is perfect prediction
print("R^2 = %0.5f" % r2_score(y_test, y_pred))
# The mean absolute error
print("MAE = %5.3f" % mean_absolute_error(y_test, y_pred))
# The mean squared error
print("MSE = %5.3f" % mean_squared_error(y_test, y_pred))
# Plot outputs
```

```
fig, ax = plt.subplots()
ax.scatter(y_test, y_pred, edgecolors = (0, 0, 0))
ax.plot([y.min(), y.max()], [y.min(), y.max()], 'k--', lw = 4)
ax.set_xlabel('Measured')
ax.set_ylabel('Predicted')
plt.title('Without Transformation')
plt.show()

# ========================================================================
# Prediction with Transformation
# ========================================================================
transformer = QuantileTransformer(output_distribution = 'normal')
regressor = DecisionTreeRegressor(random_state = 0)
regr = TransformedTargetRegressor(regressor = regressor,
            transformer = transformer)
X_train, X_test, y_train, y_test = train_test_split(X, y, random_state = 0)
regr.fit(X_train, y_train)
y_pred = regr.predict(X_test)

print('R^2 score with Transformation: {0:.2f}'.format(regr.score(X_test,
y_test)))
# Explained variance score: 1 is perfect prediction
print("R^2 = %0.5f" % r2_score(y_test, y_pred))
# The mean absolute error
print("MAE = %5.3f" % mean_absolute_error(y_test, y_pred))
# The mean squared error
print("MSE = %5.3f" % mean_squared_error(y_test, y_pred))

# Plot outputs
fig, ax = plt.subplots()
ax.scatter(y_test, y_pred, edgecolors = (0, 0, 0))
ax.plot([y.min(), y.max()], [y.min(), y.max()], 'k--', lw = 4)
ax.set_xlabel('Measured')
ax.set_ylabel('Predicted')
plt.title('With Transformation')
plt.show()
```

3.5.11 朴素贝叶斯

另一个重要的分类技术是朴素贝叶斯。首先，在早期的变量选择中会剔除高度相关的变量，因为类分离机制中这样的变量会增加相似度。这意味着可以近似地认为变量之间是相互独立的。第二，假设变量间的交互为零会导致隐式的正则化步骤，所以模型的方差减少，分类更加准确。第三，在某些情况下，当变量之间存在相关性时，最优决策面和独立假设之间会发生重叠，因此该假设对性能没有影响。第四，朴素贝叶斯模型产生的决策曲面虽然被认为是线性的，但却可以具有复杂的非线性形状。基于上述原因，朴素贝叶斯被认为是可行的分类方法。该方法非常容易构建，因为不需要任何复杂的迭代参数估计方案，对分类不熟悉的用户也能理解其分类逻辑。最后，对于某些应用来说，它可能不是最好的分类器，但是在特定的情况下，它可以做得相当好（Wu et al., 2008）。

例 3.25 以下 Python 代码通过使用 scikit-learn 库中的 API 来对朴素贝叶斯分类器进行举例说明。在本例中，我们使用了 MNIST 手写数字数据集（详见前述）。MNIST 手写数字数据集被划分为训练集和测试集，然后使用朴素贝叶斯分类器进行分类。这里计算了分类的准确率、精确度、召回率、F1 得分、Cohen kappa 得分和 Matthews 相关系数，同时给出了分类报告和混淆矩阵。请注意，本例改编自 Python-scikit-learn。

```python
# ====================================================================
# Naive Bayes example with training and testset
# ====================================================================
import time
import numpy as np
from sklearn.datasets import fetch_openml
# Import train_test_split function

from sklearn.model_selection import train_test_split
from sklearn.preprocessing import StandardScaler
from sklearn.utils import check_random_state
#Import Gaussian Naive Bayes model
from sklearn.naive_bayes import GaussianNB

print(__doc__)

# Turn down for faster convergence
train_samples = 5000
# Load data from https://www.openml.org/d/554
X, y = fetch_openml('mnist_784', version = 1, return_X_y = True)

random_state = check_random_state(0)
permutation = random_state.permutation(X.shape[0])
X = X[permutation]
y = y[permutation]
X = X.reshape((X.shape[0], -1))

# Split dataset into training set and test set
# 70% training and 30% test
Xtrain, Xtest, ytrain, ytest = train_test_split(X, y,test_size = 0.3, random_state = 0)

scaler = StandardScaler()
Xtrain = scaler.fit_transform(Xtrain)
Xtest = scaler.transform(Xtest)

#Create a Gaussian Classifier
gnb = GaussianNB()
#Train the model using the training sets
```

```
gnb.fit(Xtrain, ytrain)
#Predict the response for test dataset
ypred = gnb.predict(Xtest)

#Evaluate the Model and Print Performance Metrics
from sklearn import metrics
print('Accuracy:', np.round(metrics.accuracy_score(ytest,ypred),4))
print('Precision:', np.round(metrics.precision_score(ytest,
        ypred,average = 'weighted'),4))
print('Recall:', np.round(metrics.recall_score(ytest,ypred,
        average = 'weighted'),4))
print('F1 Score:', np.round(metrics.f1_score(ytest,ypred,
        average = 'weighted'),4))
print('Cohen    Kappa    Score:', np.round(metrics.cohen_kappa_score(ytest,
ypred),4))
print('Matthews    Corrcoef:', np.round(metrics.matthews_corrcoef(ytest,
ypred),4))
print('\t\tClassification Report:\n', metrics.classification_report(ypred,
ytest))
from sklearn.metrics import confusion_matrix
print("Confusion Matrix:\n",confusion_matrix(ytest, ypred))

#Plot Confusion Matrix
from sklearn.metrics import confusion_matrix
from io import BytesIO #neded for plot
import seaborn as sns; sns.set()
import matplotlib.pyplot as plt

mat = confusion_matrix(ytest, ypred)
sns.heatmap(mat.T, square = True, annot = True, fmt = 'd', cbar = False)
plt.xlabel('true label')
plt.ylabel('predicted label');
plt.savefig("Confusion.jpg")
# Save SVG in a fake file object.
f = BytesIO()
plt.savefig(f, format = "svg")
```

例 3.26 以下 Python 代码通过使用 scikit-learn 库中的 API 来对朴素贝叶斯分类器进行举例说明。在本例中，我们使用了 sklearn.datasets 中的 Iris 数据集（鸢尾属植物数据集）。Iris 数据集采用 10 折交叉验证，使用朴素贝叶斯分类器进行分类。这里计算了分类的准确率、精确度、召回率、F1 得分、Cohen kappa 得分和 Matthews 相关系数。请注意，本例改编自 Python-scikit-learn。

```
# =========================================================================
# Naive Bayes example with cross-validation
# =========================================================================
from sklearn.model_selection import cross_val_score
```

```
from sklearn.datasets import load_iris
#Import Gaussian Naive Bayes model
from sklearn.naive_bayes import GaussianNB
iris = load_iris()
X, y = iris.data, iris.target
#Create a Gaussian Classifier
model = GaussianNB()

CV = 10 #10-Fold Cross-Validation
#Evaluate Model Using 10-Fold Cross-Validation and Print Performance
Metrics
Acc_scores = cross_val_score(model, X, y, cv = CV)
print("Accuracy: %0.3f (+/- %0.3f)" % (Acc_scores.mean(), Acc_scores.
std() * 2))
f1_scores = cross_val_score(model, X, y, cv = CV,scoring = 'f1_macro')
print("F1 score: %0.3f (+/- %0.3f)" % (f1_scores.mean(), f1_scores.std()
 * 2))
Precision_scores = cross_val_score(model, X, y,
cv = CV,scoring = 'precision_macro')
print("Precision score: %0.3f (+/- %0.3f)" % (Precision_scores.mean(),
Precision_scores.std() * 2))
Recall_scores = cross_val_score(model, X, y, cv = CV,scoring = 'recall_macro')
print("Recall score: %0.3f (+/- %0.3f)" % (Recall_scores.mean(), Recall_
scores.std() * 2))
from sklearn.metrics import cohen_kappa_score, make_scorer
kappa_scorer = make_scorer(cohen_kappa_score)
Kappa_scores = cross_val_score(model, X, y, cv = CV,scoring = kappa_scorer)
print("Kappa score: %0.3f (+/- %0.3f)" % (Kappa_scores.mean(), Kappa_
scores.std() * 2))
```

3.5.12 集成学习

模型的组合通常也称为模型集成，它们是机器学习中最有效的方法，通常能获得比单一模型更好的性能。但模型集成会增加算法的开销以及模型的复杂度。集成学习主要有两个驱动因素：计算学习理论与统计学。著名的统计学理论指出，平均测量可以带来更可靠和稳定的估计，因为减少了单次测量中随机振荡的影响。因此，如果使用相同的训练数据构建一个集成模型，可以减少随机变化在单一模型中带来的影响。主要问题是如何实现这些不同模型的多样性。通常会使用数据的随机子集，甚至是可用特征的随机子集来训练或构建模型，以实现多样性。从学习模型的角度来研究假设语言的可学习性，由此定义其意义所在。PCA的可学习性需要这样的假设，即在大部分情况下都是大概准确的。另一种学习模型为弱可学习性，它们所需的假设是学习总是比碰运气更好。即便如此，PCA可学习性模型显然比弱可学习性模型更为严格。我们使用一个迭代算法进行证明，该算法通过提升的方式重复地创建假设目标来修正先前的假设。最后，将每个迭代中学习到的假设组合起来，形成一个集成模型（Flach, 2012）。集成模型具有以下规范：

- 通过重新采样或重新加权的方式，从训练数据的演化版本中构建不同的预测模型。
- 使用任意方式来组合这些模型的预测结果，通常使用简单的求均值或是带权重/不带权重的投票。

集成方法的思想是将精心挑选的元素产生的预测结果组合起来，构建一个更大的模型，可以实现更好的整体预测。Breiman 的随机森林是一种基于引导法的集成分类器，通过树进行集成。换句话说，随机森林是一种 bagged（bootstrap aggregated）版本的树。可见，bagging 是个很笼统的概念，ANN、SVM 或其他任何分类模型都可以用于 bagging。另一种创建集成模型的方法是按序重优化，并对结果求平均。这可以通过引导法来实现，即将一系列"弱学习器"作为集成模型。弱学习器是较差的模型，但还是能够对数据中一些关键特征进行描述。因此，在适当的弱学习器集中组合产生一个强学习器是合理的。一般来说，生成集成模型有两个主要原则：第一，组合模型意味着模型类比简单地选择其中之一更强。由此，模型集合的预测的加权和可以获得比单一预测更好的性能，因为偏差更低。集成模型组合后的预测保持独立，因为它们基于不同的假设，不容易合并，而且有各种各样的参数和不同的估计值。事实上，只有当模型产生不同的预测结果时，集成模型才会增强模型选择的方法（Clarke，Fokoue，& Zhang，2009 Clarke，Foukoue，& Zhang，2009）。最著名的两种集成算法是 bagging 和 boosting。

3.5.13 bagging 算法

bagging 或 bootstrap aggregating（装袋）是一种集成建模算法。它从训练集中随机选择数据子集进行训练，以改善模型方差。在分类和数值预测中，合适的方法分别是加权投票和加权平均。对于特定的机器学习方法，可以选择不同的测试方式以产生正确的预测结果。装袋法使用重采样的方法（有放回的采样）从原始数据集中抽取一定数量的样本，而不是使用独立的域数据集是其独特之处。它从多个样本中并行地构建不同的模型，以相同权重投票产生最终的模型。该分类器在原始训练数据中总是成功的，其预测结果大大优于实际的分类器，从未出现过差强人意的表现。它试图通过去除某些实例和重复其他实例的方式来调整原始训练数据，以抵销方差。通过避免过拟合，分类器提高了一致性，减少了偏差和方差误差。尽管如此，在训练数据因抽样而产生的变化中，偏置模型的稳健性并没有得到太大改善（Witten，Frank，& Hall，2011）。

bagging 是非常成功的集成学习器，它可以在原始数据集的不同随机样本上产生不同的模型。像这样均匀地对样本进行替换的方式称为可重复采样。可重复采样通常会包含重复的样本，因此会丢失一些原始的数据，即使可重复采样的大小与原数据集一致。这正是我们所需要的，因为可重复采样得到的不同数据集会产生不同的模型，使集成模型具有多样性。可以选择以投票的方式聚合这些不同模型的预测结果——模型预测的类别占多数的获胜——或是以求平均的方式，这更加适用于基分类器的性能分数或概率。通过投票，可以看到 bagging 会产生分段的线性决策边界，这在单一线性分类器中是不太可能的。如果把投票变成对每个模型的概率估计，可以看到决策的不同边界将样本空间分割成段，每一段可能会有不同的性能分数（Flach，2012）。

要构造模型，首先将数据集划分为训练集和测试集，然后对训练集进行重复采样，使用这些样本训练出一个模型，随机重复以上步骤若干次，组合得到最终的模型：回归采用平均值的方式，分类采用投票的方式。由此得到一个近似的抽样误差，同样，也表示为推测误差。该方法在神经网络、决策树、回归树等不平衡的学习算法中表现非常出色。然而，在一些较稳定的分类器中不太适用，如 k 近邻。缺乏对算法本身的理解是 bagging 的主要弊端，因为该策略是在无监督的群体分析背景下使用的（Kumari，2012）。

bagging 是一种能够提升模型性能的集成模型，如随机森林算法。该方法通过重复采样来保证模型能够对数据进行平均描述。一个有效的方法是寻找足够多的模型，然后生成一个具有较小错误分类误差的模型。但即使这样做了，也可能是不稳定的。不稳定方法是指其模型的选择变化较大。ANN、树、以及线性回归中的子集选择都是不稳定的，而最近邻方法则是稳定的。通常来讲，bagging 能够增强不稳定分类器的性能，使其接近最优（Clarke，Fokoue，& Zhang，2009）。

bagging 在与树型模型相结合时尤为有效，因为树型模型很容易受到训练数据变化的影响。bagging 在应用于树型模型时，常常与另一个概念结合起来。通常称为子空间采样的方法从不同的随机特征子集构建每棵树。该方法使集成模型的多样性更加丰富，并且具有减少每棵树的训练时间的额外优势。随机森林是最终的集成模型。所附示例的空间划分基本上是集合中单棵树的划分组成。因此，虽然随机森林分区比大多数树型分区更好，但理论上可以映射回单一树模型，因为交集意味着将两棵不同树的分支进行组合。这与 bagging 线性分类器不同，在 bagging 线性分类器中，集成模型有一个决策边界，而单一基础分类器是无法学习得到的。我们也可以假设树型模型的另一种学习算法是由随机森林算法实现的（Flach，2012）。

例 3.27 以下 Python 代码通过使用 scikit-learn 库中的 API 来对 bagging 集成分类器进行举例说明。在本例中，我们使用了 MNIST 手写数字数据集（详见前述）。MNIST 手写数字数据集被划分为训练集和测试集，然后使用 bagging 集成分类器进行分类。这里计算了分类的准确率、精确度、召回率、F1 得分、Cohen kappa 得分和 Matthews 相关系数，同时给出了分类报告和混淆矩阵。请注意，本例改编自 Python-scikit-learn。

```
# =====================================================================
# Bagging example with training and test set
# =====================================================================
import time
import numpy as np
from sklearn.datasets import fetch_openml
# Import train_test_split function
from sklearn.model_selection import train_test_split
```

```python
from sklearn.preprocessing import StandardScaler
from sklearn.utils import check_random_state
#Import Bagging ensemble model
from sklearn.ensemble import BaggingClassifier
#Import Tree model as a base classifier
from sklearn import tree

print(__doc__)

# Turn down for faster convergence
train_samples = 5000

# Load data from https://www.openml.org/d/554
X, y = fetch_openml('mnist_784', version = 1, return_X_y = True)

random_state = check_random_state(0)
permutation = random_state.permutation(X.shape[0])
X = X[permutation]
y = y[permutation]
X = X.reshape((X.shape[0], -1))

# Split dataset into training set and test set
# 70% training and 30% test
Xtrain, Xtest, ytrain, ytest = train_test_split(X, y,test_size = 0.3, random_state = 0)
scaler = StandardScaler()
Xtrain = scaler.fit_transform(Xtrain)
Xtest = scaler.transform(Xtest)

#Create a Bagging Ensemble Classifier
" " "BaggingClassifier(base_estimator = None, n_estimators = 10, max_samples = 1.0,
max_features = 1.0, bootstrap = True, bootstrap_features = False, oob_score = False,
warm_start = False, n_jobs = None, random_state = None, verbose = 0)" " "
bagging = BaggingClassifier(tree.DecisionTreeClassifier(),
        max_samples = 0.5, max_features = 0.5)
#Train the model using the training sets
bagging.fit(Xtrain,ytrain)
#Predict the response for test dataset
ypred = bagging.predict(Xtest)

#Evaluate the Model and Print Performance Metrics
from sklearn import metrics
print('Accuracy:', np.round(metrics.accuracy_score(ytest,ypred),4))
print('Precision:', np.round(metrics.precision_score(ytest,
        ypred,average = 'weighted'),4))
print('Recall:', np.round(metrics.recall_score(ytest,ypred,
        average = 'weighted'),4))
print('F1 Score:', np.round(metrics.f1_score(ytest,ypred,
```

```
                     average = 'weighted'),4))
print('Cohen    Kappa    Score:',   np.round(metrics.cohen_kappa_score(ytest,
ypred),4))
print('Matthews      Corrcoef:',     np.round(metrics.matthews_corrcoef(ytest,
ypred),4))
print('\t\tClassification Report:\n', metrics.classification_report(ypred,
ytest))
from sklearn.metrics import confusion_matrix
print("Confusion Matrix:\n",confusion_matrix(ytest, ypred))

#Plot Confusion Matrix
from sklearn.metrics import confusion_matrix
from io import BytesIO #neded for plot
import seaborn as sns; sns.set()
import matplotlib.pyplot as plt

mat = confusion_matrix(ytest, ypred)
sns.heatmap(mat.T, square = True, annot = True, fmt = 'd', cbar = False)
plt.xlabel('true label')
plt.ylabel('predicted label');

plt.savefig("Confusion.jpg")
# Save SVG in a fake file object.
f = BytesIO()
plt.savefig(f, format = "svg")
```

例 3.28 以下 Python 代码通过使用 scikit-learn 库中的 API 来对 bagging 集成分类器进行举例说明。在本例中，我们使用了 sklearn.datasets 中的 Iris 数据集（鸢尾属植物数据集）。Iris 数据集采用 10 折交叉验证，使用 bagging 集成分类器进行分类。这里计算了分类的准确率、精确度、召回率、F1 得分、Cohen kappa 得分和 Matthews 相关系数。请注意，本例改编自 Python-scikit-learn。

```
# =====================================================================
# Bagging example with cross-validation
# =====================================================================
print(__doc__)
from sklearn.model_selection import cross_val_score
from sklearn.datasets import load_iris
#Import Bagging ensemble model
from sklearn.ensemble import BaggingClassifier
#Import Tree model as a base classifier
from sklearn import tree
iris = load_iris()
X, y = iris.data, iris.target

"""BaggingClassifier(base_estimator = None,  n_estimators = 10,  max_sam-
ples = 1.0,
max_features = 1.0,  bootstrap = True,  bootstrap_features = False,  oob_
```

```
score = False,
warm_start = False, n_jobs = None, random_state = None, verbose = 0)" " "
#Create a Bagging Ensemble Classifier
model = BaggingClassifier(tree.DecisionTreeClassifier(),
        max_samples = 0.5, max_features = 0.5)

CV = 10 #10-Fold Cross Validation
#Evaluate Model Using 10-Fold Cross Validation and Print Performance
Metrics
Acc_scores = cross_val_score(model, X, y, cv = CV)
print("Accuracy: %0.3f (+/- %0.3f)" % (Acc_scores.mean(), Acc_scores.
std() * 2))
f1_scores = cross_val_score(model, X, y, cv = CV, scoring = 'f1_macro')
print("F1 score: %0.3f (+/- %0.3f)" % (f1_scores.mean(), f1_scores.std()
 * 2))
Precision_scores = cross_val_score(model, X, y,
cv = CV, scoring = 'precision_macro')
print("Precision score: %0.3f (+/- %0.3f)" % (Precision_scores.mean(),
Precision_scores.std() * 2))
Recall_scores = cross_val_score(model, X, y, cv = CV, scoring = 'recall_macro')
print("Recall score: %0.3f (+/- %0.3f)" % (Recall_scores.mean(), Recall_
scores.std() * 2))
from sklearn.metrics import cohen_kappa_score, make_scorer
kappa_scorer = make_scorer(cohen_kappa_score)
Kappa_scores = cross_val_score(model, X, y, cv = CV, scoring = kappa_scorer)
print("Kappa score: %0.3f (+/- %0.3f)" % (Kappa_scores.mean(), Kappa_
scores.std() * 2))
```

3.5.14 随机森林

随机森林使用决策树作为主要分类器，该集成学习方法对信息进行表征。该方法将多个训练模型组合起来，对新的实例进行分类。随机森林是一种由树型组织分类器组成的分类器。在这些分类器中，独立的随机向量被无差别地传播，而且每棵树都会为最知名的类别投票。随机向量独立于具有相同分布的前一个随机向量，并且通过训练测试的方式来创建一棵树。在随机森林中，会确定一个上界，目的是获得以下两个参数的泛化误差：单个分类器的准确率和它们之间的依赖性。随机森林中的泛化误差有两个部分，这两部分的特征为森林中各个分类器的强度以及它们之间的相关性作为原始间隔函数。在保证强度不变的前提下，需要降低相关性，以提高随机森林的准确率水平（Goel & Abhilasha, 2017）。

例 3.29 以下 Python 代码通过使用 scikit-learn 库中的 API 来对随机森林分类器进行举例说明。在本例中，我们使用了 MNIST 手写数字数据集（详见前述）。MNIST 手写数字数据集被划分为训练集和测试集，然后使用随机森林分类器进行分类。这里计算了分类的准确率、精确度、召回率、F1 得分、Cohen kappa 得分和 Matthews 相关系数，同时给出了分类报告和混淆矩阵。请注意，本例改编自 Python-scikit-learn。

```
# ==============================================================
# Random forest example with training and testset
# ==============================================================
import time
import numpy as np
from sklearn.datasets import fetch_openml
from sklearn.model_selection import train_test_split
from sklearn.preprocessing import StandardScaler
from sklearn.utils import check_random_state
from sklearn.ensemble import RandomForestClassifier

print(__doc__)

# Turn down for faster convergence
train_samples = 5000

# Load data from https://www.openml.org/d/554
X, y = fetch_openml('mnist_784', version=1, return_X_y=True)

random_state = check_random_state(0)
permutation = random_state.permutation(X.shape[0])
X = X[permutation]
y = y[permutation]
X = X.reshape((X.shape[0], -1))

Xtrain, Xtest, ytrain, ytest = train_test_split(
    X, y, train_size=train_samples, test_size=10000)

#Transform the data using Standard Scaler
scaler = StandardScaler()
Xtrain = scaler.fit_transform(Xtrain)
Xtest = scaler.transform(Xtest)

#In order to change to accuracy increase n_estimators
"""RandomForestClassifier(n_estimators = 'warn', criterion = 'gini', max_depth = None,
min_samples_split = 2, min_samples_leaf = 1, min_weight_fraction_leaf = 0.0,
max_features = 'auto', max_leaf_nodes = None, min_impurity_decrease = 0.0,
min_impurity_split = None, bootstrap = True, oob_score = False, n_jobs = None,
random_state = None, verbose = 0, warm_start = False, class_weight = None)"""
clf = RandomForestClassifier(n_estimators = 200)
#Create the Model
#Train the model with Training Dataset
clf.fit(Xtrain,ytrain)
#Test the model with Testset
ypred = clf.predict(Xtest)

#Evaluate the Model and Print Performance Metrics
from sklearn import metrics
```

```python
print('Accuracy:', np.round(metrics.accuracy_score(ytest,ypred),4))
print('Precision:', np.round(metrics.precision_score(ytest,
        ypred,average = 'weighted'),4))
print('Recall:', np.round(metrics.recall_score(ytest,ypred,
        average = 'weighted'),4))
print('F1 Score:', np.round(metrics.f1_score(ytest,ypred,
average = 'weighted'),4))
print('Cohen Kappa Score:', np.round(metrics.cohen_kappa_score(ytest,
ypred),4))
print('Matthews Corrcoef:', np.round(metrics.matthews_corrcoef(ytest,
ypred),4))
print('\t\tClassification Report:\n', metrics.classification_report(ypred,
ytest))
from sklearn.metrics import confusion_matrix
print("Confusion Matrix:\n",confusion_matrix(ytest, ypred))

#Plot Confusion Matrix
from sklearn.metrics import confusion_matrix
from io import BytesIO #neded for plot
import seaborn as sns; sns.set()
import matplotlib.pyplot as plt
mat = confusion_matrix(ytest, ypred)
sns.heatmap(mat.T, square = True, annot = True, fmt = 'd', cbar = False)
plt.xlabel('true label')
plt.ylabel('predicted label');

plt.savefig("Confusion.jpg")
# Save SVG in a fake file object.
f = BytesIO()
plt.savefig(f, format = "svg")
```

例 3.30 以下 Python 代码通过使用 scikit-learn 库中的 API 来对随机森林分类器进行举例说明。在本例中，我们使用了 sklearn.datasets 中的 Iris 数据集（鸢尾属植物数据集）。Iris 数据集采用 10 折交叉验证，使用随机森林分类器进行分类。这里计算了分类的准确率、精确度、召回率、F1 得分、Cohen kappa 得分和 Matthews 相关系数。请注意，本例改编自 Python-scikit-learn。

```python
# =====================================================================
# Random forest example with cross-validation
# =====================================================================
from sklearn.ensemble import RandomForestClassifier
from sklearn.model_selection import cross_val_score
from sklearn.datasets import load_iris
iris = load_iris()
X, y = iris.data, iris.target


#In order to change to accuracy increase n_estimators
```

```
"""RandomForestClassifier(n_estimators = 'warn', criterion = 'gini', max_
depth = None,
min_samples_split = 2, min_samples_leaf = 1, min_weight_fraction_leaf = 0.0,
max_features = 'auto', max_leaf_nodes = None, min_impurity_decrease = 0.0,
min_impurity_split = None, bootstrap = True, oob_score = False, n_jobs = None,
random_state = None, verbose = 0, warm_start = False, class_weight = None)" " "
# fit model no training data
model = RandomForestClassifier(n_estimators = 200)

CV = 10 #10-Fold Cross Validation
#Evaluate Model Using 10-Fold Cross Validation and Print Performance
Metrics
Acc_scores = cross_val_score(model, X, y, cv = CV)
print("Accuracy: %0.3f (+/- %0.3f)" % (Acc_scores.mean(), Acc_scores.
std() * 2))
f1_scores = cross_val_score(model, X, y, cv = CV,scoring = 'f1_macro')
print("F1 score: %0.3f (+/- %0.3f)" % (f1_scores.mean(), f1_scores.std()
* 2))
Precision_scores = cross_val_score(model, X, y,
cv = CV,scoring = 'precision_macro')
print("Precision score: %0.3f (+/- %0.3f)" % (Precision_scores.mean(),
Precision_scores.std() * 2))
Recall_scores = cross_val_score(model, X, y,
cv = CV,scoring = 'recall_macro')
print("Recall score: %0.3f (+/- %0.3f)" % (Recall_scores.mean(), Recall_
scores.std() * 2))
from sklearn.metrics import cohen_kappa_score, make_scorer
kappa_scorer = make_scorer(cohen_kappa_score)
Kappa_scores = cross_val_score(model, X, y, cv = CV,scoring = kappa_scorer)
print("Kappa score: %0.3f (+/- %0.3f)" % (Kappa_scores.mean(), Kappa_
scores.std() * 2))
```

例 3.31 以下 Python 代码通过使用 scikit-learn 库中的 API 来表示使用随机森林回归器进行回归中的目标转换。在本例中，我们使用了 sklearn.datasets 中的波士顿房价数据集。本例使用 TransformedTargetRegressor，利用波士顿房价数据集转换后的特征。TransformedTargetRegressor 在拟合回归模型之前对目标 y 进行转换。通过逆向转换，预测结果被映射回原始空间。用于预测的回归器和应用于目标变量的转换器可以作为参数传入 TransformedTargetRegressor。图中的直线展示了随机森林回归如何绘制出一条直线，它使数据集中观察到的值与随机森林回归预测值之间的残差平方和最小化。同时计算了相关系数 R^2、MAE 和 MSE。从本例中可以看出，目标转换改进了随机森林回归的性能。请注意，本例改编自 Python-scikit-learn。

```
# ====================================================================
# Random forest regressor example with Boston house prices dataset
# ====================================================================
```

```python
# ======================================================================
#Transforming target in regression
# ======================================================================
import matplotlib.pyplot as plt
from sklearn.compose import TransformedTargetRegressor
from sklearn.preprocessing import QuantileTransformer
from sklearn.ensemble import RandomForestRegressor
from sklearn.model_selection import train_test_split
from sklearn.metrics import mean_squared_error, r2_score
from sklearn.metrics import mean_absolute_error
from sklearn.datasets import load_boston

# Load the Boston house prices dataset
boston = load_boston()
X = boston.data
y = boston.target
X_train, X_test, y_train, y_test = train_test_split(X, y, random_state = 0)


# ======================================================================
# Prediction without transformation
# ======================================================================

# Create Regression object
regr = RandomForestRegressor(max_depth = 2, random_state = 0, n_estimators = 100)
# Train the model using the training sets
regr.fit(X_train, y_train)
# Make predictions using the testing set
y_pred = regr.predict(X_test)

print('R^2 score without Transformation: {0:.2f}'.format(regr.score(X_test, y_test)))
# Explained variance score: 1 is perfect prediction
print("R^2 = %0.5f" % r2_score(y_test, y_pred))
# The mean absolute error
print("MAE = %5.3f" % mean_absolute_error(y_test, y_pred))
# The mean squared error
print("MSE = %5.3f" % mean_squared_error(y_test, y_pred))

# Plot outputs
fig, ax = plt.subplots()
ax.scatter(y_test, y_pred, edgecolors = (0, 0, 0))
ax.plot([y.min(), y.max()], [y.min(), y.max()], 'k--', lw = 4)
ax.set_xlabel('Measured')
ax.set_ylabel('Predicted')
plt.title('Without Transformation')
plt.show()
```

```python
# ================================================================
# Prediction with transformation
# ================================================================
transformer = QuantileTransformer(output_distribution = 'normal')
regressor = RandomForestRegressor(max_depth = 2, random_state = 0,
          n_estimators = 100)
regr = TransformedTargetRegressor(regressor = regressor,
          transformer = transformer)
X_train, X_test, y_train, y_test = train_test_split(X, y, random_state = 0)
regr.fit(X_train, y_train)
y_pred = regr.predict(X_test)
print('R^2 score with Transformation: {0:.2f}'.format(regr.score(X_test,
y_test)))
# Explained variance score: 1 is perfect prediction

print("R^2 = %0.5f" % r2_score(y_test, y_pred))
# The mean absolute error
print("MAE = %5.3f" % mean_absolute_error(y_test, y_pred))
# The mean squared error
print("MSE = %5.3f" % mean_squared_error(y_test, y_pred))
# Plot outputs
fig, ax = plt.subplots()
ax.scatter(y_test, y_pred, edgecolors = (0, 0, 0))
ax.plot([y.min(), y.max()], [y.min(), y.max()], 'k--', lw = 4)
ax.set_xlabel('Measured')
ax.set_ylabel('Predicted')
plt.title('With Transformation')
plt.show()
```

3.5.15 boosting 算法

boosting（提升）算法是一种与 bagging 类似的集成学习模型，但是它使用了比重复采样更为复杂的方式来生成不同的训练集。其基本思想简单而引人入胜。假设已有一个训练好的线性分类器，我们希望通过加入另一个分类器，降低其误分类率。一个好的方法是给错误分类的实例较高的权重，并根据这些权重修改分类器。基础的线性分类器可以以加权平均的方式估计类均值（Flach，2012）。

如果学习方法较弱，在预测方面的效果可能只比随机估计略好。boosting 方法通过在数据集上迭代优化某些弱学习器，以达到更好的性能。迭代优化利用指数损失函数和数据驱动的权重序列，提高了错误分类的成本，从而使分类器的连续迭代更加敏感。事实上，迭代形成了由基础分类器产生的集成规则，可以取得更好的性能，就像通过对集成模型的加权和投票一样。

boosting 由 Schapire（Schapire，1990）提出，随后迅速得到了改进。boosting 最初是针对弱学习器提出的，弱学习器的概念为树桩。如果树桩与良好的模型拆分有关，那么它的错误率可能会很小，但大多数情况下不是这样，所以通过某种方式增强它往往是个好主意。

与 bagging 类似，boosting 倾向于通过降低其方差来提高不稳定分类器的性能。有证据表明，boosting 可能会过拟合（Clarke et al.，2009）。

有迹象表明，如果分类器已经很好（稳定），误分类率很低，那么 bagging 或 boosting 能够带来的改进甚微。原因是该分类器已经接近最优，如 LDA。此外，甚至还有迹象表明，bagging 和 boosting 会降低分类器的性能，这在样本量太小的情况下更为常见。由于缺乏数据，平均方法无法提供太多帮助，因此存在相当大的不一致性。bagging 和 boosting 被用于不同的场景，因此不容易进行对比。例如，树桩是一个弱分类器，具有高偏差但通常很稳定。在这种情况下，由于类别的限制，boosting 的优势可能不足，但 bagging 可以获得更好的效果。较大的树更适合降低方差，但偏差较小，因此它们更适合 boosting。最后，方法多种多样，鼓励大家自由应用。我们可以将 boosting 分类器用于装袋，也可以将 bagging 分类器用于提升。此外，还可以堆叠形式多样的分类器，比如说决策树、ANN、SVM 和最近邻，然后对 bagging 版本进行提升，或者将堆叠后的 ANN 和 SVM 用于提升。这些选择可能会让人精疲力竭，因此，为了改进集成方法，需要仔细决定采用哪种分类或回归方法，以及如何使用它（Clarke et al.，2009）。

1. 自适应提升

自适应提升（AdaBoost）（Freund & Schapire，1997）策略被提出用于扩展集成方法的精确度。该提升算法的基本思想是构建一系列分类器，目标是让后面的分类器更专注于上一轮中错误分类的元组。由于集成的分类器是相互补充的，由此可以创建出一个高精确度的集成分类器。boosting 被认为是一种通用策略，用于改进随机学习算法。该模型算法易于理解，而且不会出现过拟合现象。它处理二分类问题，就像机器学习领域中处理多分类问题一样。同样地，AdaBoost 也对回归问题进行了扩展。在无噪声的数据中，提升算法比 bagging 表现更好。该算法依赖于数据集，它将多个分类器进行集成，合并为一个更强的分类器，被称为连续地分类器创建（Kumari，2012）。

要开发分类器，我们需要将准备好的数据或训练集作为输入。除此之外，基础学习算法被重复地多次调用，以训练其中的权重值。一开始，所有权重的设置都是相似的，但是，在每一轮中，错误分组实例的权重都会被增大，目标是迫使弱学习器集中精力在训练数据中的难度较大的实例上。这种提升方法可以通过两种结构连接起来：采样提升和权重提升。在提升的学习方法中，加权训练集可以直接被基础学习算法所接受，由此，整个训练集被分配给基础学习算法。另外，在基于采样的提升方法中，模型的绘制是以与其权重相关的概率来替代训练数据，模型是否停止迭代由交叉验证方法控制（Kumari，2012）。

该算法不需要关于弱学习器的先验知识，因此它可以与任何发现弱假设的技术相结合。最终，伴随着一组给定充分数据的假设保证，弱学习器可以可靠地产生精确的弱假设。该算法用于学习问题，遵循两个属性。第一个是观测到的实例具有不同的难度等级。提升算法在通常情况下产生的分布与难度较大的实例有关，因此弱学习算法很难在相同的空间内展示出效率。至于第二个属性,对于不同的训练数据集或假设,算法不能反复改变(Kumari,2012)。

例 3.32 以下 Python 代码通过使用 scikit-learn 库中的 API 来对 AdaBoost 分类器进行举例说明。在本例中,我们使用了 MNIST 手写数字数据集(详见前述)。MNIST 手写数字数据集被划分为训练集和测试集,然后使用 AdaBoost 分类器进行分类。这里计算了分类的准确率、精确度、召回率、F1 得分、Cohen kappa 得分和 Matthews 相关系数,同时给出了分类报告和混淆矩阵。请注意,本例改编自 Python-scikit-learn。

```
# ====================================================================
# Adaboost example with training and test set
# ====================================================================
import time
import numpy as np
from sklearn.datasets import fetch_openml
# Import train_test_split function
from sklearn.model_selection import train_test_split
from sklearn.preprocessing import StandardScaler
from sklearn.utils import check_random_state
#Import Adaboost ensemble model
from sklearn.ensemble import AdaBoostClassifier
#Import Tree model as a base classifier
from sklearn import tree

print(__doc__)

# Turn down for faster convergence
train_samples = 5000

# Load data from https://www.openml.org/d/554
X, y = fetch_openml('mnist_784', version = 1, return_X_y = True)

random_state = check_random_state(0)
permutation = random_state.permutation(X.shape[0])
X = X[permutation]
y = y[permutation]
X = X.reshape((X.shape[0], -1))

# Split dataset into training set and test set
# 70% training and 30% test
Xtrain, Xtest, ytrain, ytest = train_test_split(X, y,test_size = 0.3, random_state = 0)
scaler = StandardScaler()
Xtrain = scaler.fit_transform(Xtrain)
Xtest = scaler.transform(Xtest)

#Create an Adaboost Ensemble Classifier
""" AdaBoostClassifier(base_estimator = None, n_estimators = 50, learning_rate = 1.0,
        algorithm = 'SAMME.R', random_state = None)" " "
```

```
clf = AdaBoostClassifier(tree.DecisionTreeClassifier(),n_estimators = 10, al-
gorithm = 'SAMME',learning_rate = 0.5)

#Train the model using the training sets
clf.fit(Xtrain,ytrain)
#Predict the response for test dataset
ypred = clf.predict(Xtest)

#Evaluate the Model and Print Performance Metrics
from sklearn import metrics
print('Accuracy:', np.round(metrics.accuracy_score(ytest,ypred),4))
print('Precision:', np.round(metrics.precision_score(ytest,
        ypred,average = 'weighted'),4))
print('Recall:', np.round(metrics.recall_score(ytest,ypred,
        average = 'weighted'),4))
print('F1 Score:', np.round(metrics.f1_score(ytest,ypred,
        average = 'weighted'),4))
print('Cohen Kappa Score:', np.round(metrics.cohen_kappa_score(ytest,
ypred),4))
print('Matthews Corrcoef:', np.round(metrics.matthews_corrcoef(ytest,
ypred),4))
print('\t\tClassification Report:\n', metrics.classification_report(ypred,
ytest))
from sklearn.metrics import confusion_matrix
print("Confusion Matrix:\n",confusion_matrix(ytest, ypred))

#Plot Confusion Matrix
from sklearn.metrics import confusion_matrix
from io import BytesIO #neded for plot
import seaborn as sns; sns.set()
import matplotlib.pyplot as plt

mat = confusion_matrix(ytest, ypred)
sns.heatmap(mat.T, square = True, annot = True, fmt = 'd', cbar = False)
plt.xlabel('true label')
plt.ylabel('predicted label');

plt.savefig("Confusion.jpg")
# Save SVG in a fake file object.
f = BytesIO()
plt.savefig(f, format = "svg")
```

例 3.33 以下 Python 代码通过使用 scikit-learn 库中的 API 来对 AdaBoost 分类器进行举例说明。在本例中，我们使用了 sklearn.datasets 中的 Iris 数据集（鸢尾属植物数据集）。Iris 数据集采用 10 折交叉验证，使用 AdaBoost 分类器进行分类。这里计算了分类的准确率、精确度、召回率、F1 得分、Cohen kappa 得分和 Matthews 相关系数。请注意，本例改编自 Python-scikit-learn。

```
# ================================================================
# Adaboost example with cross-validation
# ================================================================
print(__doc__)
from sklearn.model_selection import cross_val_score
from sklearn.datasets import load_iris
#Import Adaboost Ensemble model
from sklearn.ensemble import AdaBoostClassifier
#Import Tree model as a base classifier
from sklearn import tree
iris = load_iris()
X, y = iris.data, iris.target

#Create an Adaboost Ensemble Classifier
""" AdaBoostClassifier(base_estimator = None, n_estimators = 50, learning_rate = 1.0,
        algorithm = 'SAMME.R', random_state = None)" " "
model = clf = AdaBoostClassifier(tree.DecisionTreeClassifier(),n_estimators = 10,
        algorithm = 'SAMME',learning_rate = 0.5)
CV = 10 #10-Fold Cross Validation
#Evaluate Model Using 10-Fold Cross Validation and Print Performance Metrics
Acc_scores = cross_val_score(model, X, y, cv = CV)
print("Accuracy: %0.3f (+/- %0.3f)" % (Acc_scores.mean(), Acc_scores.std() * 2))
f1_scores = cross_val_score(model, X, y, cv = CV,scoring = 'f1_macro')
print("F1 score: %0.3f (+/- %0.3f)" % (f1_scores.mean(), f1_scores.std() * 2))
Precision_scores = cross_val_score(model, X, y, cv = CV,scoring = 'precision_macro')
print("Precision score: %0.3f (+/- %0.3f)" % (Precision_scores.mean(), Precision_scores.std() * 2))
Recall_scores = cross_val_score(model, X, y, cv = CV,scoring = 'recall_macro')
print("Recall score: %0.3f (+/- %0.3f)" % (Recall_scores.mean(), Recall_scores.std() * 2))
from sklearn.metrics import cohen_kappa_score, make_scorer
kappa_scorer = make_scorer(cohen_kappa_score)
Kappa_scores = cross_val_score(model, X, y, cv = CV,scoring = kappa_scorer)
print("Kappa score: %0.3f (+/- %0.3f)" % (Kappa_scores.mean(), Kappa_scores.std() * 2))
```

例3.34 以下Python代码通过使用scikit-learn库中的API来表示使用AdaBoost回归器进行回归中的目标转换。在本例中，我们使用了sklearn.datasets中的波士顿房价数据集。本例使用TransformedTargetRegressor，利用波士顿房价数据集转换后的特征。TransformedTargetRegressor在拟合回归模型之前对目标y进行转换。通过逆向转换，预测结果被映射回原始空间。用于预测的回归器和应用于目标变量的转换器可以作为参数传入 TransformedTargetRegressor。图中的直线展示了 AdaBoost 回归如何绘制出一条直线，它使数据集中观察到的值与 AdaBoost 回归预测值之间的残差平方和最小化。同时计算了相关系数 R^2、MAE 和 MSE。从本例中可以看出，目标转换改进了 AdaBoost 回归的性能。请注意，本例改编自 Python-scikit-learn。

```
# ===================================================================
# Adaboost regressor example with Boston house prices dataset
# ===================================================================
# ===================================================================
#Transforming target in regression
# ===================================================================
import matplotlib.pyplot as plt
from sklearn.compose import TransformedTargetRegressor
from sklearn.preprocessing import QuantileTransformer
from sklearn.ensemble import AdaBoostRegressor
from sklearn.model_selection import train_test_split
from sklearn.metrics import mean_squared_error, r2_score
from sklearn.metrics import mean_absolute_error
from sklearn.datasets import load_boston

# Load the Boston House prices dataset
boston = load_boston()
X = boston.data
y = boston.target
X_train, X_test, y_train, y_test = train_test_split(X, y, random_state = 0)
# ===================================================================
# Prediction without Transformation
# ===================================================================

# Create Regression object
regr = AdaBoostRegressor(random_state = 0, n_estimators = 100)
# Train the model using the training sets
regr.fit(X_train, y_train)
# Make predictions using the testing set
y_pred = regr.predict(X_test)

print('R^2 score without Transformation: {0:.2f}'.format(regr.score(X_test, y_test)))
# Explained variance score: 1 is perfect prediction
print("R^2 = %0.5f" % r2_score(y_test, y_pred))
# The mean absolute error
print("MAE = %5.3f" % mean_absolute_error(y_test, y_pred))
# The mean squared error
print("MSE = %5.3f" % mean_squared_error(y_test, y_pred))

# Plot outputs
fig, ax = plt.subplots()
```

```python
ax.scatter(y_test, y_pred, edgecolors = (0, 0, 0))
ax.plot([y.min(), y.max()], [y.min(), y.max()], 'k--', lw = 4)
ax.set_xlabel('Measured')
ax.set_ylabel('Predicted')
plt.title('Without Transformation')
plt.show()

# ============================================================================
# Prediction with transformation
# ============================================================================
transformer = QuantileTransformer(output_distribution = 'normal')
regressor = AdaBoostRegressor(random_state = 0, n_estimators = 100)
regr = TransformedTargetRegressor(regressor = regressor,
            transformer = transformer)
X_train, X_test, y_train, y_test = train_test_split(X, y, random_state = 0)
regr.fit(X_train, y_train)
y_pred = regr.predict(X_test)

print('R^2 score with Transformation: {0:.2f}'.format(regr.score(X_test,
y_test)))
# Explained variance score: 1 is perfect prediction
print("R^2 = %0.5f" % r2_score(y_test, y_pred))
# The mean absolute error
print("MAE = %5.3f" % mean_absolute_error(y_test, y_pred))
# The mean squared error
print("MSE = %5.3f" % mean_squared_error(y_test, y_pred))

# Plot outputs
fig, ax = plt.subplots()
ax.scatter(y_test, y_pred, edgecolors = (0, 0, 0))
ax.plot([y.min(), y.max()], [y.min(), y.max()], 'k--', lw = 4)
ax.set_xlabel('Measured')
ax.set_ylabel('Predicted')
plt.title('With Transformation')
plt.show()
```

2. 梯度提升

基于装袋过程，Breiman（Breiman，1996）引入了在函数估计过程中加入随机性的想法，以提高算法的性能。AdaBoost（Freund & Schapire，1995）的最初实现也利用了随机采样，但是，一旦基础学习器的实现不支持权重观测，该方法则被认为是确定性加权的近似，而不是一个重要的组成部分。后来，Breiman（Breiman，1999）提出了一种混合的装袋提升过程（"自适应装袋"），旨在对加性展开进行最小二乘拟合。它将常规提升过程中的基础学习器替换为相关的装袋基础学习器，并在每个提升步骤中将"袋外"残差替换为普通残差。受Breiman（Breiman，1999）的启发，对梯度提升进行了最小的修改，将随机性结合在一起作为该过程的重要部分。在每次迭代中，从完整的训练数据集中随机抽取一个训练数据的子样本（不替换）。然后利用这个随机选择的子样本，而非全样本，来对基础学习器进行拟合，并计算当前迭代的模型更新（Friedman，2002）。

例3.35 以下 Python 代码通过使用 scikit-learn 库中的 API 来对梯度提升分类器进行举例说明。在本例中，我们使用了 MNIST 手写数字数据集（详见前述）。MNIST 手写数字数据集被划分为训练集和测试集，然后使用梯度提升分类器进行分类。这里计算了分类的准确率、精确度、召回率、F1 得分、Cohen kappa 得分和 Matthews 相关系数，同时给出了分类报告和混淆矩阵。请注意，本例改编自 Python-scikit-learn。

```python
# ======================================================================
# Gradient boosting example with training and testset
# ======================================================================
import time
import numpy as np
from sklearn.datasets import fetch_openml
# Import train_test_split function
from sklearn.model_selection import train_test_split
from sklearn.preprocessing import StandardScaler
from sklearn.utils import check_random_state
#Import Gradient Boosting ensemble model
from sklearn.ensemble import GradientBoostingClassifier

print(__doc__)

# Turn down for faster convergence
train_samples = 5000

# Load data from https://www.openml.org/d/554
X, y = fetch_openml('mnist_784', version = 1, return_X_y = True)

random_state = check_random_state(0)
permutation = random_state.permutation(X.shape[0])
X = X[permutation]
y = y[permutation]
X = X.reshape((X.shape[0], -1))

# Split dataset into training set and test set
# 70% training and 30% test
Xtrain, Xtest, ytrain, ytest = train_test_split(X, y,test_size = 0.3, random_state = 0)
scaler = StandardScaler()
Xtrain = scaler.fit_transform(Xtrain)
Xtest = scaler.transform(Xtest)

#Create the Model
clf = GradientBoostingClassifier(n_estimators = 100, learning_rate = 1.0, max_depth = 1, random_state = 0).fit(Xtrain, ytrain)
```

```
clf.score(Xtest, ytest)

#Train the model using the training sets
clf.fit(Xtrain,ytrain)
#Predict the response for test dataset
ypred = clf.predict(Xtest)

#Evaluate the Model and Print Performance Metrics
from sklearn import metrics
print('Accuracy:', np.round(metrics.accuracy_score(ytest,ypred),4))
print('Precision:', np.round(metrics.precision_score(ytest,
        ypred,average = 'weighted'),4))
print('Recall:', np.round(metrics.recall_score(ytest,ypred,
        average = 'weighted'),4))
print('F1 Score:', np.round(metrics.f1_score(ytest,ypred,
        average = 'weighted'),4))
print('Cohen Kappa Score:', np.round(metrics.cohen_kappa_score(ytest,
ypred),4))
print('Matthews Corrcoef:', np.round(metrics.matthews_corrcoef(ytest,
ypred),4))
print('\t\tClassification Report:\n', metrics.classification_report(ypred,
ytest))
from sklearn.metrics import confusion_matrix
print("Confusion Matrix:\n",confusion_matrix(ytest, ypred))
```

例 3.36 以下 Python 代码通过使用 scikit-learn 库中的 API 来表示使用梯度提升回归器进行回归中的目标转换。在本例中，我们使用了 sklearn.datasets 中的波士顿房价数据集。本例使用 TransformedTargetRegressor，利用波士顿房价数据集转换后的特征。TransformedTargetRegressor 在拟合回归模型之前对目标 y 进行转换。通过逆向转换，预测结果被映射回原始空间。用于预测的回归器和应用于目标变量的转换器可以作为参数传入 TransformedTargetRegressor。图中的直线展示了梯度提升回归如何绘制出一条直线，它使数据集中观察到的值与梯度提升回归预测值之间的残差平方和最小化。同时计算了相关系数 R^2、MAE 和 MSE。从本例中可以看出，目标转换改进了梯度提升回归的性能。请注意，本例改编自 Python-scikit-learn。

```
# ==================================================================
# Gradient boosting regressor example with Boston house prices dataset
# ==================================================================
# ==================================================================
#Transforming target in regression
# ==================================================================
import matplotlib.pyplot as plt
from sklearn.compose import TransformedTargetRegressor
from sklearn.preprocessing import QuantileTransformer
from sklearn.ensemble import GradientBoostingRegressor
```

```python
from sklearn.model_selection import train_test_split
from sklearn.metrics import mean_squared_error, r2_score
from sklearn.metrics import mean_absolute_error
from sklearn.datasets import load_boston
# Load the Boston House prices dataset
boston = load_boston()
X = boston.data
y = boston.target
X_train, X_test, y_train, y_test = train_test_split(X, y, random_state = 0)

# ===================================================================
# Prediction without transformation
# ===================================================================

# Create linear regression object
# Fit regression model
params = {'n_estimators': 500, 'max_depth': 4, 'min_samples_split': 2,
          'learning_rate': 0.01, 'loss': 'ls'}
regr = GradientBoostingRegressor(**params)
# Train the model using the training sets
regr.fit(X_train, y_train)
# Make predictions using the testing set
y_pred = regr.predict(X_test)

print('R^2 score without Transformation: {0:.2f}'.format(regr.score(X_test, y_test)))
# Explained variance score: 1 is perfect prediction
print("R^2 = %0.5f" % r2_score(y_test, y_pred))
# The mean absolute error
print("MAE = %5.3f" % mean_absolute_error(y_test, y_pred))
# The mean squared error
print("MSE = %5.3f" % mean_squared_error(y_test, y_pred))

# Plot outputs
fig, ax = plt.subplots()
ax.scatter(y_test, y_pred, edgecolors = (0, 0, 0))
ax.plot([y.min(), y.max()], [y.min(), y.max()], 'k--', lw = 4)
ax.set_xlabel('Measured')
ax.set_ylabel('Predicted')
plt.title('Without Transformation')
plt.show()

# ===================================================================
# Prediction with transformation
# ===================================================================
transformer = QuantileTransformer(output_distribution = 'normal')
regressor = GradientBoostingRegressor(**params)
regr = TransformedTargetRegressor(regressor = regressor,
            transformer = transformer)
X_train, X_test, y_train, y_test = train_test_split(X, y, random_state = 0)
```

```
regr.fit(X_train, y_train)
y_pred = regr.predict(X_test)

print('R^2 score with Transformation: {0:.2f}'.format(regr.score(X_test,
y_test)))
# Explained variance score: 1 is perfect prediction
print("R^2 = %0.5f" % r2_score(y_test, y_pred))
# The mean absolute error
print("MAE = %5.3f" % mean_absolute_error(y_test, y_pred))
# The mean squared error
print("MSE = %5.3f" % mean_squared_error(y_test, y_pred))

# Plot outputs
fig, ax = plt.subplots()
ax.scatter(y_test, y_pred, edgecolors = (0, 0, 0))
ax.plot([y.min(), y.max()], [y.min(), y.max()], 'k--', lw = 4)
ax.set_xlabel('Measured')
ax.set_ylabel('Predicted')
plt.title('With Transformation')
plt.show()
```

3.5.16 其他集成方法

除了 bagging 和 boosting，还有许多其他的集成方法。关键的区别在于，基础模型预测结果的组合方式不同。一些基础分类器将其预测结果作为特征，学习产生一个元模型，用于组合其预测结果。学习线性元模型被称为 stacking。另一种可行的方式是将不同的基础模型组合成一个异构的集成模型，以实现基础模型的多样性。这样的基础模型通过采用相同的训练集，由不同的学习算法进行训练。因此，集成模型由一组基础模型和一个元模型组成，元模型经过训练得到，并决定如何组合基础模型的预测结果（Flach, 2012）。

例 3.37 以下 Python 代码通过使用 scikit-learn 库中的 API 来对 stacking 元分类器进行举例说明。在本例中，我们使用了 sklearn.datasets 中的 Iris 数据集（鸢尾属植物数据集）。请注意，本例改编自 Python-scikit-learn。

```
# ====================================================================
# Stacking metaclassifier example
# ====================================================================
#Before running you should install mlxtend.classifier for Staking using pip
install mlxtend
import numpy as np
import warnings
from sklearn import model_selection
from sklearn.linear_model import LogisticRegression
from sklearn.neighbors import KNeighborsClassifier
```

```python
from sklearn.neural_network import MLPClassifier
from sklearn.ensemble import RandomForestClassifier
from mlxtend.classifier import StackingClassifier
from sklearn.datasets import load_iris
from sklearn.model_selection import train_test_split
from sklearn import tree

#Load Iris Dataset
iris = load_iris()
X, y = iris.data, iris.target
#Create Train and Test set
Xtrain, Xtest, ytrain, ytest = train_test_split(X, y, test_size = 0.3, random_state = 0)

warnings.simplefilter('ignore')

#Create the Model
clf1 = KNeighborsClassifier(n_neighbors = 1)
clf2 = RandomForestClassifier(random_state = 1)
clf3 = MLPClassifier(hidden_layer_sizes = (100, ), learning_rate_init = 0.001,
         alpha = 1, momentum = 0.9, max_iter = 1000)
DT = tree.DecisionTreeClassifier()
sclf = StackingClassifier(classifiers = [clf1, clf2, clf3],
         meta_classifier = DT)
# =========================================================================
# # Stacking example with cross-validation
# =========================================================================
print('5-fold cross validation:\n')
for clf, label in zip([clf1, clf2, clf3, sclf],
         ['KNN',
         'Random Forest',
         'Multilayer Perceptron',
         'StackingClassifier']):

scores = model_selection.cross_val_score(clf, X, y,
cv = 5, scoring = 'accuracy')
print("Accuracy: %0.3f (+/- %0.3f) [%s]"
    % (scores.mean(), scores.std(), label))
# =========================================================================
# # Stacking example with training and test set
# =========================================================================
#Train the Model with Training dataset
sclf.fit(Xtrain, ytrain)
#Test the Model with Testing dataset
ypred = sclf.predict(Xtest)

from sklearn import metrics
print('Accuracy:', np.round(metrics.accuracy_score(ytest, ypred), 4))
```

```python
print('Precision:', np.round(metrics.precision_score(ytest,
            ypred,average = 'weighted'),4))
print('Recall:', np.round(metrics.recall_score(ytest,ypred,
            average = 'weighted'),4))
print('F1 Score:', np.round(metrics.f1_score(ytest,ypred,
            average = 'weighted'),4))
print('Cohen Kappa Score:', np.round(metrics.cohen_kappa_score(ytest,
ypred)))
print('Matthews Corrcoef:', np.round(metrics.matthews_corrcoef(ytest,
ypred)))
print('\t\tClassification Report:\n', metrics.classification_report(ypred,
ytest))

from sklearn.metrics import confusion_matrix
from io import BytesIO #neded for plot
import seaborn as sns; sns.set()
import matplotlib.pyplot as plt

mat = confusion_matrix(ytest, ypred)
sns.heatmap(mat.T, square = True, annot = True, fmt = 'd', cbar = False)
plt.xlabel('true label')
plt.ylabel('predicted label');

plt.savefig("Confusion.jpg")
# Save SVG in a fake file object.
f = BytesIO()
plt.savefig(f, format = "svg")
```

例 3.38 以下 Python 代码通过使用 scikit-learn 库中的 API 来对投票集成分类器进行举例说明。在本例中，我们使用了 sklearn.datasets 中的 Iris 数据集（鸢尾属植物数据集）。请注意，本例改编自 Python-scikit-learn。

```
# ======================================================================
# Voting ensemble classifier example
# ======================================================================
from sklearn import datasets
from sklearn.model_selection import cross_val_score
from sklearn.linear_model import LogisticRegression
from sklearn.naive_bayes import GaussianNB
from sklearn.ensemble import RandomForestClassifier
from sklearn.ensemble import VotingClassifier

#Load Iris Dataset
iris = load_iris()
X, y = iris.data, iris.target
#Create Train and Test set
Xtrain, Xtest, ytrain, ytest = train_test_split(X, y, test_size = 0.3, ran-
dom_state = 0)
```

```python
# ================================================================
# Voting example with cross-validation
# ================================================================
#Create the Model
clf1 = KNeighborsClassifier(n_neighbors = 1)
clf2 = RandomForestClassifier(random_state = 1)
clf3 = MLPClassifier(hidden_layer_sizes = (100, ), learning_rate_init = 0.001,
         alpha = 1, momentum = 0.9,max_iter = 1000)
eclf = VotingClassifier(estimators = [('kNN', clf1), ('RF', clf2), ('MLP', clf3)], voting = 'hard')

print('\nPerfromance with Cross Validation')
print('5-fold cross validation:\n')
for clf, label in zip([clf1, clf2, clf3, eclf], ['kNN', 'Random Forest',
'MLP', 'Ensemble']):
    scores = cross_val_score(clf, X, y, cv = 5, scoring = 'accuracy')
    print("Accuracy: %0.2f (+/- %0.2f) [%s]" % (scores.mean(), scores.std(), label))

# ================================================================
# Voting example with training and testset
# ================================================================
#Train the model using the training sets
eclf.fit(Xtrain,ytrain)
#Predict the response for test dataset
ypred = eclf.predict(Xtest)

#Evaluate the Model and Print Performance Metrics
print('\n\nPerfromance with Test Set')
from sklearn import metrics
print('Accuracy:', np.round(metrics.accuracy_score(ytest,ypred),4))
print('Precision:', np.round(metrics.precision_score(ytest,
       ypred,average = 'weighted'),4))
print('Recall:', np.round(metrics.recall_score(ytest,ypred,
       average = 'weighted'),4))
print('F1 Score:', np.round(metrics.f1_score(ytest,ypred,
       average = 'weighted'),4))
print('Cohen Kappa Score:', np.round(metrics.cohen_kappa_score(ytest,
ypred),4))
print('Matthews Corrcoef:', np.round(metrics.matthews_corrcoef(ytest,
ypred),4))
print('\t\tClassification Report:\n', metrics.classification_report(ypred,
ytest))
from sklearn.metrics import confusion_matrix
print("Confusion Matrix:\n",confusion_matrix(ytest, ypred))
```

例3.39 以下Python代码通过使用scikit-learn库中的API来表示使用投票回归器进行回归中的目标转换。在本例中，我们使用了sklearn.datasets中的波士顿房价数据集。图中的直线展示了投票回归如何绘制出一条直线，它使数据集中观察到的值与投票回归预测值之间的残差平方和最小化。同时计算了相关系数 R^2、MAE 和 MSE。从本例中可以看出，目标转换改进了投票回归器的性能。请注意，本例改编自 Python-scikit-learn。

```python
# ================================================================
# Voting regressor example
# ================================================================
import matplotlib.pyplot as plt
from sklearn import datasets
from sklearn.ensemble import GradientBoostingRegressor
from sklearn.ensemble import RandomForestRegressor
from sklearn.linear_model import LinearRegression
from sklearn.ensemble import VotingRegressor
from sklearn.model_selection import train_test_split
from sklearn.metrics import mean_squared_error, r2_score
from sklearn.metrics import mean_absolute_error

# Load the Boston house prices dataset
boston = datasets.load_boston()
X = boston.data
y = boston.target
X_train, X_test, y_train, y_test = train_test_split(X, y, random_state = 0)
# Training classifiers
reg1 = GradientBoostingRegressor(random_state = 1, n_estimators = 10)
reg2 = RandomForestRegressor(random_state = 1, n_estimators = 10)
reg3 = LinearRegression()
eregr = VotingRegressor(estimators = [('gb', reg1), ('rf', reg2), ('lr', reg3)])

# Train the model using the training sets
eregr.fit(X_train, y_train)
# Make predictions using the testing set
y_pred = eregr.predict(X_test)

print('R^2 score without Transformation: {0:.2f}'.format(eregr.score(X_test, y_test)))
# Explained variance score: 1 is perfect prediction
print("R^2 = %0.5f" % r2_score(y_test, y_pred))
# The mean absolute error
print("MAE = %5.3f" % mean_absolute_error(y_test, y_pred))
# The mean squared error
print("MSE = %5.3f" % mean_squared_error(y_test, y_pred))

# Plot outputs
fig, ax = plt.subplots()
ax.scatter(y_test, y_pred, edgecolors = (0, 0, 0))
ax.plot([y.min(), y.max()], [y.min(), y.max()], 'k--', lw = 4)
```

```
ax.set_xlabel('Measured')
ax.set_ylabel('Predicted')
plt.title('Voting Regressor')
plt.show()
```

3.5.17 深度学习

深度学习是对人工神经网络（ANN）的一种增强。简单神经网络由输入层、隐藏层和输出层组成。ANN 模型的参数由网络中每个连接的权重组成，有时还包括偏差参数。近来，深度学习已经成为机器学习最广为人知的特征之一。深度学习在众多领域取得了超乎寻常的精准度和普及度，尤其是在图像和音频领域（Sarkar et al., 2018）。

如果目前的线性模型不足以描述任何学习过程，那么可以添加一些由非线性输入函数产生的新特征，由此在这些特征领域建立一个线性模型。这需要确定好的基函数。创建新特征空间的方法之一是使用如 PCA 等特征提取方法。但最好的方式是，在多层感知机（MLP）的隐藏层提取这些特征。因为在监督学习中，第一层（特征提取）和第二层会一起学习被用于输出预测的合并特征。如果只有一个隐藏层，MLP 的能力有限，而如果有多个隐藏层，MLP 可以学习更复杂的输入函数。这就是深度神经网络背后的理念，每个隐藏层都结合了前一层的输出值，从网络的原始输入开始，学习更复杂的输入特征。深度网络的另一个特点是，在连续的隐藏层中使用更加抽象的描述，直到输出层，其输出通过这些无形的概念学习得到。其思想是，深度学习在一些应用中使用最少的人为干预来学习抽象程度不断上升的特征层次（Bengio, 2009）。输入中汇集的信息是不被识别的，所以任何一种依赖关系都应该在训练中自动发现。训练具有许多隐藏层的 MLP 的关键问题是，必须在所有层中连续地乘以导数，同时将误差反向传播到前面的层，直到梯度消失。这也是未展开的循环神经网络训练速度慢的原因。对于卷积神经网络来说，上述情况不会发生，因为隐藏单元的扇入和扇出可以自然地忽略不计。深度神经网络通常是逐层训练的（Hinton & Salakhutdinov, 2006）。每一层的目的是从输入的数据中提取相关特征，为此可以使用自编码器等技术。因此，从原始输入数据开始，训练一个自编码器，然后将从隐藏层学习到的编码表示作为输入来训练下一个自编码器，以此类推，直到我们到达最后一层，该层由标签数据以监督的方式训练得到。在逐层训练之后，它们被汇集到一起，整个网络都使用标签数据进行微调。如果有大量的标签数据和强大的计算能力，可以用监督的方式来训练整个深度网络，但折中的办法是，使用无监督的方法来初始化权重，其效果比随机初始化更好。因此，可以使用更少的标签数据来更快地进行训练。深度学习的方法尤为突出，因为它们只需要较少的人工干预。在很多的实现中都可以考虑抽象层，探索这样的抽象表示方式可以提供有用的信息，也可以很好地描述问题本身（Alpaydin, 2014; Subasi, 2019）。

如前所述，深度学习领域是最近相当突出的一个机器学习子领域。其关键目标是让机器学习技术更接近"制造智能机器"的真正目的。深度学习有时被认为是用于更名神经网络的华丽辞藻。在一定程度上确实如此，但深度学习肯定不只是简单的神经网络。基于深度学习的算法包括使用表征学习原理，在不同的层中学习数据的不同表征，这也被用于自动特征提取。换一种说法，深度学习方法试图通过将数据表示为一个分层的概念层次结构来构建机器智能，其中每一层的概念都是由其他更简单的层构建的。任何深度学习算法的核心组成部分之一就是分层架构本身。本质上来说，我们试图在任何简单的监督机器学习方法中学习数据样本和输出之间的映射关系，然后预测新数据样本的性能。除了学习从输入到输出的映射，表征学习还试图理解数据本身的代表的含义。这使得深度学习算法与常规方法相比极其强大，因为深度学习算法需要大量的特征提取和工程等领域的专业知识。与旧的机器学习算法相比，深度学习在性能以及数据越来越多的情况下的可扩展性方面也是非常高效的。在过去的十年中，我们观察到了深度学习的几个显著的趋势和形势（Sarkar et al., 2018），总结如下：

- 深度学习算法是建立在分布式表征学习的基础上，随着时间的推移和数据的增多，它们的性能开始变得更好
- 深度学习可以说是神经网络的一个子领域，与传统的神经网络相比，深度学习更加完善
- 更好的软件框架，如 TensorFlow、Theano、Caffe、MXNetet 和 Keras，结合更好的硬件，已经可以开发出极其复杂的、规模更大的多层深度学习模型
- 自动特征选择和监督学习为深度学习带来了多种优势，随着时间的推移，数据科学家和工程师能够解决越来越复杂的问题

以下几点定义了大多数深度学习算法的相关特征。

- 概念的分层表示
- 通过多层架构进行分布式数据表征训练
- 更复杂、高级的特征和概念都源于更简单、低级的特征
- 一个"深度"神经网络通常被认为除了输入层和输出层外，至少有一个隐藏层。通常它至少由三到四个隐藏层组成
- 深层架构具有多层结构，每层有若干个非线性处理单元。每层的输入是架构中的前一层的输出。输入通常是第一层，输出是最后一层
- 它可以进行自动提取特征、分类、异常检测等许多与机器学习相关的任务

3.5.18 深度神经网络

深度神经网络是传统人工神经网络的延伸。与传统的神经网络相比，深度神经网络主要有两个区别。具有一个或两个隐藏层的传统神经网络是很浅的，而另一方面，深度神经网络有许多隐藏层。例如，谷歌大脑项目就使用了一个由数百万神经元组成的神经网络。深度神经网络的模型种类繁多，有 DNN、CNN、RNN 和 LSTM。近来的研究甚至为我们带来了基于注意力的网络，这些网络关注深度神经网络的特定部分。网络越大，层数越多，

网络就越复杂，需要训练的资源和时间就越多。深度神经网络在基于 GPU 的架构下工作得最好，它的训练时间比传统 CPU 更短，而最近的发展则大大缩短了训练时间（Sarkar 等，2018）。

例 3.40 以下 Python 代码通过使用 TensorFlow 库，对深度神经网络分类器进行举例说明。在本例中，我们使用了 sklearn.datasets 中的 Iris 数据集（鸢尾属植物数据集）。Iris 数据集被划分为训练集和测试集，使用深度神经网络分类器进行分类。这里计算了分类的准确率、精确度、召回率、F1 得分、Cohen kappa 得分和 Matthews 相关系数，同时给出了分类报告和混淆矩阵。请注意，本例改编自 TensorFlow.org 网站。

```
# ======================================================================
# Deep neural network example with Iris dataset using TensorFlow
# ======================================================================
import numpy as np
from sklearn import datasets
from sklearn import metrics
from sklearn import model_selection

import tensorflow as tf

X_FEATURE = 'x' # Name of the input feature.
n_classes = 3
# Load dataset.
iris = datasets.load_iris()
Xtrain, Xtest, ytrain, ytest = model_selection.train_test_split(
    iris.data, iris.target, test_size = 0.3, random_state = 42)

# Build 3 layer DNN with 10, 20, 10 units respectively.
feature_columns = [
    tf.feature_column.numeric_column(
    X_FEATURE, shape = np.array(Xtrain).shape[1:])]
classifier = tf.estimator.DNNClassifier(
    feature_columns = feature_columns, hidden_units = [50, 150, 50], n_class-
es = n_classes)

# Train.
train_input_fn = tf.compat.v1.estimator.inputs.numpy_input_fn(
    x = {X_FEATURE: Xtrain}, y = ytrain, num_epochs = None, shuffle = True)
classifier.train(input_fn = train_input_fn, steps = 200)

# Predict.
test_input_fn = tf.compat.v1.estimator.inputs.numpy_input_fn(
    x = {X_FEATURE: Xtest}, y = ytest, num_epochs = 1, shuffle = False)
predictions = classifier.predict(input_fn = test_input_fn)
ypred = np.array(list(p['class_ids'] for p in predictions))
ypred = ypred.reshape(np.array(ytest).shape)
```

```python
# Score with sklearn.
score = metrics.accuracy_score(ytest, ypred)
print('Accuracy (sklearn): {0:f}'.format(score))

# Score with tensorflow.
scores = classifier.evaluate(input_fn = test_input_fn)
print('Accuracy (tensorflow): {0:f}'.format(scores['accuracy']))

#Evaluate the Model and Print Performance Metrics
from sklearn import metrics
print('Accuracy:', np.round(metrics.accuracy_score(ytest,ypred),4))
print('Precision:', np.round(metrics.precision_score(ytest,
        ypred,average = 'weighted'),4))
print('Recall:', np.round(metrics.recall_score(ytest,ypred,
        average = 'weighted'),4))
print('F1 Score:', np.round(metrics.f1_score(ytest,ypred,
        average = 'weighted'),4))
print('Cohen Kappa Score:', np.round(metrics.cohen_kappa_score(ytest,
ypred),4))
print('Matthews Corrcoef:', np.round(metrics.matthews_corrcoef(ytest,
ypred),4))
print('\t\tClassification Report:\n', metrics.classification_report(ypred,
ytest))

from sklearn.metrics import confusion_matrix
from io import BytesIO #neded for plot
import seaborn as sns; sns.set()
import matplotlib.pyplot as plt
#Plot Confusion Matrix
mat = confusion_matrix(ytest, ypred)
sns.heatmap(mat.T, square = True, annot = True, fmt = 'd', cbar = False)
plt.xlabel('true label')
```

例 3.41 以下 Python 代码通过使用 TensorFlow 库中的 Keras API，对深度神经网络分类器进行举例说明。在本例中，我们使用了 sklearn.datasets 中的 Iris 数据集（鸢尾属植物数据集）。Iris 数据集被划分为训练集和测试集，通过 Keras API 和 TensorFlow 后端，使用深度神经网络分类器进行分类。本例中，分类准确率被用作性能衡量标准。请注意，本例改编自 TensorFlow.org 网站。

```python
# =================================================================
#Deep neural network example with Iris dataset using Keras API and
TensorFlow backend
# =================================================================
from keras.models import Sequential
from keras.layers import Dense
from sklearn import datasets
```

```
#Load Iris Dataset
iris = datasets.load_iris()
X = iris.data
y = iris.target

#Encode the Target Vector
from keras.utils import np_utils
dummy_y = np_utils.to_categorical(y)

#Create Train and Test Dataset
from sklearn.model_selection import train_test_split
Xtrain, Xtest, ytrain, ytest = train_test_split(X, y, test_size = 0.3, random_state = 0)

# Initialize the constructor
model = Sequential()
# Add an input layer
model.add(Dense(50, activation = 'relu', input_shape = (4,)))
# Add two hidden layer
model.add(Dense(100, activation = 'sigmoid'))
model.add(Dense(50, activation = 'sigmoid'))
# Add an output layer
model.add(Dense(1, activation = 'sigmoid'))

# Compile model
model.compile(loss = 'binary_crossentropy', optimizer = 'adam', metrics = ['accuracy'])
# Fit model
history = model.fit(Xtrain, ytrain, validation_data = (Xtest, ytest), epochs = 500, verbose = 2)
# Evaluate the model
_, train_acc = model.evaluate(Xtrain, ytrain, verbose = 0)
_, test_acc = model.evaluate(Xtest, ytest, verbose = 0)
print('Train: %.3f, Test: %.3f' % (train_acc, test_acc))
```

3.5.19 循环神经网络

循环神经网络（RNN）是一种特殊的人工神经网络，它利用一种特殊的循环结构，持续地获取与过去知识相关的信息。它们被用于许多序列数据相关的领域，例如预测句子的下一个词。这些循环网络对输入数据序列中的每一个元素进行相同的操作和计算。RNN 具有记忆，可以协助从过去的序列中获取信息（Sarkar et al.，2018）。

RNN 最早是在 20 世纪 80 年代发展起来的，但近些年才受到关注，因为科学和硬件发展使其在训练中的计算效率得到提高。RNN 不同于前馈网络，因为它具有一种特殊类型的神经层，即所谓的循环层，它允许在网络之间保持状态。为了更好地理解 RNN 的运作方式，我们来看一个循环神经网络在经过适当的训练后是如何工作的。我们为模型创建一个新的示例，用于处理一个新的序列。通过将网络中实例的生命周期划分为离散的时间步，从而使之包含循环层。在每个时间步中，我们将输入的下一个元素投喂到模型中。前馈连接反映了信息流从一个神经元到另一个神经元，其中的数据是当前时间步计算得到的神经元激活。然而，在循环连接构成的信息流中，数据是上一个时间步中存储的神经元激活。因此，循环网络中的神经元激活反映了网络中实例的积累状态。循环层中的初始神经元激活为模型的参数，定义这些参数的最优值，就像在训练过程中定义每个连接的权重的最优值一样。事实证明，我们其实可以将 RNN 表达成一个前馈网络（虽然是不规则的结构），并给定一个固定生命周期（比如 t 个时间步）。RNN 的训练基于展开模型对梯度进行测量，它保证了所有用于前馈网络的反向传播方法也可以应用于 RNN 训练。我们要根据使用每一批训练实例后计算出的误差导数来调整权重。在展开的网络中有一组连接，它们都与原始 RNN 中的相同连接有关。然而，并不能保证为这些展开连接计算的误差导数在现实中是等价的。通过对属于同一集合的所有环节的误差导数进行平均或求和，可以克服这个问题。这使得我们可以使用一个误差导数，这个误差导数通常会考虑影响连接权重的所有动态，试图使网络构建一个准确的输出（Buduma & Locascio，2017）。

例 3.42 以下 Python 代码通过使用 TensorFlow 库中的 Keras API，对循环神经网络（RNN）分类器进行举例说明。在本例中，我们使用了 Keras 中的 IMDB 数据集（电影评论情感分类数据集）。IMDB 数据集有 50 000 条电影评论用于文本分析或自然语言处理。与之前的基准数据集相比，它包含了更多的数据用于二元情感分类。其中，25 000 条评论用于训练，25 000 条用于测试。如需了解更多数据集信息，请访问以下网站：http://ai.stanford.edu/~amaas/data/sentiment/。RNN 使用训练集和测试集对模型进行训练和测试，分别给出了训练集和测试集的分类准确率。请注意，本例改编自 Francois Chollet 编写的 *Deep Learning with Python* 一书（Chollet，2018）。

```
# ================================================================
# Recurrent neural networks example with IMDB dataset
# ================================================================
from keras.datasets import imdb
from keras.preprocessing import sequence
from keras.layers import Dense
from keras.models import Sequential
from keras.layers import Embedding, SimpleRNN

max_features = 10000 # number of words to consider as features
maxlen = 500 # cut texts after this number of words (among top max_features
most common words)
batch_size = 32

print('Loading data...')
(Xtrain, ytrain), (Xtest, ytest) = imdb.load_data(num_words = max_features)
print(len(Xtrain), 'train sequences')
```

```
print(len(Xtest), 'test sequences')

print('Pad sequences (samples x time)')
Xtrain = sequence.pad_sequences(Xtrain, maxlen = maxlen)
Xtest = sequence.pad_sequences(Xtest, maxlen = maxlen)
print('input_train shape:', Xtrain.shape)
print('input_test shape:', Xtest.shape)
#Create Model
model = Sequential()
model.add(Embedding(max_features, 32))
model.add(SimpleRNN(32))
model.add(Dense(1, activation = 'sigmoid'))

# Compile model
model.compile(optimizer = 'rmsprop', loss = 'binary_crossentropy', metrics = ['acc'])
# Fit model
history = model.fit(Xtrain, ytrain,
        epochs = 10,
        batch_size = 128,
        validation_split = 0.2)

# Evaluate the model
_, train_acc = model.evaluate(Xtrain, ytrain, verbose = 0)
_, test_acc = model.evaluate(Xtest, ytest, verbose = 0)
print('Training Accuracy: %.3f, Testing Accuracy: %.3f' % (train_acc, test_acc))
```

3.5.20 自编码器

自编码器是一种特殊类型的人工神经网络，主要用于处理无监督的机器学习任务。其主要目标是学习数据的近似、表征和编码。自编码器可用于创建生成模型、降维和异常检测（Sarkar et al., 2018）。自编码器是一种用于降维和特征提取的无监督神经网络。具体来说，自动编码器是一个前馈神经网络，它被训练用于预测输入。系统可以通过确保隐藏单元捕捉到最合适的数据特征来最小化重构误差。利用深度自编码器可以学到更有效的描述。不幸的是，使用反向传播训练这类模型的效果并不好，因为梯度信号在经过多层回传时变得太小，导致学习算法经常会陷入局部的最小值。解决该问题的一个办法是贪婪地训练一系列受限玻尔兹曼机（RBM），并利用这些机器来初始化一个自编码器。然后，整个系统就可以按照通常的方式，利用反向传播进行微调（Murphy, 2012）。

3.5.21 长短期记忆网络

RNN 擅长于处理基于序列的数据，但是随着序列的增长，算法会随着时间的推移失去序列中的历史上下文信息。然而，这正是长短期记忆网络（LSTM）起作用的地方。LSTM

可以记住相当长的序列数据的信息，并防止出现一些问题，例如通常在反向传播训练的 ANN 中出现的梯度消失的问题。LSTM 一般有三到四个门，包括输入、输出和一个特定的遗忘门。通常，输入门用于保存或消除到来的刺激和输入，以改变记忆单元的状态；输出门用于将输出值传播给其他神经元；遗忘门会控制记忆单元的自循环环节，以记忆和遗忘之前的状态（Sarkar et al., 2018）。

长短期记忆（LSTM）开发的基本原则是，构建的网络能够将重要信息有效地传输到未来的一些时间段。LSTM 模块中有几个关键部件。记忆单元是 LSTM 架构的核心部件之一，其中张量是由中心的循环定义的。记忆单元保留了它随着时间推移所学习到的关键信息，而网络的构建则是通过许多时间步来成功保存记忆单元中的有价值的信息。在三个不同的阶段中，LSTM 模型在都会根据新的信息修改记忆单元。首先，记忆单元需要确定应该保留多少从前的记忆。门的基本思想很简单。上一步内存状态张量的信息很丰富，但有些信息可能是重复的，需要删除。我们通过计算一个位张量（一个 0 和 1 的张量），来弄清楚哪些元素在内存状态张量中仍然是相关的，哪些元素是不相关的，然后将该位张量乘以之前的状态。如果位张量为 1，意味着内存单元中的位置仍然有效，应该保留；如果为 0，则表明内存单元中的位置不再相关，应予以删除。通过将当前时间步的输入和前一个时间步的 LSTM 单元的输出相加，对这个位张量进行逼近，并在结果张量上增加一个 sigmoid 层。你可能还记得，一个 sigmoid 神经元产生的值大部分情况下都接近于 0 或接近于 1，唯一的例外是当输入接近于 0 时。因此，sigmoid 层的输出值是对位张量较好的逼近，可以用它来完成门的判断。一旦弄清楚了哪些信息需要保留在旧的状态，哪些需要删除，就可以考虑要写入什么样的内存状态信息了。这主要分为两个部分：第一部分是要计算出我们想要写到状态中的信息，由 tanh 层计算创建一个中间张量；第二部分是计算出哪些组件真正需要包含在该计算张量的新状态中，而哪些组件需要在写入之前丢弃。我们通过采用在门中使用的方法（sigmoid 层）来逼近 0 和 1 的位向量来实现这一目的。通过使用中间张量乘以位向量，然后将结果相加，构建 LSTM 新的状态向量（Buduma & Locascio, 2017）。

最后，我们希望 LSTM 单元在每个时间步中提供一个输出。虽然状态向量可以被明确地看作输出，但 LSTM 系统的建立能够提供更多的通用性，它通过生成输出张量作为"解释"或外部通信，反映其状态向量。我们使用与写入门几乎相同的框架：tanh 层从状态向量中创建一个中间张量，sigmoid 层使用当前输入和之前的输出创建一个位张量掩码，中间张量乘以位张量产生最终输出。那么，为什么这比使用单个的原始 RNN 更好呢？这里主要考虑的点是展开 LSTM 模块时，信息如何在网络中随时间传播。由最上层可以看到，状态向量的传播过程中，其关系大多恒定不变。因此，与当前输出相关的梯度由几个时间步构成，不会像在标准 RNN 模型中那样显著衰减。这确保了 LSTM 可以比原始的 RNN 更容易学习长期关系相关的知识。最终，我们想知道使用 LSTM 单元创建任意的网络结构有多简单。LSTM 如何才能做到"可组合"？为了使用 LSTM 单元而不是普通 RNN，必须牺牲灵活性吗？就像通过堆叠 RNN 层来构建容量更大的功能模型一样，我们也可以类似地堆叠

LSTM 单元：第二个单元的输入是第一个单元的输出，第三个单元的输入是第二个单元的输出，以此类推。这确保可以很容易地在任何使用普通 RNN 层的地方使用 LSTM 系统进行替换。既然已经解决了梯度消失的问题，并知晓了 LSTM 单元的内部工作原理，为深入研究模型的实现做好了准备（Buduma & Locascio，2017）。

例 3.43　以下 Python 代码通过使用 TensorFlow 库中的 Keras API，对长短期记忆网络（LSTM）分类器进行举例说明。在本例中，我们使用了 Keras 中的 IMDB 数据集（电影评论情感分类数据集）。IMDB 数据集有 50 000 条电影评论用于文本分析或自然语言处理。与之前的基准数据集相比，它包含了更多的数据用于二元情感分类。其中，25 000 条评论用于训练，25 000 条用于测试。如需了解更多数据集信息，请访问以下网站：http://ai.stanford.edu/~amaas/data/sentiment/。LSTM 使用训练集和测试集对模型进行训练和测试，分别给出了训练集和测试集的分类准确率。请注意，本例改编自 Francois Chollet 编写的 *Deep Learning with Python* 一书（Chollet，2018）。

```
# ======================================================================
# LSTM example in Keras using IMDB
# ======================================================================
from keras.datasets import imdb
from keras.preprocessing import sequence
from keras.layers import Dense
from keras.models import Sequential
from keras.layers import Embedding, LSTM

max_features = 10000 # number of words to consider as features
maxlen = 500 # cut texts after this number of words (among top max_features
most common words)
batch_size = 32

print('Loading data...')
(Xtrain, ytrain), (Xtest, ytest) = imdb.load_data(num_words = max_features)
print(len(Xtrain), 'train sequences')
print(len(Xtest), 'test sequences')

print('Pad sequences (samples x time)')
Xtrain = sequence.pad_sequences(Xtrain, maxlen = maxlen)
Xtest = sequence.pad_sequences(Xtest, maxlen = maxlen)
print('input_train shape:', Xtrain.shape)
print('input_test shape:', Xtest.shape)

#Create Model
model = Sequential()
model.add(Embedding(max_features, 32))
model.add(LSTM(32))
model.add(Dense(1, activation = 'sigmoid'))

# Compile model
```

```
model.compile(optimizer = 'rmsprop', loss = 'binary_crossentropy', met-
rics = ['acc'])
# Fit model
history = model.fit(Xtrain, ytrain,
          epochs = 2,
          batch_size = 128,
          validation_split = 0.2)

# Evaluate the model
_, train_acc = model.evaluate(Xtrain, ytrain, verbose = 0)
_, test_acc = model.evaluate(Xtest, ytest, verbose = 0)
print('Training Accuracy: %.3f, Testing Accuracy: %.3f' % (train_acc,
test_acc))
```

3.5.22 卷积神经网络

卷积神经网络（CNN）是对人工神经网络的一种改进，注重于模仿我们视觉皮层的功能和行为。隐藏单元的目的是学习原始输入的非线性变化，这称为特征提取或特征创建。然后将这些隐藏特征传递到最终的广义线性模型（GLM）中作为输入。该方法对于原始输入特征不具备独立信息的情况特别有用。例如，图像中的每一个像素是不包含太多信息的，而像素的组合向我们展示了图像中存在的对象。另一方面，对于使用词袋法表示的任务（例如文本分类）来说，其特征（词频）本身就具有信息量，因此提取高阶特征的意义不大。因此，大多数神经网络的研究都是出自对视觉模式的识别，这并不奇怪，即使它们也被用于某些数据类型的任务，包括文本处理。卷积神经网络是多层感知机（MLP）的一种类型，特别适用于一维信号（如语音、生物医学信号或文本）以及二维信号，如图像。它包含了隐藏单元的局部感知区域（类似初级视觉皮层），以及在整个图像中添加和共享的权重，以减少变量的数量。当然，这种固定空间参数的结果是，在图像的某一部分"发现"的任何信息特征都可以在其他地方被复制，而不需要单独学习。由此产生的网络显示出平移不变性，这意味着可以对出现在输入图像中任何地方的模式进行分类（Murphy，2012）。

CNN 不仅是具有多个隐藏层的深度神经网络，还是模拟和理解刺激的大型网络，如同大脑的视觉皮层处理。因此，在初次接触时，即使是神经网络专家也仍然难以理解这个术语。这就是 CNN 与以往神经网络在理论和实践上的差异。CNN 的输出层通常使用神经网络进行多分类。但是，无论采用何种识别方法，直接利用原始图像进行图像识别效果都很差，应该对图像进行处理以作为特征。否则，识别过程会产生很差的结果。为此，人们创建了不同的方法来提取图像特征。特征提取器由特定领域的专家进行开发，因此，需要相当大的成本和时间，而且性能水平不一致。这样的特征提取器独立于机器学习算法。CNN 在训练过程中使用特征提取器，而不是手动实现。CNN 的特征提取器由特殊类型的神经网络组成，通过训练过程确定权重。CNN 的主要特点和优势在于它将人工特征提取过程转变为自动提取。当 CNN 的神经网络特征提取变得更深（包含更多的层数）时，可以提供更好的图像识别效果，但代价是学习方法复杂，这使得 CNN 在一段时间内由于效率低下被忽视。CNN 是一个用于提取输入图像特征的神经网络，而另一个神经网络用于对图像特征进行分类。输入图像被用于特征提取的网络使用，提取的特征信号被神经网络用于分类。神经网络根据图像特征进行分类，并产生输出。用于特征提取的神经网络包括卷积层堆和池化层集。顾名思义，卷积层利用卷积的过程对图像进行变换，它可以被描述为一系列数字滤波器。池化层将相邻像素转化为单个像素，降低了图像的维度。由于 CNN 主要关注的是图像，因此直观上来说，卷积层和池化层过程是在二维平面上进行的。这也是 CNN 与其他神经网络的区别之一（Kim，2017）。

1. 卷积层

卷积层产生的新图像称为特征映射，特征映射表示了原始图像的唯一特征。与其他神经网络结构相反，卷积层的工作方式与众不同。卷积层不使用连接权重和加权和，而是使用图像转换过滤器，称为卷积过滤器。特征映射是通过将图像输入到卷积过滤器中产生的。卷积是一个非常难以用文字解释的过程，因为它是在二维平面上发生的。但卷积的理论和度量步骤却很简单。特征映射中的元素值取决于图像矩阵是否与卷积滤波器相匹配，类似于第一个卷积过程。卷积滤波器产生的特征映射经激活函数处理后，再产生输出。卷积层的激活函数与传统神经网络激活函数类似。虽然在目前的许多实现方案中使用了 ReLU 函数，但 sigmoid 函数和 tanh 函数也经常使用。

2. 池化层

池化层将图像区域的相邻像素组合成单一的表示值，从而减小图像的尺寸。池化是一种流行的方法，已经被许多其他图像处理技术所采用。我们必须决定如何从图像中选择池化像素，如何设置表示值以实现池化层的操作。相邻的像素通常从方阵中选取，不同问题组合的像素数不同。

例 3.44 以下 Python 代码通过使用 TensorFlow 库中的 Keras API，对深度神经网络分类器进行举例说明。在本例中，我们使用了 keras.datasets 中的 MNIST 手写数字数据集。MNIST 数据集包含 6000 个训练图像和 10 000 个测试图像。使用卷积神经网络对这些图像进行分类，分类准确率被用作性能衡量标准。请注意，本例改编自 TensorFlow.org 网站。

```
# ===================================================================
# Convolutional neural network example
# ===================================================================
from keras import models
from keras import layers
from keras.datasets import mnist
(train_images, train_labels), (test_images, test_labels) = mnist.load_
data()
```

```
train_images = train_images.reshape((60000, 28, 28, 1))
test_images = test_images.reshape((10000, 28, 28, 1))

train_images, test_images = train_images / 255.0, test_images / 255.0

classes = [0,1,2,3,4,5,6,7,8,9]
#Built the Model
model = models.Sequential()
model.add(layers.Conv2D(32, (3, 3), activation = 'relu', input_shape = (28,
28, 1)))
model.add(layers.MaxPooling2D((2, 2)))
model.add(layers.Conv2D(64, (3, 3), activation = 'relu'))
model.add(layers.MaxPooling2D((2, 2)))
model.add(layers.Conv2D(64, (3, 3), activation = 'relu'))
model.add(layers.Flatten())
model.add(layers.Dense(64, activation = 'relu'))
model.add(layers.Dense(10, activation = 'softmax'))
model.compile(optimizer = 'adam',
      loss = 'sparse_categorical_crossentropy',
      metrics = ['accuracy'])
      #Fit the model and Test it
model.fit(x = train_images,
      y = train_labels,
      epochs = 5,
      validation_data = (test_images, test_labels))
```

3.6 无监督学习

在监督学习中，其目的是将输入映射到由监督者提供的正确输出。而在无监督学习中，没有这样的监督者，有的只是输入数据。其目标是发现输入中存在的一致性。输入空间的某种配置使得特定的模式比其他模式出现得更频繁，我们想找出通常什么会发生，什么不会。在统计学中，这就是所谓的密度估计。聚类是密度估计的一种方法，其目标是找到输入的聚类或组。在公司拥有客户历史数据的情况下，其中客户数据包括人口统计信息以及过去与公司的交易情况，公司需要了解客户的资料分布，找出常见的客户类型。在该案例中，可以使用聚类模型将属性相近的客户分到同一组中，为公司提供自然的客户分组，这就是所谓的客户细分。一旦发现了这些分组，公司可以针对不同组选择有关服务和产品的策略，这就是所谓的客户关系管理。这样的分组方法还可以发现有别于其他客户的离群点（Alpaydin，2014）。

众多著名的聚类算法包括k均值都可以用于聚类预测。它们可以从训练数据中学习聚类模型，然后利用这个模型将新样本分配到所属聚类中。这里保留了对随机数据进行聚类和从训练数据中学习聚类模型的区分。尽管如此，这种区分对于描述性的聚类方法来说并不是特别合适。事实上，问题变成了为给定数据学习一个合适的聚类模型。一个好的聚类在于，能够将数据划分到一致的聚类或组。这需要一些估计相似性的方法，或者，通常更合适的方式是，估计任意一对实例的相异性或距离。如果是数值特征，广泛使用的距离度量方法是欧氏距离，但也有其他的选择。大多数基于距离的聚类方法都是为任意实例集定义一个"质心"或示例点，然后使得示例点与集合中的所有实例之间的距离（即散度）最小化。一个值得注意的问题是如何评估聚类方法。在缺乏标签数据的情况下，不能像分类或回归中那样用测试集进行评估。然而，我们可以将类内散度作为衡量聚类性能的标准 (Flach, 2012)。

在某些情况下，我们不能自由灵活地使用预标记的数据进行训练。但无论如何，需要从信息中提取有用的知识或实例。在这种情况下，无监督学习技术是不可思议的。这些技术称为无监督是因为模型或算法试图从给定的信息中学习基本的潜在结构、关系和模式，而不需要任何辅助或监督，如提供标记结果或输出形式的信息。在机器学习任务中，无监督学习方法可以被分为以下几个主要领域 (Sarkar et al., 2018)。

- 聚类
- 降维
- 异常检测
- 关联规则挖掘

聚类是一种机器学习方法，它试图在数据集中找到数据样本之间的相似模式和关系，然后将这些样本聚成不同的组，使每一组或每一聚类的数据样本都具有某种基于其实际属性或特征的相关性。这类方法是完全无监督的，因为它们通过观察数据特征进行聚类，而不需要事先进行训练、指导或是具备数据特征、关系和关联的知识 (Sarkar et al., 2018)。

聚类方法有不同的形式，可分为以下几种：

- 基于中心点的方法，如 k 中心点和 k 均值算法
- 层次聚类方法，如分裂式和凝聚式
- 基于分布的聚类方法，如高斯混合模型
- 基于密度的聚类方法，如光学 dbscan 方法

3.6.1 k 均值算法

在 k 均值问题中，我们没有有效的解决方案来识别全局最小值，因此需要利用启发式算法。可以看出，k 均值的迭代可能永远无法改善聚类的类内散度，导致算法接近一个稳定点，无法进一步提升。值得注意的是，即使在最简单的数据集中也可能存在几个稳定点。通常来讲，虽然 k 均值能够在有限时间内收敛到一个稳定点，但无论离它多远，都无法确定这个收敛点是否真的是全局最小点 (Flach, 2012)。

例 3.45 以下 Python 代码通过使用 scikit-learn 库中的 API，利用 k 均值聚类寻找乳腺癌数据的聚类中心。在本例中，我们使用了 sklearn.datasets 中的乳腺癌数据集，并

绘制出了聚类中心点。请注意，本例改编自 scikit-learn。

```python
# ================================================================
# K-means clustering example
# ================================================================
from sklearn import datasets
import matplotlib.pyplot as plt
import numpy as np
# ################################################################
# Import some data to play with
Breast_Cancer = datasets.load_breast_cancer()
X = Breast_Cancer.data
y = Breast_Cancer.target

# Plot the original data points
plt.scatter(X[:, 0], X[:, 1], c = y, cmap = plt.cm.Set1,
     edgecolor = 'k')
plt.xlabel('Attribute I')
plt.ylabel('Attribute II')
plt.title('Original data Scatter')
plt.xticks(())
plt.yticks(())
#%%
from sklearn.cluster import KMeans

"""
sklearn.cluster.KMeans(n_clusters = 2, init = 'k-means++', n_init = 10, max_iter = 300, tol = 0.0001,
   precompute_distances = 'auto', verbose = 0, random_state = None,
   copy_x = True, n_jobs = None,
   algorithm = 'auto')
"""

#Find Cluster Centers
kmeans = KMeans(n_clusters = 2)
kmeans.fit(X)
y_kmeans = kmeans.predict(X)

#Plot the Cluster Centers
plt.scatter(X[:, 0], X[:, 1], c = y_kmeans, s = 50, cmap = 'viridis')
centers = kmeans.cluster_centers_
plt.scatter(centers[:, 0], centers[:, 1], c = 'black', s = 200, alpha = 0.5);
plt.xlabel('Attribute I')
plt.ylabel('Attribute II')
plt.title('Cluster Centers')
plt.xticks(())
plt.yticks(())
```

3.6.2 轮廓系数

如何检测一个聚类算法的质量好坏？轮廓系数是一种有效的方法。对每个以聚类进行分组的例子，会将轮廓系数 $s(x)$ 进行排序和绘制。在这种情况下，欧氏距离的平方被用于构造轮廓系数，但也可以使用其他距离度量方法。可以清楚地看到，第一个聚类比第二个聚类强得多。除了用图形表示以外，我们还可以估计出每个聚类以及整个数据集的平均轮廓值（Flach，2012）。

例 3.46 以下 Python 代码通过使用 scikit-learn 库中的 API，利用 k 均值聚类寻找模拟数据的轮廓系数。在本例中，我们使用了由 scikit-learn 库中的 make_blobs 方法生成的样本数据，并分别绘制出了轮廓和聚类中心点。请注意，本例改编自 scikit-learn。

```
# ======================================================================
#Silhouettes example
# ======================================================================
from sklearn.datasets import make_blobs
from sklearn.cluster import KMeans
from sklearn.metrics import silhouette_samples, silhouette_score

import matplotlib.pyplot as plt
import matplotlib.cm as cm
import numpy as np

print(__doc__)

# Generating the sample data from make_blobs
# This particular setting has one distinct cluster and 3 clusters placed close
# together.
X, y = make_blobs(n_samples = 500,
        n_features = 2,
        centers = 4,
        cluster_std = 1,
        center_box = (-10.0, 10.0),
        shuffle = True,
        random_state = 1) # For reproducibility

range_n_clusters = [2, 3, 4, 5, 6]

for n_clusters in range_n_clusters:
  # Create a subplot with 1 row and 2 columns
  fig, (ax1, ax2) = plt.subplots(1, 2)
  fig.set_size_inches(18, 7)

  # The 1st subplot is the silhouette plot
  # The silhouette coefficient can range from -1, 1 but in this example all
  # lie within [-0.1, 1]
```

```python
ax1.set_xlim([-0.1, 1])
# The (n_clusters+1)*10 is for inserting blank space between silhouette
# plots of individual clusters, to demarcate them clearly.
ax1.set_ylim([0, len(X) + (n_clusters + 1) * 10])

# Initialize the clusterer with n_clusters value and a random generator
# seed of 10 for reproducibility.
clusterer = KMeans(n_clusters = n_clusters, random_state = 10)
cluster_labels = clusterer.fit_predict(X)

# The silhouette_score gives the average value for all the samples.
# This gives a perspective into the density and separation of the formed

# clusters
silhouette_avg = silhouette_score(X, cluster_labels)
print("For n_clusters =", n_clusters,
"The average silhouette_score is :", silhouette_avg)

# Compute the silhouette scores for each sample
sample_silhouette_values = silhouette_samples(X, cluster_labels)

y_lower = 10
for i in range(n_clusters):
# Aggregate the silhouette scores for samples belonging to
# cluster i, and sort them
    ith_cluster_silhouette_values = \
    sample_silhouette_values[cluster_labels == i]

    ith_cluster_silhouette_values.sort()
size_cluster_i = ith_cluster_silhouette_values.shape[0]
y_upper = y_lower + size_cluster_i

color = cm.nipy_spectral(float(i) / n_clusters)
ax1.fill_betweenx(np.arange(y_lower, y_upper),
0, ith_cluster_silhouette_values,
facecolor = color, edgecolor = color, alpha = 0.7)

# Label the silhouette plots with their cluster numbers at the middle
ax1.text(-0.05, y_lower + 0.5 * size_cluster_i, str(i))

# Compute the new y_lower for next plot
y_lower = y_upper + 10 # 10 for the 0 samples

ax1.set_title("The silhouette plot for the various clusters.")
ax1.set_xlabel("The silhouette coefficient values")
ax1.set_ylabel("Cluster label")

# The vertical line for average silhouette score of all the values
ax1.axvline(x = silhouette_avg, color = "red", linestyle = "--")
```

```
ax1.set_yticks([]) # Clear the yaxis labels / ticks
ax1.set_xticks([-0.1, 0, 0.2, 0.4, 0.6, 0.8, 1])

# 2nd Plot showing the actual clusters formed
colors = cm.nipy_spectral(cluster_labels.astype(float) / n_clusters)
ax2.scatter(X[:, 0], X[:, 1], marker = '.', s = 30, lw = 0, alpha = 0.7,
    c = colors, edgecolor = 'k')

# Labeling the clusters
centers = clusterer.cluster_centers_
# Draw white circles at cluster centers
ax2.scatter(centers[:, 0], centers[:, 1], marker = 'o',
c = "white", alpha = 1, s = 200, edgecolor = 'k')

for i, c in enumerate(centers):
    ax2.scatter(c[0], c[1], marker = '$%d$' % i, alpha = 1,
    s = 50, edgecolor = 'k')

ax2.set_title("The visualization of the clustered data.")
ax2.set_xlabel("Feature space for the 1st feature")
ax2.set_ylabel("Feature space for the 2nd feature")

plt.suptitle(("Silhouette analysis for KMeans clustering on sample data "
    "with n_clusters = %d" % n_clusters),
    fontsize = 14, fontweight = 'bold')
plt.show()
```

3.6.3 异常检测

异常检测技术也称为离群点检测。在异常检测中，我们从历史数据样本中发现通常不会出现的异常事件和异常点。异常情况通常很少发生，导致不寻常的现象，也可能在不同的情况下并不罕见，但是在很短的时间内突然出现，因此异常点具有独特的模式。将无监督学习方法用于异常检测，该方法在训练数据集中使用正常的、非异常的数据样本来训练算法。一旦算法在正常样本中学习到必要的数据解释、模式和属性关系，便可以利用这些信息判断一个新样本是属于异常还是普通数据点。基于异常的检测技术在现实生活中非常受欢迎，例如安全攻击或入侵检测、储蓄卡欺诈、制造异常、网络问题等（Sarkar et al., 2018）。

例 3.47 以下 Python 代码通过使用 scikit-learn 库中的 API，介绍了二维数据集上各种异常检测算法的特征。数据集包括一种或两种模式，以说明算法处理多模式数据的能力。对于每个数据集，将生成 15% 的样本作为随机均匀噪声。该比例是为 OneClassSVM 的 nu 参数和其他离群值检测算法的污染参数提供的值。离群值的决策边界以黑色显示，但局部离群值因子（LOF）除外，因为当用于离群值检测时，它没有适用于新数据的预测方法。在本例中，我们使用了由 scikit-learn 库中的 make_moons 和 make_blobs 方法生成的样本数据。sklearn.svm.OneClassSVM 类对离群点很敏感，因此在离群点检测中表现得不是很好；sklearn.covariance.EllipticEnvelope 类假定数据服从高斯分布而且要学习一个椭圆（ellipse），但它不受离群点的影响；sklearn.ensemble.IsolationForest 和 sklearn.neighbors.LocalOutlierFactor 在多模态数据集中表现得相当好。请注意，本例来自 scikit-learn。

```
# ======================================================================
# Anomaly detection
# ======================================================================
# Author: Alexandre Gramfort <alexandre.gramfort@inria.fr>
# Albert Thomas <albert.thomas@telecom-paristech.fr>
# License: BSD 3 clause

import time

import numpy as np
import matplotlib
import matplotlib.pyplot as plt
from sklearn import svm
from sklearn.datasets import make_moons, make_blobs
from sklearn.covariance import EllipticEnvelope
from sklearn.ensemble import IsolationForest
from sklearn.neighbors import LocalOutlierFactor

print(__doc__)

matplotlib.rcParams['contour.negative_linestyle'] = 'solid'

# Example settings
n_samples = 300
outliers_fraction = 0.15
n_outliers = int(outliers_fraction * n_samples)
n_inliers = n_samples - n_outliers

# define outlier/anomaly detection methods to be compared
anomaly_algorithms = [
    ("Robust covariance",
EllipticEnvelope(contamination=outliers_fraction)),
    ("One-Class SVM", svm.OneClassSVM(nu=outliers_fraction, kernel="rbf",
            gamma=0.1)),
    ("Isolation Forest", IsolationForest(behaviour='new',
            contamination=outliers_fraction,
            random_state=42)),
    ("Local Outlier Factor", LocalOutlierFactor(
      n_neighbors=35, contamination=outliers_fraction))]
```

```python
# Define datasets
blobs_params = dict(random_state = 0, n_samples = n_inliers, n_features = 2)
datasets = [
    make_blobs(centers = [[0, 0], [0, 0]], cluster_std = 0.5,
        **blobs_params)[0],
    make_blobs(centers = [[2, 2], [-2, -2]], cluster_std = [0.5, 0.5],
        **blobs_params)[0],
    make_blobs(centers = [[2, 2], [-2, -2]], cluster_std = [1.5, .3],
        **blobs_params)[0],
    4. * (make_moons(n_samples = n_samples, noise = .05, random_state = 0)[0] -
        np.array([0.5, 0.25])),
    14. * (np.random.RandomState(42).rand(n_samples, 2) - 0.5)]
# Compare given classifiers under given settings
xx, yy = np.meshgrid(np.linspace(-7, 7, 150),
        np.linspace(-7, 7, 150))

plt.figure(figsize = (len(anomaly_algorithms) * 2 + 3, 12.5))
plt.subplots_adjust(left = .02, right = .98, bottom = .001, top = .96,
wspace = .05,
        hspace = .01)

plot_num = 1
rng = np.random.RandomState(42)

for i_dataset, X in enumerate(datasets):
    # Add outliers
    X = np.concatenate([X, rng.uniform(low = -6, high = 6,
    size = (n_outliers, 2))], axis = 0)

    for name, algorithm in anomaly_algorithms:
    t0 = time.time()
    algorithm.fit(X)
    t1 = time.time()
    plt.subplot(len(datasets), len(anomaly_algorithms), plot_num)
    if i_dataset == 0:
    plt.title(name, size = 18)

    # fit the data and tag outliers
    if name == "Local Outlier Factor":
    y_pred = algorithm.fit_predict(X)
    else:
    y_pred = algorithm.fit(X).predict(X)

    # plot the levels lines and the points
    if name != "Local Outlier Factor": # LOF does not implement predict
    Z = algorithm.predict(np.c_[xx.ravel(), yy.ravel()])
    Z = Z.reshape(xx.shape)
    plt.contour(xx, yy, Z, levels = [0], linewidths = 2, colors = 'black')

    colors = np.array(['#377eb8', '#ff7f00'])
    plt.scatter(X[:, 0], X[:, 1], s = 10, color = colors[(y_pred + 1) // 2])
```

```
    plt.xlim(-7, 7)
    plt.ylim(-7, 7)
    plt.xticks(())
    plt.yticks(())
    plt.text(.99, .01, ('%.2fs' % (t1 - t0)).lstrip('0'),
    transform = plt.gca().transAxes, size = 15,
    horizontalalignment = 'right')
    plot_num += 1
plt.show()
```

3.6.4 关联规则挖掘

关联规则挖掘是一种数据挖掘的方法，用于探索和解释大型的交易数据集，以识别其中独特的模式和规则。在交易过程中，这些模式定义了不同商品之间有趣的关系和相互作用。此外，关联规则挖掘也通常称为购物篮分析，用于分析客户的购买习惯。关联规则通过使用由交易相关属性训练得到的信息识别和预测交易行为。利用该方法可以回答一些问题，比如人们倾向于同时购买什么货物，说明这样的货物搭配容易被卖出。我们还可以将产品和商品进行关联（Sarkar et al., 2018）。

3.7 强化学习

在一些实现中，机器的输出是一系列的行动。在这种情况下，单一的行动显得不重要，重要的是策略，即采用一系列正确的行动以实现目标。任何的中间状态都不存在特殊干预项，所以如果某个行动是一个好的策略步骤，它就足够了。在这种情况下，为了创建策略，机器学习算法必须能够检验策略的有效性，并从过去的行动序列中学习。这种通过学习进行强化的方法称为强化学习算法。一个合适的比喻是玩游戏，单一的行动不是很重要，而一系列正确的行动才更有意义。当某个行动是一系列正确行动的一部分时，才显得恰当。游戏是机器学习和人工智能的重要研究领域，这是因为游戏比较容易描述，同时要想玩好游戏也相当有挑战性。像象棋这样的游戏，规则有限但非常复杂，因为每一种状态下可能的棋步数量巨大，一盘棋涉及的棋步数量也很大。一旦我们有了强有力的算法能够学习如何玩好游戏，就可以将其用到更有经济效益的应用中。强化学习的另一个实现是机器人领域，即机器人在现实环境中导航，寻找目标区域。机器人会随时向多个可能的方向移动。经过多次试验后，它需要从初始状态中学习到实现目标所需的正确行为序列，要尽可能快，并且不能碰到任何障碍（Alpaydin, 2014）。

解决机器学习问题的主要目标在于，通过学习创建智能的程序或智能体（agent），对不断变化的环境做出反应。学习器和软件程序智能体能够从机器学习与显式的环境交互中获益，这模仿了人类学习的方法。此外，即使无法获取环境的全部信息或模型，智能体也能够学习。智能体为每个行动提供反馈作为奖励或惩罚。在学习过程中，这些反馈条件映射为环境中的行动。强化学习算法会对与环境交互过程中获得的激励进行某种程度的优化，并建立行动的状态映射，作为决策的策略。算法可以直接选择策略，或是根据环境的改变做出调整。强化学习不同于监督学习，后者是在外部专家监督器的帮助下，从提供的样例中学习。强化学习是一种训练参数化函数逼近的方法。然而，仅从交互中学习是不够的。这更像是从外部指令中学习，指引产生于特定的情景和环境。在交互式问题中，往往很难实现所期望的既准确又具有代表性的行动，使智能体能够在所有情况下做出的反应。人们会做这样的推测，学习在未知领域是最有用的方法，智能体应该能够从自己的经验以及环境中学习。由此，强化学习将监督学习与动态编程相结合，建立了一个机器学习的框架，该框架与人类学习的方式非常相似。探索与利用之间的权衡是强化学习的挑战之一，在其他类型的学习中并不存在这个问题。为了获得激励，强化学习应该选择从前实施过的、被认为能有效提供激励的行动。但同时它也需要尝试没有实施过的行动进行探索。机器应该利用已经学习到的优势来获得激励，但它也必须学习，以便在未来做出更好的选择。两难的是，没有失败的任务，就没有发现和发展。机器会探寻一系列的行动，并逐步选择那些看起来最好的。为了获得对预测奖励的可靠估计，每个行动都应该在大量的实例中进行随机地尝试。正如通常所说的那样，在整个监督学习中，不存在平衡探索与利用的挑战。因此，专家们有责任在监督学习中进行探索发明。强化学习与更广泛的监督学习有许多不同之处，主要区别在于没有输入输出对的表示。在强化学习中，机器在采取某项行动之后会得到即时的奖励以及条件结果，但这并不能说明该行动会对长期利益产生怎样的影响。重要的是，机器能够从潜在的条件、转换、行动以及从激励中获得宝贵的经验，从而有效地采取行动（Kulkarni，2012）。

3.8 基于实例的学习

有多种方法可以通过对输入数据进行泛化，以构建机器学习模型。基于实例的学习包括一些机器学习算法和技术，它们使用原始数据本身来确定那些较新的、从未见过的数据样本的结果，而不是在训练数据上构建一个特定的模型，然后进行测试。一个简单的例子是 k 近邻方法。机器学习从数据的特征、维度以及位置等信息中识别出对数据的解释。它利用相似性度量方法（如欧式距离）对新数据点进行测量，从而定位离该新数据点最近的三个输入数据点。对于一个新的数据点，其预测结果大部分取决于这三个训练因子，并以此作为该数据点的响应/标签。因此，基于实例的学习的工作方法是对输入数据进行探索，并使用相似性度量来预测和泛化新的数据点（Sarkar et al., 2018）。

3.9 本章小结

本章介绍了一组精选的机器学习软件包，它们经常用于数据的存储、分析和建模。这些库和框架是构成数据科学家工具箱的关键要素。然而，本章所涵盖的软件包清单还远远不够完整，我们强烈建议你通过阅读文档和相关视频来熟悉这些软件包。在随后的章节中，我们将继续阐明这些框架的其他重要特征，以及其他方面的内容。本章的例子能够让你良好地掌握对机器学习的理解，了解如何以简洁明了的方式解决问题，以及相关的理论信息。通常情况下，利用各种数据学习模型的过程就是重复这些简单步骤和原则。在接下来的章节中，你将学习如何使用这套工具来解决更复杂的数据处理、整理、分析和可视化问题。

3.10　参考文献

Alpaydin, E. (2014). Introduction to machine learning. Cambridge, MA, London, England: The MIT Press.

Bengio, Y. (2009). Learning deep architectures for AI. Foundations and Trends® in Machine Learning, 2(1), 1–127.

Breiman, L., Friedman, J. H., Olshen, R. A., & Stone, C. J. (1984). Classification and regression trees. Boca Raton: Hall/CRC.

Breiman, L. (1996). Bagging predictors. Machine Learning, 24(2), 123–140.

Breiman, L. (1999). Using adaptive bagging to debias regressions. Statistics Dept., University of California, Berkeley: (Technical report 547).

Buduma, N., & Locascio, N. (2017). Fundamentals of deep learning: Designing next-generation machine intelligence algorithms. Sebastopol, CA: O'Reilly Media, Inc.

Clarke, B., Fokoue, E., & Zhang, H. H. (2009). Principles and theory for data mining and machine learning. New York, NY: Springer Science & Business Media, LLC.

Flach, P. (2012). Machine learning: The art and science of algorithms that make sense of data. Cambridge, UK: Cambridge University Press.

Chollet, F. (2018). Deep learning with Python. Shelter Island, NY: Manning Publications Co.

Freund, Y., & Schapire, R. E. (1995). *A decision-theoretic generalization of on-line learning and an application to boosting*. In P. Vitányi (Ed.), *Computational Learning Theory. EuroCOLT*. Berlin, Heidelberg: Springer Lecture Notes in Computer Science (Lecture Notes in Artificial Intelligence), vol. 904.

Freund, Y., & Schapire, R. E. (1997). A decision-theoretic generalization of on-line learning and an application to boosting. Journal of Computer and System Sciences, 55(1), 119–139.

Friedman, J. H. (2002). Stochastic gradient boosting. Computational Statistics & Data Analysis, 38(4), 367–378.

Goel, E., & Abhilasha, E. (2017). Random forest: A review. *International Journal of Advanced Research in Computer Science and Software Engineering, 7*(2), 251–257 https://doi.org/10.23956/ijarcsse/V7I1/01113.

Han, J., Pei, J., & Kamber, M. (2011). Data mining: Concepts and techniques. San Francisco, CA: Elsevier.

Hapke, H. M., Lane, H., & Howard, C. (2019). Natural language processing in action. Shelter Island, NY: Manning Publications Co.

Hastie, T., Tibshirani, R., Friedman, J., & Franklin, J. (2005). The elements of statistical learning: Data mining, inference and prediction. The Mathematical Intelligencer, 27(2), 83–85.

Hinton, G. E., & Salakhutdinov, R. R. (2006). Reducing the dimensionality of data with neural networks. Science, 313(5786), 504–507.

Kim, P. (2017). Convolutional neural network. *MATLAB deep learning*. New York, NY: Springer Science+Business Media New York 121-147.

Kulkarni, P. (2012). *Reinforcement and systemic machine learning for decision making (Vol. 1)*. Hoboken, NJ: John Wiley & Sons Inc.

Kumari, G. T. P. (2012). A study of bagging and boosting approaches to develop meta-classifier. Engineering Science and Technology: An International Journal, 2(5), 850–855.

Mohri, M., Rostamizadeh, A., & Talwalkar, A. (2018). Foundations of machine learning. Cambridge, MA: MIT press.

Murphy, K. P. (2012). Machine learning: A probabilistic perspective. MIT press, Cambridge, MA.

Pedregosa, F., Varoquaux, G., Gramfort, A., Michel, V., Thirion, B., Grisel, O., & Dubourg, V. (2011). Scikit-learn: Machine learning in Python. Journal of Machine Learning Research, 12(Oct), 2825–2830.

Quinlan, J.R. (1993). Improved use of continuous attributes in C4.5. 77–90.

Quinlan, J. R. (1986). Induction of decision trees. Machine Learning, 1(1), 81–106.

Raschka, S. (2015). Python machine learning. Birmingham, UK: Packt Publishing Ltd.

Ray, P. K., Mohanty, A., & Panigrahi, T. (2019). Power quality analysis in solar PV integrated microgrid using independent component analysis and support vector machine. Optik, 180, 691–698 https://doi.org/10.1016/j.

ijleo.2018.11.041.

Rosenblatt, F. (1958). The perceptron: A probabilistic model for information storage and organization in the brain. Psychological Review, 65(6), 386–408.

Sanei, S. (2013). *Adaptive processing of brain signals*. United Kingdom: John Wiley & Sons.

Sarkar, D., Bali, R., & Sharma, T. (2018). *Practical machine learning with Python: A problem-solver's guide to building real-world intelligent systems*. New York, NY: Apress, Springer Science+Business Media New York.

Schapire, R. E. (1990). The strength of weak learnability. Machine Learning, 5(2), 197–227.

Subasi, A. (2019). Practical Guide for Biomedical Signals Analysis Using Machine Learning Techniques, A MATLAB Based Approach (1st Edition). United Kingdom: Academic Press.

Veit, D. (2012). 2—Neural networks and their application to textile technology. In D. Veit (Ed.), Simulation in Textile Technology (pp. 9-71). https://doi.org/10.1533/9780857097088.9.

Witten, I. H., Frank, E., & Hall, M. A. (2011). *Data mining: Practical machine learning tools and techniques*. Retrieved from https://books.google.com.sa/books?id=bDtLM8CODsQC.

Wu, X., Kumar, V., Ross Quinlan, J., Ghosh, J., Yang, Q., Motoda, H., & Steinberg, D. (2008). Top 10 algorithms in data mining. Knowledge and Information Systems, 14(1), 1–37 https://doi.org/10.1007/s10115-007-0114-2.

第 4 章

医疗保健分类示例

4.1 简介

一般来说,生物医学数据分类过程可分为四个阶段,即数据采集和分段、数据预处理、特征提取/降维以及识别和分类。如图 4.1 所示,生物医学数据从人体采集,然后进行预处理。数据预处理是一种将原始数据转换为有用且有效格式的技术。数据可能包含噪声,即许多不相关和缺失的部分,这些都应该被消除。然后,从处理后得到的数据中提取特征,并转换为特征向量。特征向量描述了原始数据中的相关结构。下一步,使用降维来消除特征向量中的冗余信息,得到降维后的特征向量。最后一步,分类器将使用降维后的特征向量进行分类(Subasi, 2019c)。

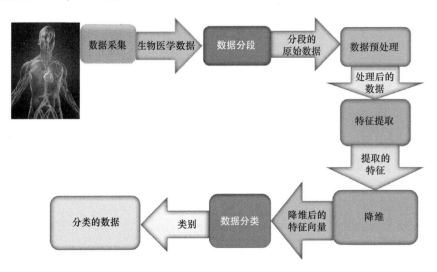

图 4.1 生物医学数据分类的一般框架。该图改编自(Subasi, 2019c)

4.2 脑电图信号分析

大脑产生的电信号描述了大脑的功能和整个身体的状况。它们为使用数字信号处理技术处理从人脑获得的脑电图（EEG）提供了灵感。大脑活动的生理特征与原始来源及其真实模式的特征有关。在处理这些信号以进行识别时，结合信号产生和获取的基本工作原理来了解大脑的神经生理特性和神经元功能是非常有用的。脑电图引入了人体中多种神经系统疾病和异常的诊断方法。这表明，脑电图在使用先进的信号处理方法来支持临床医生的决策方面具有巨大的潜力（Sanei & Chambers, 2013; Subasi, 2019c）。

EEG 信号是一种非侵入性的医疗工具，可用于分析多种脑部疾病，以及更好地理解人脑。但是，任何单个数学或生物学模型都无法完全解释 EEG 模式的多样性。因此，理解 EEG 信号本质上仍属于现象学医学学科的范畴（Barlow, 1993）。脑电图记录的内容是大脑产生的电势，通常小于 300 μV。一名脑电图专家，也就是可以在相当长的脑电图记录中定性地区分正常和异常的脑电图活动的人，可能需要多年的训练才能对脑电图进行视觉分析，而临床医生和研究人员看到的只有一堆脑电图记录。功能强大的现代计算机和相关技术的出现为利用不同技术量化 EEG 信号开辟了全新的可能性（Bronzino, 1999）。通过选择具有不同目的的数字信号处理技术，如降噪和特征提取，可大大加快分析速度，这些是视觉分析无法实现的。脑电图是分析多种疾病（如癫痫、睡眠障碍和痴呆症）极其有效的工具。此外，EEG 信号对于实时监测脑病患者或昏迷患者也很重要（Sörnmo & Laguna, 2005; Subasi, 2019c）。EEG 信号分析的一般框架如图 4.2 所示。

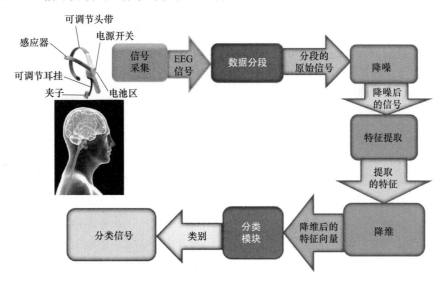

图 4.2　EEG 信号分析的一般框架。该图改编自（Subasi, 2019a）

4.2.1 癫痫症的预测和检测

随着数学和机器学习技术的广泛使用,脑电图分析已大大增强。机器学习技术让 EEG 的模式分类得以改善,从而提高了识别能力,使 EEG 信号可用于识别脑部疾病和关键病理。因此,人们对神经系统疾病与 EEG 信号特征的相关性进行了大量研究(Begg, Lai, & Palaniswami, 2008)。癫痫是一种神经系统疾病,全世界有超过五千万人受其影响,是除中风之外最常见的神经系统疾病,这种严重的疾病被描述为大脑生物电功能的暂时变化。这些变化会导致异常的神经元同步和癫痫症,从而影响意识、感觉或运动。癫痫症是由大批神经元特定同步的活动触发的,是一组大脑皮层神经元突然爆发的强烈电活动。由于多个大脑区域电活动的焦点(起源)位置的不同,癫痫症可能表现为不同的方式(Sörnmo & Laguna, 2005; Subasi, 2019c)。

癫痫发作的 EEG 信号包括健康专业人员用来区分正常(非癫痫)EEG 信号的标准模式。因此,它们的识别可用于应对即将到来或正在发生的癫痫。此外,自动化识别技术已经被评估,以通过使用较少数据量,更快、更准确地检测出表征癫痫发作的病理性 EEG 波形(Begg et al., 2008)。通常,有必要长时间记录以捕捉突发的 EEG。在这种情况下,受试者通常需要在医院进行长达几天的视频录制。因此,神经学家可以在脑电图和视觉记录之间找到相似之处,以改善他们的评估。这种记录形式称为视频脑电图。另一种记录形式需要至少在家中进行一天,称为动态脑电图,是使用小型数字录制设备在日常活动中录制的。这种技术相较于视频脑电图更加便宜,包括睡眠阶段和清醒阶段。如果受试者挠了自己的头,脑电图记录中则会引入噪声。神经学家利用发作期(癫痫发作)EEG 波形的不同模式,以区别于发作间期(非癫痫发作)EEG 波形。任何形式的长期脑电图监测都会产生大量数据。此数据需要大量时间才能进行适当的分析。高效的癫痫发作预测算法可以提醒佩戴移动记录设备的患者,在癫痫发作前要考虑适当的安全预防措施(Sörnmo & Laguna, 2005; Subasi, 2019c)。

最近,生物医学设备和医疗保健技术有了很多改进,可以帮助满足当前医疗保健诊断和治疗的需求。虽然医疗中心的创新医疗设备可提供快速、准确的分析,但仍有必要持续监测癫痫等慢性病患者的状态(Chiauzzi, Rodarte, & DasMahapatra, 2015)。图 4.3 显示了基于云的移动患者监测可用于癫痫发作的预测。在这种框架下,移动设备可以持续收集 EEG 信号,从智能耳机传感器和采集设备中获取少量信息,通过互联网在紧急情况下通知紧急救援人员、临床医生和病人家属(Hsieh & Hsu, 2012)。通过这种方法,可以将不同的信号处理和机器学习技术用于分析 EEG 信号以预测癫痫发作。癫痫发作随机且难以预测(Menshawy, Benharref, & Serhani, 2015),正确预测癫痫发作的方法是持续监测患者。由于智能可穿戴传感器、移动传感设备和智能手机以及无线和蜂窝通信网络的创新发展,基于云的移动患者监测系统的发展成为可能。智能传感器可以轻松与智能手机进行集成,成为发展基于云技术的移动健康监测系统的重要一环。智能传感器可直接准确地从患者的大脑接收 EEG 信号，并将收集到的 EEG 信号传输到智能手机。然后智能手机将收集到的 EEG 信号发送到云端，以持续跟踪患者。如果预测会有癫痫发作，则云服务器上的应用程序将发送此信息给急诊科、临床医生和患者家属（Serhani, El Menshawy, & Benharref, 2016）。如图 4.3 所示的癫痫发作预测框架由数据采集与传输模块、特征提取模块和分类模块组成。其中，智能手机通过适当的通信协议连接到云端（Subasi, Bandic, & Qaisar, 2020）。

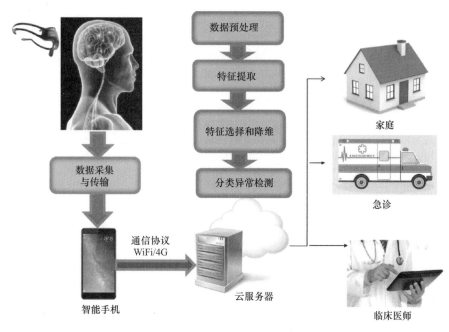

图 4.3　一种基于云的移动癫痫患者监测框架。该图改编自（Subasi et al., 2020a）

例 4.1　以下 Python 代码首先使用离散小波变换（DWT）对数据进行变换，从 EEG 信号中提取特征，然后使用 DWT 子带的统计值。接下来，通过单独的训练和测试数据集，使用不同的分类器对这些数据进行分类。本例计算了分类的准确率、精确度、召回率、F1 得分、Cohen kappa 得分和 Matthews 相关系数，并且给出了分类报告和混淆矩阵。

数据集信息：该癫痫数据集是一种广泛使用的 EEG 数据集，具有记为 A～E 的 5 组信息，每组包含 100 个 23.6 秒持续时间的单通道 EEG 段。A 组和 B 组由 5 名健康志愿者的脑电图记录组成，采用标准化电极放置方案。受试者处于放松的清醒状态，分别为睁眼（A-Z.zip）和闭眼（B-O.zip）。集合 C-N.zip、D-F.zip 和 E-S.zip 取自癫痫病人术前诊断的 EEG 存档。EEG 信号是从切除一个海马结构后获得完全癫痫控制的受试者身上采集的，该海马结构被正确识别为癫痫发生区。D 组的记录来自癫痫发生区，C 组的记录来自大脑对侧半球的海马区。C 组和 D 组仅包含癫痫发作间歇期记录的活动，E 组仅包含癫痫发作活动。这里的片段是从所有显示发作活动的记录中选择的。经过 12 位模数转换后，数据以173.61Hz进行采样。带通滤波器设置为0.53～40 Hz（12 dB/oct）。可以从以下网站下载数据：http://epileptologie-bonn.de/cms/front_content.php?idcat=193&lang=3&changelang=3。

```
"""
Created on Thu May 9 12:18:30 2019
@author: asubasi
"""
# descriptive statistics
import scipy as sp
import scipy.io as sio
import pywt
import numpy as np
import scipy.stats as stats
from sklearn.metrics import classification_report
from sklearn.metrics import confusion_matrix
from sklearn import metrics
from io import BytesIO #needed for plot
import seaborn as sns; sns.set()
import matplotlib.pyplot as plt
import seaborn as sns
#Mother Wavelet db1
waveletname='db1'
level=6
#Load mat file
mat_contents = sio.loadmat('AS_BONN_ALL_EEG_DATA_1024.mat')
sorted(mat_contents.keys())
EpilepticZone_Interictal=mat_contents['EpilepticZone_Interictal']
Epileptic_Ictal=mat_contents['Epileptic_Ictal']
NonEpilepticZone_Interictal=mat_contents['NonEpilepticZone_Interictal_']
Normal_Eyes_Closed=mat_contents['Normal_Eye_Closed']
Normal_Eyes_Open=mat_contents['Normal_Eyes_Open']

Labels = [] #Empty List For Labels
Length = 1024; # Length of signal
Nofsignals=len(Normal_Eyes_Open[0]) ; #Total Number of Signal for each class
NofClasses=3 #Number of Classes
numfeatures =48 #Number of features extracted from DWT decomposition
#Create Empty Array For Features
Extracted_Features=np.ndarray(shape=(NofClasses*Nofsignals,numfeatures), dtype=float, order='F')
# ================================================================
# Define utility functions
# ================================================================
def print_confusion_matrix(y_test, y_pred):
    matrix = confusion_matrix(y_test, y_pred)
    plt.figure(figsize=(10, 8))
```

```python
    sns.heatmap(matrix,cmap='coolwarm',linecolor='white',linewidths=1,
                annot=True,
                fmt='d')
    plt.title('Confusion Matrix')
    plt.ylabel('True Label')
    plt.xlabel('Predicted Label')
    plt.show()
def print_performance_metrics(y_test, y_pred):
    print('Accuracy:', np.round(metrics.accuracy_score(y_test, y_pred),4))
    print('Precision:', np.round(metrics.precision_score(y_test,
                    y_pred,average='weighted'),4))
    print('Recall:', np.round(metrics.recall_score(y_test, y_pred,
                    average='weighted'),4))
    print('F1 Score:', np.round(metrics.f1_score(y_test, y_pred,
                    average='weighted'),4))
    print('Cohen Kappa Score:', np.round(metrics.cohen_kappa_score(y_test,
    y_pred),4))
    print('Matthews Corrcoef:', np.round(metrics.matthews_corrcoef(y_test,
    y_pred),4))
    print('\t\tClassification  Report:\n', metrics.classification_report(y_
    test, y_pred))

def print_confusion_matrix_and_save(y_test, y_pred):
    mat = confusion_matrix(y_test, y_pred)
    sns.heatmap(mat, square=True, annot=True, fmt='d', cbar=False)
    plt.title('Confusion Matrix')
    plt.ylabel('True Label')
    plt.xlabel('Predicted Label')
    plt.show()

    plt.savefig("Confusion.jpg")
    # Save SVG in a fake file object.
    f = BytesIO()
    plt.savefig(f, format="svg")

def plot_history(history):
    accuracy = history.history['accuracy']
    val_accuracy = history.history['val_accuracy']
    loss = history.history['loss']
    val_loss = history.history['val_loss']
    x = range(1, len(accuracy) + 1)

    plt.figure(figsize=(12, 5))
    plt.subplot(1, 2, 1)
    plt.plot(x, accuracy, 'b', label='Training acc')
    plt.plot(x, val_accuracy, 'r', label='Validation acc')
    plt.title('Training and validation accuracy')
    plt.legend()
    plt.subplot(1, 2, 2)
    plt.plot(x, loss, 'b', label='Training loss')
```

```python
    plt.plot(x, val_loss, 'r', label='Validation loss')
    plt.title('Training and validation loss')
    plt.legend()
# ==================================================================
# Feature extraction using the statistical values of discrete wavelet
transform
# ==================================================================
def DWT_Feature_Extraction(signal, i, wname, level):
    coeff = pywt.wavedec(signal, wname, level=level)
    cA6,cD6,cD5,cD4, cD3, cD2, cD1=coeff
    #Mean Values of each subbands
    Extracted_Features[i,0]=sp.mean(abs(cD1[:]))
    Extracted_Features[i,1]=sp.mean(abs(cD2[:]))
    Extracted_Features[i,2]=sp.mean(abs(cD3[:]))
    Extracted_Features[i,3]=sp.mean(abs(cD4[:]))
    Extracted_Features[i,4]=sp.mean(abs(cD5[:]))
    Extracted_Features[i,5]=sp.mean(abs(cD6[:]))
    Extracted_Features[i,6]=sp.mean(abs(cA6[:]))
    #Standart Deviation of each subbands
Extracted_Features[i,7]=sp.std(cD1[:]);
Extracted_Features[i,8]=sp.std(cD2[:]);
Extracted_Features[i,9]=sp.std(cD3[:]);
Extracted_Features[i,10]=sp.std(cD4[:]);
Extracted_Features[i,11]=sp.std(cD5[:]);
Extracted_Features[i,12]=sp.std(cD6[:]);
Extracted_Features[i,13]=sp.std(cA6[:]);
#Skewness of each subbands
Extracted_Features[i,14]=stats.skew(cD1[:]);
Extracted_Features[i,15]=stats.skew(cD2[:]);
Extracted_Features[i,16]=stats.skew(cD3[:]);
Extracted_Features[i,17]=stats.skew(cD4[:]);
Extracted_Features[i,18]=stats.skew(cD5[:]);
Extracted_Features[i,19]=stats.skew(cD6[:]);
Extracted_Features[i,20]=stats.skew(cA6[:]);
#Kurtosis of each subbands
Extracted_Features[i,21]=stats.kurtosis(cD1[:]);
Extracted_Features[i,22]=stats.kurtosis(cD2[:]);
Extracted_Features[i,23]=stats.kurtosis(cD3[:]);
Extracted_Features[i,24]=stats.kurtosis(cD4[:]);
Extracted_Features[i,25]=stats.kurtosis(cD5[:]);
Extracted_Features[i,26]=stats.kurtosis(cD6[:]);
Extracted_Features[i,27]=stats.kurtosis(cA6[:]);
#Median Values of each subbands
Extracted_Features[i,28]=sp.median(cD1[:]);
Extracted_Features[i,29]=sp.median(cD2[:]);
Extracted_Features[i,30]=sp.median(cD3[:]);
Extracted_Features[i,31]=sp.median(cD4[:]);
Extracted_Features[i,32]=sp.median(cD5[:]);
Extracted_Features[i,33]=sp.median(cD6[:]);
Extracted_Features[i,34]=sp.median(cA6[:]);
```

```python
#RMS Values of each subbands
Extracted_Features[i,35]=np.sqrt(np.mean(cD1[:]**2));
Extracted_Features[i,36]=np.sqrt(np.mean(cD2[:]**2));
Extracted_Features[i,37]=np.sqrt(np.mean(cD3[:]**2));
Extracted_Features[i,38]=np.sqrt(np.mean(cD4[:]**2));
Extracted_Features[i,39]=np.sqrt(np.mean(cD5[:]**2));
Extracted_Features[i,40]=np.sqrt(np.mean(cD6[:]**2));
Extracted_Features[i,41]=np.sqrt(np.mean(cA6[:]**2));
#Ratio of subbands
Extracted_Features[i,42]=sp.mean(abs(cD1[:]))/sp.mean(abs(cD2[:]))
Extracted_Features[i,43]=sp.mean(abs(cD2[:]))/sp.mean(abs(cD3[:]))
Extracted_Features[i,44]=sp.mean(abs(cD3[:]))/sp.mean(abs(cD4[:]))
Extracted_Features[i,45]=sp.mean(abs(cD4[:]))/sp.mean(abs(cD5[:]))
Extracted_Features[i,46]=sp.mean(abs(cD5[:]))/sp.mean(abs(cD6[:]))
    Extracted_Features[i,47]=sp.mean(abs(cD6[:]))/sp.mean(abs(cA6[:]))
    return Extracted_Features
# ================================================================
# Feature extraction from normal EEG signal
# ================================================================
for i in range(Nofsignals):
    DWT_Feature_Extraction(Normal_Eyes_Open[:,i], i, waveletname, level)
    Labels.append("NORMAL")
# ================================================================
# Feature extraction from interictal EEG signal
# ================================================================
for i in range(Nofsignals, 2*Nofsignals):
    DWT_Feature_Extraction(EpilepticZone_Interictal[:,i-Nofsignals],    i,
    waveletname, level)
    Labels.append("INTERICTAL")

# ================================================================
# Feature extraction from ictal EEG signal
# ================================================================
for i in range(2*Nofsignals, 3*Nofsignals):
    DWT_Feature_Extraction(Epileptic_Ictal[:,i-2*Nofsignals], i, wavelet-
    name, level)
    Labels.append("ICTAL")
#%%
# ================================================================
# Classification
# ================================================================
X = Extracted_Features
y = Labels
# Import train_test_split function
from sklearn.model_selection import train_test_split
# Split dataset into training set and test set
# 70% training and 30% test
Xtrain, Xtest, ytrain, ytest = train_test_split(X, y,test_size=0.3, ran-
dom_state=0)
#%%
```

```python
# ================================================================
# LDA classification with training and test set
# ================================================================
#Import LDA model
from sklearn.discriminant_analysis import LinearDiscriminantAnalysis
#Create a LDA Classifier
clf = LinearDiscriminantAnalysis(solver='lsqr', shrinkage=None)
#Train the model using the training sets
clf.fit(Xtrain,ytrain)
#Predict the response for test dataset
ypred = clf.predict(Xtest)
#Evaluate the Model and Print Performance Metrics
print_performance_metrics(ytest, ypred)

#Plot Confusion Matrix
print_confusion_matrix_and_save(ytest, ypred)

#%%
# ================================================================
# Naive Bayes classification with training and test set
# ================================================================
#Import Gaussian Naive Bayes model
from sklearn.naive_bayes import GaussianNB
#Create a Gaussian Classifier
gnb = GaussianNB()
#Train the model using the training sets
gnb.fit(Xtrain, ytrain)
#Predict the response for test dataset
ypred = gnb.predict(Xtest)

#Evaluate the Model and Print Performance Metrics
print_performance_metrics(ytest, ypred)

#Plot Confusion Matrix
print_confusion_matrix_and_save(ytest, ypred)

#%%
# ================================================================
# Quadratic discriminant analysis (QDA) example
# ================================================================
import numpy as np
from sklearn.discriminant_analysis import QuadraticDiscriminantAnalysis
from sklearn.model_selection import train_test_split
Xtrain, Xtest, ytrain, ytest = train_test_split(X, y, test_size=0.3, random_state=1)

# Quadratic Discriminant Analysis
clf = QuadraticDiscriminantAnalysis(store_covariance=True)
clf.fit(Xtrain,ytrain)
ypred = clf.predict(Xtest)
```

```python
#Evaluate the Model and Print Performance Metrics
print_performance_metrics(ytest, ypred)

#Plot Confusion Matrix
print_confusion_matrix_and_save(ytest, ypred)

#%%
from sklearn.neural_network import MLPClassifier
#Create Train and Test set
Xtrain, Xtest, ytrain, ytest = train_test_split(X, y, test_size=0.3, random_state=0)

"""mlp=MLPClassifier(hidden_layer_sizes=(100, ),   activation='relu', solver='adam',
      alpha=0.0001, batch_size='auto', learning_rate='constant',
      learning_rate_init=0.001, power_t=0.5, max_iter=200,
      shuffle=True, random_state=None, tol=0.0001, verbose=False,
      warm_start=False, momentum=0.9, nesterovs_momentum=True,
      early_stopping=False, validation_fraction=0.1, beta_1=0.9,
      beta_2=0.999, epsilon=1e-08, n_iter_no_change=10)"""
#Create the Model
mlp = MLPClassifier(hidden_layer_sizes=(50, ), learning_rate_init=0.001,
      alpha=1, momentum=0.7,max_iter=1000)
#Train the Model with Training dataset
mlp.fit(Xtrain,ytrain)
#Test the Model with Testing dataset
ypred = mlp.predict(Xtest)

#Evaluate the Model and Print Performance Metrics
print_performance_metrics(ytest, ypred)

#Plot Confusion Matrix
print_confusion_matrix_and_save(ytest, ypred)

#%%
# ====================================================================
# k-NN example with training and test set
# ====================================================================
from sklearn.neighbors import KNeighborsClassifier

#Create Train and Test set
Xtrain, Xtest, ytrain, ytest = train_test_split(X, y, test_size=0.3, random_state=0)
#Create the Model
clf = KNeighborsClassifier(n_neighbors=1)
#Train the model with Training Dataset
clf.fit(Xtrain,ytrain)
#Test the model with Testset
ypred = clf.predict(Xtest)
```

```python
#Evaluate the Model and Print Performance Metrics
print_performance_metrics(ytest, ypred)

#Plot Confusion Matrix
print_confusion_matrix_and_save(ytest, ypred)

#%%
# ====================================================================
# SVM example with training and test set
# ====================================================================
from sklearn import svm
Xtrain, Xtest, ytrain, ytest = train_test_split(X, y, test_size=0.3, random_state=0)

""" The parameters and kernels of SVM classifierr can be changed as follows
C = 10.0 # SVM regularization parameter
svm.SVC(kernel='linear', C=C)
svm.LinearSVC(C=C, max_iter=10000)
svm.SVC(kernel='rbf', gamma=0.7, C=C)
svm.SVC(kernel='poly', degree=3, gamma='auto', C=C)
"""
C = 10.0 # SVM regularization parameter
#Create the Model
clf =svm.SVC(kernel='poly', degree=4, gamma='auto', C=C)
#Train the model with Training Dataset
clf.fit(Xtrain,ytrain)
#Test the model with Testset
ypred = clf.predict(Xtest)

#Evaluate the Model and Print Performance Metrics
print_performance_metrics(ytest, ypred)

#Plot Confusion Matrix
print_confusion_matrix_and_save(ytest, ypred)

#%%
# ====================================================================
# Decision tree example with training and test set
# ====================================================================
from sklearn import tree
Xtrain, Xtest, ytrain, ytest = train_test_split(X, y, test_size=0.3, random_state=0)
#Create the Model
clf = tree.DecisionTreeClassifier()
#Train the Model with Training dataset
clf.fit(Xtrain,ytrain)
#Test the Model with Testing dataset
ypred = clf.predict(Xtest)

#Evaluate the Model and Print Performance Metrics
```

```
print_performance_metrics(ytest, ypred)

#Plot Confusion Matrix
print_confusion_matrix_and_save(ytest, ypred)

#%%
# ======================================================================
# Extra trees classification example with training and test set
# ======================================================================
#Import Extra Trees model
from sklearn.ensemble import ExtraTreesClassifier

# Split dataset into training set and test set
# 70% training and 30% test
Xtrain, Xtest, ytrain, ytest = train_test_split(X, y,test_size=0.3, random_state=0)

#Create the Model
clf = ExtraTreesClassifier(n_estimators=100, max_features=48)

#Train the model using the training sets
clf.fit(Xtrain,ytrain)
#Predict the response for test dataset
ypred = clf.predict(Xtest)

#Evaluate the Model and Print Performance Metrics
print_performance_metrics(ytest, ypred)

#Plot Confusion Matrix
print_confusion_matrix_and_save(ytest, ypred)

#%%
# ======================================================================
# Bagging example with training and test set
# ======================================================================
#Import Bagging ensemble model
from sklearn.ensemble import BaggingClassifier
#Import Tree model as a base classifier
from sklearn import tree

# Split dataset into training set and test set
# 70% training and 30% test
Xtrain, Xtest, ytrain, ytest = train_test_split(X, y,test_size=0.3, random_state=0)

#Create a Bagging Ensemble Classifier
"""BaggingClassifier(base_estimator=None,   n_estimators=10,    max_samples=1.0,
max_features=1.0,    bootstrap=True,    bootstrap_features=False,    oob_score=False,
```

```
warm_start=False, n_jobs=None, random_state=None, verbose=0)"""
bagging = BaggingClassifier(tree.DecisionTreeClassifier(),
            max_samples=0.5, max_features=0.5)
#Train the model using the training sets
bagging.fit(Xtrain,ytrain)
#Predict the response for test dataset
ypred = bagging.predict(Xtest)

#Evaluate the Model and Print Performance Metrics
print_performance_metrics(ytest, ypred)

#Plot Confusion Matrix
print_confusion_matrix_and_save(ytest, ypred)

#%%
# =====================================================================
# Random forest example with training and test set
# =====================================================================
from sklearn.ensemble import RandomForestClassifier

#In order to change to accuracy increase n_estimators
"""RandomForestClassifier(n_estimators='warn',    criterion='gini',    max_
depth=None,
min_samples_split=2, min_samples_leaf=1, min_weight_fraction_leaf=0.0,
max_features='auto', max_leaf_nodes=None, min_impurity_decrease=0.0,
min_impurity_split=None, bootstrap=True, oob_score=False, n_jobs=None,
random_state=None, verbose=0, warm_start=False, class_weight=None)"""
clf = RandomForestClassifier(n_estimators=200)
#Create the Model
#Train the model with Training Dataset
clf.fit(Xtrain,ytrain)
#Test the model with Testset
ypred = clf.predict(Xtest)
#Evaluate the Model and Print Performance Metrics
print_performance_metrics(ytest, ypred)

#Plot Confusion Matrix
print_confusion_matrix_and_save(ytest, ypred)

#%%
# =====================================================================
# AdaBoost example with training and test set
# =====================================================================
#Import AdaBoost ensemble model
from sklearn.ensemble import AdaBoostClassifier
#Import Tree model as a base classifier
from sklearn import tree

# Split dataset into training set and test set
# 70% training and 30% test
```

```python
Xtrain, Xtest, ytrain, ytest = train_test_split(X, y,test_size=0.3, random_state=0)

#Create an AdaBoost Ensemble Classifier
""" AdaBoostClassifier(base_estimator=None, n_estimators=50, learning_rate=1.0,
            algorithm='SAMME.R', random_state=None)"""
clf=AdaBoostClassifier(tree.DecisionTreeClassifier(),n_estimators=100,
algorithm='SAMME',learning_rate=0.5)

#Train the model using the training sets
clf.fit(Xtrain,ytrain)
#Predict the response for test dataset
ypred = clf.predict(Xtest)

#Evaluate the Model and Print Performance Metrics
print_performance_metrics(ytest, ypred)

#Plot Confusion Matrix
print_confusion_matrix_and_save(ytest, ypred)

#%%
# =====================================================================
# Gradient boosting example with training and test set
# =====================================================================
#Import Gradient Boosting ensemble model
from sklearn.ensemble import GradientBoostingClassifier

# Split dataset into training set and test set
# 70% training and 30% test
Xtrain, Xtest, ytrain, ytest = train_test_split(X, y,test_size=0.3, random_state=0)

#Create the Model
clf = GradientBoostingClassifier(n_estimators=100, learning_rate=1.0,
    max_depth=1, random_state=0)

#Train the model using the training sets
clf.fit(Xtrain,ytrain)
#Predict the response for test dataset
ypred = clf.predict(Xtest)

#Evaluate the Model and Print Performance Metrics
print_performance_metrics(ytest, ypred)

#Plot Confusion Matrix
print_confusion_matrix_and_save(ytest, ypred)

#%%
# =====================================================================
```

```python
# Stacking meta classifier example
# ==================================================================
#Before running you should install mlxtend.classifier for Staking using pip install mlxtend
import numpy as np
import warnings
from sklearn import model_selection
from sklearn.linear_model import LogisticRegression
from sklearn.neighbors import KNeighborsClassifier
from sklearn.neural_network import MLPClassifier
from sklearn.ensemble import RandomForestClassifier
from mlxtend.classifier import StackingClassifier
from sklearn.model_selection import train_test_split
from sklearn import tree

#Create Train and Test set
Xtrain, Xtest, ytrain, ytest = train_test_split(X, y, test_size=0.3, random_state=0)

warnings.simplefilter('ignore')
#Create the Model
clf1 = KNeighborsClassifier(n_neighbors=1)
clf2 = RandomForestClassifier(random_state=1)
clf3 = MLPClassifier(hidden_layer_sizes=(100, ), learning_rate_init=0.001,
          alpha=1, momentum=0.9,max_iter=1000)
DT = tree.DecisionTreeClassifier()
sclf = StackingClassifier(classifiers=[clf1, clf2, clf3],
          meta_classifier=DT)
#Evaluate the Model and Print Performance Metrics
print_performance_metrics(ytest, ypred)

#Plot Confusion Matrix
print_confusion_matrix_and_save(ytest, ypred)

#%%
# ==================================================================
# Voting example with training and test set
# ==================================================================
from sklearn import datasets
from sklearn.model_selection import cross_val_score
from sklearn.linear_model import LogisticRegression
from sklearn.naive_bayes import GaussianNB
from sklearn.ensemble import RandomForestClassifier
from sklearn.ensemble import VotingClassifier
from sklearn.model_selection import train_test_split

#Create Train and Test set
Xtrain, Xtest, ytrain, ytest = train_test_split(X, y, test_size=0.3, random_state=0)
```

```python
#Create the Models
clf1 = KNeighborsClassifier(n_neighbors=1)
clf2 = RandomForestClassifier(random_state=1)
clf3 = MLPClassifier(hidden_layer_sizes=(100, ), learning_rate_init=0.001,
        alpha=1, momentum=0.9,max_iter=1000)
eclf = VotingClassifier(estimators=[('kNN', clf1), ('RF', clf2), ('MLP', clf3)], voting='hard')

#Train the model using the training sets
eclf.fit(Xtrain,ytrain)
#Predict the response for test dataset
ypred = eclf.predict(Xtest)

#Evaluate the Model and Print Performance Metrics
print_performance_metrics(ytest, ypred)
#Plot Confusion Matrix
print_confusion_matrix_and_save(ytest, ypred)

#%%
# ==================================================================
# Deep neural network example using Keras
# ==================================================================
from keras.models import Sequential
from keras.layers import Dense, Activation
from keras.utils.np_utils import to_categorical
from keras.utils.vis_utils import plot_model
from sklearn import preprocessing
lb = preprocessing.LabelBinarizer()
y=lb.fit_transform( Labels)
X = Extracted_Features
# Import train_test_split function
from sklearn.model_selection import train_test_split
# Split dataset into training set and test set
# 70% training and 30% test
Xtrain, Xtest, ytrain, ytest = train_test_split(X, y,test_size=0.3, random_state=0)
InputDataDimension=numfeatures
#NofClasses
# ==================================================================
# Create a model
# ==================================================================
model = Sequential()
model.add(Dense(128,  input_dim=InputDataDimension,  init='uniform', activation='relu'))
model.add(Dense(32, activation='relu'))
model.add(Dense(NofClasses, activation='softmax'))
#%%
# ==================================================================
# Compile the model
# ==================================================================
```

```python
model.compile(loss='categorical_crossentropy',         optimizer='adam',
metrics=['accuracy'])
# ================================================================
# Train and validate the model
# ================================================================
history = model.fit(Xtrain, ytrain, validation_split=0.33, epochs=50,
batch_size=25,verbose=2)
#%%
# ================================================================
# Evaluate the model
# ================================================================
test_loss, test_acc = model.evaluate(Xtest, ytest, verbose=0)
print('\nTest accuracy:', test_acc)

#%%
# ================================================================
# Plot the history
# ================================================================
#Plot the Model Accuracy and Loss for Training and Validation dataset
plot_history(history)

#%%
#Test the Model with testing data
ypred_test = model.predict(Xtest,)
# Round the test predictions
max_y_pred_test = np.round(ypred_test)
#Convert binary Labels back to numbers
ypred=max_y_pred_test.argmax(axis=1)
ytest=ytest.argmax(axis=1)
#%%
#Print the Confusion Matrix
print_confusion_matrix(ytest,ypred)
#%%
#Evaluate the Model and Print Performance Metrics
print_performance_metrics(ytest,ypred)
#%%
#Print and Save the Confusion Matrix
print_confusion_matrix_and_save(ytest,ypred)
#%%
from keras.models import Sequential
from keras.layers import Dense, Activation
from keras.utils.np_utils import to_categorical
from keras.utils.vis_utils import plot_model
from sklearn import preprocessing
lb = preprocessing.LabelBinarizer()
y=lb.fit_transform( Labels)
X = Extracted_Features
# Import train_test_split function
from sklearn.model_selection import train_test_split
# Split dataset into training set and test set
```

```python
# 70% training and 30% test
Xtrain, Xtest, ytrain, ytest = train_test_split(X, y,test_size=0.3, random_state=0)
InputDataDimension=numfeatures
# ================================================================
# Build a deep model using Keras
# ================================================================
model = Sequential()
model.add(Dense(128, input_dim=InputDataDimension, init='uniform', activation='relu'))
model.add(Dense(64, activation='relu'))
model.add(Dense(32, activation='relu'))
model.add(Dense(NofClasses, activation='softmax'))
#%%
# ================================================================
# Compile the model
# ================================================================
model.compile(loss='categorical_crossentropy', optimizer='adam', metrics=['accuracy'])
# ================================================================
# Train and validate the model
# ================================================================
history = model.fit(Xtrain, ytrain, validation_split=0.33, epochs=50, batch_size=25,verbose=2)
#%%
# ================================================================
# Evaluate the model
# ================================================================
test_loss, test_acc = model.evaluate(Xtest, ytest, verbose=0)
print('\nTest accuracy:', test_acc)

#%%
# ================================================================
# Plot the history
# ================================================================
#Plot the model accuracy and loss for training and validation dataset
plot_history(history)

#%%
#Test the Model with testing data
ypred_test = model.predict(Xtest,)
# Round the test predictions
max_y_pred_test = np.round(ypred_test)
#Convert binary Labels back to numbers
ypred=max_y_pred_test.argmax(axis=1)
ytest=ytest.argmax(axis=1)
#%%
#Print the Confusion Matrix
print_confusion_matrix(ytest,ypred)
#%%
```

```
#Evaluate the Model and Print Performance Metrics
print_performance_metrics(ytest,ypred)
#%%
#Print and Save the Confusion Matrix
print_confusion_matrix_and_save(ytest,ypred)

#%%
"""Adapted From Scikit Learn"""
# =====================================================================
# ROC curves for the multiclass problem
# =====================================================================
import numpy as np
import matplotlib.pyplot as plt
from itertools import cycle
from sklearn.discriminant_analysis import LinearDiscriminantAnalysis
from sklearn import svm
from sklearn.metrics import roc_curve, auc
from sklearn.model_selection import train_test_split
from sklearn.preprocessing import label_binarize
from sklearn.multiclass import OneVsRestClassifier
from scipy import interp
X = Extracted_Features
y = Labels
# Import train_test_split function
from sklearn.model_selection import train_test_split
# Split dataset into training set and test set
# 70% training and 30% test
Xtrain, Xtest, ytrain, ytest = train_test_split(X, y,test_size=0.3, random_state=0)
# Binarize the output
y = label_binarize(y, classes=['NORMAL','INTERICTAL','ICTAL' ])
n_classes = y.shape[1]

#%%
# shuffle and split training and test sets
random_state = np.random.RandomState(0)
X_train, X_test, y_train, y_test = train_test_split(X, y, test_size=.5,
                                    random_state=0)

# Learn to predict each class against the other
"""classifier = OneVsRestClassifier(svm.SVC(kernel='linear',
probability=True,
                 random_state=random_state))"""
classifier = OneVsRestClassifier(LinearDiscriminantAnalysis(solver='lsqr',
shrinkage=None))
y_score = classifier.fit(X_train, y_train).decision_function(X_test)

# Compute ROC curve and ROC area for each class
fpr = dict()
tpr = dict()
```

```python
roc_auc = dict()
for i in range(n_classes):
    fpr[i], tpr[i], _ = roc_curve(y_test[:, i], y_score[:, i])
    roc_auc[i] = auc(fpr[i], tpr[i])

# Compute micro-average ROC curve and ROC area
fpr["micro"], tpr["micro"], _ = roc_curve(y_test.ravel(), y_score.ravel())
roc_auc["micro"] = auc(fpr["micro"], tpr["micro"])

##################################################################
# Plot of a ROC curve for a specific class
plt.figure()
lw = 2
plt.plot(fpr[2], tpr[2], color='darkorange',
         lw=lw, label='ROC curve (area = %0.2f)' % roc_auc[2])
plt.plot([0, 1], [0, 1], color='navy', lw=lw, linestyle='--')
plt.xlim([0.0, 1.0])
plt.ylim([0.0, 1.05])
plt.xlabel('False Positive Rate')
plt.ylabel('True Positive Rate')
plt.title('Receiver operating characteristic')
plt.legend(loc="lower right")
plt.show()
##################################################################
# Plot ROC curves for the multiclass problem

# Compute macro-average ROC curve and ROC area

# First aggregate all false positive rates
all_fpr = np.unique(np.concatenate([fpr[i] for i in range(n_classes)]))

# Then interpolate all ROC curves at this points
mean_tpr = np.zeros_like(all_fpr)
for i in range(n_classes):
    mean_tpr += interp(all_fpr, fpr[i], tpr[i])
# Finally average it and compute AUC
mean_tpr /= n_classes

fpr["macro"] = all_fpr
tpr["macro"] = mean_tpr
roc_auc["macro"] = auc(fpr["macro"], tpr["macro"])

# Plot all ROC curves
plt.figure()
plt.plot(fpr["micro"], tpr["micro"],
         label='micro-average ROC curve (area = {0:0.2f})'
               ''.format(roc_auc["micro"]),
         color='deeppink', linestyle=':', linewidth=4)
```

```python
plt.plot(fpr["macro"], tpr["macro"],
    label='macro-average ROC curve (area = {0:0.2f})'
        ''.format(roc_auc["macro"]),
    color='navy', linestyle=':', linewidth=4)

colors = cycle(['aqua', 'darkorange', 'cornflowerblue'])
for i, color in zip(range(n_classes), colors):
    plt.plot(fpr[i], tpr[i], color=color, lw=lw,
        label='ROC curve of class {0} (area = {1:0.2f})'
        ''.format(i, roc_auc[i]))

plt.plot([0, 1], [0, 1], 'k--', lw=lw)
plt.xlim([0.0, 1.0])
plt.ylim([0.0, 1.05])
plt.xlabel('False Positive Rate')
plt.ylabel('True Positive Rate')
plt.title('Some extension of Receiver operating characteristic to multi-class')
plt.legend(loc="lower right")
plt.show()
```

4.2.2 情绪识别

情绪是由人类有意识和（或）不自觉的感知触发的心理生理过程。人的敏锐感知通常与气质、性格、个性、情绪和动机有关。情绪在人类交流中起着至关重要的作用，可以通过情绪词汇进行口头表达，也可以通过诸如语音语调、面部表情和手势等非语言提示进行表达（Koelstra et al., 2012）。研究人员对更多常见的情绪进行了研究，例如愤怒、幸福、惊奇、恐惧、羞耻、厌恶、悲伤、鄙视、痛苦、爱、紧张和欢乐。他们通过分析大脑信号和图像来对情绪进行理解、控制和调节。EEG 信号处理和分析技术已被用于情绪检测和识别。工程、计算机科学、神经科学和心理学方面的研究旨在开发识别、监测和模拟情绪的设备。在计算机科学中，高效评估是应用机器学习研究的一个分支，涉及处理、解释和识别人类情绪的系统和设备的设计。因此，通过自然的互动和对话可以间接产生新的技术来评估压力、沮丧和情绪。此外，使计算机更具有情绪智能，特别是以减少负面情绪的方式对人的挫折做出反应，可以成为情绪识别和调节的重要研究方向。脑电图有助于不同情绪下大脑的连接、定位和同步，以提取和处理高价值的情绪相关信息。情绪化 EEG 信号的分类和识别通常很复杂，需要复杂的信号处理技术（Sanei, 2013; Subasi, 2019c）。情绪识别的一般框架如图 4.4 所示。

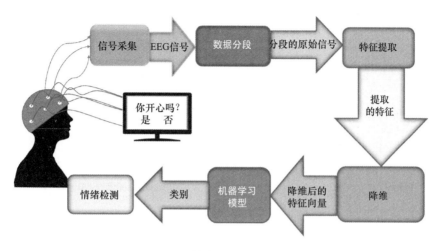

图 4.4 情绪识别的一般框架

例 4.2 以下 Python 代码使用平稳小波变换（SWT）从 EEG 信号中提取能够显示正、负以及中性情绪的特征。接下来，它使用 SWT 子带的统计值，然后使用随机森林分类器通过训练和测试数据集将这些数据分类。本例计算了分类准确率、精确度、召回率、F1 得分、Cohen kappa 得分以及 Matthews 相关系数。并给出了分类报告和混淆矩阵。本例使用了 SEED 情绪识别数据。

数据集信息：SEED 数据库包含 15 名受试者的 EEG 信号，这些信号是在受试者观看情感电影片段时记录的。每个受试者被要求在三个时段内进行实验。这个数据集总共有 45 个实验。这些是在实验中被选择作为刺激物的情感电影片段，这些影片剪辑的选择标准是：整个实验的时间不应过长，以免受试者产生疲劳感；视频应该不需要解释就能理解，并且应该引起一个单一的预期目标情绪。我们选择了不同的电影片段（积极的、中立的和消极的情绪），用来获得对不同参与者的最佳匹配。每个视频长度约为 4min。每个电影片段都是精心设计的，以产生一致的情绪，并最大限度地提高情绪本身。每个实验总共有 15 次试验。每个片段前有 15 秒的提示，每个片段后有 10 秒的反馈。演示的顺序是这样安排的：针对同一情绪的两个电影片段不会连续呈现。在反馈方面，参与者被要求在观看每个片段后立即回答问卷，以确定他们对每个电影片段的情绪回应。每个受试者的脑电图信号都被记录为单独的文件，其中包含受试者的名字和日期。15 名受试者（7 名男性和 8 名女性，平均年龄为 23.27，标准差为 2.37）参加了实验。这些文件包含脑电图数据的预处理、下采样和分段的版本。数据被下采样到 200Hz，并且使用了 0 至 75Hz 的带通频率滤波器。之后，与每部电影相关的 EEG 片段被提取出来。每个实验都有一个，总共有 45 个 .mat 文件。每个人在一周内进行三次实验。每个主题文件包括 16 个数组，其中 15 个数组包括同一实验中的 15 次试验的预处理和分段的脑电数据。名为 LABELS 的数组包含相应的情绪标签（-1 代表负数、0 代表中性、+1 代表正）。脑电图是根据国际 10-20 系统采集的，有 62 个通道。可以从以下网站下载数据：http://bcmi.sjtu.edu.cn/~seed/index.html。

```
"""
Created on Thu May 9 12:18:30 2019
@author: asubasi
"""

# ================================================================
# Feature extraction using the statistical values of stationary wavelet
transform
# ================================================================

import scipy.io as sio
# descriptive statistics
import scipy as sp
import pywt
import matplotlib.pyplot as plt
import numpy as np
import scipy.stats as stats
waveletname='db1'
level=6 #Decomposition Level
#Load mat file
mat_contents = sio.loadmat('EMOTIONSDAT.mat')
sorted(mat_contents.keys())
#Load each datset separately
NEGATIVE=mat_contents['NEGATIVE']
NEUTRAL=mat_contents['NEUTRAL']
POSITIVE=mat_contents['POSITIVE']

Labels = [] #Empty List For Labels
Length = 4096; # Length of signal
Nofsignal=100; #Total Number of Signal for each class
numrows =83 #Number of features extracted from Wavelet Packet
Decomposition
#Create Empty Array For Features
Extracted_Features=np.ndarray(shape=(3*Nofsignal,numrows),
dtype=float, order='F')
# ================================================================
# Define utility functions
# ================================================================
def SWT_Feature_Extraction(signal, i, wname, level):
    coeffs = pywt.swt(signal, wname, level=level)
    #Mean Values of each subbands
    Extracted_Features[i,0]=sp.mean(abs(coeffs[0][0]))
    Extracted_Features[i,1]=sp.mean(abs(coeffs[1][0]))
    Extracted_Features[i,2]=sp.mean(abs(coeffs[2][0]))
    Extracted_Features[i,3]=sp.mean(abs(coeffs[3][0]))
```

```
Extracted_Features[i,4]=sp.mean(abs(coeffs[4][0]))
Extracted_Features[i,5]=sp.mean(abs(coeffs[5][0]))
Extracted_Features[i,6]=sp.mean(abs(coeffs[0][1]))
Extracted_Features[i,7]=sp.mean(abs(coeffs[1][1]))
Extracted_Features[i,8]=sp.mean(abs(coeffs[2][1]))
Extracted_Features[i,9]=sp.mean(abs(coeffs[3][1]))
Extracted_Features[i,10]=sp.mean(abs(coeffs[4][1]))
Extracted_Features[i,11]=sp.mean(abs(coeffs[5][1]))
#Standart Deviation of each subbands
Extracted_Features[i,12]=sp.std(coeffs[0][0])
Extracted_Features[i,13]=sp.std(coeffs[1][0])
Extracted_Features[i,14]=sp.std(coeffs[2][0])
Extracted_Features[i,15]=sp.std(coeffs[3][0])
Extracted_Features[i,16]=sp.std(coeffs[4][0])
Extracted_Features[i,17]=sp.std(coeffs[5][0])
Extracted_Features[i,18]=sp.std(coeffs[0][1])
Extracted_Features[i,19]=sp.std(coeffs[1][1])
Extracted_Features[i,20]=sp.std(coeffs[2][1])
Extracted_Features[i,21]=sp.std(coeffs[3][1])
Extracted_Features[i,22]=sp.std(coeffs[4][1])
Extracted_Features[i,23]=sp.std(coeffs[5][1])
#Median Values of each subbands
Extracted_Features[i,24]=sp.median(coeffs[0][0])
Extracted_Features[i,25]=sp.median(coeffs[1][0])
Extracted_Features[i,26]=sp.median(coeffs[2][0])
Extracted_Features[i,27]=sp.median(coeffs[3][0])
Extracted_Features[i,28]=sp.median(coeffs[4][0])
Extracted_Features[i,29]=sp.median(coeffs[5][0])
Extracted_Features[i,30]=sp.median(coeffs[0][1])
Extracted_Features[i,31]=sp.median(coeffs[1][1])
Extracted_Features[i,32]=sp.median(coeffs[2][1])
Extracted_Features[i,33]=sp.median(coeffs[3][1])
Extracted_Features[i,34]=sp.median(coeffs[4][1])
Extracted_Features[i,35]=sp.median(coeffs[5][1])
#Skewness of each subbands
Extracted_Features[i,36]=stats.skew(coeffs[0][0])
Extracted_Features[i,37]=stats.skew(coeffs[1][0])
Extracted_Features[i,38]=stats.skew(coeffs[2][0])
Extracted_Features[i,39]=stats.skew(coeffs[3][0])
Extracted_Features[i,40]=stats.skew(coeffs[4][0])
Extracted_Features[i,41]=stats.skew(coeffs[5][0])
Extracted_Features[i,42]=stats.skew(coeffs[0][1])
Extracted_Features[i,43]=stats.skew(coeffs[1][1])
Extracted_Features[i,44]=stats.skew(coeffs[2][1])
Extracted_Features[i,45]=stats.skew(coeffs[3][1])
Extracted_Features[i,46]=stats.skew(coeffs[4][1])
Extracted_Features[i,47]=stats.skew(coeffs[5][1])
#Kurtosis of each subbands
Extracted_Features[i,48]=stats.kurtosis(coeffs[0][0])
Extracted_Features[i,49]=stats.kurtosis(coeffs[1][0])
```

```
Extracted_Features[i,50]=stats.kurtosis(coeffs[2][0])
Extracted_Features[i,51]=stats.kurtosis(coeffs[3][0])
Extracted_Features[i,52]=stats.kurtosis(coeffs[4][0])
Extracted_Features[i,53]=stats.kurtosis(coeffs[5][0])
Extracted_Features[i,54]=stats.kurtosis(coeffs[0][1])
Extracted_Features[i,55]=stats.kurtosis(coeffs[1][1])
Extracted_Features[i,56]=stats.kurtosis(coeffs[2][1])
Extracted_Features[i,57]=stats.kurtosis(coeffs[3][1])
Extracted_Features[i,58]=stats.kurtosis(coeffs[4][1])
Extracted_Features[i,59]=stats.kurtosis(coeffs[5][1])
#RMS Values of each subbands
Extracted_Features[i,60]=np.sqrt(np.mean(coeffs[0][0]**2))
Extracted_Features[i,61]=np.sqrt(np.mean(coeffs[1][0]**2))
Extracted_Features[i,62]=np.sqrt(np.mean(coeffs[2][0]**2))
Extracted_Features[i,63]=np.sqrt(np.mean(coeffs[3][0]**2))
Extracted_Features[i,64]=np.sqrt(np.mean(coeffs[4][0]**2))
Extracted_Features[i,65]=np.sqrt(np.mean(coeffs[5][0]**2))
Extracted_Features[i,66]=np.sqrt(np.mean(coeffs[0][1]**2))
Extracted_Features[i,67]=np.sqrt(np.mean(coeffs[1][1]**2))
Extracted_Features[i,68]=np.sqrt(np.mean(coeffs[2][1]**2))
Extracted_Features[i,69]=np.sqrt(np.mean(coeffs[3][1]**2))
Extracted_Features[i,70]=np.sqrt(np.mean(coeffs[4][1]**2))
Extracted_Features[i,71]=np.sqrt(np.mean(coeffs[5][1]**2))
#Ratio of subbands
Extracted_Features[i,72]=sp.mean(abs(coeffs[0][0]))/sp.mean(abs    (coeffs
[1][0]))
Extracted_Features[i,73]=sp.mean(abs(coeffs[1][0]))/sp.mean(abs(coeffs
[2][0]))
Extracted_Features[i,74]=sp.mean(abs(coeffs[2][0]))/sp.mean(abs    (coeffs
[3][0]))
Extracted_Features[i,75]=sp.mean(abs(coeffs[3][0]))/sp.mean(abs(coeffs
[4][0]))
Extracted_Features[i,76]=sp.mean(abs(coeffs[4][0]))/sp.mean (abs(coeffs[5]
[0]))
Extracted_Features[i,77]=sp.mean(abs(coeffs[5][0]))/sp.mean(abs
(coeffs [0][1]))
Extracted_Features[i,78]=sp.mean(abs(coeffs[0][1]))/sp.mean(abs
(coeffs[1][1]))
Extracted_Features[i,79]=sp.mean(abs(coeffs[1][1]))/sp.mean(abs(coeffs
[2][1]))
Extracted_Features[i,80]=sp.mean(abs(coeffs[2][1]))/sp.mean(abs
(coeffs[3][1]))
Extracted_Features[i,81]=sp.mean(abs(coeffs[3][1]))/sp.mean(abs(coeffs
[4][1]))
Extracted_Features[i,82]=sp.mean(abs(coeffs[4][1]))/sp.mean(abs
(coeffs[5][1]))
#%%
# ======================================================================
# Feature extraction from negative emotion EEG signal
# ======================================================================
```

```python
for i in range(Nofsignal):
    SWT_Feature_Extraction(NEGATIVE[i,:], i, waveletname, level)
    Labels.append("NEGATIVE")

# ====================================================================
# Feature extraction from neutral emotion EEG signal
# ====================================================================
for i in range(Nofsignal, 2*Nofsignal):
    SWT_Feature_Extraction(NEUTRAL[i-Nofsignal,:], i, waveletname, level)
    Labels.append("NEUTRAL")

# ====================================================================
# Feature extraction from positive emotion EEG signal
# ====================================================================
for i in range(2*Nofsignal, 3*Nofsignal):
    SWT_Feature_Extraction(POSITIVE[i-2*Nofsignal,:], i, waveletname, level)
    Labels.append("POSITIVE")
#%%
# ====================================================================
# Classification
# ====================================================================
X = Extracted_Features
y = Labels
# Import train_test_split function
from sklearn.model_selection import train_test_split
# Split dataset into training set and test set
# 70% training and 30% test
Xtrain, Xtest, ytrain, ytest = train_test_split(X, y,test_size=0.3, random_state=1)

#%%
# ====================================================================
# Random forest example with training and test set
# ====================================================================
from sklearn.ensemble import RandomForestClassifier

#In order to change to accuracy increase n_estimators
"""RandomForestClassifier(n_estimators='warn', criterion='gini', max_depth=None,
min_samples_split=2, min_samples_leaf=1, min_weight_fraction_leaf=0.0,
max_features='auto', max_leaf_nodes=None, min_impurity_decrease=0.0,
min_impurity_split=None, bootstrap=True, oob_score=False, n_jobs=None,
random_state=None, verbose=0, warm_start=False, class_weight=None)"""
clf = RandomForestClassifier(n_estimators=200)
#Create the Model
 #Train the model with Training Dataset
clf.fit(Xtrain,ytrain)
#Test the model with Testset
ypred = clf.predict(Xtest)
```

```python
#Evaluate the Model and Print Performance Metrics
from sklearn import metrics
print('Accuracy:', np.round(metrics.accuracy_score(ytest,ypred),4))
print('Precision:', np.round(metrics.precision_score(ytest,
    ypred,average='weighted'),4))
print('Recall:', np.round(metrics.recall_score(ytest,ypred,
    average='weighted'),4))
print('F1 Score:', np.round(metrics.f1_score(ytest,ypred,
average='weighted'),4))
print('Cohen  Kappa  Score:',  np.round(metrics.cohen_kappa_score(ytest,
ypred),4))
print('Matthews Corrcoef:', np.round(metrics.matthews_corrcoef(ytest,
ypred),4))
print('\t\tClassification Report:\n', metrics.classification_report(ypred,
ytest))

#Plot Confusion Matrix
from sklearn.metrics import confusion_matrix
print("Confusion Matrix:\n",confusion_matrix(ytest, ypred))

#%%
# ====================================================================
# Random forest example with cross-validation
# ====================================================================
from sklearn.ensemble import RandomForestClassifier
from sklearn.model_selection import cross_val_score
#In order to change to accuracy increase n_estimators
# fit model no training data
model = RandomForestClassifier(n_estimators=200)

CV=10 #10-Fold Cross Validation
#Evaluate Model Using 10-Fold Cross Validation and Print Performance
Metrics
Acc_scores = cross_val_score(model, X, y, cv=CV)
print("Accuracy: %0.3f (+/- %0.3f)" % (Acc_scores.mean(), Acc_scores.
std() * 2))
f1_scores = cross_val_score(model, X, y, cv=CV,scoring='f1_macro')
print("F1 score: %0.3f (+/- %0.3f)" % (f1_scores.mean(), f1_scores.std()
* 2))
Precision_scores = cross_val_score(model, X, y, cv=CV,scoring='precision_
macro')
print("Precision score: %0.3f (+/- %0.3f)" % (Precision_scores.mean(),
Precision_scores.std() * 2))
Recall_scores = cross_val_score(model, X, y, cv=CV,scoring='recall_macro')
print("Recall score: %0.3f (+/- %0.3f)" % (Recall_scores.mean(), Recall_
scores.std() * 2))
from sklearn.metrics import cohen_kappa_score, make_scorer
kappa_scorer = make_scorer(cohen_kappa_score)
Kappa_scores = cross_val_score(model, X, y, cv=CV,scoring=kappa_scorer)
```

```
print("Kappa score: %0.3f (+/- %0.3f)" % (Kappa_scores.mean(), Kappa_
scores.std() * 2))
```

4.2.3 局灶性和非局灶性癫痫 EEG 信号的分类

大脑分为多个区域，这些区域会产生突触电流或局部磁场。来自脑电图的脑信号源的定位是近年来的活跃研究领域。这对于研究生理、精神、病理和功能异常以及各种身体残疾，并最终确定如肿瘤和癫痫等异常来源很有帮助。尽管通常情况下脑源的定位是一项棘手的任务，但在一些简单的情况下，可以简化定位工作（Sanei & Chambers, 2013;Subasi, 2019c）。

癫痫病的病灶是由临床诊断来确定的，该诊断基于产生了异常的信号的脑电信号通道。因此，脑电图是癫痫评估的决定性方法。神经外科的主要问题是无创的癫痫初始放电定位，确定包含异常活动源的大脑区域。尽管癫痫的诊断取决于个人病史，但脑电图对于检测和诊断依然至关重要（Subasi, 2019c）。

大脑记录的 EEG 信号有助于我们了解大脑功能。这些记录的目的是对癫痫发作开始的大脑区域进行定位，并评估病人是否可能从这些大脑区域的神经外科切除手术中受益。因此这些类型的癫痫患者的颅内记录是分析信号的挑战性应用领域。其中涉及两种不同类型的信号：第一种类型是从注意到发作性脑电信号变化的大脑区域记录的（"局灶性信号"），第二种类型是从发作开始时未包括的大脑区域记录的（"非局灶性信号"）。发作区在发病时可能会产生发作间期的尖峰，因此覆盖大多数尖峰的聚类开始于最活跃的区域。这个区域的定位和由这些尖峰聚类决定的大脑活动焦点可以为外科评估致痫组织的确切位置提供证据（Subasi, 2019c）。

例 4.3 以下 Python 代码用于从局灶性和非局灶性 EEG 信号中提取特征，使用了小波包分解（WPD）和 WPD 子带的统计值。接下来，它使用各种分类器和 10 折交叉验证对这些数据进行分类。本例计算了分类准确率、精确度、召回率、F1 得分、Cohen kappa 得分和 Matthews 相关系数，并使用了癫痫病源定位数据。

数据集信息：有两种类型的文件——F 和 N，分别涉及局灶性和非局灶性信号对。每个压缩文件包含 750 个独立的文本文件。文件名称中的数字与这个文件中包含的信号对的索引有关。每个文本文件包括一个不同的信号对。第一列涉及 x 信号，第二列涉及 y 信号。这两列由逗号分隔。所有文件都有 10 240 行。后续的行与后续的样本有关。这些文件没有头文件（Andrzejak, Schindler, & Rummel, 2012）。

```python
"""
Created on Thu May 9 12:18:30 2019
@author: asubasi
"""
# descriptive statistics
import scipy as sp
import scipy.io as sio
import pywt
import numpy as np
import scipy.stats as stats
from sklearn.metrics import cohen_kappa_score, make_scorer
from sklearn.model_selection import cross_val_score
wname = pywt.Wavelet('db1')
level=6 #Number of decomposition level
#Load mat file
mat_contents = sio.loadmat('FOCAL_NFOCAL.mat')
sorted(mat_contents.keys())

FOCAL=mat_contents['focal_5000']
NONFOCAL=mat_contents['nfocal_5000']

Labels = [] #Empty List For Labels
NofClasses=2 #Number of Classes
Length = 4096; # Length of signal
Nofsignal=100; #Number of Signal for each Class
numfeatures =83 #Number of features extracted from Wavelet Packet
Decomposition
#Create Empty Array For Features
Extracted_Features=np.ndarray(shape=(NofClasses*Nofsignal,numfeatures),
dtype=float, order='F')
# ==================================================================
# Define utility functions
# ==================================================================
def kFold_Cross_Validation_Metrics(model,CV):
    Acc_scores = cross_val_score(model, X, y, cv=CV)
    print("Accuracy: %0.3f (+/- %0.3f)" % (Acc_scores.mean(), Acc_scores.
    std() * 2))
    f1_scores = cross_val_score(model, X, y, cv=CV,scoring='f1_macro')
    print("F1 score: %0.3f (+/- %0.3f)" % (f1_scores.mean(), f1_scores.std()
    * 2))
    Precision_scores = cross_val_score(model, X, y, cv=CV,scoring='precision_
    macro')
    print("Precision score: %0.3f (+/- %0.3f)" % (Precision_scores.mean(),
    Precision_scores.std() * 2))
    Recall_scores = cross_val_score(model, X, y, cv=CV,scoring='recall_macro')
    print("Recall score: %0.3f (+/- %0.3f)" % (Recall_scores.mean(), Re-
    call_scores.std() * 2))
    kappa_scorer = make_scorer(cohen_kappa_score)
    Kappa_scores = cross_val_score(model, X, y, cv=CV,scoring=kappa_scorer)
    print("Kappa score: %0.3f (+/- %0.3f)" % (Kappa_scores.mean(), Kappa_
```

```python
        scores.std() * 2))
# ==============================================================
# Feature extraction using the statistical values of wavelet packet
transform
# ==============================================================
def WPD_Feature_Extraction(signal, i, wname, level):
    #Mean Values of each subbands
    wp= pywt.WaveletPacket(signal, wname, mode='symmetric', maxlevel=level)
    Extracted_Features[i,0]=sp.mean(abs(wp['a'].data))
    Extracted_Features[i,1]=sp.mean(abs(wp['aa'].data))
    Extracted_Features[i,2]=sp.mean(abs(wp['aaa'].data))
    Extracted_Features[i,3]=sp.mean(abs(wp['aaaa'].data))
    Extracted_Features[i,4]=sp.mean(abs(wp['aaaaa'].data))
    Extracted_Features[i,5]=sp.mean(abs(wp['aaaaaa'].data))
    Extracted_Features[i,6]=sp.mean(abs(wp['d'].data))
    Extracted_Features[i,7]=sp.mean(abs(wp['dd'].data))
    Extracted_Features[i,8]=sp.mean(abs(wp['ddd'].data))
    Extracted_Features[i,9]=sp.mean(abs(wp['dddd'].data))
    Extracted_Features[i,10]=sp.mean(abs(wp['ddddd'].data))
    Extracted_Features[i,11]=sp.mean(abs(wp['dddddd'].data))
    #Standart Deviation of each subbands
    Extracted_Features[i,12]=sp.std(wp['a'].data)
    Extracted_Features[i,13]=sp.std(wp['aa'].data)
    Extracted_Features[i,14]=sp.std(wp['aaa'].data)
    Extracted_Features[i,15]=sp.std(wp['aaaa'].data)
    Extracted_Features[i,16]=sp.std(wp['aaaaa'].data)
    Extracted_Features[i,17]=sp.std(wp['aaaaaa'].data)
    Extracted_Features[i,18]=sp.std(wp['d'].data)
    Extracted_Features[i,19]=sp.std(wp['dd'].data)
    Extracted_Features[i,20]=sp.std(wp['ddd'].data)
    Extracted_Features[i,21]=sp.std(wp['dddd'].data)
    Extracted_Features[i,22]=sp.std(wp['ddddd'].data)
    Extracted_Features[i,23]=sp.std(wp['dddddd'].data)
    #Median Values of each subbands
    Extracted_Features[i,24]=sp.median(wp['a'].data)
    Extracted_Features[i,25]=sp.median(wp['aa'].data)
    Extracted_Features[i,26]=sp.median(wp['aaa'].data)
    Extracted_Features[i,27]=sp.median(wp['aaaa'].data)
    Extracted_Features[i,28]=sp.median(wp['aaaaa'].data)
    Extracted_Features[i,29]=sp.median(wp['aaaaaa'].data)
    Extracted_Features[i,30]=sp.median(wp['d'].data)
    Extracted_Features[i,31]=sp.median(wp['dd'].data)
    Extracted_Features[i,32]=sp.median(wp['ddd'].data)
    Extracted_Features[i,33]=sp.median(wp['dddd'].data)
    Extracted_Features[i,34]=sp.median(wp['ddddd'].data)
    Extracted_Features[i,35]=sp.median(wp['dddddd'].data)
    #Skewness of each subbands
    Extracted_Features[i,36]=stats.skew(wp['a'].data)
    Extracted_Features[i,37]=stats.skew(wp['aa'].data)
    Extracted_Features[i,38]=stats.skew(wp['aaa'].data)
```

```
Extracted_Features[i,39]=stats.skew(wp['aaaa'].data)
Extracted_Features[i,40]=stats.skew(wp['aaaaa'].data)
Extracted_Features[i,41]=stats.skew(wp['aaaaaa'].data)
Extracted_Features[i,42]=stats.skew(wp['d'].data)
Extracted_Features[i,43]=stats.skew(wp['dd'].data)
Extracted_Features[i,44]=stats.skew(wp['ddd'].data)
Extracted_Features[i,45]=stats.skew(wp['dddd'].data)
Extracted_Features[i,46]=stats.skew(wp['ddddd'].data)
Extracted_Features[i,47]=stats.skew(wp['dddddd'].data)
#Kurtosis of each subbands
Extracted_Features[i,48]=stats.kurtosis(wp['a'].data)
Extracted_Features[i,49]=stats.kurtosis(wp['aa'].data)
Extracted_Features[i,50]=stats.kurtosis(wp['aaa'].data)
Extracted_Features[i,51]=stats.kurtosis(wp['aaaa'].data)
Extracted_Features[i,52]=stats.kurtosis(wp['aaaaa'].data)
Extracted_Features[i,53]=stats.kurtosis(wp['aaaaaa'].data)
Extracted_Features[i,54]=stats.kurtosis(wp['d'].data)
Extracted_Features[i,55]=stats.kurtosis(wp['dd'].data)
Extracted_Features[i,56]=stats.kurtosis(wp['ddd'].data)
Extracted_Features[i,57]=stats.kurtosis(wp['dddd'].data)
Extracted_Features[i,58]=stats.kurtosis(wp['ddddd'].data)
Extracted_Features[i,59]=stats.kurtosis(wp['dddddd'].data)
#RMS Values of each subbands
Extracted_Features[i,60]=np.sqrt(np.mean(wp['a'].data**2))
Extracted_Features[i,61]=np.sqrt(np.mean(wp['aa'].data**2))
Extracted_Features[i,62]=np.sqrt(np.mean(wp['aaa'].data**2))
Extracted_Features[i,63]=np.sqrt(np.mean(wp['aaaa'].data**2))
Extracted_Features[i,64]=np.sqrt(np.mean(wp['aaaaa'].data**2))
Extracted_Features[i,65]=np.sqrt(np.mean(wp['aaaaaa'].data**2))
Extracted_Features[i,66]=np.sqrt(np.mean(wp['d'].data**2))
Extracted_Features[i,67]=np.sqrt(np.mean(wp['dd'].data**2))
Extracted_Features[i,68]=np.sqrt(np.mean(wp['ddd'].data**2))
Extracted_Features[i,69]=np.sqrt(np.mean(wp['dddd'].data**2))
Extracted_Features[i,70]=np.sqrt(np.mean(wp['ddddd'].data**2))
Extracted_Features[i,71]=np.sqrt(np.mean(wp['dddddd'].data**2))
#Ratio of subbands
Extracted_Features[i,72]=sp.mean(abs(wp['a'].data))/sp.mean(abs(wp['aa'].
data))
Extracted_Features[i,73]=sp.mean(abs(wp['aa'].data))/sp.mean(abs(wp['aaa'].
data))
Extracted_Features[i,74]=sp.mean(abs(wp['aaa'].data))/sp.mean(abs(wp
['aaaa'].data))
Extracted_Features[i,75]=sp.mean(abs(wp['aaaa'].data))/sp.mean(abs
(wp['aaaaa'].data))
Extracted_Features[i,76]=sp.mean(abs(wp['aaaaa'].data))/sp.mean(abs(wp
['aaaaaa'].data))
Extracted_Features[i,77]=sp.mean(abs(wp['aaaaaa'].data))/sp.mean(abs
(wp['d'].data))
Extracted_Features[i,78]=sp.mean(abs(wp['d'].data))/sp.mean(abs(wp
['dd'].data))
```

```python
    Extracted_Features[i,79]=sp.mean(abs(wp['dd'].data))/sp.mean(abs(wp
['ddd'].data))
    Extracted_Features[i,80]=sp.mean(abs(wp['ddd'].data))/sp.mean(abs
(wp['dddd'].data))
    Extracted_Features[i,81]=sp.mean(abs(wp['dddd'].data))/sp.mean(abs(wp
['ddddd'].data))
    Extracted_Features[i,82]=sp.mean(abs(wp['ddddd'].data))/sp.mean(abs(wp
['dddddd'].data))
# ==========================================================================
# Feature extraction from focal EEG signal
# ==========================================================================
for i in range(Nofsignal):
    WPD_Feature_Extraction(FOCAL[:,i], i, wname, level)
    Labels.append("FOCAL")
# ==========================================================================
# Feature extraction from nonfocal EEG signal
# ==========================================================================
for i in range(Nofsignal, 2*Nofsignal):
    WPD_Feature_Extraction(NONFOCAL[:,i-Nofsignal], i, wname, level)
    Labels.append("NONFOCAL")

#%%
# ==========================================================================
# Classification
# ==========================================================================
X = Extracted_Features
y = Labels
#To prevent warnings
import warnings
warnings.filterwarnings("ignore")
#%%
# ==========================================================================
# Logistic regression example with cross-validation
# ==========================================================================
from sklearn.linear_model import LogisticRegressionCV
CV=10 #10-Fold Cross Validation
#Create the Model
model = LogisticRegressionCV(cv=CV, random_state=0,
            multi_class='multinomial').fit(X, y)
#Evaluate Model Using 10-Fold Cross Validation and Print Performance
Measures
kFold_Cross_Validation_Metrics(model,CV)
#%%
# ==========================================================================
# LDA example with cross-validation
# ==========================================================================
from sklearn.discriminant_analysis import LinearDiscriminantAnalysis
# fit model no training data
#lda = LinearDiscriminantAnalysis(solver="svd", store_covariance=True)
#model = LinearDiscriminantAnalysis(solver='lsqr', shrinkage='auto')
```

```
model = LinearDiscriminantAnalysis(solver='lsqr', shrinkage=None)
CV=10 #10-Fold Cross Validation
#Evaluate Model Using 10-Fold Cross Validation
kFold_Cross_Validation_Metrics(model,CV)
#%%
# ====================================================================
# Naive Bayes example with cross-validation
# ====================================================================
#Import Gaussian Naive Bayes model
from sklearn.naive_bayes import GaussianNB
#Create a Gaussian Classifier
model = GaussianNB()
CV=10 #10-Fold Cross Validation
#Evaluate Model Using 10-Fold Cross Validation and Print Performance
Metrics
kFold_Cross_Validation_Metrics(model,CV)
#%%
# ====================================================================
# MLP example with cross-validation
# ====================================================================
from sklearn.neural_network import MLPClassifier
#Create the Model
model = MLPClassifier(hidden_layer_sizes=(60, ), learning_rate_init=0.001,
         alpha=1, momentum=0.9,max_iter=1000)
CV=10 #10-Fold Cross Validation
#Evaluate Model Using 10-Fold Cross Validation and Print Performance
Metrics
kFold_Cross_Validation_Metrics(model,CV)
#%%
from sklearn.linear_model import LogisticRegression
from elm import ELMClassifier
from random_hidden_layer import RBFRandomHiddenLayer
from random_hidden_layer import SimpleRandomHiddenLayer

nh = 75 #Number of Hidden Layer

# pass user defined transfer func
sinsq = (lambda x: np.power(np.sin(x), 2.0))
srhl_sinsq = SimpleRandomHiddenLayer(n_hidden=nh,
                     activation_func=sinsq,
                     random_state=0)

# use internal transfer funcs
srhl_tanh = SimpleRandomHiddenLayer(n_hidden=nh,
                    activation_func='tanh',
                    random_state=0)
srhl_tribas = SimpleRandomHiddenLayer(n_hidden=nh,
                      activation_func='tribas',
                      random_state=0)
```

```python
srhl_hardlim = SimpleRandomHiddenLayer(n_hidden=nh,
                    activation_func='hardlim',
                    random_state=0)

# use gaussian RBF
#In order to get better accuracy decrease gamma=0.0001
srhl_rbf = RBFRandomHiddenLayer(n_hidden=nh*2, gamma=0.1, random_state=0)

log_reg = LogisticRegression()

#ELMClassifier(srhl_tanh)
#ELMClassifier(srhl_tanh, regressor=log_reg)
#ELMClassifier(srhl_sinsq)
#ELMClassifier(srhl_tribas)
#ELMClassifier(srhl_hardlim)
#ELMClassifier(srhl_rbf)
# fit model no training data
model = ELMClassifier(srhl_rbf)
CV=10 #10-Fold Cross Validation
#Evaluate Model Using 10-Fold Cross Validation and Print Performance Metrics
kFold_Cross_Validation_Metrics(model,CV)
#%%
# ======================================================================
# k-NN example with cross-validation
# ======================================================================
from sklearn.neighbors import KNeighborsClassifier
#Create a Model
model = KNeighborsClassifier(n_neighbors=5)

CV=10 #10-Fold Cross Validation
#Evaluate Model Using 10-Fold Cross Validation and Print Performance Metrics
kFold_Cross_Validation_Metrics(model,CV)

#%%
# ======================================================================
# SVM example with cross-validation
# ======================================================================
from sklearn import svm
""" The parameters and kernels of SVM classifierr can be changed as follows
C = 10.0 # SVM regularization parameter
svm.SVC(kernel='linear', C=C)
svm.LinearSVC(C=C, max_iter=10000)
svm.SVC(kernel='rbf', gamma=0.7, C=C)
svm.SVC(kernel='poly', degree=3, gamma='auto', C=C))
"""
C = 50.0 # SVM regularization parameter
# fit model no training data
model = svm.SVC(kernel='linear', C=C)
```

```
CV=10 #10-Fold Cross Validation
#Evaluate Model Using 10-Fold Cross Validation and Print Performance
Metrics
kFold_Cross_Validation_Metrics(model,CV)
#%%
# ====================================================================
# Decision tree example with cross-validation
# ====================================================================
from sklearn import tree
# fit model no training data
model = tree.DecisionTreeClassifier()

CV=10 #10-Fold Cross Validation
#Evaluate Model Using 10-Fold Cross Validation and Print Performance
Metrics
kFold_Cross_Validation_Metrics(model,CV)

#%%
# ====================================================================
# Extra trees example with cross-validation
# ====================================================================
#Import Extra Trees model
from sklearn.ensemble import ExtraTreesClassifier
# fit model no training data
model = ExtraTreesClassifier(n_estimators=100, max_features=83)

CV=10 #10-Fold Cross Validation
#Evaluate Model Using 10-Fold Cross Validation and Print Performance
Metrics
kFold_Cross_Validation_Metrics(model,CV)
#%%
# ====================================================================
# Bagging example with cross-validation
# ====================================================================
#Import Bagging ensemble model
from sklearn.ensemble import BaggingClassifier
#Import Tree model as a base classifier
from sklearn import tree

"""BaggingClassifier(base_estimator=None,   n_estimators=10,    max_sam-
ples=1.0,
max_features=1.0,   bootstrap=True,   bootstrap_features=False,    oob_
score=False,
warm_start=False, n_jobs=None, random_state=None, verbose=0)"""
#Create a Bagging Ensemble Classifier
model = BaggingClassifier(tree.DecisionTreeClassifier(),
           max_samples=0.5, max_features=0.5)

CV=10 #10-Fold Cross Validation
```

```python
#Evaluate Model Using 10-Fold Cross Validation and Print Performance 
Metrics
kFold_Cross_Validation_Metrics(model,CV)
#%%
# =====================================================================
# Random forest example with cross-validation
# =====================================================================
from sklearn.ensemble import RandomForestClassifier
#In order to change to accuracy increase n_estimators
# fit model no training data
model = RandomForestClassifier(n_estimators=200)

CV=10 #10-Fold Cross Validation
#Evaluate Model Using 10-Fold Cross Validation and Print Performance 
Metrics
kFold_Cross_Validation_Metrics(model,CV)

#%%
#Import Adaboost Ensemble model
from sklearn.ensemble import AdaBoostClassifier
#Import Tree model as a base classifier
from sklearn import tree
#Create an Adaboost Ensemble Classifier
""" AdaBoostClassifier(base_estimator=None, n_estimators=50, learning_
rate=1.0,
            algorithm='SAMME.R', random_state=None)"""
model = clf=AdaBoostClassifier(tree.DecisionTreeClassifier(),
n_estimators=10,
            algorithm='SAMME',learning_rate=0.5)
CV=10 #10-Fold Cross Validation
#Evaluate Model Using 10-Fold Cross Validation and Print Performance 
Metrics
kFold_Cross_Validation_Metrics(model,CV)
#%%
# =====================================================================
# Gradient boosting example with cross-validation
# =====================================================================
#Import Gradient Boosting ensemble model
from sklearn.ensemble import GradientBoostingClassifier
# fit model no training data
model = GradientBoostingClassifier(n_estimators=100, learning_rate=1.0,
                max_depth=1, random_state=0)

CV=10 #10-Fold Cross Validation
#Evaluate Model Using 10-Fold Cross Validation and Print Performance 
Metrics
kFold_Cross_Validation_Metrics(model,CV)
#%%
# =====================================================================
# Voting example with cross-validation
```

```python
# ================================================================
from sklearn.linear_model import LogisticRegression
from sklearn.naive_bayes import GaussianNB
from sklearn.neighbors import KNeighborsClassifier
from sklearn.neural_network import MLPClassifier
from sklearn.ensemble import RandomForestClassifier
from sklearn.ensemble import VotingClassifier
#Create the Model
clf1 = KNeighborsClassifier(n_neighbors=1)
clf2 = RandomForestClassifier(random_state=1)
clf3 = MLPClassifier(hidden_layer_sizes=(100, ), learning_rate_init=0.001,
        alpha=1, momentum=0.9,max_iter=1000)
eclf = VotingClassifier(estimators=[('kNN', clf1), ('RF', clf2), ('MLP',
clf3)], voting='hard')
CV=10 #10-Fold Cross Validation
#Evaluate Model Using 10-Fold Cross Validation and Print Performance
Metrics
kFold_Cross_Validation_Metrics(model,CV)

#%%
from keras.models import Sequential
from keras.layers import Dense, Activation
from keras.utils.np_utils import to_categorical
from keras.utils.vis_utils import plot_model
from keras.wrappers.scikit_learn import KerasClassifier
from sklearn import preprocessing
lb = preprocessing.LabelBinarizer()
y=lb.fit_transform(Labels)
X = Extracted_Features

#from keras.utils import to_categorical
y_binary = to_categorical(y)
#%%
# ================================================================
# Keras DNN example with cross-validation
# ================================================================
# define a Keras model
InputDataDimension=numfeatures
def my_model():
        # create model
        model = Sequential()
        model.add(Dense(128, input_dim=InputDataDimension,
        activation='relu'))
        model.add(Dense(NofClasses, activation='softmax'))
        # Compile model
        model.compile(loss='categorical_crossentropy',
        optimizer='adam', metrics=['accuracy'])
        return model
```

```python
estimator = KerasClassifier(build_fn=my_model, epochs=100, batch_size=5, verbose=1)
from sklearn.model_selection import KFold
kfold = KFold(n_splits=5, shuffle=True)
from sklearn.model_selection import cross_val_score
results = cross_val_score(estimator, X, y_binary, cv=kfold)
print("Accuracy: %.2f%% (%.2f%%)" % (results.mean()*100, results.std()*100))
#%%
# =====================================================================
# ROC analysis for binary classification
# =====================================================================
from sklearn.discriminant_analysis import LinearDiscriminantAnalysis
from sklearn.metrics import roc_curve, auc
from sklearn.model_selection import StratifiedKFold
from scipy import interp
import matplotlib.pyplot as plt
random_state = np.random.RandomState(0)
X = Extracted_Features
y = Labels
######################################################################
from sklearn.preprocessing import label_binarize
y = label_binarize(y, classes=['FOCAL','NONFOCAL' ])
n_classes = y.shape[1]
# Run classifier with cross-validation and plot ROC curves
cv = StratifiedKFold(n_splits=5)
classifier = LinearDiscriminantAnalysis(solver='lsqr', shrinkage=None)

tprs = []
aucs = []
mean_fpr = np.linspace(0, 1, 100)

i = 0
for train, test in cv.split(X, y):
    probas_ = classifier.fit(X[train], y[train]).predict_proba(X[test])
    # Compute ROC curve and area the curve
    fpr, tpr, thresholds = roc_curve(y[test], probas_[:, 1])
    tprs.append(interp(mean_fpr, fpr, tpr))
    tprs[-1][0] = 0.0
    roc_auc = auc(fpr, tpr)
    aucs.append(roc_auc)
    plt.plot(fpr, tpr, lw=1, alpha=0.3,
        label='ROC fold %d (AUC = %0.2f)' % (i, roc_auc))

    i += 1
plt.plot([0, 1], [0, 1], linestyle='--', lw=2, color='r',
 label='Chance', alpha=.8)

mean_tpr = np.mean(tprs, axis=0)
mean_tpr[-1] = 1.0
```

```
mean_auc = auc(mean_fpr, mean_tpr)
std_auc = np.std(aucs)
plt.plot(mean_fpr, mean_tpr, color='b',
label=r'Mean ROC (AUC = %0.2f $\pm$ %0.2f)' % (mean_auc, std_
auc),
lw=2, alpha=.8)
std_tpr = np.std(tprs, axis=0)
tprs_upper = np.minimum(mean_tpr + std_tpr, 1)
tprs_lower = np.maximum(mean_tpr - std_tpr, 0)
plt.fill_between(mean_fpr,    tprs_lower,    tprs_upper,
color='grey', alpha=.2,
 label=r'$\pm$ 1 std. dev.')
plt.xlim([-0.05, 1.05])
plt.ylim([-0.05, 1.05])
plt.xlabel('False Positive Rate')
plt.ylabel('True Positive Rate')
plt.title('Receiver operating characteristic example')
plt.legend(loc="lower right")
plt.show()
```

4.2.4 偏头痛检测

偏头痛是一种持续的神经系统疾病。其重大指标包括：大脑一侧或两侧的搏动性疼痛及对光线的敏感性。值得注意的是，它是第三大普遍的疾病，每七个人中就有一个受到影响。值得一提的是，偏头痛一般被评为第七大致残性疾病，在神经系统疾病中排名第一。但偏头痛常常会被误诊，因为它的症状与其他疾病如紧张性头痛、癫痫和中风相似。在过去的几十年里，许多研究都围绕高度精确地识别偏头痛而进行。这些研究中的一项特定技术已呈现出良好的表现，这就是闪光刺激。该方法的基础是分析受试者的神经反应，在不同频率的闪光刺激下，用多通道 EEG 记录，时间长短不一。文献中的一项研究发现闪光频率为 4Hz 是最准确的（Akben, Subasi, & Tuncel, 2012）（Subasi, Ahmed, Aličković, & Rashik Hassan, 2019）。偏头痛检测的一般框架如图 4.5 所示。

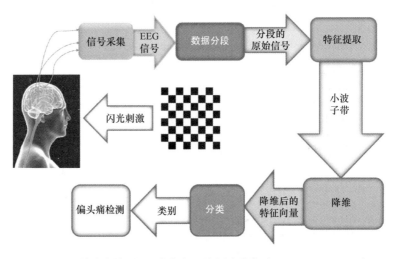

图 4.5 偏头痛检测的一般框架。该图改编自（Subasi et al., 2019d）

例 4.4 以下 Python 代码通过采用小波包分解（WPD）从健康和偏头痛脑电信号中提取特征。接下来，它使用 WPD 子带的统计值，通过单独的训练和测试数据集，利用额外的树分类器对这些数据进行分类。本例计算了分类准确率、精确度、召回率、F1 得分、Cohen kappa 得分和 Matthews 相关系数，还给出了分类报告和混淆矩阵。

数据集信息：数据由一台 18 通道（10-20 系统）的 Nicolet One 机器来记录。脑电图数据是由 Kahramanmaras Sutcu Imam University 的神经内科系召集的偏头痛患者和健康受试者提供的。数据集可以描述如下：

- 共有 15 位偏头痛患者（2 位男性，13 位女性，无先兆），年龄 20 ~ 34 岁（平均年龄 ± 标准差为 27 ± 4.4 岁，根据国际头痛协会（IHS）提出的标准诊断，还有 15 位（男性 5 位，女性 10 位）年龄 19 ~ 35 岁（平均年龄 ± 标准差为 26 ± 5.3 岁）的对照受试者参加了实验
- 实验是在昏暗的房间里进行的
- 记录前没有参与者服用任何药物
- 所有参与者都处于发作间歇期（无痛）状态，同时处于伏卧位
- 使用 10-20 EEG 系统收集 EEG 数据
- 以 256 Hz 的采样频率收集 EEG 信号
- 每次刺激的时间为 30 秒，频率为 4Hz

脑电图频率范围从 0 到 100Hz（Niedermeyer & da Silva，2005）。此外，脑电图频段 delta（1 ~ 4Hz）、theta（4 ~ 8Hz）、alpha（8 ~ 13Hz）、beta（13 ~ 30Hz）和 gamma（30 ~ 100Hz）都在 0 ~ 100Hz 范围内。根据 Nyquist 定理，为了从脑电图信号中获取所有有用的信息，采样率应该至少为 200Hz。请注意，本研究中使用的采样率是 256Hz，明显高于所需的 200Hz。Akben 等人（2012）和 Subasi 等人（2019d）使用了这个数据集，其中考虑了不同的分析策略。

```
"""
Created on Thu May 9 12:18:30 2019
@author: asubasi
"""
# =================================================================
# Feature extraction using the statistical values of wavelet packet
transform
# =================================================================
# descriptive statistics
import scipy as sp
import scipy.io as sio
import pywt
import numpy as np
import scipy.stats as stats
```

```python
wname = pywt.Wavelet('db1')
level=6 #Number of decomposition level
#Load mat file
mat_contents = sio.loadmat('Migraine.mat')
sorted(mat_contents.keys())
HEALTHY=mat_contents['Healthy']
MIGRAINE=mat_contents['Migraineur']
Labels = [] #Empty List For Labels
NofClasses=2 #Number of Classes
Length = 768; # Length of signal
Nofsignal=135; #Number of Signal for each Class
numrows =83 #Number of features extracted from Wavelet Packet Decomposition
#Create Empty Array For Features
Extracted_Features=np.ndarray(shape=(NofClasses*Nofsignal,numrows),
dtype=float, order='F')
# =================================================================
# Utility function for feature extraction
# =================================================================
def WPD_Feature_Extraction(signal, i, wname, level):
#Mean Values of each subbands
  wp= pywt.WaveletPacket(signal, wname, mode='symmetric', maxlevel=level)
  Extracted_Features[i,0]=sp.mean(abs(wp['a'].data))
  Extracted_Features[i,1]=sp.mean(abs(wp['aa'].data))
  Extracted_Features[i,2]=sp.mean(abs(wp['aaa'].data))
  Extracted_Features[i,3]=sp.mean(abs(wp['aaaa'].data))
  Extracted_Features[i,4]=sp.mean(abs(wp['aaaaa'].data))
  Extracted_Features[i,5]=sp.mean(abs(wp['aaaaaa'].data))
  Extracted_Features[i,6]=sp.mean(abs(wp['d'].data))
  Extracted_Features[i,7]=sp.mean(abs(wp['dd'].data))
  Extracted_Features[i,8]=sp.mean(abs(wp['ddd'].data))
  Extracted_Features[i,9]=sp.mean(abs(wp['dddd'].data))
  Extracted_Features[i,10]=sp.mean(abs(wp['ddddd'].data))
  Extracted_Features[i,11]=sp.mean(abs(wp['dddddd'].data))
  #Standart Deviation of each subbands
  Extracted_Features[i,12]=sp.std(wp['a'].data)
  Extracted_Features[i,13]=sp.std(wp['aa'].data)
  Extracted_Features[i,14]=sp.std(wp['aaa'].data)
  Extracted_Features[i,15]=sp.std(wp['aaaa'].data)
  Extracted_Features[i,16]=sp.std(wp['aaaaa'].data)
  Extracted_Features[i,17]=sp.std(wp['aaaaaa'].data)
  Extracted_Features[i,18]=sp.std(wp['d'].data)
  Extracted_Features[i,19]=sp.std(wp['dd'].data)
  Extracted_Features[i,20]=sp.std(wp['ddd'].data)
  Extracted_Features[i,21]=sp.std(wp['dddd'].data)
  Extracted_Features[i,22]=sp.std(wp['ddddd'].data)
  Extracted_Features[i,23]=sp.std(wp['dddddd'].data)
  #Median Values of each subbands
  Extracted_Features[i,24]=sp.median(wp['a'].data)
  Extracted_Features[i,25]=sp.median(wp['aa'].data)
```

```
Extracted_Features[i,26]=sp.median(wp['aaa'].data)
Extracted_Features[i,27]=sp.median(wp['aaaa'].data)
Extracted_Features[i,28]=sp.median(wp['aaaaa'].data)
Extracted_Features[i,29]=sp.median(wp['aaaaaa'].data)
Extracted_Features[i,30]=sp.median(wp['d'].data)
Extracted_Features[i,31]=sp.median(wp['dd'].data)
Extracted_Features[i,32]=sp.median(wp['ddd'].data)
Extracted_Features[i,33]=sp.median(wp['dddd'].data)
Extracted_Features[i,34]=sp.median(wp['ddddd'].data)
Extracted_Features[i,35]=sp.median(wp['dddddd'].data)
#Skewness of each subbands
Extracted_Features[i,36]=stats.skew(wp['a'].data)
Extracted_Features[i,37]=stats.skew(wp['aa'].data)
Extracted_Features[i,38]=stats.skew(wp['aaa'].data)
Extracted_Features[i,39]=stats.skew(wp['aaaa'].data)
Extracted_Features[i,40]=stats.skew(wp['aaaaa'].data)
Extracted_Features[i,41]=stats.skew(wp['aaaaaa'].data)
Extracted_Features[i,42]=stats.skew(wp['d'].data)
Extracted_Features[i,43]=stats.skew(wp['dd'].data)
Extracted_Features[i,44]=stats.skew(wp['ddd'].data)
Extracted_Features[i,45]=stats.skew(wp['dddd'].data)
Extracted_Features[i,46]=stats.skew(wp['ddddd'].data)
Extracted_Features[i,47]=stats.skew(wp['dddddd'].data)
#Kurtosis of each subbands
Extracted_Features[i,48]=stats.kurtosis(wp['a'].data)
Extracted_Features[i,49]=stats.kurtosis(wp['aa'].data)
Extracted_Features[i,50]=stats.kurtosis(wp['aaa'].data)
Extracted_Features[i,51]=stats.kurtosis(wp['aaaa'].data)
Extracted_Features[i,52]=stats.kurtosis(wp['aaaaa'].data)
Extracted_Features[i,53]=stats.kurtosis(wp['aaaaaa'].data)
Extracted_Features[i,54]=stats.kurtosis(wp['d'].data)
Extracted_Features[i,55]=stats.kurtosis(wp['dd'].data)
Extracted_Features[i,56]=stats.kurtosis(wp['ddd'].data)
Extracted_Features[i,57]=stats.kurtosis(wp['dddd'].data)
Extracted_Features[i,58]=stats.kurtosis(wp['ddddd'].data)
Extracted_Features[i,59]=stats.kurtosis(wp['dddddd'].data)
#RMS Values of each subbands
Extracted_Features[i,60]=np.sqrt(np.mean(wp['a'].data**2))
Extracted_Features[i,61]=np.sqrt(np.mean(wp['aa'].data**2))
Extracted_Features[i,62]=np.sqrt(np.mean(wp['aaa'].data**2))
Extracted_Features[i,63]=np.sqrt(np.mean(wp['aaaa'].data**2))
Extracted_Features[i,64]=np.sqrt(np.mean(wp['aaaaa'].data**2))
Extracted_Features[i,65]=np.sqrt(np.mean(wp['aaaaaa'].data**2))
Extracted_Features[i,66]=np.sqrt(np.mean(wp['d'].data**2))
Extracted_Features[i,67]=np.sqrt(np.mean(wp['dd'].data**2))
Extracted_Features[i,68]=np.sqrt(np.mean(wp['ddd'].data**2))
Extracted_Features[i,69]=np.sqrt(np.mean(wp['dddd'].data**2))
Extracted_Features[i,70]=np.sqrt(np.mean(wp['ddddd'].data**2))
Extracted_Features[i,71]=np.sqrt(np.mean(wp['dddddd'].data**2))
#Ratio of subbands
```

```python
    Extracted_Features[i,72]=sp.mean(abs(wp['a'].data))/sp.mean(abs(wp['aa'].
    data))
    Extracted_Features[i,73]=sp.mean(abs(wp['aa'].data))/sp.mean(abs
    (wp['aaa'].data))
    Extracted_Features[i,74]=sp.mean(abs(wp['aaa'].data))/sp.mean(abs(wp
    ['aaaa'].data))
    Extracted_Features[i,75]=sp.mean(abs(wp['aaaa'].data))/sp.mean(abs(wp
    ['aaaaa'].data))
    Extracted_Features[i,76]=sp.mean(abs(wp['aaaaa'].data))/sp.mean(abs(wp
    ['aaaaaa'].data))
    Extracted_Features[i,77]=sp.mean(abs(wp['aaaaaa'].data))/sp.mean(abs
    (wp['d'].data))
    Extracted_Features[i,78]=sp.mean(abs(wp['d'].data))/sp.mean(abs(wp
    ['dd'].data))
    Extracted_Features[i,79]=sp.mean(abs(wp['dd'].data))/sp.mean(abs(wp
    ['ddd'].data))
    Extracted_Features[i,80]=sp.mean(abs(wp['ddd'].data))/sp.mean(abs
    (wp['dddd'].data))
    Extracted_Features[i,81]=sp.mean(abs(wp['dddd'].data))/sp.mean(abs(wp
    ['ddddd'].data))
    Extracted_Features[i,82]=sp.mean(abs(wp['ddddd'].data))/sp.mean(abs(wp
    ['dddddd'].data))
#%%
# =================================================================
# Feature extraction from healthy EEG signal
# =================================================================
for i in range(Nofsignal):
    WPD_Feature_Extraction(HEALTHY[:,i], i, wname, level)
    Labels.append("HEALTHY")
# =================================================================
# Feature extraction from migraine EEG signal
# =================================================================
for i in range(Nofsignal, 2*Nofsignal):
    WPD_Feature_Extraction(MIGRAINE[:,i-Nofsignal], i, wname, level)
    Labels.append("MIGRAINE")
#%%
# =================================================================
# Classification
# =================================================================
X = Extracted_Features
y = Labels
# Import train_test_split function
from sklearn.model_selection import train_test_split
# Split dataset into training set and test set
# 70% training and 30% test
Xtrain, Xtest, ytrain, ytest = train_test_split(X, y,test_size=0.3,
random_state=1)
#%%
# =================================================================
```

```
# Extra trees classification example with training and test set
# =====================================================================
#Import Extra Trees model
from sklearn.ensemble import ExtraTreesClassifier
# Split dataset into training set and test set
# 70% training and 30% test
Xtrain, Xtest, ytrain, ytest = train_test_split(X, y,test_size=0.3,
random_state=0)
#Create the Model
clf = model = ExtraTreesClassifier(n_estimators=100, max_features=48)
#Train the model using the training sets
clf.fit(Xtrain,ytrain)
#Predict the response for test dataset
ypred = clf.predict(Xtest)

#Evaluate the Model and Print Performance Metrics
from sklearn import metrics
print('Accuracy:', np.round(metrics.accuracy_score(ytest,ypred),4))
print('Precision:', np.round(metrics.precision_score(ytest,
                  ypred,average='weighted'),4))
print('Recall:', np.round(metrics.recall_score(ytest,ypred,
                  average='weighted'),4))
print('F1 Score:', np.round(metrics.f1_score(ytest,ypred,
                  average='weighted'),4))
print('Cohen Kappa Score:', np.round(metrics.cohen_kappa_score(ytest,
ypred),4))
print('Matthews Corrcoef:', np.round(metrics.matthews_corrcoef(ytest,
ypred),4))
print('\t\tClassification Report:\n', metrics.classification_report(ypred,
ytest))
#Plot Confusion Matrix
from sklearn.metrics import confusion_matrix
print("Confusion Matrix:\n",confusion_matrix(ytest, ypred))
```

4.3 EMG 信号分析

人类骨骼肌系统的主要职责是提供进行一些活动所需的力量。这个系统包含两个子系统——神经系统和肌肉系统,它们共同构成了神经肌肉系统(Begg et al., 2008)。肢体的运动和安排是由肌肉和外周及中枢神经系统之间来回传播的电信号控制的(Bronzino, 1999)。神经可以理解为传导电流的导线。神经头(核)开始于脊柱,它们的长轴突体扩大到远处和深处,刺激各种肌肉中的单个运动单元。骨骼肌系统由通过肌腱连接到骨骼的肌肉群组成,一旦神经产生肌肉收缩和放松信号,吸引或排斥骨骼,运动就完成了(Begg et al., 2008;Subasi, 2019c)。

肌电图(EMG)信号代表骨骼肌的电状态,并包括与肌肉结构相关的数据(用于激活各个身体部位)。EMG 信号携带与中枢和外周神经系统对肌肉的控制作用有关的数据。当然,

EMG 信号对神经肌肉系统的描述是非常有用的，因为不管是出现在神经系统还是肌肉中的各种病理迹象，都可以通过信号特征的变化看到。近几十年来，随着新的信号处理技术的发展，用以了解肌肉电生理学的 EMG 信号分析质量有了很大的提高（Sörnmo & Laguna, 2005）。EMG 分析在确定异常肌肉纤维类型方面比临床试验更准确。即使病人仍有正常的运动-神经传导，EMG 分析依然可以显示不规则的感觉-神经现象。此外，EMG 分析可以帮助医生诊断疾病，而不需要进行肌肉活检（Begg et al., 2008）。工程技术的进步，使得肌电图检查可以考虑常规诊断程序之外进行不同领域的应用，例如康复学、人体工程学、运动分析、运动生理学、生物反馈以及假肢的肌电控制（Sörnmo & Laguna, 2005;Subasi, 2019c）。

4.3.1 神经肌肉疾病的诊断

神经肌肉疾病是最初发生在神经系统、神经肌肉连接处和肌肉纤维中的异常疾病。这些异常的严重程度不同，从可以忽略的肌肉损伤到由神经元或肌肉死亡引起的截肢。对于更严重的疾病，如肌萎缩侧索硬化症（ALS），通常会导致死亡。尽早和正确的诊断对于改善预后和提高完全康复的概率至关重要。在大多数情况下，临床试验不足以识别和防止疾病的蔓延，因为一个特定的症状可能会导致很多不同的异常情况。因此，对疾病的准确诊断具有极其重要的意义，这样才能建立起更有决定性的治疗。目前，电诊断研究包括神经传导研究，EMG 用于评估和识别神经肌肉疾病患者。EMG 最初被用作检查在整个肌肉活动过程中细胞活动能力形成的神经肌肉状态。EMG 波形的特点（平均 0.01 ~ 10 mV 和 10 ~ 2000 Hz）可以表明位置、病因和解剖学类型。例如，EMG 信号间隔显示了肌肉纤维的位置和网络代谢状态（Emly, Gilmore, & Roy, 1992），不规则的峰值会显示肌肉疾病。另一方面，电诊断方法有助于医生进行疾病诊断，但对确诊几乎没有帮助，在复杂的情况下，更具侵入性的工具（肌肉活检）或更先进的成像方法（如超声波或核磁共振）不可或缺。EMG 的解释通常由训练有素的专业神经学家完成，他们除了检查 EMG 波形外，还积极应用与针传导研究和肌肉声学有关的方法。当几乎没有专家满足病人的需求时，问题就出现了，因此，开发基于 EMG 读数的自动诊断系统非常重要。相关的机器学习技术可用于基于 EMG 处理的神经肌肉疾病的检测和分类。这些智能系统将帮助医生检测神经肌肉系统的异常情况。智能诊断和人工控制神经肌肉系统的目的是对原始 EMG 信号进行预处理，从而提取特征数据或特征。提取的特征包括时域和频域数据、傅里叶系数、自回归系数、小波系数以及从其他信号处理方法得出的各种量。然后，这些数据可以作为决策树和支持向量机等分类器的输入数据，对神经肌肉疾病进行分类。神经肌肉疾病通常是周围神经系统的异常，它们可以根据疾病的位置和原因进行分类。两种主要的疾病是神经病变和肌肉疾病（Begg et al., 2008;Subasi, 2019c）。

神经病变是一个用以识别引起疼痛和残疾的神经紊乱的术语。神经性疾病的原因是不同的，包括受伤、感染、糖尿病、酗酒和癌症化疗。肌肉疾病通常是与骨骼肌有关的疾病，由肌肉群损伤或一些基因突变引起的。肌肉疾病阻碍了受影响肌肉的正常功能。因此，患有肌肉疾病的病人肌肉无力，根据疾病的严重程度，或在执行日常生活任务中有问题，或者发现不利用受影响的肌肉就无法做任何动作（Begg et al., 2008; Subasi, 2019c）。

针式肌电图是用于诊断神经肌肉病理的典型临床记录方法。例如，当病人因肌肉无力去看医生时，医生会记录特定肌肉收缩时的针状肌电图。这些数据可能有助于识别在肌肉疼痛、手臂和腿部神经受伤、神经受压和肌肉萎缩等情况下发生的不规则活动。针式肌电图也可以与神经损伤一起检查，并可用于确定损伤是否正在恢复或是否已恢复到正常状态而具有完整的肌肉反应性，例如，通过分析特定时间段内运动单位成就的改变。诊断性 EMG 包括检查意外的运动动作，这可能发生在肌肉放松期间。在正常情况下，肌肉放松时在电方面是安静的，否则会产生不规则的自发波形和波形模式，这与自发的肌肉活动有关（Sörnmo & Laguna, 2005;Subasi, 2019c）。使用 EMG 信号诊断神经肌肉疾病的一般框架如图 4.6 所示。

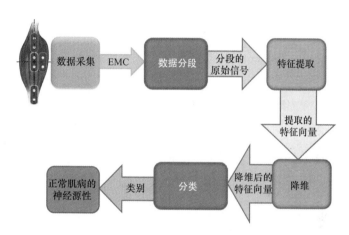

图 4.6　基于 EMG 信号诊断神经肌肉疾病的一般框架

例 4.5　下面的 Python 代码用于通过平稳小波变换（SWT）和利用 SWT 子带的统计值从 EMG 信号中提取特征。然后使用具有单独的训练和测试数据集的 SVM 分类器对这些数据进行分类。本例计算了分类准确率、精确度、召回率、F1 得分、Cohen kappa 得分和 Matthews 相关系数，还给出了分类报告、混淆矩阵和 ROC 区域。

数据集信息：EMG 信号取自 EMGLAB 网站（http://www.emglab.net/）。临床 EMG 信号是在正常条件下获取的，用于 MUAP 分析。EMG 信号是在低度自愿和恒定收缩水平下用标准的同心针电极记录的。EMG 信号在 2Hz 和 10kHz 之间被过滤，包括一个对照组和一组 ALS 和肌病患者。对照组有 10 名正常受试者（4 名女性和 6 名男性），年龄在 21～37 岁。ALS 组有 8 名患者（4 名女性和 4 名男性），年龄在 35～67 岁。肌病组有 7 名患者（2 名女性和 5 名男性），年龄在 19～63 岁（Nikolic, 2001）。可以从以下网站下载数据：http://www.emglab.net/emglab/Signals/signals.php。

```
"""
Created on Thu May 9 12:18:30 2019
```

```
@author: asubasi
"""
# ====================================================================
# Feature extraction using the statistical values of stationary wavelet
transform
# ====================================================================
import scipy.io as sio
# descriptive statistics
import scipy as sp
import pywt
import matplotlib.pyplot as plt
import numpy as np
import scipy.stats as stats
waveletname='db1'
level=6 #Decomposition Level
#Load mat file
mat_contents = sio.loadmat('EMGDAT.mat')
sorted(mat_contents.keys())

CONTROL=mat_contents['CON']
ALS=mat_contents['ALS']
MYOPATHIC=mat_contents['MYO']

Labels = [] #Empty List For Labels
Length = 8192; # Length of signal
Nofsignal=200; #Number of Signal
NofClasses=3; #Number of Classes
numrows =83 #Number of features extracted from Wavelet Packet
Decomposition
#Create Empty Array For Features
Extracted_Features=np.ndarray(shape=(NofClasses*Nofsignal,numrows),
dtype=float, order='F')
# ====================================================================
# Define utility functions for feature extraction
# ====================================================================
def SWT_Feature_Extraction(signal, i, wname, level):
    coeffs = pywt.swt(signal, wname, level=level)
    #Mean Values of each subbands
    Extracted_Features[i,0]=sp.mean(abs(coeffs[0][0]))
    Extracted_Features[i,1]=sp.mean(abs(coeffs[1][0]))
    Extracted_Features[i,2]=sp.mean(abs(coeffs[2][0]))
    Extracted_Features[i,3]=sp.mean(abs(coeffs[3][0]))
    Extracted_Features[i,4]=sp.mean(abs(coeffs[4][0]))
    Extracted_Features[i,5]=sp.mean(abs(coeffs[5][0]))
    Extracted_Features[i,6]=sp.mean(abs(coeffs[0][1]))
    Extracted_Features[i,7]=sp.mean(abs(coeffs[1][1]))
    Extracted_Features[i,8]=sp.mean(abs(coeffs[2][1]))
    Extracted_Features[i,9]=sp.mean(abs(coeffs[3][1]))
    Extracted_Features[i,10]=sp.mean(abs(coeffs[4][1]))
    Extracted_Features[i,11]=sp.mean(abs(coeffs[5][1]))
```

```
#Standart Deviation of each subbands
Extracted_Features[i,12]=sp.std(coeffs[0][0])
Extracted_Features[i,13]=sp.std(coeffs[1][0])
Extracted_Features[i,14]=sp.std(coeffs[2][0])
Extracted_Features[i,15]=sp.std(coeffs[3][0])
Extracted_Features[i,16]=sp.std(coeffs[4][0])
Extracted_Features[i,17]=sp.std(coeffs[5][0])
Extracted_Features[i,18]=sp.std(coeffs[0][1])
Extracted_Features[i,19]=sp.std(coeffs[1][1])
Extracted_Features[i,20]=sp.std(coeffs[2][1])
Extracted_Features[i,21]=sp.std(coeffs[3][1])
Extracted_Features[i,22]=sp.std(coeffs[4][1])
Extracted_Features[i,23]=sp.std(coeffs[5][1])
#Median Values of each subbands
Extracted_Features[i,24]=sp.median(coeffs[0][0])
Extracted_Features[i,25]=sp.median(coeffs[1][0])
Extracted_Features[i,26]=sp.median(coeffs[2][0])
Extracted_Features[i,27]=sp.median(coeffs[3][0])
Extracted_Features[i,28]=sp.median(coeffs[4][0])
Extracted_Features[i,29]=sp.median(coeffs[5][0])
Extracted_Features[i,30]=sp.median(coeffs[0][1])
Extracted_Features[i,31]=sp.median(coeffs[1][1])
Extracted_Features[i,32]=sp.median(coeffs[2][1])
Extracted_Features[i,33]=sp.median(coeffs[3][1])
Extracted_Features[i,34]=sp.median(coeffs[4][1])
Extracted_Features[i,35]=sp.median(coeffs[5][1])
#Skewness of each subbands
Extracted_Features[i,36]=stats.skew(coeffs[0][0])
Extracted_Features[i,37]=stats.skew(coeffs[1][0])
Extracted_Features[i,38]=stats.skew(coeffs[2][0])
Extracted_Features[i,39]=stats.skew(coeffs[3][0])
Extracted_Features[i,40]=stats.skew(coeffs[4][0])
Extracted_Features[i,41]=stats.skew(coeffs[5][0])
Extracted_Features[i,42]=stats.skew(coeffs[0][1])
Extracted_Features[i,43]=stats.skew(coeffs[1][1])
Extracted_Features[i,44]=stats.skew(coeffs[2][1])
Extracted_Features[i,45]=stats.skew(coeffs[3][1])
Extracted_Features[i,46]=stats.skew(coeffs[4][1])
Extracted_Features[i,47]=stats.skew(coeffs[5][1])
#Kurtosis of each subbands
Extracted_Features[i,48]=stats.kurtosis(coeffs[0][0])
Extracted_Features[i,49]=stats.kurtosis(coeffs[1][0])
Extracted_Features[i,50]=stats.kurtosis(coeffs[2][0])
Extracted_Features[i,51]=stats.kurtosis(coeffs[3][0])
Extracted_Features[i,52]=stats.kurtosis(coeffs[4][0])
Extracted_Features[i,53]=stats.kurtosis(coeffs[5][0])
Extracted_Features[i,54]=stats.kurtosis(coeffs[0][1])
Extracted_Features[i,55]=stats.kurtosis(coeffs[1][1])
Extracted_Features[i,56]=stats.kurtosis(coeffs[2][1])
Extracted_Features[i,57]=stats.kurtosis(coeffs[3][1])
```

```
    Extracted_Features[i,58]=stats.kurtosis(coeffs[4][1])
    Extracted_Features[i,59]=stats.kurtosis(coeffs[5][1])
    #RMS Values of each subbands
    Extracted_Features[i,60]=np.sqrt(np.mean(coeffs[0][0]**2))
    Extracted_Features[i,61]=np.sqrt(np.mean(coeffs[1][0]**2))
    Extracted_Features[i,62]=np.sqrt(np.mean(coeffs[2][0]**2))
    Extracted_Features[i,63]=np.sqrt(np.mean(coeffs[3][0]**2))
    Extracted_Features[i,64]=np.sqrt(np.mean(coeffs[4][0]**2))
    Extracted_Features[i,65]=np.sqrt(np.mean(coeffs[5][0]**2))
    Extracted_Features[i,66]=np.sqrt(np.mean(coeffs[0][1]**2))
    Extracted_Features[i,67]=np.sqrt(np.mean(coeffs[1][1]**2))
    Extracted_Features[i,68]=np.sqrt(np.mean(coeffs[2][1]**2))
    Extracted_Features[i,69]=np.sqrt(np.mean(coeffs[3][1]**2))
    Extracted_Features[i,70]=np.sqrt(np.mean(coeffs[4][1]**2))
    Extracted_Features[i,71]=np.sqrt(np.mean(coeffs[5][1]**2))
    #Ratio of subbands
    Extracted_Features[i,72]=sp.mean(abs(coeffs[0][0]))/sp.mean(abs(coeffs[1][0]))
    Extracted_Features[i,73]=sp.mean(abs(coeffs[1][0]))/sp.mean(abs(coeffs[2][0]))
    Extracted_Features[i,74]=sp.mean(abs(coeffs[2][0]))/sp.mean(abs(coeffs[3][0]))
    Extracted_Features[i,75]=sp.mean(abs(coeffs[3][0]))/sp.mean(abs(coeffs[4][0]))
    Extracted_Features[i,76]=sp.mean(abs(coeffs[4][0]))/sp.mean(abs(coeffs[5][0]))
    Extracted_Features[i,77]=sp.mean(abs(coeffs[5][0]))/sp.mean(abs(coeffs[0][1]))
    Extracted_Features[i,78]=sp.mean(abs(coeffs[0][1]))/sp.mean(abs(coeffs[1][1]))
    Extracted_Features[i,79]=sp.mean(abs(coeffs[1][1]))/sp.mean(abs(coeffs[2][1]))
    Extracted_Features[i,80]=sp.mean(abs(coeffs[2][1]))/sp.mean(abs(coeffs[3][1]))
    Extracted_Features[i,81]=sp.mean(abs(coeffs[3][1]))/sp.mean(abs(coeffs[4][1]))
    Extracted_Features[i,82]=sp.mean(abs(coeffs[4][1]))/sp.mean(abs(coeffs[5][1]))
#%%
# ==================================================================
# Feature extraction from control EMG signal
# ==================================================================
for i in range(Nofsignal):
    SWT_Feature_Extraction(CONTROL[:,i], i, waveletname, level)
    Labels.append("CONTROL")
# ==================================================================
# Feature extraction from ALS EMG signal
# ==================================================================
for i in range(Nofsignal, 2*Nofsignal):
    SWT_Feature_Extraction(ALS[:,i-Nofsignal], i, waveletname, level)
```

```python
    Labels.append("ALS")
# ================================================================
# Feature extraction from myopathic EMG signal
# ================================================================
for i in range(2*Nofsignal, 3*Nofsignal):
    SWT_Feature_Extraction(MYOPATHIC[:,i-2*Nofsignal], i, waveletname, level)
    Labels.append("MYOPATHIC")
#%%
# ================================================================
# Classification
# ================================================================
X = Extracted_Features
y = Labels
# Import train_test_split function
from sklearn.model_selection import train_test_split
# Split dataset into training set and test set
# 70% training and 30% test
Xtrain, Xtest, ytrain, ytest = train_test_split(X, y,test_size=0.3,
random_state=1)
#%%
# ================================================================
# SVM example with training and test set
# ================================================================
from sklearn import svm
    """ The parameters and kernels of SVM classifierr can be changed as
follows
C = 10.0 # SVM regularization parameter
svm.SVC(kernel='linear', C=C)
svm.LinearSVC(C=C, max_iter=10000)
svm.SVC(kernel='rbf', gamma=0.7, C=C)
svm.SVC(kernel='poly', degree=3, gamma='auto', C=C)
"""
C = 10.0 # SVM regularization parameter
#Create the Model
clf =svm.SVC(kernel='linear', C=C)
#Train the model with Training Dataset
clf.fit(Xtrain,ytrain)
#Test the model with Testset
ypred = clf.predict(Xtest)
#Evaluate the Model and Print Performance Metrics
from sklearn import metrics
print('Accuracy:', np.round(metrics.accuracy_score(ytest,ypred),4))
print('Precision:', np.round(metrics.precision_score(ytest,
                    ypred,average='weighted'),4))
print('Recall:', np.round(metrics.recall_score(ytest,ypred,
                        average='weighted'),4))
print('F1 Score:', np.round(metrics.f1_score(ytest,ypred,
                        average='weighted'),4))
print('Cohen Kappa Score:', np.round(metrics.cohen_kappa_score(ytest,
ypred),4))
```

```python
print('Matthews    Corrcoef:',    np.round(metrics.matthews_corrcoef(ytest,
ypred),4))
print('\t\tClassification Report:\n', metrics.classification_report(ypred,
ytest))
#Plot Confusion Matrix
from sklearn.metrics import confusion_matrix
print("Confusion Matrix:\n",confusion_matrix(ytest, ypred))
#%%
# ==================================================================
# ROC curves for the multiclass problem
# ==================================================================
import numpy as np
import matplotlib.pyplot as plt
from itertools import cycle
from sklearn.metrics import roc_curve, auc
from sklearn.model_selection import train_test_split
from sklearn.preprocessing import label_binarize
from sklearn.multiclass import OneVsRestClassifier
from scipy import interp
# Binarize the output
y = label_binarize(y, classes=['CONTROL','ALS','MYOPATHIC' ])
n_classes = y.shape[1]
# shuffle and split training and test sets
Xtrain, Xtest, ytrain, ytest = train_test_split(X, y,test_size=0.3,
random_state=1)
# Learn to predict each class against the other
classifier = OneVsRestClassifier(svm.SVC(kernel='linear', C=C))
yscore = classifier.fit(Xtrain, ytrain).decision_function(Xtest)
# Compute ROC curve and ROC area for each class
fpr = dict()
tpr = dict()
roc_auc = dict()
for i in range(n_classes):
    fpr[i], tpr[i], _ = roc_curve(ytest[:, i], yscore[:, i])
    roc_auc[i] = auc(fpr[i], tpr[i])
# Compute micro-average ROC curve and ROC area
fpr["micro"], tpr["micro"], _ = roc_curve(ytest.ravel(), yscore.ravel())
roc_auc["micro"] = auc(fpr["micro"], tpr["micro"])
################################################################
# Plot of a ROC curve for a specific class
plt.figure()
lw = 2
plt.plot(fpr[2], tpr[2], color='darkorange',
      lw=lw, label='ROC curve (area = %0.2f)' % roc_auc[2])
plt.plot([0, 1], [0, 1], color='navy', lw=lw, linestyle='--')
plt.xlim([0.0, 1.0])
plt.ylim([0.0, 1.05])
plt.xlabel('False Positive Rate')
plt.ylabel('True Positive Rate')
plt.title('Receiver operating characteristic')
```

```python
plt.legend(loc="lower right")
plt.show()
###################################################################
# Plot ROC curves for the multiclass problem
# Compute macro-average ROC curve and ROC area
# First aggregate all false positive rates
all_fpr = np.unique(np.concatenate([fpr[i] for i in range(n_classes)]))
# Then interpolate all ROC curves at this points
mean_tpr = np.zeros_like(all_fpr)
for i in range(n_classes):
    mean_tpr += interp(all_fpr, fpr[i], tpr[i])
# Finally average it and compute AUC
mean_tpr /= n_classes
fpr["macro"] = all_fpr
tpr["macro"] = mean_tpr
roc_auc["macro"] = auc(fpr["macro"], tpr["macro"])
# Plot all ROC curves
plt.figure()
plt.plot(fpr["micro"], tpr["micro"],
         label='micro-average ROC curve (area = {0:0.2f})'
               ".format(roc_auc["micro"]),
         color='deeppink', linestyle=':', linewidth=4)
plt.plot(fpr["macro"], tpr["macro"],
         label='macro-average ROC curve (area = {0:0.2f})'
               ".format(roc_auc["macro"]),
         color='navy', linestyle=':', linewidth=4)
colors = cycle(['aqua', 'darkorange', 'cornflowerblue'])
for i, color in zip(range(n_classes), colors):
    plt.plot(fpr[i], tpr[i], color=color, lw=lw,
             label='ROC curve of class {0} (area = {1:0.2f})'
                   ".format(i, roc_auc[i]))
plt.plot([0, 1], [0, 1], 'k--', lw=lw)
plt.xlim([0.0, 1.0])
plt.ylim([0.0, 1.05])
plt.xlabel('False Positive Rate')
plt.ylabel('True Positive Rate')
plt.title('Detailed Receiver operating characteristic')
plt.legend(loc="lower right")
plt.show()
```

4.3.2 假体控制中的 EMG 信号

肌电控制的假体可以被上肢缺失或是截肢的人使用。肌电信号从位于肌肉上的表面电极记录下来，并被带到假肢上，在那里对其特性进行分析和解释，以激活必要的功能。考虑到假肢的类别，调节数据从仅由一块肌肉产生的简单开/关指令到由肌肉组合产生的复杂多功能指令。单一肌肉的调节器通常基于 EMG 振幅，其方式是不同强度和不同振幅的肌肉收缩可以区分手的关闭和打开或肘部的弯曲和伸展。多功能假体在不同的肌肉组合上混合使用几个电极，应用先进的信号处理技术来增加数据量，从而能够提取相对活跃肌肉状态的数据。多功能假肢通过分析基于收缩的瞬时信号模式，应用时间、频率或时频方法，实现了对用户意图的更精确把握。实时控制是假肢控制算法发展的重点（Sörnmo & Laguna, 2005）。人类之手之所以重要，是因为它们可以抓取和操纵许多物体。即使失去一只手也会影响人类活动，假手是为无臂对象配备的解决方案。肌肉信号控制手部假体，并在手部截肢后可以进行控制的原因是因为手臂残端有相当数量的肌肉可以控制假体（Kurzynski, Krysmann, Trajdos, & Wolczowski, 2016）(Subasi, 2019c)。

生物医学信号是描述感兴趣的物理变量的一组身体信号，如表面 EMG（sEMG）被用来控制假体的运动。上肢假体结构主要是基于肌电控制，表征皮肤表面肌肉收缩时产生的 EMG 信号。因为大部分产生手指运动的肌肉在手部截肢后都留在残肢上，这些肌肉可以被利用来控制假肢（Wojtczak, Amaral, Dias, Wolczowski, & Kurzynski, 2009）。由于 EMG 信号的分析被普遍用于医疗诊断、体育、康复和假肢控制，EMG 信号的识别为人类任务的自动化提供了相当大的支持。sEMG 信号分类的新应用是个体骨骼肌收缩，相关的 EMG 信号在分类后被用来控制机器动作。但面临的挑战是假手控制，其中手可以实现不同的运动，从而可以通过抓取和操纵几个物体来实现运动甚至演奏乐器。失去一只手可能会降低人的完整功能，而失去两只手则基本上降低了独立性。假手的目的是相对地重建失去的肢体的功能，特别是其工作功能，使其能够进行各种运动和实现不同的手指配置。控制假手的肌肉信号与健康的手和手指运动有关，应通过设置在皮肤上肌肉上方的合适传感器无创地获得。因此，sEMG 控制的假手涉及由手部残肢肌肉激活的表面信号，然后通过分类识别预定的假体动作类别（Subasi, 2019c; Wołczowski & Zdunek, 2017;）。

手是人类最关键的组成部分之一，被用作感觉的基本元素。在现实生活的经验中，手被用来感受表面和执行基本的提升功能。例如，通过控制肌肉运动，手臂的基本功能是抓握、抬起、挥动和进行手臂的其他旋转运动。对于截肢者来说，假手是一种用于替代缺失部分的人工装置。假手可以被改进和操纵，以实现手的不同功能。例如，肌电臂利用由电荷产生的受控肌肉收缩来转移和加强控制中心。以这种方式，通过控制运动，截肢者可以执行手臂的正常功能，如抓握、感觉和挥舞等与手有关的运动（Subasi, 2019c）。使用 sEMG 信号进行假体控制的一般框架如图 4.7 所示。

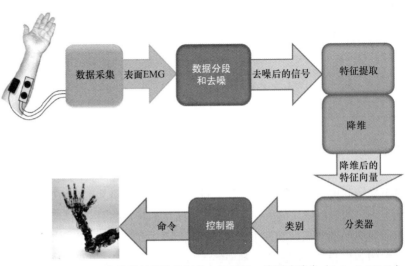

图 4.7　基于 sEMG 信号进行假体控制的一般框架。该图改编自（Subasi, 2019b）

例 4.6 以下 Python 代码被用来从 sEMG 信号中使用小波包分解（WPD）提取特征，并采用 WPD 子带的统计值。然后，它使用多层感知器分类器通过训练和测试数据集对这些数据进行分类。本例计算了分类准确率、精确度、召回率、F1 得分、Cohen kappa 得分和 Matthews 相关系数，还给出了分类报告和混淆矩阵。在这个例子中，将使用基本手部运动的表面 EMG 数据集。

数据集信息：数据以 500 Hz 采样，使用 Butterworth 带通滤波器进行滤波，分别在 15 Hz 和 500 Hz 时具有低和高截止频率，在 50Hz 时用陷波滤波器消除线路干扰。该信号是从两个差分 EMG 传感器记录的。信号由 Delsys Bagnoli Handheld 传导至两通道 EMG 系统。实验包括自由反复地抓取不同的物品，这些物品对进行手部运动至关重要。力量和速度特意由受试者的意愿决定。受试者前臂表面有两个 EMG 电极，用松紧带固定在尺桡肌和伸肌上，参考电极固定在中间，以收集关于肌肉激活的信息。五名年龄大致相同（20～22 岁）的健康受试者（两男三女）分别完成 6 个抓握动作，每个动作 30 次。测量时间为 6 秒。每个受试者都有一个 mat 文件。受试者被要求反复进行的六个日常抓握动作为：

1. 抓握球形：抓握球形工具
2. 抓握尖锐物：抓握小工具
3. 手掌反射：用手掌面对物体抓握
4. 侧面抓握：抓握薄而扁平的物体
5. 抓握圆柱形：抓握圆柱形工具
6. 吊钩：承重。

可以在 Sapsanis 等人的论文中找到更多信息（Sapsanis et al., 2013 Sapsanis, Georgoulas, & Tzes, 2013）。可以从以下网站下载数据：https://archive.ics.uci.edu/ml/datasets/sEMG+for+Basic+Hand+movements#。

```
"""
Created on Thu May 9 12:18:30 2019
@author: absubasi
"""
# ======================================================================
# Feature extraction using the statistical values of wavelet packet
transform
# ======================================================================
# descriptive statistics
import scipy as sp
import scipy.io as sio
import pywt
import numpy as np
import scipy.stats as stats
```

```python
wname = pywt.Wavelet('db1')
level=6 #Number of decomposition level
#Load mat file
mat_contents = sio.loadmat('sEMG_UCI_BHM.mat')
sorted(mat_contents.keys())
CYLINDIRICAL=mat_contents['F1_cyl_ch1']
HOOK=mat_contents['F1_hook_ch1']
LATERAL=mat_contents['F1_lat_ch1']
PALMAR=mat_contents['F1_palm_ch1']
SPHERICAL=mat_contents['F1_spher_ch1']
TIP=mat_contents['F1_tip_ch1']
Labels = [] #Empty List For Labels
NofClasses=6 #Number of Classes
Length = 3000; # Length of signal
Nofsignal=30; #Number of Signal for each Class
numrows =83 #Number of features extracted from Wavelet Packet
Decomposition
#Create Empty Array For Features
Extracted_Features=np.ndarray(shape=(NofClasses*Nofsignal,numrows),
dtype=float, order='F')
# ====================================================================
# Feature extraction using the statistical values of wavelet packet
transform
# ====================================================================
def WPD_Feature_Extraction(signal, i, wname, level):
  #Mean Values of each subbands
  wp= pywt.WaveletPacket(signal, wname, mode='symmetric', maxlevel=level)
  Extracted_Features[i,0]=sp.mean(abs(wp['a'].data))
  Extracted_Features[i,1]=sp.mean(abs(wp['aa'].data))
  Extracted_Features[i,2]=sp.mean(abs(wp['aaa'].data))
  Extracted_Features[i,3]=sp.mean(abs(wp['aaaa'].data))
  Extracted_Features[i,4]=sp.mean(abs(wp['aaaaa'].data))
  Extracted_Features[i,5]=sp.mean(abs(wp['aaaaaa'].data))
  Extracted_Features[i,6]=sp.mean(abs(wp['d'].data))
  Extracted_Features[i,7]=sp.mean(abs(wp['dd'].data))
  Extracted_Features[i,8]=sp.mean(abs(wp['ddd'].data))
  Extracted_Features[i,9]=sp.mean(abs(wp['dddd'].data))
  Extracted_Features[i,10]=sp.mean(abs(wp['ddddd'].data))
  Extracted_Features[i,11]=sp.mean(abs(wp['dddddd'].data))
  #Standart Deviation of each subbands
  Extracted_Features[i,12]=sp.std(wp['a'].data)
  Extracted_Features[i,13]=sp.std(wp['aa'].data)
  Extracted_Features[i,14]=sp.std(wp['aaa'].data)
  Extracted_Features[i,15]=sp.std(wp['aaaa'].data)
  Extracted_Features[i,16]=sp.std(wp['aaaaa'].data)
  Extracted_Features[i,17]=sp.std(wp['aaaaaa'].data)
  Extracted_Features[i,18]=sp.std(wp['d'].data)
  Extracted_Features[i,19]=sp.std(wp['dd'].data)
  Extracted_Features[i,20]=sp.std(wp['ddd'].data)
  Extracted_Features[i,21]=sp.std(wp['dddd'].data)
```

```
Extracted_Features[i,22]=sp.std(wp['ddddd'].data)
Extracted_Features[i,23]=sp.std(wp['dddddd'].data)
#Median Values of each subbands
Extracted_Features[i,24]=sp.median(wp['a'].data)
Extracted_Features[i,25]=sp.median(wp['aa'].data)
Extracted_Features[i,26]=sp.median(wp['aaa'].data)
Extracted_Features[i,27]=sp.median(wp['aaaa'].data)
Extracted_Features[i,28]=sp.median(wp['aaaaa'].data)
Extracted_Features[i,29]=sp.median(wp['aaaaaa'].data)
Extracted_Features[i,30]=sp.median(wp['d'].data)
Extracted_Features[i,31]=sp.median(wp['dd'].data)
Extracted_Features[i,32]=sp.median(wp['ddd'].data)
Extracted_Features[i,33]=sp.median(wp['dddd'].data)
Extracted_Features[i,34]=sp.median(wp['ddddd'].data)
Extracted_Features[i,35]=sp.median(wp['dddddd'].data)
#Skewness of each subbands
Extracted_Features[i,36]=stats.skew(wp['a'].data)
Extracted_Features[i,37]=stats.skew(wp['aa'].data)
Extracted_Features[i,38]=stats.skew(wp['aaa'].data)
Extracted_Features[i,39]=stats.skew(wp['aaaa'].data)
Extracted_Features[i,40]=stats.skew(wp['aaaaa'].data)
Extracted_Features[i,41]=stats.skew(wp['aaaaaa'].data)
Extracted_Features[i,42]=stats.skew(wp['d'].data)
Extracted_Features[i,43]=stats.skew(wp['dd'].data)
Extracted_Features[i,44]=stats.skew(wp['ddd'].data)
Extracted_Features[i,45]=stats.skew(wp['dddd'].data)
Extracted_Features[i,46]=stats.skew(wp['ddddd'].data)
Extracted_Features[i,47]=stats.skew(wp['dddddd'].data)
#Kurtosis of each subbands
Extracted_Features[i,48]=stats.kurtosis(wp['a'].data)
Extracted_Features[i,49]=stats.kurtosis(wp['aa'].data)
Extracted_Features[i,50]=stats.kurtosis(wp['aaa'].data)
Extracted_Features[i,51]=stats.kurtosis(wp['aaaa'].data)
Extracted_Features[i,52]=stats.kurtosis(wp['aaaaa'].data)
Extracted_Features[i,53]=stats.kurtosis(wp['aaaaaa'].data)
Extracted_Features[i,54]=stats.kurtosis(wp['d'].data)
Extracted_Features[i,55]=stats.kurtosis(wp['dd'].data)
Extracted_Features[i,56]=stats.kurtosis(wp['ddd'].data)
Extracted_Features[i,57]=stats.kurtosis(wp['dddd'].data)
Extracted_Features[i,58]=stats.kurtosis(wp['ddddd'].data)
Extracted_Features[i,59]=stats.kurtosis(wp['dddddd'].data)
#RMS Values of each subbands
Extracted_Features[i,60]=np.sqrt(np.mean(wp['a'].data**2))
Extracted_Features[i,61]=np.sqrt(np.mean(wp['aa'].data**2))
Extracted_Features[i,62]=np.sqrt(np.mean(wp['aaa'].data**2))
Extracted_Features[i,63]=np.sqrt(np.mean(wp['aaaa'].data**2))
Extracted_Features[i,64]=np.sqrt(np.mean(wp['aaaaa'].data**2))
Extracted_Features[i,65]=np.sqrt(np.mean(wp['aaaaaa'].data**2))
Extracted_Features[i,66]=np.sqrt(np.mean(wp['d'].data**2))
Extracted_Features[i,67]=np.sqrt(np.mean(wp['dd'].data**2))
```

```
Extracted_Features[i,68]=np.sqrt(np.mean(wp['ddd'].data**2))
Extracted_Features[i,69]=np.sqrt(np.mean(wp['dddd'].data**2))
Extracted_Features[i,70]=np.sqrt(np.mean(wp['ddddd'].data**2))
Extracted_Features[i,71]=np.sqrt(np.mean(wp['dddddd'].data**2))
#Ratio of subbands
Extracted_Features[i,72]=sp.mean(abs(wp['a'].data))/sp.mean(abs
(wp['aa'].data))
Extracted_Features[i,73]=sp.mean(abs(wp['aa'].data))/sp.mean(abs(wp
['aaa'].data))
Extracted_Features[i,74]=sp.mean(abs(wp['aaa'].data))/sp.mean(abs
(wp['aaaa'].data))
Extracted_Features[i,75]=sp.mean(abs(wp['aaaa'].data))/sp.mean(abs(wp
['aaaaa'].data))
Extracted_Features[i,76]=sp.mean(abs(wp['aaaaa'].data))/sp.mean(abs
(wp['aaaaaa'].data))
Extracted_Features[i,77]=sp.mean(abs(wp['aaaaaa'].data))/sp.mean(abs
(wp['d'].data))
Extracted_Features[i,78]=sp.mean(abs(wp['d'].data))/sp.mean(abs
(wp['dd'].data))
Extracted_Features[i,79]=sp.mean(abs(wp['dd'].data))/sp.mean(abs(wp
['ddd'].data))
Extracted_Features[i,80]=sp.mean(abs(wp['ddd'].data))/sp.mean(abs(wp
['dddd'].data))
Extracted_Features[i,81]=sp.mean(abs(wp['dddd'].data))/sp.mean(abs
(wp['ddddd'].data))
Extracted_Features[i,82]=sp.mean(abs(wp['ddddd'].data))/sp.mean(abs(wp
['dddddd'].data))

#%%
# =============================================================
# Feature extraction from cylindirical sEMG signal
# =============================================================
for i in range(Nofsignal):
    WPD_Feature_Extraction(CYLINDIRICAL[i,:], i, wname, level)
    Labels.append("CYLINDIRICAL")
# =============================================================
# Feature extraction from hook sEMG signal
# =============================================================
for i in range(Nofsignal, 2*Nofsignal):
    WPD_Feature_Extraction(HOOK[i-Nofsignal,:], i, wname, level)
    Labels.append("HOOK")
# =============================================================
# Feature extraction from lateral sEMG signal
# =============================================================
for i in range(2*Nofsignal, 3*Nofsignal):
    WPD_Feature_Extraction(LATERAL[i-2*Nofsignal,:], i, wname, level)
    Labels.append("LATERAL")
# =============================================================
# Feature extraction from palmar sEMG signal
# =============================================================
```

```python
for i in range(3*Nofsignal, 4*Nofsignal):
    WPD_Feature_Extraction(PALMAR[i-3*Nofsignal,:], i, wname, level)
    Labels.append("PALMAR")
# ================================================================
# Feature extraction from spherical sEMG signal
# ================================================================
for i in range(4*Nofsignal, 5*Nofsignal):
    WPD_Feature_Extraction(SPHERICAL[i-4*Nofsignal,:], i, wname, level)
    Labels.append("SPHERICAL")
# ================================================================
# Feature extraction from tip sEMG signal
# ================================================================
for i in range(5*Nofsignal, 6*Nofsignal):
    WPD_Feature_Extraction(TIP[i-5*Nofsignal,:], i, wname, level)
    Labels.append("TIP")
#%%
# ================================================================
# Classification
# ================================================================
from sklearn.model_selection import cross_val_score
from sklearn.metrics import cohen_kappa_score, make_scorer
X = Extracted_Features
y = Labels
#To prevent warnings
import warnings
warnings.filterwarnings("ignore")
# Import train_test_split function
from sklearn.model_selection import train_test_split
# Split dataset into training set and test set
# 70% training and 30% test
Xtrain, Xtest, ytrain, ytest = train_test_split(X, y,test_size=0.3,
random_state=1)
#%%
from sklearn.neural_network import MLPClassifier
#Create Train and Test set
Xtrain, Xtest, ytrain, ytest = train_test_split(X, y, test_size=0.3,
random_state=1)
"""mlp=MLPClassifier(hidden_layer_sizes=(100,    ),    activation='relu',
solver='adam',
        alpha=0.0001, batch_size='auto', learning_rate='constant',
        learning_rate_init=0.001, power_t=0.5, max_iter=200,
        shuffle=True, random_state=None, tol=0.0001, verbose=False,
        warm_start=False, momentum=0.9, nesterovs_momentum=True,
        early_stopping=False, validation_fraction=0.1, beta_1=0.9,
        beta_2=0.999, epsilon=1e-08, n_iter_no_change=10)"""
#Create the Model
mlp = MLPClassifier(hidden_layer_sizes=(50, ), learning_rate_init=0.001,
        alpha=1, momentum=0.7,max_iter=1000)
#Train the Model with Training dataset
mlp.fit(Xtrain,ytrain)
```

```
#Test the Model with Testing dataset
ypred = mlp.predict(Xtest)
#Evaluate the Model and Print Performance Metrics
from sklearn import metrics
print('Accuracy:', np.round(metrics.accuracy_score(ytest,ypred),4))
print('Precision:', np.round(metrics.precision_score(ytest,
                    ypred,average='weighted'),4))
print('Recall:', np.round(metrics.recall_score(ytest,ypred,
                    average='weighted'),4))
print('F1 Score:', np.round(metrics.f1_score(ytest,ypred,
                    average='weighted'),4))
print('Cohen Kappa Score:', np.round(metrics.cohen_kappa_score(ytest,
ypred),4))
print('Matthews Corrcoef:', np.round(metrics.matthews_corrcoef(ytest,
ypred),4))
print('\t\tClassification Report:\n', metrics.classification_report(ypred,
ytest))
#Plot Confusion Matrix
from sklearn.metrics import confusion_matrix
print("Confusion Matrix:\n",confusion_matrix(ytest, ypred))
```

4.3.3 康复机器人中的 EMG 信号

EMG 控制的辅助设备也被用于强化治疗环境中的中风治疗，以帮助康复。在这种情况下，患者的活动目标可以通过 sEMG 来实现（Lum, Burgar, Shor, Majmundar, & Van der Loos, 2002; Riener, Nef, & Colombo, 2005）。随着最近的技术改进，主动外骨骼机器人在康复应用、人类动力增强、辅助机器人技术、损伤评估以及虚拟和远程操作环境中的触觉交流中得到许多应用。为了协助人类，这些机器人必须理解人类的身体指令。因此，需要获取 EMG 信号并对其进行分析，以控制外骨骼机器人（Sasaki et al., 2005）。外骨骼机器人有两个控制器同时工作——机器人控制器和人类肌肉。由于人类操作指令，而控制系统则采用这些指令作为其决策组件的一部分，因此上肢外骨骼机器人的控制方式与传统的工业和野外机器人不同。外骨骼准确地实现了人类操作员的真实决策。然而，根据机器人用户的运动意图识别做出决策仍然存在挑战（Abdullah, Subasi, & Qaisar, 2017; Subasi, 2019c）。

设计上肢外骨骼机器人控制器的最佳方法是专注于控制器的输入信息。在现代技术中，输入包括人类生物医学信号和与平台无关的控制信号。在不同的领域中，许多不同的策略已经尝试使用。EMG 信号已经成功地作为人类生物医学信号输入到一些外骨骼发展中，如上肢外骨骼机器人（Lo & Xie, 2012）。例如，Gopura 等人（2009）的文献提出了基于 EMG 控制的肌肉模型来控制具有七个自由度的上肢外骨骼机器人。用户可以调整该方法使得大多数上肢残疾的人可以操作它。使用 EMG 的上肢外骨骼机器人的控制方法大多是二进制（开 - 关）性质的（Lenzi et al., 2009）。一个好的设计可以高度准确地识别即使是身体虚弱、不能正确进行日常动作的人的运动意向（Abdullah et al., 2017; Subasi, 2019c）。

例 4.7 以下 Python 代码被用来从与各种物理动作有关的 sEMG 信号中提取特征，使用 WPD 和采用 WPD 子带的统计值。然后使用具有单独训练和测试数据集的随机森林分类器对这些数据进行分类。计算分类准确率、精确度、召回率、F1 得分、Cohen kappa 得分和 Matthews 相关系数。给出了分类报告和混淆矩阵。本例将使用表面 EMG 物理动作数据集。

数据集信息：实验对象为一名女性和三名男性受试者（年龄在 25～30 岁之间），他们都曾经历过身体上的攻击行为。每个受试者都要通过 20 个不同的实验实施 10 个正常活动和 10 个攻击性活动。Essex 机器人竞技场是进行数据收集的主要实验厅。受试者的表现是由 Delsys EMG 仪器收集的，该仪器将人类活动与肌电收缩联系起来。基于这种情况，数据记录程序包括位于上臂（肱二头肌和肱三头肌）和上腿（大腿和腿筋）的八个皮肤表面电极。它们对应于每个肌肉通道产生八个输入时间序列。每个时间序列包含 10 000 个样本（每个受试者每个实验时段有 15 个动作）。可以从以下网站下载数据：https://archive.ics.uci.edu/ml/datasets/EMG+Physical+Action+Data+Set。

```
"""
Created on Thu May 9 12:18:30 2019
@author: absubasi
"""
# =====================================================================
# Feature extraction using the statistical values of wavelet packet
transform
# =====================================================================
# descriptive statistics
import scipy as sp
import scipy.io as sio
import pywt
import numpy as np
import scipy.stats as stats

wname = pywt.Wavelet('db1')
level=6 #Number of decomposition level
#Load mat file
mat_contents = sio.loadmat('sEMG_UCI_PA_NOR.mat')
sorted(mat_contents.keys())
BOWING=mat_contents['Bow']
CLAPPING=mat_contents['Cla']
HANDSHAKING=mat_contents['Han']
HUGGING=mat_contents['Hug']
JUMPING=mat_contents['Jum']
RUNNING=mat_contents['Run']
SEATING=mat_contents['Sea']
STANDING=mat_contents['Sta']
WALKING=mat_contents['Wal']
WAVING=mat_contents['Wav']
```

```
Labels=[] #Empty List For Labels
NofClasses=10 #Number of Classes
Length = 512; # Length of signal
Nofsignal=72; #Number of Signal for each Class
Ch=1 #Channel To be used
numrows =83 #Number of features extracted from Wavelet Packet
Decomposition
#Create Empty Array For Features
Extracted_Features=np.ndarray(shape=(NofClasses*Nofsignal,numrows),
dtype=float, order='F')
# ========================================================================
# Utility function for feature extraction using the statistical values
of WPD
# ========================================================================
def WPD_Feature_Extraction(signal, i, wname, level):
    #Mean Values of each subbands
    wp= pywt.WaveletPacket(signal, wname, mode='symmetric', maxlevel=level)
    Extracted_Features[i,0]=sp.mean(abs(wp['a'].data))
    Extracted_Features[i,1]=sp.mean(abs(wp['aa'].data))
    Extracted_Features[i,2]=sp.mean(abs(wp['aaa'].data))
    Extracted_Features[i,3]=sp.mean(abs(wp['aaaa'].data))
    Extracted_Features[i,4]=sp.mean(abs(wp['aaaaa'].data))
    Extracted_Features[i,5]=sp.mean(abs(wp['aaaaaa'].data))
    Extracted_Features[i,6]=sp.mean(abs(wp['d'].data))
    Extracted_Features[i,7]=sp.mean(abs(wp['dd'].data))
    Extracted_Features[i,8]=sp.mean(abs(wp['ddd'].data))
    Extracted_Features[i,9]=sp.mean(abs(wp['dddd'].data))
    Extracted_Features[i,10]=sp.mean(abs(wp['ddddd'].data))
    Extracted_Features[i,11]=sp.mean(abs(wp['dddddd'].data))
    #Standart Deviation of each subbands
    Extracted_Features[i,12]=sp.std(wp['a'].data)
    Extracted_Features[i,13]=sp.std(wp['aa'].data)
    Extracted_Features[i,14]=sp.std(wp['aaa'].data)
    Extracted_Features[i,15]=sp.std(wp['aaaa'].data)
    Extracted_Features[i,16]=sp.std(wp['aaaaa'].data)
    Extracted_Features[i,17]=sp.std(wp['aaaaaa'].data)
    Extracted_Features[i,18]=sp.std(wp['d'].data)
    Extracted_Features[i,19]=sp.std(wp['dd'].data)
    Extracted_Features[i,20]=sp.std(wp['ddd'].data)
    Extracted_Features[i,21]=sp.std(wp['dddd'].data)
    Extracted_Features[i,22]=sp.std(wp['ddddd'].data)
    Extracted_Features[i,23]=sp.std(wp['dddddd'].data)
    #Median Values of each subbands
    Extracted_Features[i,24]=sp.median(wp['a'].data)
    Extracted_Features[i,25]=sp.median(wp['aa'].data)
    Extracted_Features[i,26]=sp.median(wp['aaa'].data)
    Extracted_Features[i,27]=sp.median(wp['aaaa'].data)
    Extracted_Features[i,28]=sp.median(wp['aaaaa'].data)
    Extracted_Features[i,29]=sp.median(wp['aaaaaa'].data)
    Extracted_Features[i,30]=sp.median(wp['d'].data)
```

```
Extracted_Features[i,31]=sp.median(wp['dd'].data)
Extracted_Features[i,32]=sp.median(wp['ddd'].data)
Extracted_Features[i,33]=sp.median(wp['dddd'].data)
Extracted_Features[i,34]=sp.median(wp['ddddd'].data)
Extracted_Features[i,35]=sp.median(wp['dddddd'].data)
#Skewness of each subbands
Extracted_Features[i,36]=stats.skew(wp['a'].data)
Extracted_Features[i,37]=stats.skew(wp['aa'].data)
Extracted_Features[i,38]=stats.skew(wp['aaa'].data)
Extracted_Features[i,39]=stats.skew(wp['aaaa'].data)
Extracted_Features[i,40]=stats.skew(wp['aaaaa'].data)
Extracted_Features[i,41]=stats.skew(wp['aaaaaa'].data)
Extracted_Features[i,42]=stats.skew(wp['d'].data)
Extracted_Features[i,43]=stats.skew(wp['dd'].data)
Extracted_Features[i,44]=stats.skew(wp['ddd'].data)
Extracted_Features[i,45]=stats.skew(wp['dddd'].data)
Extracted_Features[i,46]=stats.skew(wp['ddddd'].data)
Extracted_Features[i,47]=stats.skew(wp['dddddd'].data)
#Kurtosis of each subbands
Extracted_Features[i,48]=stats.kurtosis(wp['a'].data)
Extracted_Features[i,49]=stats.kurtosis(wp['aa'].data)
Extracted_Features[i,50]=stats.kurtosis(wp['aaa'].data)
Extracted_Features[i,51]=stats.kurtosis(wp['aaaa'].data)
Extracted_Features[i,52]=stats.kurtosis(wp['aaaaa'].data)
Extracted_Features[i,53]=stats.kurtosis(wp['aaaaaa'].data)
Extracted_Features[i,54]=stats.kurtosis(wp['d'].data)
Extracted_Features[i,55]=stats.kurtosis(wp['dd'].data)
Extracted_Features[i,56]=stats.kurtosis(wp['ddd'].data)
Extracted_Features[i,57]=stats.kurtosis(wp['dddd'].data)
Extracted_Features[i,58]=stats.kurtosis(wp['ddddd'].data)
Extracted_Features[i,59]=stats.kurtosis(wp['dddddd'].data)
#RMS Values of each subbands
Extracted_Features[i,60]=np.sqrt(np.mean(wp['a'].data**2))
Extracted_Features[i,61]=np.sqrt(np.mean(wp['aa'].data**2))
Extracted_Features[i,62]=np.sqrt(np.mean(wp['aaa'].data**2))
Extracted_Features[i,63]=np.sqrt(np.mean(wp['aaaa'].data**2))
Extracted_Features[i,64]=np.sqrt(np.mean(wp['aaaaa'].data**2))
Extracted_Features[i,65]=np.sqrt(np.mean(wp['aaaaaa'].data**2))
Extracted_Features[i,66]=np.sqrt(np.mean(wp['d'].data**2))
Extracted_Features[i,67]=np.sqrt(np.mean(wp['dd'].data**2))
Extracted_Features[i,68]=np.sqrt(np.mean(wp['ddd'].data**2))
Extracted_Features[i,69]=np.sqrt(np.mean(wp['dddd'].data**2))
Extracted_Features[i,70]=np.sqrt(np.mean(wp['ddddd'].data**2))
Extracted_Features[i,71]=np.sqrt(np.mean(wp['dddddd'].data**2))
#Ratio of subbands
Extracted_Features[i,72]=sp.mean(abs(wp['a'].data))/sp.mean(abs
(wp['aa'].data))
Extracted_Features[i,73]=sp.mean(abs(wp['aa'].data))/sp.mean(abs
(wp['aaa'].data))
Extracted_Features[i,74]=sp.mean(abs(wp['aaa'].data))/sp.mean(abs
```

```python
                    (wp['aaaa'].data))
    Extracted_Features[i,75]=sp.mean(abs(wp['aaaa'].data))/sp.mean(abs(wp
                    ['aaaaa'].data))
    Extracted_Features[i,76]=sp.mean(abs(wp['aaaaa'].data))/sp.mean(abs
                    (wp['aaaaaa'].data))
    Extracted_Features[i,77]=sp.mean(abs(wp['aaaaaa'].data))/sp.mean(abs
                    (wp['d'].data))
    Extracted_Features[i,78]=sp.mean(abs(wp['d'].data))/sp.mean(abs(wp
                    ['dd'].data))
    Extracted_Features[i,79]=sp.mean(abs(wp['dd'].data))/sp.mean(abs
                    (wp['ddd'].data))
    Extracted_Features[i,80]=sp.mean(abs(wp['ddd'].data))/sp.mean(abs(wp
                    ['dddd'].data))
    Extracted_Features[i,81]=sp.mean(abs(wp['dddd'].data))/sp.mean(abs
                    (wp['ddddd'].data))
    Extracted_Features[i,82]=sp.mean(abs(wp['ddddd'].data))/sp.mean(abs(wp
                    ['dddddd'].data))
#%%
# ======================================================================
# Feature extraction from bowing sEMG signal
# ======================================================================
for i in range(Nofsignal):
    WPD_Feature_Extraction(BOWING[i,:, Ch], i, wname, level)
    Labels.append("BOWING")
# ======================================================================
# Feature extraction from clapping sEMG signal
# ======================================================================
for i in range(Nofsignal, 2*Nofsignal):
    WPD_Feature_Extraction(CLAPPING[i-Nofsignal,:, Ch], i, wname, level)
    Labels.append("CLAPPING")
# ======================================================================
# Feature extraction from handshaking sEMG signal
# ======================================================================
for i in range(2*Nofsignal, 3*Nofsignal):
    WPD_Feature_Extraction(HANDSHAKING[i-2*Nofsignal,:,Ch], i, wname, level)
    Labels.append("HANDSHAKING")
# ======================================================================
# Feature extraction from hugging sEMG signal
# ======================================================================
for i in range(3*Nofsignal, 4*Nofsignal):
    WPD_Feature_Extraction(HUGGING[i-3*Nofsignal,:, Ch], i, wname, level)
    Labels.append("HUGGING")
# ======================================================================
# Feature extraction from jumping sEMG signal
# ======================================================================
for i in range(4*Nofsignal, 5*Nofsignal):
    WPD_Feature_Extraction(JUMPING[i-4*Nofsignal,:,Ch], i, wname, level)
    Labels.append("JUMPING")
# ======================================================================
# Feature extraction from running sEMG signal
```

```python
# =================================================================
for i in range(5*Nofsignal, 6*Nofsignal):
    WPD_Feature_Extraction(RUNNING[i-5*Nofsignal,:,Ch], i, wname, level)
    Labels.append("RUNNING")
# =================================================================
# Feature extraction from seating sEMG signal
# =================================================================
for i in range(6*Nofsignal, 7*Nofsignal):
    WPD_Feature_Extraction(SEATING[i-6*Nofsignal,:,Ch], i, wname, level)
    Labels.append("SEATING")
# =================================================================
# Feature extraction from standing sEMG signal
# =================================================================
for i in range(7*Nofsignal, 8*Nofsignal):
    WPD_Feature_Extraction(STANDING[i-7*Nofsignal,:,Ch], i, wname, level)
    Labels.append("STANDING")
# =================================================================
# Feature extraction from walking sEMG signal
# =================================================================
for i in range(8*Nofsignal, 9*Nofsignal):
    WPD_Feature_Extraction(WALKING[i-8*Nofsignal,:,Ch], i, wname, level)
    Labels.append("WALKING")
# =================================================================
# Feature extraction from waving sEMG signal
# =================================================================
for i in range(9*Nofsignal, 10*Nofsignal):
    WPD_Feature_Extraction(WAVING[i-9*Nofsignal,:,Ch], i, wname, level)
    Labels.append("WAVING")
#%%
# =================================================================
# Classification
# =================================================================
from sklearn.model_selection import cross_val_score
from sklearn.metrics import cohen_kappa_score, make_scorer
X = Extracted_Features
y = Labels

#To prevent warnings
import warnings
warnings.filterwarnings("ignore")
# Import train_test_split function
from sklearn.model_selection import train_test_split
# Split dataset into training set and test set
# 70% training and 30% test
Xtrain, Xtest, ytrain, ytest = train_test_split(X, y,test_size=0.3,
random_state=1)
#%%
# =================================================================
# Random forest example with training and test set
# =================================================================
```

```python
from sklearn.ensemble import RandomForestClassifier

#In order to change to accuracy increase n_estimators
"""RandomForestClassifier(n_estimators='warn',   criterion='gini',   max_
depth=None,
min_samples_split=2, min_samples_leaf=1, min_weight_fraction_leaf=0.0,
max_features='auto', max_leaf_nodes=None, min_impurity_decrease=0.0,
min_impurity_split=None, bootstrap=True, oob_score=False, n_jobs=None,
random_state=None, verbose=0, warm_start=False, class_weight=None)"""
clf = RandomForestClassifier(n_estimators=200)
#Create the Model
#Train the model with Training Dataset
clf.fit(Xtrain,ytrain)
#Test the model with Testset
ypred = clf.predict(Xtest)

#Evaluate the Model and Print Performance Metrics
from sklearn import metrics
print('Accuracy:', np.round(metrics.accuracy_score(ytest,ypred),4))
print('Precision:', np.round(metrics.precision_score(ytest,
                  ypred,average='weighted'),4))
print('Recall:', np.round(metrics.recall_score(ytest,ypred,
                           average='weighted'),4))
print('F1 Score:', np.round(metrics.f1_score(ytest,ypred,
                         average='weighted'),4))
print('Cohen Kappa Score:', np.round(metrics.cohen_kappa_score(ytest,
ypred),4))
print('Matthews Corrcoef:', np.round(metrics.matthews_corrcoef(ytest,
ypred),4))
print('\t\tClassification Report:\n', metrics.classification_report(ypred,
ytest))

#Plot Confusion Matrix
from sklearn.metrics import confusion_matrix
print("Confusion Matrix:\n",confusion_matrix(ytest, ypred))
```

4.4 心电图信号分析

心电图（ECG）是对源自心脏的体表电活动的记录。为了追踪心电图的波形，在体表的两个点之间进行差分记录。在默认情况下，每个差分记录被称为一个导联。Einthoven定义了三个导联，用罗马数字Ⅰ、Ⅱ和Ⅲ命名。任何两个部位的电压差都由心电图记录。心电图信号通常在 ±2 mV 的范围内，需要 0.05 至 150 Hz 的记录带宽。12 导联心电图用于有限模式的记录活动，如磁带记录的门诊心电图（通常是两个导联）、床边的重症监护（通常是一个或两个导联）或在医院各区域遥测的无约束病人（一个导联）。现代心电图设备完全集成了一个模拟前端、一个 12 至 16 位的模数（A/D）转换器、一个计算型微处理器和专用的输入-输出（I/O）处理器。这些系统从 12 个导联信号中找到一个维度矩阵，并使用一套规则来检查这个矩阵，以达到最终的一套解释声明。更好的医院系统会记录这些变化，并保存在一个大的数据库，使得所有的心电图数据都可以通过不同的参数组合被访问到，例如，所有 30 岁以上患有先天性心脏病的女性（Abdullah et al., 2017; Subasi, 2019c）。

针对每张心电图进行具体的诊断可以使用很多种不同的示范性方法，但 ECG 仅仅使用了五六种主要分类集。心电图分析的最初步骤需要计算心房和心室的速率和节律。这包括在不同心室之间的连接或心室本身的任何传导不稳定。然后进行特征识别，这将与是否有心肌梗塞导致的瘢痕有关。心电图一直是评估心室大小或生长的主要方法，但人们可能会认为，这一领域更精确的数据可以通过无创成像技术获得（Berbari, 2000; Subasi, 2019c）。ECG 信号分类的一般框架如图 4.8 所示。

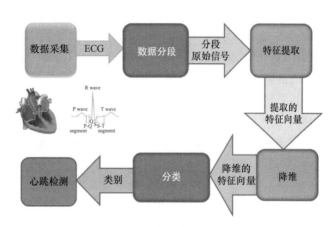

图 4.8　ECG 信号分类的一般框架

心律失常的诊断

心血管疾病（CVD）是全球导致死亡的主要原因之一。精确的设计和快速的技术以自动进行诸如心律不齐等心电信号分类至关重要，这些技术可以用于各种 CVD 的临床诊断（Thaler, 2017）。心律失常的特点是一组情况，其中不规则的电活动来自心脏，并通过心电图的跳动或模式来识别（De Chazal, O'Dwyer, & Reilly, 2004; Pan &Tompkins, 1985）。心电图是一种高效、简单、无创的心脏疾病诊断技术。医生根据几个波形的特点（振幅、极性等）进行检查，并在此基础上进行诊断和治疗（Subasi, 2019c）。

例 4.8　以下 Python 代码用于从正常的房性早搏（APC）、室性早搏（PCV）、左束支阻滞（LBBB）和右束支阻滞（RBBB）的心电信号中提取特征，采用平稳小波变换（SWT）和 SWT 子带的统计值。然后使用随机森林（RF）和单独的训练和测试数据集对这些数据进行分类。计算分类准确率、精确度、召回率、F1 得分、Cohen kappa 得分和 Matthews 相关系数。给出了分类报告和混淆矩阵。

数据集信息：波士顿的贝斯以色列医院和麻省理工学院一直保持着心律失常分析和相关课题的研究。这项工作的首批主要成就之一是 MIT-BIH 心律失常数据库，该数据库于 1980 年发布。它是第一个可公开访问的用于评估心律失常检测器的标准测试材料数据集，并已用于这一目的以及全球500多个地点的心脏动力学基础研究。MIT-BIH心律失常数据库包括48个半小时的双通道动态心电图记录，取自BIH心律失常实验室在1975年至1979年间研究的47名受试者。该数据集是由波士顿贝斯以色列医院的住院病人（约60%）和门诊病人（约40%）的4000个24小时动态心电图信号中随机选择23个信号，以及其余的25个从同一集合中选择的记录组成，从而包含不太常见但在临床上很重要的心律失常。这些非常见的心律失常在一个小的随机样本中不会得到很好的体现。心电图信号的采样频率为每通道360Hz，分辨率为11位，范围为10mV。两个或两个以上的心脏病专家分别解释每个记录以解决分歧，为每个节拍获得了计算机可读的参考注释（总共约11万条），包括在数据库中。该数据集包含许多心律失常类型，包括房性早搏（APC）、室性早搏（PVC）、左束支阻滞（LBBB）和右束支阻滞（RBBB）。自1999年9月PhysioNet成立以来，整个MIT-BIH心律失常数据库可以免费访问。可以从以下网站下载数据：https://www.physionet.org/physiobank/database/mitdb/。

```
"""
Created on Thu May 9 12:18:30 2019
@author: asubasi
"""
# ==================================================================
# Feature extraction using the statistical values of wavelet packet
transform
# ==================================================================
# descriptive statistics
import scipy as sp
import scipy.io as sio
import pywt
import numpy as np
import scipy.stats as stats

wname = pywt.Wavelet('db1')
level=6 #Number of decomposition level
#Load mat file
mat_contents = sio.loadmat('MITBIH_ECG.mat')
sorted(mat_contents.keys())
ECGN=mat_contents['ECGN']
ECGAPC=mat_contents['ECGAPC']
ECGPVC=mat_contents['ECGPVC']
ECGLBBB=mat_contents['ECGLBBB']
ECGRBBB=mat_contents['ECGRBBB']

Labels = [] #Empty List For Labels
NofClasses=5 #Number of Classes
Length = 320; # Length of signal
Nofsignal=300; #Number of Signal for each Class
numrows =83 #Number of features extracted from Wavelet Packet Decomposition
```

```python
#Create Empty Array For Features
Extracted_Features=np.ndarray(shape=(NofClasses*Nofsignal,numrows),
dtype=float, order='F')
# ======================================================================
# Utility function for feature extraction using the statistical values
of WPD
# ======================================================================
def WPD_Feature_Extraction(signal, i, wname, level):
  #Mean Values of each subbands
  wp= pywt.WaveletPacket(signal, wname, mode='symmetric', maxlevel=level)
  Extracted_Features[i,0]=sp.mean(abs(wp['a'].data))
  Extracted_Features[i,1]=sp.mean(abs(wp['aa'].data))
  Extracted_Features[i,2]=sp.mean(abs(wp['aaa'].data))
  Extracted_Features[i,3]=sp.mean(abs(wp['aaaa'].data))
  Extracted_Features[i,4]=sp.mean(abs(wp['aaaaa'].data))
  Extracted_Features[i,5]=sp.mean(abs(wp['aaaaaa'].data))
  Extracted_Features[i,6]=sp.mean(abs(wp['d'].data))
  Extracted_Features[i,7]=sp.mean(abs(wp['dd'].data))
  Extracted_Features[i,8]=sp.mean(abs(wp['ddd'].data))
  Extracted_Features[i,9]=sp.mean(abs(wp['dddd'].data))
  Extracted_Features[i,10]=sp.mean(abs(wp['ddddd'].data))
  Extracted_Features[i,11]=sp.mean(abs(wp['dddddd'].data))
  #Standart Deviation of each subbands
  Extracted_Features[i,12]=sp.std(wp['a'].data)
  Extracted_Features[i,13]=sp.std(wp['aa'].data)
  Extracted_Features[i,14]=sp.std(wp['aaa'].data)
  Extracted_Features[i,15]=sp.std(wp['aaaa'].data)
  Extracted_Features[i,16]=sp.std(wp['aaaaa'].data)
  Extracted_Features[i,17]=sp.std(wp['aaaaaa'].data)
  Extracted_Features[i,18]=sp.std(wp['d'].data)
  Extracted_Features[i,19]=sp.std(wp['dd'].data)
  Extracted_Features[i,20]=sp.std(wp['ddd'].data)
  Extracted_Features[i,21]=sp.std(wp['dddd'].data)
  Extracted_Features[i,22]=sp.std(wp['ddddd'].data)
  Extracted_Features[i,23]=sp.std(wp['dddddd'].data)
  #Median Values of each subbands
  Extracted_Features[i,24]=sp.median(wp['a'].data)
  Extracted_Features[i,25]=sp.median(wp['aa'].data)
  Extracted_Features[i,26]=sp.median(wp['aaa'].data)
  Extracted_Features[i,27]=sp.median(wp['aaaa'].data)
  Extracted_Features[i,28]=sp.median(wp['aaaaa'].data)
  Extracted_Features[i,29]=sp.median(wp['aaaaaa'].data)
  Extracted_Features[i,30]=sp.median(wp['d'].data)
  Extracted_Features[i,31]=sp.median(wp['dd'].data)
  Extracted_Features[i,32]=sp.median(wp['ddd'].data)
  Extracted_Features[i,33]=sp.median(wp['dddd'].data)
  Extracted_Features[i,34]=sp.median(wp['ddddd'].data)
  Extracted_Features[i,35]=sp.median(wp['dddddd'].data)
  #Skewness of each subbands
  Extracted_Features[i,36]=stats.skew(wp['a'].data)
```

```
Extracted_Features[i,37]=stats.skew(wp['aa'].data)
Extracted_Features[i,38]=stats.skew(wp['aaa'].data)
Extracted_Features[i,39]=stats.skew(wp['aaaa'].data)
Extracted_Features[i,40]=stats.skew(wp['aaaaa'].data)
Extracted_Features[i,41]=stats.skew(wp['aaaaaa'].data)
Extracted_Features[i,42]=stats.skew(wp['d'].data)
Extracted_Features[i,43]=stats.skew(wp['dd'].data)
Extracted_Features[i,44]=stats.skew(wp['ddd'].data)
Extracted_Features[i,45]=stats.skew(wp['dddd'].data)
Extracted_Features[i,46]=stats.skew(wp['ddddd'].data)
Extracted_Features[i,47]=stats.skew(wp['dddddd'].data)
#Kurtosis of each subbands
Extracted_Features[i,48]=stats.kurtosis(wp['a'].data)
Extracted_Features[i,49]=stats.kurtosis(wp['aa'].data)
Extracted_Features[i,50]=stats.kurtosis(wp['aaa'].data)
Extracted_Features[i,51]=stats.kurtosis(wp['aaaa'].data)
Extracted_Features[i,52]=stats.kurtosis(wp['aaaaa'].data)
Extracted_Features[i,53]=stats.kurtosis(wp['aaaaaa'].data)
Extracted_Features[i,54]=stats.kurtosis(wp['d'].data)
Extracted_Features[i,55]=stats.kurtosis(wp['dd'].data)
Extracted_Features[i,56]=stats.kurtosis(wp['ddd'].data)
Extracted_Features[i,57]=stats.kurtosis(wp['dddd'].data)
Extracted_Features[i,58]=stats.kurtosis(wp['ddddd'].data)
Extracted_Features[i,59]=stats.kurtosis(wp['dddddd'].data)
#RMS Values of each subbands
Extracted_Features[i,60]=np.sqrt(np.mean(wp['a'].data**2))
Extracted_Features[i,61]=np.sqrt(np.mean(wp['aa'].data**2))
Extracted_Features[i,62]=np.sqrt(np.mean(wp['aaa'].data**2))
Extracted_Features[i,63]=np.sqrt(np.mean(wp['aaaa'].data**2))
Extracted_Features[i,64]=np.sqrt(np.mean(wp['aaaaa'].data**2))
Extracted_Features[i,65]=np.sqrt(np.mean(wp['aaaaaa'].data**2))
Extracted_Features[i,66]=np.sqrt(np.mean(wp['d'].data**2))
Extracted_Features[i,67]=np.sqrt(np.mean(wp['dd'].data**2))
Extracted_Features[i,68]=np.sqrt(np.mean(wp['ddd'].data**2))
Extracted_Features[i,69]=np.sqrt(np.mean(wp['dddd'].data**2))
Extracted_Features[i,70]=np.sqrt(np.mean(wp['ddddd'].data**2))
Extracted_Features[i,71]=np.sqrt(np.mean(wp['dddddd'].data**2))
#Ratio of subbands
Extracted_Features[i,72]=sp.mean(abs(wp['a'].data))/sp.mean(abs(wp['aa'].data))
Extracted_Features[i,73]=sp.mean(abs(wp['aa'].data))/sp.mean(abs(wp['aaa'].data))
Extracted_Features[i,74]=sp.mean(abs(wp['aaa'].data))/sp.mean(abs(wp['aaaa'].data))
Extracted_Features[i,75]=sp.mean(abs(wp['aaaa'].data))/sp.mean(abs(wp['aaaaa'].data))
Extracted_Features[i,76]=sp.mean(abs(wp['aaaaa'].data))/sp.mean(abs(wp['aaaaaa'].data))
Extracted_Features[i,77]=sp.mean(abs(wp['aaaaaa'].data))/sp.mean(abs(wp['d'].data))
```

```
    Extracted_Features[i,78]=sp.mean(abs(wp['d'].data))/sp.mean(abs(wp
    ['dd'].data))
    Extracted_Features[i,79]=sp.mean(abs(wp['dd'].data))/sp.mean(abs
    (wp['ddd'].data))
    Extracted_Features[i,80]=sp.mean(abs(wp['ddd'].data))/sp.mean(abs(wp
    ['dddd'].data))
    Extracted_Features[i,81]=sp.mean(abs(wp['dddd'].data))/sp.mean(abs
    (wp['ddddd'].data))
    Extracted_Features[i,82]=sp.mean(abs(wp['ddddd'].data))/sp.mean(abs
    (wp['dddddd'].data))
#%%
# =====================================================================
# Feature extraction from normal ECG signal
# =====================================================================
for i in range(Nofsignal):
    WPD_Feature_Extraction(ECGN[:,i], i, wname, level)
    Labels.append("NORMAL")
# =====================================================================
# Feature extraction from APC ECG signal
# =====================================================================
for i in range(Nofsignal, 2*Nofsignal):
    WPD_Feature_Extraction(ECGAPC[:,i-Nofsignal], i, wname, level)
    Labels.append("APC")
# =====================================================================
# Feature extraction from PVC ECG signal
# =====================================================================
for i in range(2*Nofsignal, 3*Nofsignal):
    WPD_Feature_Extraction(ECGPVC[:,i-2*Nofsignal], i, wname, level)
    Labels.append("PVC")
# =====================================================================
# Feature extraction from LBBB ECG signal
# =====================================================================
for i in range(3*Nofsignal, 4*Nofsignal):
    WPD_Feature_Extraction(ECGLBBB[:,i-3*Nofsignal], i, wname, level)
    Labels.append("LBBB")
# =====================================================================
# Feature extraction from RBBB ECG signal
# =====================================================================
for i in range(4*Nofsignal, 5*Nofsignal):
    WPD_Feature_Extraction(ECGRBBB[:,i-4*Nofsignal], i, wname, level)
    Labels.append("RBBB")
#%%
# =====================================================================
# Classification using random forest
# =====================================================================
from sklearn.model_selection import cross_val_score
from sklearn.metrics import cohen_kappa_score, make_scorer
from matplotlib import pyplot as plt
#To prevent warnings
import warnings
```

```
warnings.filterwarnings("ignore")
X = Extracted_Features
y = Labels

from sklearn.model_selection import train_test_split
# Split dataset into training set and test set
# 70% training and 30% test
Xtrain, Xtest, ytrain, ytest = train_test_split(X, y,test_size=0.3, ran-
dom_state=0)

from sklearn.ensemble import RandomForestClassifier
#In order to change to accuracy increase n_estimators
"""RandomForestClassifier(n_estimators='warn',   criterion='gini',   max_
depth=None,
min_samples_split=2, min_samples_leaf=1, min_weight_fraction_leaf=0.0,
max_features='auto', max_leaf_nodes=None, min_impurity_decrease=0.0,
min_impurity_split=None, bootstrap=True, oob_score=False, n_jobs=None,
random_state=None, verbose=0, warm_start=False, class_weight=None)"""
clf = RandomForestClassifier(n_estimators=200)
clf.fit(Xtrain,ytrain)
ypred = clf.predict(Xtest)

from sklearn import metrics
print('Accuracy:', np.round(metrics.accuracy_score(ytest,ypred),4))
print('Precision:', np.round(metrics.precision_score(ytest,
                    ypred,average='weighted'),4))
print('Recall:', np.round(metrics.recall_score(ytest,ypred,
                            average='weighted'),4))
print('F1 Score:', np.round(metrics.f1_score(ytest,ypred,
                            average='weighted'),4))
print('Cohen Kappa Score:', np.round(metrics.cohen_kappa_score(ytest,
ypred)))
print('Matthews Corrcoef:', np.round(metrics.matthews_corrcoef(ytest,
ypred)))
print('\t\tClassification Report:\n', metrics.classification_report(ypred,
ytest))

from sklearn.metrics import confusion_matrix
from io import BytesIO #neded for plot
import seaborn as sns; sns.set()

mat = confusion_matrix(ytest, ypred)
sns.heatmap(mat.T, square=True, annot=True, fmt='d', cbar=False)
plt.xlabel('true label')
plt.ylabel('predicted label');

plt.savefig("Confusion.jpg")
# Save SVG in a fake file object.
f = BytesIO()
plt.savefig(f, format="svg")
```

```python
#%%
"""The following PYTHON code is adapted from Scikit Learn to find the ROC
Area"""
# ================================================================
# ROC curves for the multiclass problem
# ================================================================
import numpy as np
import matplotlib.pyplot as plt
from itertools import cycle
from sklearn.discriminant_analysis import LinearDiscriminantAnalysis
from sklearn import svm
from sklearn.metrics import roc_curve, auc
from sklearn.model_selection import train_test_split
from sklearn.preprocessing import label_binarize
from sklearn.multiclass import OneVsRestClassifier
from scipy import interp
X = Extracted_Features
y = Labels
# Binarize the output
y = label_binarize(y, classes=['NORMAL','APC','PVC', 'LBBB','RBBB' ])
n_classes = y.shape[1]

#%%
# Import train_test_split function
from sklearn.model_selection import train_test_split
# Split dataset into training set and test set
# 70% training and 30% test
random_state = np.random.RandomState(1)
X_train, X_test, y_train, y_test = train_test_split(X, y,test_size=0.3,
random_state=0)
# Learn to predict each class against the other
"""classifier = OneVsRestClassifier(svm.SVC(kernel='linear',
probability=True,
                 random_state=random_state))"""
classifier = OneVsRestClassifier(LinearDiscriminantAnalysis(solver='lsqr',
shrinkage=None))
y_score = classifier.fit(X_train, y_train).decision_function(X_test)

# Compute ROC curve and ROC area for each class
fpr = dict()
tpr = dict()
roc_auc = dict()
for i in range(n_classes):
    fpr[i], tpr[i], _ = roc_curve(y_test[:, i], y_score[:, i])
    roc_auc[i] = auc(fpr[i], tpr[i])

# Compute micro-average ROC curve and ROC area
fpr["micro"], tpr["micro"], _ = roc_curve(y_test.ravel(), y_score.ravel())
```

```python
roc_auc["micro"] = auc(fpr["micro"], tpr["micro"])

###################################################################
# Plot of a ROC curve for a specific class
plt.figure()
lw = 2
plt.plot(fpr[2], tpr[2], color='darkorange',
         lw=lw, label='ROC curve (area = %0.2f)' % roc_auc[2])
plt.plot([0, 1], [0, 1], color='navy', lw=lw, linestyle='--')
plt.xlim([0.0, 1.0])
plt.ylim([0.0, 1.05])
plt.xlabel('False Positive Rate')
plt.ylabel('True Positive Rate')
plt.title('Receiver operating characteristic')
plt.legend(loc="lower right")
plt.show()

###################################################################
# Plot ROC curves for the multiclass problem
# Compute macro-average ROC curve and ROC area
# First aggregate all false positive rates
all_fpr = np.unique(np.concatenate([fpr[i] for i in range(n_classes)]))

# Then interpolate all ROC curves at this points
mean_tpr = np.zeros_like(all_fpr)
for i in range(n_classes):
    mean_tpr += interp(all_fpr, fpr[i], tpr[i])
# Finally average it and compute AUC
mean_tpr /= n_classes

fpr["macro"] = all_fpr
tpr["macro"] = mean_tpr
roc_auc["macro"] = auc(fpr["macro"], tpr["macro"])

# Plot all ROC curves
plt.figure()
plt.plot(fpr["micro"], tpr["micro"],
         label='micro-average ROC curve (area = {0:0.2f})'
               ''.format(roc_auc["micro"]),
         color='deeppink', linestyle=':', linewidth=4)

plt.plot(fpr["macro"], tpr["macro"],
         label='macro-average ROC curve (area = {0:0.2f})'
               ''.format(roc_auc["macro"]),
         color='navy', linestyle=':', linewidth=4)

colors = cycle(['aqua', 'darkorange', 'cornflowerblue'])
for i, color in zip(range(n_classes), colors):
    plt.plot(fpr[i], tpr[i], color=color, lw=lw,
```

```
            label='ROC curve of class {0} (area = {1:0.2f})'
            ''.format(i, roc_auc[i]))
plt.plot([0, 1], [0, 1], 'k--', lw=lw)
plt.xlim([0.0, 1.0])
plt.ylim([0.0, 1.05])
plt.xlabel('False Positive Rate')
plt.ylabel('True Positive Rate')
plt.title('Receiver Operating Characteristic to multi-class')
plt.legend(loc="lower right")
plt.show()
```

4.5 人类活动识别

大量的老年人口忍受着与年龄有关的健康问题。这些并发症，再加上老年人身体上明显发生的认知能力的逐渐减弱，使他们无法独自生活。近年来，信息和通信技术（ICT）的发展以及智能传感器和智能手机的进步，带来了智能环境的快速发展。智能医疗似乎是解决日益增长的老龄人口挑战的一个有希望的解决方案。为了满足日益增长的人群需求，我们正在提供智能医疗服务。特别是，使用智能医疗系统监测和评估老年人在日常活动中的任何关键健康状况的服务。智能医疗结构不仅可以让老年人独立生活，还可以通过减少老年人和依赖者对医疗系统的负担，提供更可持续的医疗服务。智能医疗监控系统（SHMS）已被开发为提供适合受试者实际需求的智能医疗服务的出色方法。为了解决这些结构的不同特点，我们已经实现了各种解决方案和方法。这些解决方案的主要目的是提供一个智能环境，让系统监测和分析受试者的健康状况，并为他们提供及时、智能的健康服务（Mshali, Lemlouma, Moloney, & Magoni, 2018）（Subasi, Khateeb, Brahimi, & Sarirete, 2020）。图 4.9 显示了 SHMS 的一般框架。

图 4.9　SHMS 的一般实验设定，该图改编自（Subasi et al., 2020b）

可穿戴设备的使用把医疗专业人员和病人一起带到了现代医疗保健系统中，从而可以对老年人的活动进行智能和自动的常规监测。智能可穿戴式传感器的集成促进了医疗保健智能监测系统的发展。在这个意义上，人们对在人类活动识别（HAR）中发挥重要作用的机器学习算法的发展有很大热情。智能医疗监控系统通过使用机器学习方法，准确有效地模拟和识别日常生活活动（ADL），提供了自动化的人类活动识别。在此过程中，手机传感器或可穿戴的身体传感器可被用于可靠和精确的人类活动识别（Subasi et al., 2020b）。

4.5.1 基于传感器的人类活动识别

随着无线网络技术的出现，可穿戴式人体传感器被越来越多地引入到各个领域的智能日常活动监测中，如紧急援助、认知援助和安全等（Majumder et al., 2017; Neves, Stachyra, & Rodrigues, 2008）。我们使用 HAR 从对使用可穿戴传感器的人活动的观察集合中识别不同的人类动作和手势。可穿戴传感器的优势可以通过使用 HAR 得以体现。在定义复杂活动方面，通常数据驱动的策略受到便携性、扩展性和感知问题的影响，而基于知识的处理复杂时间数据的解决方案通常很差（Liu, Peng, Liu, & Huang, 2015）。机器学习的目标之一是最大限度地减少大量数据之间的联系，使它们可以简单地反映整个数据而无须流通（Xu et al., 2017）。机器学习增加了可穿戴设备的可用性，改进了普适计算中的 HAR 工具（Liu et al., 2015）。机器学习算法可以处理从可穿戴式传感器获得的信息。由于 HAR 是一个发展迅速的科学领域，因此具有广泛的医疗应用、生活辅助、个人健身助手、家庭监护仪和恐怖分子的侦查。在智能家庭保健系统中，HAR 技术可被更改用以增强和建立临床康复过程（Hassan, Uddin, Mohamed, & Almogren, 2018）。因此，为了向老年人提供预防性帮助，护理人员可以使用这些设备来监测和解读他们的日常生活活动。（Liu, Nie, Liu, & Rosenblum, 2016）。这可以使老年人留在自己的家中。然而，有时我们仍需面对的困难之一是如何更可靠地监测日常生活活动（Subasi et al., 2018）。

例 4.9 以下 Python 代码用于使用 SVM 分类器对基于传感器的人类活动识别（HAR）数据进行分类，并采用了单独的训练和测试数据集。计算了分类准确率、精确度、召回率、F1 得分、Cohen kappa 得分和 Matthews 相关系数。还给出了分类报告和混淆矩阵。

数据集信息：该数据集从 UCI 机器学习资源库下载（UCI, 2018a）。REALDISP（REAListic sensor DISPlacement）数据集最初是为了研究现实世界中活动识别过程中传感器位移的影响而收集的。该数据集包括一系列广泛的物理活动和传感器模式。这些设置在数据集中得到了检验，该数据集采用了 17 个参与者的 9 个惯性传感器单元，考虑了 33 项健身活动。该数据集包含表示整个身体运动的简单活动（例如，行走或跳跃），而其他部分则侧重于个别部位的训练。采样率为 50Hz，这足以满足锻炼的需要。其中

8个传感器通常位于肢体的中部。另一个位于背部，略低于肩胛骨（Baños et al., 2012; Banos, Toth, Damas, Pomares, & Rojas, 2014）。实验中采用了不同的传感器位移设置以评估活动识别问题的复杂性对系统稳健性的影响。

本例中使用的是原始数据集的缩减版，有7项活动。为了确保这个数据集所记录的各种类型活动的平均分布，我们选择包括了身体几个部分在内的7项活动。

```
"""
REALDISP Activity Recognition IDEAL Dataset
Created on Tue Jun 18 18:14:15 2019
@author: asubasi
"""
from sklearn.model_selection import train_test_split
import numpy as np
import pandas as pd

# load data
dataset = pd.read_csv("REALDISPActivityRecognitionDataset.csv")
# split data into X and Y
X = dataset.iloc[:,0:117]
y = dataset.iloc[:, 117]
class_names = dataset.iloc[:, 117]
# split data into train and test sets
Xtrain, Xtest, ytrain, ytest = train_test_split(X, y, test_size=0.3,
random_state=7)
from sklearn import svm
""" The parameters and kernels of SVM classifierr can be changed as follows
C = 10.0 # SVM regularization parameter
svm.SVC(kernel='linear', C=C)
svm.LinearSVC(C=C, max_iter=10000)
svm.SVC(kernel='rbf', gamma=0.7, C=C)
svm.SVC(kernel='poly', degree=3, gamma='auto', C=C))
"""
C = 10.0 # SVM regularization parameter
clf = svm.SVC(kernel='linear', C=C)
clf.fit(Xtrain,ytrain)
ypred = clf.predict(Xtest)

from sklearn import metrics
print('Accuracy:', np.round(metrics.accuracy_score(ytest,ypred),4))
print('Precision:', np.round(metrics.precision_score(ytest,
                  ypred,average='weighted'),4))
print('Recall:', np.round(metrics.recall_score(ytest,ypred,
                  average='weighted'),4))
print('F1 Score:', np.round(metrics.f1_score(ytest,ypred,
                  average='weighted'),4))
print('Cohen Kappa Score:', np.round(metrics.cohen_kappa_score(ytest,
ypred)))
```

```
print('Matthews   Corrcoef:', np.round(metrics.matthews_corrcoef(ytest,
ypred)))
print('\t\tClassification Report:\n', metrics.classification_report(ypred,
ytest))

from sklearn.metrics import confusion_matrix
from io import BytesIO #neded for plot
import seaborn as sns; sns.set()
import matplotlib.pyplot as plt

mat = confusion_matrix(ytest, ypred)
sns.heatmap(mat.T, square=True, annot=True, fmt='d', cbar=False)
plt.xlabel('true label')
plt.ylabel('predicted label');

plt.savefig("SVM_Confusion.jpg")
# Save SVG in a fake file object.
f = BytesIO()
plt.savefig(f, format="svg")
```

4.5.2 基于智能手机的人类活动识别

信息和通信技术的进步促进了智能手机应用的更广泛使用。在现代医疗应用中，智能手机技术的使用将医生和病人联系起来，进行实时监控和医疗管理。此外，参与医疗领域的智能手机还推出了移动医疗和智能医疗监控系统等智能应用。移动医疗（m-healthcare）是这场革命前沿的一个重要改进特征。为了监测个人健康治疗和福祉，移动设备正在被广泛使用。多年来，移动电话的惊人发展已经极大地改变了人们的行为。随着移动设备和智能手机的增长，创新空间越来越大，提供的广泛的发展前景和更多的机会。智能手机有很多优点，比如作为便携式设备，在实验中不一定影响用户的生活方式（Boulos, Wheeler, Tavares, & Jones, 2011）。世界上有非常高比例的人口可以使用智能手机，而且可以上网的智能手机使用量正在急剧增加。HAR 是另一个具有广泛的创新可能，其与智能手机的结合值得进一步研究的领域。HAR 已被证明有助于检测各种健康问题的实例并保持健康的生活方式。人类活动的识别已经成为一个重要的计算领域，因为它对一般人，特别是对医疗保健服务有重大影响。需要长期照顾和监护的老人数量明显在增加。因此，更多的研究人员对老年病例感兴趣，将其作为一个研究和发展的领域（Reyes-Ortiz, Oneto, Sama, Parra, & Anguita, 2016）。过去的几年里，医疗服务经历了相当大的变化。移动医疗是这场革命前端的主要驱动力。移动医疗（mHealth）是一个由移动和物联网支持的医疗基础设施，包括使用通用分组无线电服务（GPRS）、4G 系统、全球定位系统（GPS）和蓝牙技术的手机（Kay, Santos, & Takane, 2011; Subasi, Fllatah, Alzobidi, Brahimi, & Sarirete, 2019）。基于智能手机的 HAR 一般框架如图 4.10 所示。

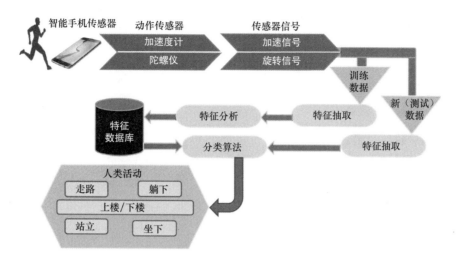

图 4.10 基于智能手机的 HAR 一般框架，该图改编自（Subasi et al., 2020b）

例 4.10 以下 Python 代码用于使用 LSTM 分类器对基于智能手机的 HAR 数据进行分类，该分类器采用了单独的训练和测试数据集。计算了分类准确率、精确度、召回率、F1 得分、Cohen kappa 得分和 Matthews 相关系数。还给出了分类报告和混淆矩阵。这个例子改编自 https://machinelearningmastery.com/how-toevelop-rnn-models-for-human-activity-recognition-timeseries-classification/。

数据集信息：该数据集从 UCI 资料库下载（UCI，2018b）。实验是在一组 30 名 19～48 岁的志愿者中进行的。他们完成了由六项基本活动组成的活动方案：三种静态姿势（站、坐、卧）和三种动态活动（走、下楼、上楼）。实验还涉及静态姿势之间的姿势转换。这些姿势是：站到坐、坐到站、坐到躺、躺到坐、站到躺和躺到站。在实验实施过程中，所有参与者的腰上都带着一部智能手机（三星 Galaxy S II）。我们利用设备的嵌入式加速度计和陀螺仪，以 50Hz 的恒定速率捕捉 3 轴线性加速度和 3 轴角速度。实验是用视频记录的，然后手动标注数据。实验获得的数据集被随机分为两组，其中 70% 的志愿者被选来生成训练数据，30% 被选来生成测试数据。

传感器信号（加速度计和陀螺仪）通过利用噪声滤波器进行预处理，然后在固定宽度的滑动窗口中进行采样，窗口宽度为 2.56 秒，重叠度为 50%（128 个读数/窗口）。传感器的加速度信号有重力和身体运动分量，用 Butterworth 低通滤波器分离成身体加速度和重力。重力被认为只有低频分量，因此采用了 0.3Hz 截止频率的滤波器。从每个窗口中，通过计算时域和频域变量，得到 561 个特征向量（Reyes-Ortiz et al., 2016; Subasi, et al., 2019e）。

```python
"""
Created on Sun Nov 24 15:50:51 2019
@author: asubasi
"""
# lstm model
from numpy import mean
from numpy import std
from numpy import dstack
from pandas import read_csv
from keras.models import Sequential
from keras.layers import Dense
from keras.layers import Flatten
from keras.layers import Dropout
from keras.layers import LSTM
from keras.utils import to_categorical
from keras import callbacks
from matplotlib import pyplot

# load a single file as a numpy array
def load_file(filepath):
    dataframe = read_csv(filepath, header=None, delim_whitespace=True)
    return dataframe.values

# load a list of files and return as a 3d numpy array
def load_group(filenames, prefix=''):
    loaded = list()
    for name in filenames:
        data = load_file(prefix + name)
        loaded.append(data)
    # stack group so that features are the 3rd dimension
    loaded = dstack(loaded)
    return loaded

# load a dataset group, such as train or test
def load_dataset_group(group, prefix=''):
    filepath = prefix + group + '/Inertial Signals/'
    # load all 9 files as a single array
    filenames = list()
    # total acceleration
    filenames += ['total_acc_x_'+group+'.txt',
        'total_acc_y_'+group+'.txt', 'total_acc_z_'+group+'.txt']
    # body acceleration
    filenames += ['body_acc_x_'+group+'.txt',
        'body_acc_y_'+group+'.txt', 'body_acc_z_'+group+'.txt']
    # body gyroscope
    filenames += ['body_gyro_x_'+group+'.txt',
        'body_gyro_y_'+group+'.txt', 'body_gyro_z_'+group+'.txt']
    # load input data
    X = load_group(filenames, filepath)
```

```python
            # load class output
            y = load_file(prefix + group + '/y_'+group+'.txt')
            return X, y
# load the dataset, returns train and test X and y elements
def load_dataset(prefix=""):
            # load all train
            Xtrain, ytrain = load_dataset_group('train', prefix + 'UCI HAR
            Dataset/')
            print(Xtrain.shape, ytrain.shape)
            # load all test
            Xtest, ytest = load_dataset_group('test', prefix + 'UCI HAR
            Dataset/')
            print(Xtest.shape, ytest.shape)
            # zero-offset class values
            ytrain = ytrain - 1
            ytest = ytest - 1
            # one hot encode y
            ytrain = to_categorical(ytrain)
            ytest = to_categorical(ytest)
            print(Xtrain.shape, ytrain.shape, Xtest.shape, ytest.shape)
            return Xtrain, ytrain, Xtest, ytest
# load data
Xtrain, ytrain, Xtest, ytest = load_dataset()
target_names = ['Walking',
           'Upstairs',
           'Downstairs',
           'Sitting',
           'Standing',
           'Laying' ]
#%%
# ===================================================================
# Create an LSTM model
# ===================================================================
verbose, epochs, batch_size = 1, 5, 64
n_timesteps, n_features, n_outputs = Xtrain.shape[1], Xtrain.shape[2],
ytrain.shape[1]
model = Sequential()
model.add(LSTM(100, input_shape=(n_timesteps,n_features)))
model.add(Dropout(0.5))
model.add(Dense(100, activation='relu'))
model.add(Dense(n_outputs, activation='softmax'))
# ===================================================================
# Compile the model
# ===================================================================
model.compile(loss='categorical_crossentropy', optimizer='adam',
metrics=['accuracy'])
#%%
# ===================================================================
# Enable validation to use ModelCheckpoint and EarlyStopping callbacks
# ===================================================================
```

```python
callbacks_list = [
    callbacks.ModelCheckpoint(
        filepath='best_model.{epoch:02d}-{val_loss:.2f}.h5',
        monitor='val_loss', save_best_only=True),]
# ================================================================
# Train the model
# ================================================================
history = model.fit(Xtrain,
          ytrain,epochs=epochs, batch_size=batch_size,
          callbacks=callbacks_list,
          validation_split=0.3,
          verbose=1)
#%%
# ================================================================
# Evaluate the model
# ================================================================
test_loss, test_acc = model.evaluate(Xtest, ytest, batch_size=batch_size,
verbose=verbose)
print('\nTest accuracy:', test_acc)
#%%
from matplotlib import pyplot as plt
plt.figure(figsize=(6, 4))
plt.plot(history.history['accuracy'], 'r', label='Accuracy of training
data')
plt.plot(history.history['val_accuracy'], 'b', label='Accuracy of
validation data')
plt.plot(history.history['loss'], 'r--', label='Loss of training data')
plt.plot(history.history['val_loss'], 'b--', label='Loss of validation
data')
#plt.plot(history.history['val_accuracy'], label = 'val_accuracy')
plt.title('Model Accuracy and Loss')
plt.ylabel('Accuracy and Loss')
plt.xlabel('Training Epoch')
plt.ylim(0)
plt.legend()
plt.show()
#%%
from sklearn.metrics import classification_report
import numpy as np
# Print confusion matrix for training data
y_pred_test = model.predict(Xtest,)
# Take the class with the highest probability from the train predictions
max_y_pred_test = np.round(y_pred_test)
#max_y_train = np.argmax(testy, axis=1)
#max_y_pred_train = np.argmax(y_pred_train, axis=1)
print(classification_report(ytest, max_y_pred_test))
#%%
from sklearn.metrics import confusion_matrix
import seaborn as sns
matrix = confusion_matrix(ytest.argmax(axis=1), max_y_pred_test.
```

```
argmax(axis=1))
plt.figure(figsize=(6, 4))
sns.heatmap(matrix,cmap='coolwarm',linecolor='white',linewidths=1,
            xticklabels=target_names,
            yticklabels=target_names,
            annot=True,
            fmt='d')
plt.title('Confusion Matrix')
plt.ylabel('True Label')
plt.xlabel('Predicted Label')
plt.show()
#%%
from sklearn import metrics
print('Accuracy:', np.round(metrics.accuracy_score(ytest,
max_y_pred_test),4))
print('Precision:', np.round(metrics.precision_score(ytest,
                  max_y_pred_test,average='weighted'),4))
print('Recall:', np.round(metrics.recall_score(ytest, max_y_pred_test,
                  average='weighted'),4))
print('F1 Score:', np.round(metrics.f1_score(ytest, max_y_pred_test,
                  average='weighted'),4))
print('Cohen Kappa Score:', np.round(metrics.cohen_kappa_score(ytest.
argmax(axis=1), max_y_pred_test.argmax(axis=1)),4))
print('Matthews Corrcoef:', np.round(metrics.matthews_corrcoef(ytest.
argmax(axis=1), max_y_pred_test.argmax(axis=1)),4))

print('\t\tClassification Report:\n', metrics.classification_report(ytest,
max_y_pred_test))

#%%
from sklearn.metrics import confusion_matrix
from io import BytesIO #neded for plot
import seaborn as sns; sns.set()
import matplotlib.pyplot as plt

#Convert the binary labels back
#confusion_matrix(y_test.values.argmax(axis=1), predictions.
argmax(axis=1))
mat = confusion_matrix(ytest.argmax(axis=1), max_y_pred_test.
argmax(axis=1))
sns.heatmap(mat.T, square=True, annot=True, fmt='d', cbar=False)
plt.title('Confusion Matrix')
plt.ylabel('True Label')
plt.xlabel('Predicted Label')
plt.show()

plt.savefig("Confusion.jpg")
# Save SVG in a fake file object.
f = BytesIO()
plt.savefig(f, format="svg")
```

4.6 用于癌症检测的微阵列基因表达数据分类

癌症是世界各地死亡的主要原因之一。许多不同类型的癌症已经在各种组织和器官中被诊断出来。因为它与细胞的基因异常有关，允许同时测量基因表达水平的 DNA 微阵列已用于描述肿瘤细胞的基因表达谱。因此，这些测量可以显示出检测细胞中的异常情况（Chen, Li, & Wei, 2007）。微阵列技术还可以对癌基因进行标准化的临床评估逻辑诊断和预后。发现通常引发癌症的基因，在理解癌症的生物学机制中具有重要意义（Rojas-Galeano, Hsieh, Agranoff, Krishna, & Fernandez-Reyes, 2008）。从微阵列数据分析中获得准确的结果，找到信息丰富的基因至关重要（Xu & Zhang, 2006）。尽管如此，这些研究都是针对有限的癌症类型，而且由于未在广泛的病人年龄组中得到充分验证，其结果的用途有限。尽管数据检索过程具有挑战性，但近几十年来，微阵列技术已被作为一个令人鼓舞的工具来利用，以提高癌症的诊断和治疗。另一方面，从基因表达中产生的数据包括高水平的噪音和相对于可用样本数量的大量基因。因此，用微阵列数据进行分类和统计技术是一个巨大的挑战。发现在癌症中被普遍调控的基因可能对理解癌症的共同生物学机制有重要意义。在过去的几年里，许多科学家分析了不同癌症类型的全球基因表达谱。基于机器学习的决策支持系统有助于医生和临床医生的诊断和预后过程（Vural & Subasi, 2015）。

例 4.11 以下 Python 代码使用极限树分类器和 10 折交叉验证对微阵列基因表达数据（白血病）进行分类。计算了分类准确率、精确度、召回率、F1 得分、Cohen kappa 得分和 Matthews 相关系数。

数据集信息：我们将该方法应用于广泛使用的公开微阵列数据集，其中包括急性髓系白血病（AML）- 急性淋巴细胞白血病（ALL）和混合系白血病（MLL）基因（http://portals.broadinstitute.org/cgi-bin/cancer/publications/pub_paper.cgi?mode=view&paper_id=63）。这个数据集包括从外周血和骨髓中提取的白血病患者样本的测量。携带涉及混合系白血病基因（MLL、ALL1、HRX）的染色体易位的 ALL，预后特别差。Armstrong 等人（Armstrong et al., 2002）提出它们构成一种独特的疾病，表示为 MLL，并表明基因表达的差异足以将白血病正确分类为 MLL、ALL 或 AML。确立 MLL 是一个独特的实体是至关重要的，因为它要求对选择性表达的基因进行检查，以寻找急需的分子目标。测量结果对应于急性淋巴细胞白血病（ALL）、急性髓细胞白血病（AML）和混合系白血病基因（MLL），分别包括 24、28 和 20 个样本。它们使用由 11 224 个基因组成的 Affymetrix 微阵列进行分析。

```
"""
Created on Tue Jun 18 18:14:15 2019
@author: asubasi
"""
import scipy.io as sio
```

```python
from sklearn.model_selection import cross_val_score
from sklearn.metrics import cohen_kappa_score, make_scorer
import numpy as np
# import file into a dictionary
mat_contents = sio.loadmat('Leukemia2.mat')
sorted(mat_contents.keys())
# read in the structure
data = mat_contents['data']
# get the fields
# split data into X and y
X = data[:,1:11225]
y = data[:,0]
#%%
# =====================================================================
# Extra trees example with cross-validation
# =====================================================================
#Import Extra Trees model
from sklearn.ensemble import ExtraTreesClassifier
# fit model no training data
model = ExtraTreesClassifier(n_estimators=100, max_features=11200)

CV=10 #10-Fold Cross Validation
#Evaluate Model Using 10-Fold Cross Validation and Print Performance Metrics
Acc_scores = cross_val_score(model, X, y, cv=CV)
print("Accuracy: %0.3f (+/- %0.3f)" % (Acc_scores.mean(), Acc_scores.std() * 2))
f1_scores = cross_val_score(model, X, y, cv=CV,scoring='f1_macro')
print("F1 score: %0.3f (+/- %0.3f)" % (f1_scores.mean(), f1_scores.std() * 2))
Precision_scores = cross_val_score(model, X, y, cv=CV,scoring='precision_macro')
print("Precision score: %0.3f (+/- %0.3f)" % (Precision_scores.mean(), Precision_scores.std() * 2))
Recall_scores = cross_val_score(model, X, y, cv=CV,scoring='recall_macro')
print("Recall score: %0.3f (+/- %0.3f)" % (Recall_scores.mean(), Recall_scores.std() * 2))
from sklearn.metrics import cohen_kappa_score, make_scorer
kappa_scorer = make_scorer(cohen_kappa_score)
Kappa_scores = cross_val_score(model, X, y, cv=CV,scoring=kappa_scorer)
print("Kappa score: %0.3f (+/- %0.3f)" % (Kappa_scores.mean(), Kappa_scores.std() * 2))
```

4.7 乳腺癌检测

乳腺癌是妇女乳腺组织内恶性肿瘤的一种可能病变。它是妇女中常见的病症之一。乳腺癌是世界各地的妇女癌症死亡率的最重要的原因之一。随着生物医学和计算机技术的不断发展，与乳腺癌有关的各种临床因素已经被记录下来。为了应对乳腺癌的急剧增长，许多研究人员考虑利用病人的临床记录来预测病人是否患有乳腺癌。高效的乳腺癌诊断仍然是一个重大的挑战，早期诊断对于防止疾病的发展是极为重要的（Hassan, Hossain, Begg, Ramamohanarao, & Morsi, 2010）。乳腺癌是全世界女性中最常见的癌症，患者占了所有女性癌症的15%。尽管有一定的风险，但通过预防可以减少患病人数，这些方法不能减少大多数晚期诊断的乳腺癌患者。因此，早期检测是控制乳腺癌以提高乳腺癌生存率的基石。乳房X光检查和细针穿刺细胞学（FNAC）是主要的诊断方法，但这些方法没有足够的诊断精确度。毫无疑问，对患者的数据评估和医生的决定是诊断中最关键的因素。与乳腺造影和FNAC一起，使用不同的方法可以作为医生诊断的决策支持工具，从而获得更好的诊断系统。就上述要求而言，可以利用机器学习方法来促进诊断系统的改进。通过使用基于机器学习的自动诊断系统，可以消除医生潜在的诊断错误，并且可以详细检查医疗数据（Aličković & Subasi, 2017）。

例4.12 以下Python代码被用来对从UCI下载的乳腺癌数据集（WDBC）进行分类，使用Keras深度学习模型，并进行10折交叉验证。计算分类准确率。

数据集信息：乳腺癌是一种由乳腺细胞产生的恶性肿瘤。尽管一些风险因素（即遗传风险因素、衰老、肥胖、家族史、未生育和月经期）会增加女性患乳腺癌的几率，但目前还不知道是什么原因导致大多数患者罹患乳腺癌，以及不同的因素是如何激发细胞癌变的。为了进一步了解情况，研究人员已经做了许多研究。其中，在确定DNA的具体变化如何导致健康的乳腺细胞癌变方面取得了很大进展（Aličković & Subasi, 2017; Jerez-Aragonés, Gómez-Ruiz, Ramos-Jiménez, Muñoz-Pérez, & Alba-Conejo, 2003; Marcano-Cedeño, Quintanilla-Domínguez, Andina, 2011）。

在UCI机器学习资源库中，有两个不同的威斯康星州乳腺癌数据集（UCI, 2019a）。第一个数据集是威斯康星州乳腺癌诊断（WBCD）数据集。它包括569个不同的实例和32个属性：良性357个，恶性212个。所有的属性都是从病人的乳腺组织细针穿刺（FNA）的数字化图像中计算出来的。乳腺组织中的所有细胞核由10个实值特征定义，并计算所有这些特征的平均值、标准误差和"最差值"（三个最大值的平均值）。因此，每张图像有30个属性（Aličković & Subasi, 2017）。

- 半径（从中心到周边各点距离的平均值）
- 纹理（灰度值的标准差）
- 周长
- 面积
- 平滑度（半径长度的局部变化）
- 紧凑度（周长^2/面积-1.0）
- 凹度（轮廓线凹陷部分的严重程度）

- 凹陷点（轮廓线凹陷部分的数量）
- 对称性
- 分形维数（"海岸线近似值"-1）

第二个数据集是威斯康星州的乳腺癌原始数据集，包括699个取自乳腺组织的样本。然后，从数据集中剔除了缺失值的数据，结果剩下683个样例。这个数据库中的每条记录都有9个属性，所有的值都是1～10之间的整数，并且在良性和恶性之间变化明显。测量的9个属性是（Aličković & Subasi, 2017; UCI, 2019a）：

- 肿块厚度
- 细胞大小的均匀性
- 细胞形状的均匀性
- 边缘粘附
- 单个上皮细胞大小
- 裸露的细胞核
- 平淡的染色质
- 正常细胞核
- 有丝分裂

```
"""
Created on Tue Jun 18 18:14:15 2019
@author: asubasi
"""
from sklearn.model_selection import cross_val_score
from sklearn.metrics import cohen_kappa_score, make_scorer
import numpy as np
import pandas as pd
# load data
dataset = pd.read_csv("wdbc.csv")
# split data into X and Y
X = dataset.iloc[:,1:31]
y = dataset.iloc[:, 0]
class_names = dataset.iloc[:, 0]
#%%
# ===================================================================
# Classification with Keras deep learning model with cross-validation
# ===================================================================
from keras.models import Sequential
from keras.layers import Dense
from keras.wrappers.scikit_learn import KerasClassifier
from keras.utils import np_utils
from sklearn.model_selection import cross_val_score
from sklearn.model_selection import KFold
from sklearn.preprocessing import LabelEncoder
# encode class values as integers
```

```
encoder = LabelEncoder()
encoder.fit(y)
encoded_Y = encoder.transform(y)
# convert integers to dummy variables (i.e. one hot encoded)
dummy_y = np_utils.to_categorical(encoded_Y)
lenOfCoded=dummy_y.shape[1] # Dimension of binary coded output data
InputDataDimension=30
# define baseline model
def baseline_model():
            # create model
            model = Sequential()
            model.add(Dense(20, input_dim=InputDataDimension,
            activation='relu'))
            model.add(Dense(lenOfCoded, activation='softmax'))
            # Compile model
            model.compile(loss='categorical_crossentropy',
            optimizer='adam', metrics=['accuracy'])
            return model
estimator = KerasClassifier(build_fn=baseline_model, epochs=50, batch_size=5, verbose=1)
kfold = KFold(n_splits=10, shuffle=True)
results = cross_val_score(estimator, X, dummy_y, cv=kfold)
print("Accuracy: %.2f%% (%.2f%%)" % (results.mean()*100, results.std()*100))
```

4.8 预测胎儿风险的心电图数据分类

在妊娠周期中，胎儿心率（FHR）是关于胎儿的最重要的指标之一。产科医生利用持续胎心电子监护（CTG）来获取信息，其中包括与胎儿有关的FHR和子宫收缩（UC）。CTG不仅能获得FHR，还能协助观察母亲的宫缩和其他类型的胎儿监测。一般来说，产前CTG的适当时期是妊娠28周，即第三个妊娠期。这种测试可以通过内部或外部技术来进行。在内部测试中，一个膨胀至一定阶段的导管将被放在子宫中。在外部测试中，一对传感器节点被连接到母亲的腹部。CTG数据一般表示两条线。上边的线记录了以每分钟为单位的FHR。下线则记录子宫收缩。为了在CTG的基础上发现胎儿的风险，机器学习技术是一个日益增长的趋势，在医学上创造了决策支持系统。各种研究都对CTG数据进行了分类（Sahin & Subasi, 2015）。从CTG中获取的信息可以用于早期发现病理状态，进而帮助产科医生预测未来的问题并阻止对胎儿的永久性损害。在整个分娩过程中，表现出缺氧的婴儿会造成暂时的损伤或死亡。对FHR记录的错误诊断和对胎儿采用的不适当的治疗可以造成一半以上的死亡（Ayres-de-Campos et al., 2005; Cesarelli, Romano, & Bifulco, 2009; Gribbin & Thornton, 2006; Grivell, Alfirevic, Gyte, & Devane, 2015）。虽然它很实用，但CTG监测的成功率可能有一些不一致，尤其是在低危妊娠中。如果评估不正确，则可能导致治疗无效或错误的胎儿健康状况结果，导致胎儿无法得到必要的治疗（Subasi, Kadasa, & Kremic,2019; Van Geijn, Jongsma, de Haan, & Eskes, 1980）。

例 4.13 以下 Python 代码用于对心电图数据集进行分类，采用了线性判别分析（LDA）分类器和 10 折交叉验证。计算了分类准确率、精确度、召回率、F1 得分、Cohen kappa 得分和 Matthews 相关系数。

数据集信息：利用从 UCI 下载的心电图（CTG）数据集（https://archive.ics.uci.edu/ml/datasets/cardiotocography）（Bache & Lichman, 2013）对学习器的表现进行了评估。共有 2126 张胎儿心电图（CTG）被自动处理，并测量各自的诊断特征。CTG 数据由三位产科专家检查，通过检查胚胎状态确定为正常或病态。CTG 数据有 3 个等级（N= 正常, S= 可疑, P= 病理）和 21 个特征，其中 13 个是离散的, 8 个是连续的。该数据有以下属性：LB, FHR 基线（每分钟心跳数）；AC, 每秒加速次数；FM, 每秒胎动次数；UC, 每秒子宫收缩次数；DL, 每秒轻度减速次数；DS, 每秒严重减速次数；DP, 每秒长时间减速次数；ASTV, 短期变化异常的时间百分比；MSTV, 短期变化的平均值；ALTV, 长期变化异常的时间百分比；MLTV, 长期变异性的平均值；width, FHR 直方图的宽度；min, FHR 直方图的最小值；max, FHR 直方图的最大值；Nmax, 直方图峰值数；Nzeros, 直方图零点的数量；mode, 直方图模式；平均数, 直方图平均数；中位数, 直方图中位数；方差, 直方图方差；趋势, 直方图趋势；类, FHR 模式类代码（1 至 10）；NSP, 胎儿状态类代码（N= 正常、S= 可疑、P= 病理）。

```
"""
Created on Thu May 9 12:18:30 2019
@author: absubasi
"""
from sklearn.metrics import cohen_kappa_score, make_scorer
import scipy as sp
import numpy as np
import pandas as pd
#To prevent warnings
import warnings
warnings.filterwarnings("ignore")


# load data
dataset = pd.read_csv("CTG3CLASS.csv")
# split data into X and y
X = dataset.iloc[:,0:21]
y = dataset.iloc[:, 21]
class_names = dataset.iloc[:, 21]
X=X.to_numpy() #Convert Pandas Dataframe into Numpy
```

```python
# ================================================================
# LDA classification with training and test set
# ================================================================
# Import train_test_split function
from sklearn.model_selection import train_test_split
# Split dataset into training set and test set
# 70% training and 30% test
Xtrain, Xtest, ytrain, ytest = train_test_split(X, y,test_size=0.3, random_state=1)
#Import LDA model
from sklearn.discriminant_analysis import LinearDiscriminantAnalysis
#Create a LDA Classifier
clf = LinearDiscriminantAnalysis(solver='lsqr', shrinkage=None)
#Train the model using the training sets
clf.fit(Xtrain,ytrain)
#Predict the response for test dataset
ypred = clf.predict(Xtest)
#Evaluate the Model and Print Performance Metrics
from sklearn import metrics
print('Accuracy:', np.round(metrics.accuracy_score(ytest,ypred),4))
print('Precision:', np.round(metrics.precision_score(ytest,
                   ypred,average='weighted'),4))
print('Recall:', np.round(metrics.recall_score(ytest,ypred,
                            average='weighted'),4))
print('F1 Score:', np.round(metrics.f1_score(ytest,ypred,
                            average='weighted'),4))
print('Cohen Kappa Score:', np.round(metrics.cohen_kappa_score(ytest,ypred),4))
print('Matthews Corrcoef:', np.round(metrics.matthews_corrcoef(ytest,ypred),4))
print('\n\t\tClassification Report:\n', metrics.classification_report(ypred, ytest))

#Plot Confusion Matrix
from sklearn.metrics import confusion_matrix
print("Confusion Matrix:\n",confusion_matrix(ytest, ypred))

#%%
# ================================================================
# LDA example with cross-validation
# ================================================================
from sklearn.discriminant_analysis import LinearDiscriminantAnalysis
from sklearn.model_selection import cross_val_score
# fit model no training data
#model = LinearDiscriminantAnalysis(solver='lsqr', shrinkage='auto')
model = LinearDiscriminantAnalysis(solver='lsqr', shrinkage=None)

CV=10 #10-Fold Cross Validation
#Evaluate Model Using 10-Fold Cross Validation
Acc_scores = cross_val_score(model, X, y, cv=CV)
```

```
print("Accuracy: %0.3f (+/- %0.3f)" % (Acc_scores.mean(), Acc_scores.
std() * 2))
f1_scores = cross_val_score(model, X, y, cv=CV,scoring='f1_macro')
print("F1 score: %0.3f (+/- %0.3f)" % (f1_scores.mean(), f1_scores.std()
* 2))
Precision_scores = cross_val_score(model, X, y, cv=CV,scoring='precision_
macro')
print("Precision score: %0.3f (+/- %0.3f)" % (Precision_scores.mean(),
Precision_scores.std() * 2))
Recall_scores = cross_val_score(model, X, y, cv=CV,scoring='recall_macro')
print("Recall score: %0.3f (+/- %0.3f)" % (Recall_scores.mean(), Recall_
scores.std() * 2))
kappa_scorer = make_scorer(cohen_kappa_score)
Kappa_scores = cross_val_score(model, X, y, cv=CV,scoring=kappa_scorer)
print("Kappa score: %0.3f (+/- %0.3f)" % (Kappa_scores.mean(), Kappa_
scores.std() * 2))
```

4.9 糖尿病检测

糖尿病是人类普遍存在的一种生理性疾病。当一个人由于缺乏胰岛素而无法分解葡萄糖时，被称为糖尿病患者。人体器官胰腺负责分泌激素胰岛素，它是一种调节人体血液中糖分水平的重要酶。它利用糖为人体产生能量，如果没有足够的胰岛素，身体细胞就不能获得所需要的能量。因此血液中的糖分水平变得过高，身体就会出现一些问题。糖尿病不是一种可以治愈的疾病，幸运的是，它是可以治疗的。在当代医疗保健中，预测和准确治疗疾病已成为医学预测学科的首要任务。糖尿病的治疗完全是人工完成的，通常由医生推荐。Smith 等人（Smith, Everhart, Dickson, Knowler, Johannes, 1988）利用了基于感知器的算法——称为自适应学习例程（ADAP）的算法，是一种早期的神经网络模型，用于预测糖尿病的发作。本实验使用了皮马印第安人（Pima Indians）糖尿病（PID）数据集。该数据集取自 UCI 机器学习库，包含年龄在 20 岁以上的遗传皮马印第安人血统的妇女，在研究时居住在美国。输出的二元变量有 0、1 两个取值，其中 1 表示糖尿病测试阳性，0 是测试阴性。总共有 268 个（34.9%）病例测试阳性（1 类），500 个（65.1%）病例测试阴性（0 类）。该数据集有八种临床属性（Ashiquzzaman et al., 2018）。

例 4.14 以下 Python 代码用于使用随机森林分类器对糖尿病数据集进行分类。计算分类准确率、精确度、召回率、F1 得分、Cohen kappa 得分和 Matthews 相关系数。也计算混淆矩阵和 ROC 区域。

数据集信息：该数据集包括 768 个不同的实例，数据集中的所有病人都是至少 21 岁的女性。输出的二元变量取值 0 或 1，其中，0 意味着糖尿病检测为阴性，1 表示检测为阳性。标识为 0 的有 500 例，1 的有 268 例。数据集的采样人口是亚利桑那州凤凰城附近的皮马印第安人。自1965年以来，生活在亚利桑那州南部吉拉河印第安社区的皮马印第安人为糖尿病及其影响的纵向研究做出了贡献。这个社区的糖尿病发生率是世界上最高的（35岁时占50%）。皮马印第安人的糖尿病与胰岛素依赖性、酮症酸中毒或胰岛细胞抗体无关，因此即使发生在年轻人身上，也是2型糖尿病。糖尿病肾病是这一人群中普遍存在的肾脏疾病，其临床特征和常规病理特征与其他人群中描述的相似。它经常导致终末期肾病，近15%患糖尿病的皮马印第安人在20年后会出现这种情况（Mercaldo, Nardone, & Santone, 2017）。该数据集的属性如下：

1. 怀孕次数
2. 口服葡萄糖耐量试验（GTIT）中2小时的血浆葡萄糖浓度
3. 舒张压（mm Hg）
4. 三头肌皮肤褶皱厚度（mm）
5. 2小时血清胰岛素（μU/ml）
6. 体重指数
7. 糖尿病谱系功能
8. 年龄（岁）

```python
"""
Created on Thu May 9 12:18:30 2019
@author: absubasi
"""

import scipy as sp
import numpy as np
import pandas as pd
#To prevent warnings
import warnings
warnings.filterwarnings("ignore")

# load data
from numpy import loadtxt
dataset = loadtxt('pima-indians-diabetes.csv', delimiter=",")

# split data into X and y
X = dataset[:,0:8]
y = dataset[:,8]

# Import train_test_split function
from sklearn.model_selection import train_test_split
# Split dataset into training set and test set
# 70% training and 30% test
Xtrain, Xtest, ytrain, ytest = train_test_split(X, y,test_size=0.3, random_state=1)
#%%
```

```python
# ================================================================
# Random forest example with training and test set
# ================================================================
from sklearn.ensemble import RandomForestClassifier

#In order to change to accuracy increase n_estimators
"""RandomForestClassifier(n_estimators='warn', criterion='gini', max_depth=None,
min_samples_split=2, min_samples_leaf=1, min_weight_fraction_leaf=0.0,
max_features='auto', max_leaf_nodes=None, min_impurity_decrease=0.0,
min_impurity_split=None, bootstrap=True, oob_score=False, n_jobs=None,
random_state=None, verbose=0, warm_start=False, class_weight=None)"""
clf = RandomForestClassifier(n_estimators=200)
#Create the Model
#Train the model with Training Dataset
clf.fit(Xtrain,ytrain)
#Test the model with Testset
ypred = clf.predict(Xtest)

#Evaluate the Model and Print Performance Metrics
from sklearn import metrics
print('Accuracy:', np.round(metrics.accuracy_score(ytest,ypred),4))
print('Precision:', np.round(metrics.precision_score(ytest,
                    ypred,average='weighted'),4))
print('Recall:', np.round(metrics.recall_score(ytest,ypred,
                    average='weighted'),4))
print('F1 Score:', np.round(metrics.f1_score(ytest,ypred,
                    average='weighted'),4))
print('Cohen Kappa Score:', np.round(metrics.cohen_kappa_score(ytest,
ypred),4))
print('Matthews Corrcoef:', np.round(metrics.matthews_corrcoef(ytest,
ypred),4))
print('\t\tClassification Report:\n', metrics.classification_report(ypred,
ytest))

#Plot Confusion Matrix
from sklearn.metrics import confusion_matrix
print("Confusion Matrix:\n",confusion_matrix(ytest, ypred))

#%%
# ================================================================
# Random forest example with cross-validation
# ================================================================
from sklearn.ensemble import RandomForestClassifier
from sklearn.model_selection import cross_val_score
#In order to change to accuracy increase n_estimators
# fit model no training data
model = RandomForestClassifier(n_estimators=200)

CV=10 #10-Fold Cross Validation
```

```python
#Evaluate Model Using 10-Fold Cross Validation and Print Performance
Metrics
Acc_scores = cross_val_score(model, X, y, cv=CV)
print("Accuracy: %0.3f (+/- %0.3f)" % (Acc_scores.mean(), Acc_scores.
std() * 2))
f1_scores = cross_val_score(model, X, y, cv=CV,scoring='f1_macro')
print("F1 score: %0.3f (+/- %0.3f)" % (f1_scores.mean(), f1_scores.std()
* 2))
Precision_scores = cross_val_score(model, X, y, cv=CV,scoring='precision_
macro')
print("Precision score: %0.3f (+/- %0.3f)" % (Precision_scores.mean(),
Precision_scores.std() * 2))
Recall_scores = cross_val_score(model, X, y, cv=CV,scoring='recall_macro')
print("Recall score: %0.3f (+/- %0.3f)" % (Recall_scores.mean(), Recall_
scores.std() * 2))
from sklearn.metrics import cohen_kappa_score, make_scorer
kappa_scorer = make_scorer(cohen_kappa_score)
Kappa_scores = cross_val_score(model, X, y, cv=10,scoring=kappa_scorer)
print("Kappa score: %0.3f (+/- %0.3f)" % (Kappa_scores.mean(), Kappa_
scores.std() * 2))

#%%
# =====================================================================
# # Classification and ROC analysis for binary classification
# =====================================================================
from sklearn.ensemble import RandomForestClassifier
from sklearn.metrics import roc_curve, auc
from sklearn.model_selection import StratifiedKFold
from scipy import interp
import matplotlib.pyplot as plt
random_state = np.random.RandomState(0)
###################################################################
# Run classifier with cross-validation and plot ROC curves
cv = StratifiedKFold(n_splits=5)
classifier = RandomForestClassifier(n_estimators=200)

tprs = []
aucs = []
mean_fpr = np.linspace(0, 1, 100)

i = 0
for train, test in cv.split(X, y):
    probas_ = classifier.fit(X[train], y[train]).predict_proba(X[test])
    # Compute ROC curve and area the curve
    fpr, tpr, thresholds = roc_curve(y[test], probas_[:, 1])
    tprs.append(interp(mean_fpr, fpr, tpr))
    tprs[-1][0] = 0.0
    roc_auc = auc(fpr, tpr)
    aucs.append(roc_auc)
    plt.plot(fpr, tpr, lw=1, alpha=0.3,
```

```
            label='ROC fold %d (AUC = %0.2f)' % (i, roc_auc))
    i += 1
plt.plot([0, 1], [0, 1], linestyle='--', lw=2, color='r',
         label='Chance', alpha=.8)

mean_tpr = np.mean(tprs, axis=0)
mean_tpr[-1] = 1.0
mean_auc = auc(mean_fpr, mean_tpr)
std_auc = np.std(aucs)
plt.plot(mean_fpr, mean_tpr, color='b',
         label=r'Mean ROC (AUC = %0.2f $\pm$ %0.2f)' % (mean_auc, std_auc),
         lw=2, alpha=.8)

std_tpr = np.std(tprs, axis=0)
tprs_upper = np.minimum(mean_tpr + std_tpr, 1)
tprs_lower = np.maximum(mean_tpr - std_tpr, 0)
plt.fill_between(mean_fpr, tprs_lower, tprs_upper, color='grey',
                 alpha=.2,
         label=r'$\pm$ 1 std. dev.')
plt.xlim([-0.05, 1.05])
plt.ylim([-0.05, 1.05])
plt.xlabel('False Positive Rate')
plt.ylabel('True Positive Rate')
plt.title('Receiver operating characteristic example')
plt.legend(loc="lower right")
plt.show()
```

4.10 心脏病检测

心脏是人体中最重要的器官之一。它是循环系统的中心。如果心脏不能正常工作，其他多个器官将停止工作。根据美国心脏协会 2015 年心脏病和中风统计更新，心血管疾病是世界范围内主要的死因，每年死亡 1730 万人，到 2030 年估计将增加到 2360 万以上。并且，每年死于心血管疾病的总人数正在急剧上升。如果能在早期阶段准确识别和诊断心脏病，并提供适当的后续治疗，那么就可以挽救相当多的生命，降低死亡率。心脏病的诊断是一个复杂的过程，医生根据他们的知识和治疗患有类似问题和症状的病人时所遇到的经验得出结论。这一过程可能导致不正确的假设，因为有些因素与各种器官有关。本书所介绍的工作旨在实现医疗诊断过程的自动化，并开发一个预测系统，使用机器学习以更高的准确率检测心脏病。为了诊断一种疾病，要对病人进行一系列的测试。在机器学习技术的帮助下对疾病进行预测，可以减少测试的数量，从而节省时间，并以合理的成本提供优质服务（Dembla & Bhatia, 2016）。

例 4.15 以下 Python 代码利用包括训练和测试数据集的随机森林分类器对 Cleveland 心脏数据集进行分类。计算分类准确率、精确度、召回率、F1 得分、Cohen kappa 得分和 Matthews 相关系数。给出了分类报告和混淆矩阵。由于分类器的准确率较低，我们采用了有数据规范化和无数据规范化两种不同的情况。通过使用数据规范化，分类器的性能得到了改善。

数据集信息：随机森林分类器的性能利用 Cleveland 心脏数据集进行评估。Cleveland 心脏数据集可以从 UCI 机器学习数据集下载（https://archive.ics.uci.edu/ml/datasets/Heart+Disease），由 Detrano 提供。该数据集包括 303 个实例和 14 个属性，被分为两类疾病。该数据有以下属性：age，年龄，以岁为单位；sex，性别（1= 男性，0= 女性）；cp，胸痛类型（1：典型心绞痛，2：非典型心绞痛，3：非心绞痛，4：无症状）；trestbpsm，静息血压（入院时毫米汞柱）；chol，血清胆固醇，毫克/分升；fbs，空腹血糖 >120 毫克/分升（1= 真；0= 假）；restecg，静息心电图结果（0：正常，1：有 ST-T 波异常，2：按 Estes 标准显示可能或明确的左心室肥厚）；thalach，达到的最大心率；exang，运动性心绞痛（1= 是，0= 否）；oldpeak，相对于休息时，运动诱发的 ST 段压低；slope，运动 ST 段峰值的斜率（1：上斜，2：平坦，3：下斜）；ca，荧光染色的主要血管数（0～3）；thal，3= 正常，6= 固定缺陷，7= 可逆缺陷；num，心脏病的诊断（血管造影疾病状态，0：<50% 直径狭窄，1：>50% 直径狭窄）。

```
"""
Created on Tue Jun 18 18:14:15 2019
@author: asubasi
"""
import numpy as np
import pandas as pd
# load data
dataset = pd.read_csv('cleveland_heart.csv')
# split data into X and Y
X = dataset.iloc[:,0:13]
y = dataset.iloc[:, 13]
class_names = dataset.iloc[:, 13]
#%%
# =====================================================================
# Evaluation without data normalization
# =====================================================================
from sklearn.ensemble import RandomForestClassifier
from sklearn.model_selection import train_test_split
# split data into train and test sets
Xtrain, Xtest, ytrain, ytest = train_test_split(X, y, test_size=0.33, random_state=7)
#In order to change to accuracy increase n_estimators
"""RandomForestClassifier(n_estimators='warn',    criterion='gini',    max_depth=None,
min_samples_split=2, min_samples_leaf=1, min_weight_fraction_leaf=0.0,
max_features='auto', max_leaf_nodes=None, min_impurity_decrease=0.0,
min_impurity_split=None, bootstrap=True, oob_score=False, n_jobs=None,
```

```python
random_state=None, verbose=0, warm_start=False, class_weight=None)"""
clf = RandomForestClassifier(n_estimators=200)
#Create the Model
#Train the model with Training Dataset
clf.fit(Xtrain,ytrain)
#Test the model with Testset
ypred = clf.predict(Xtest)
#Evaluate the Model and Print Performance Metrics
from sklearn import metrics
print('Accuracy:', np.round(metrics.accuracy_score(ytest,ypred),4))
print('Precision:', np.round(metrics.precision_score(ytest,
                 ypred,average='weighted'),4))
print('Recall:', np.round(metrics.recall_score(ytest,ypred,
                          average='weighted'),4))
print('F1 Score:', np.round(metrics.f1_score(ytest,ypred,
                            average='weighted'),4))
print('Cohen Kappa Score:', np.round(metrics.cohen_kappa_score(ytest,
ypred),4))
print('Matthews  Corrcoef:',  np.round(metrics.matthews_corrcoef(ytest,
ypred),4))
print('\t\tClassification Report:\n', metrics.classification_report(ypred,
ytest))
#Plot Confusion Matrix
from sklearn.metrics import confusion_matrix
print("Confusion Matrix:\n",confusion_matrix(ytest, ypred))
#%%
# ====================================================================
# Evaluation with data normalization
# ====================================================================
from sklearn.ensemble import RandomForestClassifier
from sklearn.model_selection import train_test_split
from sklearn import preprocessing

X_normalized = preprocessing.normalize(X, norm='l2')
# split data into train and test sets
Xtrain, Xtest, ytrain, ytest = train_test_split(X_normalized, y, test_size=0.33, random_state=7)
#In order to change to accuracy increase n_estimators
"""RandomForestClassifier(n_estimators='warn',   criterion='gini',   max_depth=None,
min_samples_split=2, min_samples_leaf=1, min_weight_fraction_leaf=0.0,
max_features='auto', max_leaf_nodes=None, min_impurity_decrease=0.0,
min_impurity_split=None, bootstrap=True, oob_score=False, n_jobs=None,
random_state=None, verbose=0, warm_start=False, class_weight=None)"""
clf = RandomForestClassifier(n_estimators=200)
#Create the Model
#Train the model with Training Dataset
clf.fit(Xtrain,ytrain)
#Test the model with Testset
ypred = clf.predict(Xtest)
```

```
#Evaluate the Model and Print Performance Metrics
from sklearn import metrics
print('Accuracy:', np.round(metrics.accuracy_score(ytest,ypred),4))
print('Precision:', np.round(metrics.precision_score(ytest,
              ypred,average='weighted'),4))
print('Recall:', np.round(metrics.recall_score(ytest,ypred,
              average='weighted'),4))
print('F1 Score:', np.round(metrics.f1_score(ytest,ypred,
              average='weighted'),4))
print('Cohen Kappa Score:', np.round(metrics.cohen_kappa_score(ytest,
ypred),4))
print('Matthews Corrcoef:', np.round(metrics.matthews_corrcoef(ytest,
ypred),4))
print('\t\tClassification Report:\n', metrics.classification_report(ypred,
ytest))
#Plot Confusion Matrix
from sklearn.metrics import confusion_matrix
print("Confusion Matrix:\n",confusion_matrix(ytest, ypred))
```

4.11 慢性肾脏病的诊断

慢性肾脏病（CKD）是一种慢性医疗保健问题，几乎影响了全球10%的人口（Cueto-Manzano et al., 2014; Pérez-Sáez et al., 2015）。在现实生活中，CKD可以发现与心血管疾病和肾脏功能逐渐丧失导致的入院、发病和死亡风险增加有关的病例。被诊断为CKD的受试者有很高的风险受到动脉硬化和其他类型疾病的影响。这些疾病对他们的生活质量有很大影响。CKD发现的关键后果是肾脏损害（Levin & Stevens, 2014）。许多适应症或风险因素也与CKD的进展有关，因此这些因素可能会极大地影响CKD的识别。通过监测CKD的进展性，建立在机器学习概念基础上的诊断计算模型新理解为改善CKD的诊断提供了巨大的潜力（Chen, Zhang, Zhu, Xiang, & Harrington, 2016）。众多研究提出了用于CKD诊断的模型（Chen et al., 2016; Muthukumar & Krishnan, 2016; Subasi et al., 2017）。

例4.16 以下Python代码用于对慢性肾脏病（CKD）数据集（https://archive.ics.uci.edu/ml/datasets/Chronic_Kidney_Disease#）采用LDA分类器进行分类。计算分类准确率、精确度、召回率、F1得分、Cohen kappa得分和Matthews相关系数。同时还计算了混淆矩阵和ROC区域。

数据集信息：本例中使用的CKD数据集是从UCI机器学习库中下载的（UCI, 2019b）。该数据由Soundarapandian等人捐赠，在近2个月的时间内收集，共包括400个样本，由14个数字属性和10个标称属性以及一个类描述符表示。在400个样本中，250个样本属于CKD组，另外150个样本属于非CKD组。更多细节请见Chen等人（2016）。

数据集包含以下属性：age，年龄；bp，血压；sg，比重；al，白蛋白；su，糖；rbc，红细胞；pc，脓细胞；pcc，脓细胞团块；ba，细菌；bgr，随机血糖；bu，血液尿素；sc，血清肌酐；sod，钠；pot，钾；hemo，血红蛋白；pcv，红细胞压积；wc，白细胞计数；rc，红细胞计数；htn，高血压；dm，糖尿病；cad，冠状动脉疾病；appet，食欲；pe，脚部水肿；ane，贫血；class，CKD/NOCKD。

```
"""
Created on Thu May 9 12:18:30 2019
@author: absubasi
"""
import numpy as np
import pandas as pd
from sklearn.preprocessing import LabelEncoder
#To prevent warnings
import warnings
warnings.filterwarnings("ignore")

# load data
dataset = pd.read_csv("chronic_kidney_disease.csv")
#Encode string data
for col in dataset.columns:
    if dataset[col].dtype == "object":
        encoded = LabelEncoder()
        encoded.fit(dataset[col])
        dataset[col] = encoded.transform(dataset[col])
#%%
# split data into X and y
X = dataset.iloc[:,0:24]
y = dataset.iloc[:, 24]
#%%
# Import train_test_split function
from sklearn.model_selection import train_test_split
# Split dataset into training set and test set
# 70% training and 30% test
Xtrain, Xtest, ytrain, ytest = train_test_split(X, y,test_size=0.3, random_state=1)
#%%
# ================================================================
# LDA example with training and test set
# ================================================================
from sklearn.discriminant_analysis import LinearDiscriminantAnalysis
clf = LinearDiscriminantAnalysis(solver='lsqr', shrinkage=None)
#Create the Model
#Train the model with Training Dataset
clf.fit(Xtrain,ytrain)
#Test the model with Testset
ypred = clf.predict(Xtest)
#Evaluate the Model and Print Performance Metrics
```

```python
from sklearn import metrics
print('Accuracy:', np.round(metrics.accuracy_score(ytest,ypred),4))
print('Precision:', np.round(metrics.precision_score(ytest,
                ypred,average='weighted'),4))
print('Recall:', np.round(metrics.recall_score(ytest,ypred,
                        average='weighted'),4))
print('F1 Score:', np.round(metrics.f1_score(ytest,ypred,
                        average='weighted'),4))
print('Cohen Kappa Score:', np.round(metrics.cohen_kappa_score(ytest,
ypred),4))
print('Matthews   Corrcoef:',  np.round(metrics.matthews_corrcoef(ytest,
ypred),4))
print('\t\tClassification Report:\n', metrics.classification_report(ypred,
ytest))
#Plot Confusion Matrix
from sklearn.metrics import confusion_matrix
print("Confusion Matrix:\n",confusion_matrix(ytest, ypred))

from io import BytesIO #neded for plot
import seaborn as sns; sns.set()
import matplotlib.pyplot as plt

mat = confusion_matrix(ytest, ypred)
sns.heatmap(mat.T, square=True, annot=True, fmt='d', cbar=False)
plt.xlabel('true label')
plt.ylabel('predicted label');
plt.savefig("Confusion.jpg")
# Save SVG in a fake file object.
f = BytesIO()
plt.savefig(f, format="svg")


#%%
# ====================================================================
# LDA example with cross-validation
# ====================================================================
from sklearn.discriminant_analysis import LinearDiscriminantAnalysis
from sklearn.model_selection import cross_val_score
#In order to change to accuracy increase n_estimators
# fit model no training data
model = LinearDiscriminantAnalysis(solver='lsqr', shrinkage=None)

CV=10 #10-Fold Cross Validation
#Evaluate Model Using 10-Fold Cross Validation and Print Performance
Metrics
Acc_scores = cross_val_score(model, X, y, cv=CV)
print("Accuracy: %0.3f (+/- %0.3f)" % (Acc_scores.mean(), Acc_scores.
std() * 2))
f1_scores = cross_val_score(model, X, y, cv=CV,scoring='f1_macro')
print("F1 score: %0.3f (+/- %0.3f)" % (f1_scores.mean(), f1_scores.std()
* 2))
```

```
Precision_scores = cross_val_score(model, X, y, cv=CV,scoring='precision_
macro')
print("Precision score: %0.3f (+/- %0.3f)" % (Precision_scores.mean(),
Precision_scores.std() * 2))
Recall_scores = cross_val_score(model, X, y, cv=CV,scoring='recall_macro')
print("Recall score: %0.3f (+/- %0.3f)" % (Recall_scores.mean(), Recall_
scores.std() * 2))
from sklearn.metrics import cohen_kappa_score, make_scorer
kappa_scorer = make_scorer(cohen_kappa_score)
Kappa_scores = cross_val_score(model, X, y, cv=10,scoring=kappa_scorer)
print("Kappa score: %0.3f (+/- %0.3f)" % (Kappa_scores.mean(), Kappa_
scores.std() * 2))
```

4.12 本章小结

机器学习是一个相对较新的领域，也可能是计算机科学中最活跃的领域之一。鉴于数字化数据的广泛可用性及其众多应用，我们可以假设，在未来的几十年里，它将继续以非常快的速度增长。有许多不同的学习问题，有些是由数据规模的大幅增长驱动的——在一些应用中已经涉及数十亿条记录的处理，有些则与全新学习系统的实施有关——这些问题可能会带来新的研究挑战并需要新的算法解决方案。在所有情况下，学习理论、算法和实现都是计算机科学和数学的一个迷人的领域，我们希望这本书至少能在一定程度上解释清楚这些（Mohri, Rostamizadeh, & Talwalkar, 2018）。本章中，我们定义了医疗保健数据分类中广泛的机器学习算法和技术，以及它们不同的实现方式。每一节末尾的例子将帮助读者对所提到的策略和原则以及单独提供的完整解决方案更有经验。其中一些也可以作为学术工作和研究新问题的一个切入点。在算法的具体实现过程中，本章讨论了几种机器学习算法及其变体，这些算法可以直接用于提取现实世界学习问题的有效解决方案。本书对所提出算法的详细描述和分析将有助于读者对于算法的实现和将其改进并运用于其他场景。

4.13 参考文献

Abdullah, A.A., Subasi, A., & Qaisar, S.M. (2017). *Surface EMG Signal Classification by Using WPD and Ensemble Tree Classifiers*. 62, 475. Springer.

Akben, S. B., Subasi, A., & Tuncel, D. (2012). Analysis of repetitive flash stimulation frequencies and record periods to detect migraine using artificial neural network. Journal of Medical Systems, 36(2), 925–931.

Aličković, E., & Subasi, A. (2017). Breast cancer diagnosis using GA feature selection and Rotation Forest. Neural Computing and Applications, 28(4), 753–763.

Andrzejak, R. G., Schindler, K., & Rummel, C. (2012). Nonrandomness, nonlinear dependence, and nonstationarity of electroencephalographic recordings from epilepsy patients. Physical Review E, 86(4), 046206.

Armstrong, S. A., Staunton, J. E., Silverman, L. B., Pieters, R., den Boer, M. L., Minden, M. D., & Korsmeyer, S. J. (2002). MLL translocations specify a distinct gene expression profile that distinguishes a unique leukemia. Nature Genetics, 30(1), 41–47.

Ashiquzzaman, A., Tushar, A. K., Islam, M. R., Shon, D., Im, K., Park, J. -H., & Kim, J. (2018). Reduction of overfitting in diabetes prediction using deep learning neural network. In *IT Convergence and Security 2017*. Berlin: Springer 35–43.

Ayres-de-Campos, D., Costa-Santos, C., & Bernardes, J. SisPorto® Multicentre Validation Study Group. (2005). Prediction of neonatal state by computer analysis of fetal heart rate tracings: The antepartum arm of the SisPorto® multicentre validation study. *European Journal of Obstetrics & Gynecology and Reproductive Biology, 118*(1), 52–60.

Bache, K., & Lichman, M. (2013). UCI Machine Learning Repository [http://archive. Ics. Uci. Edu/ml]. University of California, School of Information and Computer Science. Irvine, CA.

Baños, O., Damas, M., Pomares, H., Rojas, I., Tóth, M. A., & Amft, O. (2012). A benchmark dataset to evaluate sensor displacement in activity recognition. *UbiComp '12: Proceedings of the 2012 ACM Conference on Ubiquitous Computing*, 1026–1035.

Banos, O., Toth, M. A., Damas, M., Pomares, H., & Rojas, I. (2014). Dealing with the effects of sensor displacement in wearable activity recognition. *Sensors, 14*(6), 9995–10023.

Barlow, J. S. (1993). *The electroencephalogram: Its patterns and origins*. Cambridge, MA: MIT press.

Begg, R., Lai, D. T., & Palaniswami, M. (2008). *Computational intelligence in biomedical engineering*. Boca Raton, FL: CRC Press.

Berbari, E. J. (2000). Principles of electrocardiography. *The Biomedical Engineering Handbook, 1*, 1 13–11.

Boulos, M. N. K., Wheeler, S., Tavares, C., & Jones, R. (2011). How smartphones are changing the face of mobile and participatory healthcare: An overview, with example from eCAALYX. *Biomedical Engineering Online, 10*(1), 24.

Bronzino, J. D. (1999). *Biomedical engineering handbook* (Vol. 2). Boca Raton, FL: CRC press.

Cesarelli, M., Romano, M., & Bifulco, P. (2009). Comparison of short term variability indexes in cardiotocographic foetal monitoring. *Computers in Biology and Medicine, 39*(2), 106–118.

Chen, Z., Zhang, Z., Zhu, R., Xiang, Y., & Harrington, P. B. (2016). Diagnosis of patients with chronic kidney disease by using two fuzzy classifiers. *Chemometrics and Intelligent Laboratory Systems, 153*, 140–145.

Chen, Z., Li, J., & Wei, L. (2007). A multiple kernel support vector machine scheme for feature selection and rule extraction from gene expression data of cancer tissue. *Artificial Intelligence in Medicine, 41*(2), 161–175.

Chiauzzi, E., Rodarte, C., & DasMahapatra, P. (2015). Patient-centered activity monitoring in the self-management of chronic health conditions. *BMC Medicine, 13*(1), 77.

Cueto-Manzano, A. M., Cortés-Sanabria, L., Martínez-Ramírez, H. R., Rojas-Campos, E., Gómez-Navarro, B., & Castillero-Manzano, M. (2014). Prevalence of chronic kidney disease in an adult population. *Archives of Medical Research, 45*(6), 507–513.

De Chazal, P., O'Dwyer, M., & Reilly, R. B. (2004). Automatic classification of heartbeats using ECG morphology and heartbeat interval features. *IEEE Transactions on Biomedical Engineering, 51*(7), 1196–1206.

Dembla, P. (2016). *Multiclass Diagnosis Model for Heart Disease using PSO based SVM*. MS Thesis, Thapar University

Emly, M., Gilmore, L. D., & Roy, S. H. (1992). Electromyography. *IEEE Potentials, 11*(2), 25–28.

Gopura, R.A. R. C., Kiguchi, K., & Li, Y. (2009). SUEFUL-7: A 7DOF upper-limb exoskeleton robot with muscle-model-oriented EMG-based control. 1126–1131. IEEE.

Gribbin, C., & Thornton, J. (2006). Critical evaluation of fetal assessment methods. *High risk pregnancy management options*. Amsterdam: Elsevier.

Grivell, R. M., Alfirevic, Z., Gyte, G. M., & Devane, D. (2015). Antenatal cardiotocography for fetal assessment. *Cochrane Database of Systematic Reviews*, 9.

Hassan, M. M., Uddin, M. Z., Mohamed, A., & Almogren, A. (2018). A robust human activity recognition system using smartphone sensors and deep learning. *Future Generation Computer Systems, 81*, 307–313.

Hassan, M. R., Hossain, M. M., Begg, R. K., Ramamohanarao, K., & Morsi, Y. (2010). Breast-cancer identification using HMM-fuzzy approach. *Computers in Biology and Medicine, 40*(3), 240–251.

Hsieh, J., & Hsu, M. -W. (2012). A cloud computing based 12-lead ECG telemedicine service. *BMC Medical Informatics and Decision Making, 12*(1), 77.

Jerez-Aragonés, J. M., Gómez-Ruiz, J. A., Ramos-Jiménez, G., Muñoz-Pérez, J., & Alba-Conejo, E. (2003). A combined neural network and decision trees model for prognosis of breast cancer relapse. *Artificial Intelligence in Medicine, 27*(1), 45–63.

Kay, M., Santos, J., & Takane, M. (2011). mHealth: New horizons for health through mobile technologies. *World Health Organization, 64*(7), 66–71.

Koelstra, S., Muhl, C., Soleymani, M., Lee, J. -S., Yazdani, A., Ebrahimi, T., & Patras, I. (2012). Deap: A database for emotion analysis; using physiological signals. *IEEE Transactions on Affective Computing, 3*(1), 18–31.

Kurzynski, M., Krysmann, M., Trajdos, P., & Wolczowski, A. (2016). Multiclassifier system with hybrid learning applied to the control of bioprosthetic hand. *Computers in Biology and Medicine, 69*, 286–297. doi: 10.1016/j.compbiomed.2015.04.023.

Lenzi, T., De Rossi, S., Vitiello, N., Chiri, A., Roccella, S., Giovacchini, F., & Carrozza, M. C. (2009). The neuro-robotics paradigm: NEURARM, NEUROExos, HANDEXOS. *2009 Annual International Conference of the IEEE Engineering in Medicine and Biology Society*, 2430–2433 Minneapolis.

Levin, A., & Stevens, P. E. (2014). Summary of KDIGO 2012 CKD Guideline: Behind the scenes, need for guidance, and a framework for moving forward. Kidney International, 85(1), 49–61.

Liu, L., Peng, Y., Liu, M., & Huang, Z. (2015). Sensor-based human activity recognition system with a multilayered model using time series shapelets. Knowledge-Based Systems, 90, 138–152.

Liu, Y., Nie, L., Liu, L., & Rosenblum, D. S. (2016). From action to activity: Sensor-based activity recognition. Neurocomputing, 181, 108–115.

Lo, H. S., & Xie, S. Q. (2012). Exoskeleton robots for upper-limb rehabilitation: State of the art and future prospects. Medical Engineering and Physics, 34(3), 261–268.

Lum, P. S., Burgar, C. G., Shor, P. C., Majmundar, M., & Van der Loos, M. (2002). Robot-assisted movement training compared with conventional therapy techniques for the rehabilitation of upper-limb motor function after stroke. Archives of Physical Medicine and Rehabilitation, 83(7), 952–959.

Majumder, S., Aghayi, E., Noferesti, M., Memarzadeh-Tehran, H., Mondal, T., Pang, Z., & Deen, M. J. (2017). Smart homes for elderly healthcare—recent advances and research challenges. Sensors, 17(11), 2496.

Marcano-Cedeño, A., Quintanilla-Domínguez, J., & Andina, D. (2011). WBCD breast cancer database classification applying artificial metaplasticity neural network. *Expert Systems with Applications*, 38(8), 9573–9579.

Menshawy, M. E., Benharref, A., & Serhani, M. (2015). An automatic mobile-health based approach for EEG epileptic seizures detection. Expert Systems with Applications, 42(20), 7157–7174.

Mercaldo, F., Nardone, V., & Santone, A. (2017). Diabetes mellitus affected patients classification and diagnosis through machine learning techniques. Procedia Computer Science, 112, 2519–2528.

Mohri, M., Rostamizadeh, A., & Talwalkar, A. (2018). Foundations of machine learning. Cambridge MA: MIT press.

Mshali, H., Lemlouma, T., Moloney, M., & Magoni, D. (2018). A survey on health monitoring systems for health smart homes. International Journal of Industrial Ergonomics, 66, 26–56.

Muthukumar, P., & Krishnan, G. S. S. (2016). A similarity measure of intuitionistic fuzzy soft sets and its application in medical diagnosis. Applied Soft Computing, 41, 148–156.

Neves, P., Stachyra, M., & Rodrigues, J. (2008). Application of wireless sensor networks to healthcare promotion. *Journal of Communications Software and Systems*, 4(3).

Niedermeyer, E., & da Silva, F. L. (2005). Electroencephalography: Basic principles, clinical applications and related fields. Philadelphia: Lippincott Williams & Wilkins.

Nikolic, M. (2001). Findings and firing pattern analysis in controls and patients with myopathy and amytrophic lateral sclerosis. Copenhagen: University of Copenhagen.

Pan, J., & Tompkins, W. J. (1985). A real-time QRS detection algorithm. IEEE Transactions on Biomedical Engineering, 32(3), 230–236.

Pérez-Sáez, M. J., Prieto-Alhambra, D., Barrios, C., Crespo, M., Redondo, D., Nogués, X., & Pascual, J. (2015). Increased hip fracture and mortality in chronic kidney disease individuals: The importance of competing risks. Bone, 73, 154–159.

Reyes-Ortiz, J. -L., Oneto, L., Sama, A., Parra, X., & Anguita, D. (2016). Transition-aware human activity recognition using smartphones. Neurocomputing, 171, 754–767.

Riener, R., Nef, T., & Colombo, G. (2005). Robot-aided neurorehabilitation of the upper extremities. Medical and Biological Engineering and Computing, 43(1), 2–10.

Rojas-Galeano, S., Hsieh, E., Agranoff, D., Krishna, S., & Fernandez-Reyes, D. (2008). Estimation of relevant variables on high-dimensional biological patterns using iterated weighted kernel functions. PloS One, 3(3), e1806.

Sahin, H., & Subasi, A. (2015). Classification of the cardiotocogram data for anticipation of fetal risks using machine learning techniques. Applied Soft Computing, 33, 231–238.

Sanei, S. (2013). *Adaptive processing of brain signals*. West Sussex, United Kingdom: John Wiley & Sons.

Sanei, S., & Chambers, J. A. (2013). EEG signal processing. Hoboken: John Wiley & Sons.

Sapsanis, C., Georgoulas, G., & Tzes, A. (2013). EMG based classification of basic hand movements based on time-frequency features. *21st Mediterranean Conference on Control and Automation*, 716–722.

Sasaki, D., Noritsugu, T., & Takaiwa, M. (2005). Development of active support splint driven by pneumatic soft actuator (ASSIST). *Proceedings of the 2005 IEEE International Conference on Robotics and Automation*, 520–525 Barcelona, Spain.

Serhani, M. A., El Menshawy, M., & Benharref, A. (2016). SME2EM: Smart mobile end-to-end monitoring architecture for life-long diseases. Computers in Biology and Medicine, 68, 137–154.

Smith, J. W., Everhart, J., Dickson, W., Knowler, W., & Johannes, R. (1988). Using the ADAP learning algorithm to forecast the onset of diabetes mellitus. *Proceedings of the Annual Symposium on Computer Application in Medical Care*, 261–265 American Medical Informatics Association.

Sörnmo, L., & Laguna, P. (2005). *Bioelectrical signal processing in cardiac and neurological applications (Vol. 8)*. San Diego, CA: Academic Press.

Subasi, A. (2019a). Electroencephalogram-controlled assistive devices. In K. Pal, H.-B. Kraatz, A. Khasnobish, S. Bag, I. Banerjee, & U. Kuruganti (Eds.), Bioelectronics and Medical Devices (pp. 261–284). https://doi.org/10.1016/B978-0-08-102420-1.00016-9.

Subasi, A. (2019b). Electromyogram-controlled assistive devices. *Bioelectronics and medical devices*. Amsterdam: Elsevier 285 - 311.

Subasi, A. (2019c). *Practical guide for biomedical signals analysis using machine learning techniques, a MATLAB based approach (1st ed.)*. Cambridge, MA: Academic Press.

Subasi, A., Ahmed, A., Aličković, E., & Rashik Hassan, A. (2019d). Effect of photic stimulation for migraine detection using random forest and discrete wavelet transform. Biomedical Signal Processing and Control, 49, 231–239. doi:10.1016/j.bspc.2018.12.011.

Subasi, A., Alickovic, E., & Kevric, J. (2017). Diagnosis of chronic kidney disease by using random forest. In CMBEBIH 2017 (pp. 589–594). Springer.

Subasi, A., Bandic, L., & Qaisar, S. M. (2020a). Cloud-based health monitoring framework using smart sensors and smartphone. *Innovation in health informatics*. Amsterdam: Elsevier 217 -L 243.

Subasi, A., Dammas, D. H., Alghamdi, R. D., Makawi, R. A., Albiety, E. A., Brahimi, T., & Sarirete, A. (2018). Sensor based human activity recognition using Adaboost ensemble classifier. *Procedia Computer Science*, 140, 104–111. doi:10.1016/j.procs.2018.10.298.

Subasi, A., Fllatah, A., Alzobidi, K., Brahimi, T., & Sarirete, A. (2019e). Smartphone-based human activity recognition using bagging and boosting. *Presented at the the 16th International Learning and Technology Conference* Jeddah, Saudi Arabia.

Subasi, A., Kadasa, B., & Kremic, E. (2019). Classification of the cardiotocogram data for anticipation of fetal risks using bagging ensemble classifier. *Presented at the Complex Adaptive Systems Conference*. Malvern: PA November 13.

Subasi, A., Khateeb, K., Brahimi, T., & Sarirete, A. (2020b). Human activity recognition using machine learning methods in a smart healthcare environment. *Innovation in Health Informatics*. Amsterdam: Elsevier 123 –L 144.

Thaler, M. (2017). The only EKG book you'll ever need. Philadelphia: Lippincott Williams & Wilkins.

UCI Machine Learning Repository: Breast Cancer Wisconsin (Diagnostic) Data Set. (n.d.). Retrieved November 17, 2019, from https://archive.ics.uci.edu/ml/datasets/Breast+Cancer+Wisconsin+(Diagnostic).

UCI Machine Learning Repository: Chronic_Kidney_Disease Data Set. (n.d.). Retrieved November 19, 2019, from https://archive.ics.uci.edu/ml/datasets/Chronic_Kidney_Disease#.

UCI Machine Learning Repository: REALDISP Activity Recognition Dataset Data Set. (n.d.). Retrieved June 24, 2018, from https://archive.ics.uci.edu/ml/datasets/REALDISP+Activity+Recognition+Dataset#.

UCI Machine Learning Repository: Smartphone-Based Recognition of Human Activities and Postural Transitions Data Set. (n.d.). Retrieved November 22, 2018, from https://archive.ics.uci.edu/ml/datasets/Smartphone-Based+Recognition+of+Human+Activities+and+Postural+Transitions#.

Van Geijn, H. P., Jongsma, H. W., de Haan, J., & Eskes, T. K. (1980). Analysis of heart rate and beat-to-beat variability: Interval difference index. American Journal of Obstetrics and Gynecology, 138(3), 246–252.

Vural, H., & Subasi, A. (2015). Data-mining techniques to classify microarray gene expression data using gene selection by SVD and information gain. Modeling of Artificial Intelligence, 6, 171–182.

Wojtczak, P., Amaral, T. G., Dias, O. P., Wolczowski, A., & Kurzynski, M. (2009). Hand movement recognition based on biosignal analysis. Engineering Applications of Artificial Intelligence, 22(4), 608–615.

Wołczowski, A., & Zdunek, R. (2017). Electromyography and mechanomyography signal recognition: Experimental analysis using multi-way array decomposition methods. Biocybernetics and Biomedical Engineering, 37(1), 103–113 https://doi.org/10.1016/j.bbe.2016.09.004.

Xu, X., & Zhang, A. (2006). *Boost feature subset selection: A new gene selection algorithm for microarray dataset*. Berlin: Springer 670 – 677.

Xu, Y., Shen, Z., Zhang, X., Gao, Y., Deng, S., Wang, Y., & Chang, C. (2017). Learning multi-level features for sensor-based human action recognition. Pervasive and Mobile Computing, 40, 324–338.

CHAPTER 5

第 5 章

其他分类示例

5.1 入侵检测

对信息技术（IT）基础设施的攻击是企业网络面临的最大威胁之一。特别是对那些以分布式方式运营的公司，IT 管理者需要将安全性扩展到企业骨干网之外，并且将潜在的漏洞考虑在内，包括但不仅限于互联网连接、远程和本地公司的办公连接，以及和受信任的合作伙伴之间的链接。然而只关注企业资源和内部流量的安全并不能帮助我们识别出刻意的攻击或发现整个公司的潜在漏洞。通常入侵检测系统（IDS）的必要性是受到质疑的。为什么防火墙的安全性不够？对安全威胁进行分类则可以看出两者的差异。攻击来自公司外部网络还是内部网络？防火墙往往扮演着公司内部网络和外部网络（互联网）之间的防护墙的角色。任何经过防火墙的流量都要根据预先定义的安全策略进行过滤。可惜不是所有互联网上的访问都会经过防火墙。有些人也许会使用未授权的连接从内部网络连接互联网。同时所有的威胁并非都发生在防火墙之外。事实上，大量的安全事件和损失来源于内部人员。此外，防火墙也是攻击者们的重点。如果你不考虑以上问题，防火墙是足够的。连接公域网络的公司无法完全依赖单一形式的安全策略。所以，防火墙必须辅以入侵检测系统（IDS）（Bace, 1999）。

IDS 可以检测计算机攻击和（或）系统滥用，一旦检测到攻击就会向相关人员发出警报。每个 IDS 工具或系统都有三个关键的安全功能：监控、检测和响应来自公司网络内部或者外部的任何类型的未经授权的活动。任何 IDS 程序都会利用策略来识别入侵事件。一旦事件被触发，就会根据策略作出适当的响应。一些 IDS 程序可以向系统管理员发出警告。某些 IDS 程序还可以作为一种预防措施并在安全威胁发生时采取行动。它们可以对某些攻击类型作出响应，如禁用账户、注销用户和运行某些脚本。因此，IDS 程序是一种实用的调查企业网络内部和外部的攻击的方法。入侵可能是为了经济、政治、军事利益或某种个人原因而发起的（Innella & McMillan, 2017）。

IDS 是一种跟踪系统网络或操作的软件，它可以检查可疑的活动，并在发现未经授权的行为时发出警告（Tiwari, Kumar, Bharti, & Kishan, 2017）。IDS 使用机器学习算法来区分网络中的正常和异常行为。机器学习可以通过判断采集的数据间的关系来优化可用的知识，从而有效地用于最小化冗余信息和提高决策水平（Govindarajan & Chandrasekaran, 2011）。入侵检测是一种对主动破坏网络的攻击事件的检测机制。为了实现更好的保护并减少损害和（或）其他将来的攻击，记录这些事件至关重要。安全专业人员会利用布置在防火墙前和防火墙后的传感器并比较两种传感器获得的结果（Sahu, Mishra, Das, & Mishra, 2014）。由于城市化和物联网（IoT）的快速发展，许多安全问题和挑战也随之上升。网络攻击容易发生在基于传感器的物联网设备以及用于收集和分析数据的计算机上。机器学习算法可以有效检测入侵和恶意活动（Goel & Hong, 2015）。

> **例 5.1** 以下 Python 代码使用决策树分类器对入侵检测数据进行分类，采用 10 折交叉验证并划分训练集和测试集。该代码计算了分类准确率、精确度、召回率、F1 得分和熵，同时给出了混淆矩阵。
>
> 数据集信息：数据集下载自 KDD Cup 1999 数据网站（http://kdd.ics.uci.edu/databases/kddcup99/kddcup99.html）。这是与 KDD-99 第五届知识发现和数据挖掘国际会议同时进行的第三届国际知识发现和数据挖掘工具竞赛所采用的数据集。比赛任务是开发一个网络入侵检测系统，一个能够区分"异常"连接（称为入侵或攻击）和正常连接的预测模型。这个数据库涉及一组标准的待审计数据，其中包含了在军事网络环境中模拟的各种入侵行为。
>
> ```
> #==================================================================
> # Anomaly detection with KDD dataset using decision tree
> #==================================================================
> #import the libraries
> import numpy as np
> import matplotlib.pyplot as plt
> import pandas as pd
> #Load the dataset
> dataset = pd.read_csv('kddcup.data_10_percent_corrected')
> dataset['normal.'] = dataset['normal.'].replace(['back.', 'buffer_
> overflow.', 'ftp_write.', 'guess_passwd.', 'imap.', 'ipsweep.',
> 'land.', 'loadmodule.', 'multihop.', 'neptune.', 'nmap.', 'perl.',
> 'phf.', 'pod.', 'portsweep.', 'rootkit.', 'satan.', 'smurf.', 'spy.',
> 'teardrop.', 'warezclient.', 'warezmaster.'], 'attack')
> x = dataset.iloc[:, :-1].values
> y = dataset.iloc[:, 41].values
> #%%
> #encoding categorical data
> from sklearn.preprocessing import LabelEncoder, OneHotEncoder
> ```

```python
labelencoder_x_1 = LabelEncoder()
labelencoder_x_2 = LabelEncoder()
labelencoder_x_3 = LabelEncoder()
x[:, 1] = labelencoder_x_1.fit_transform(x[:, 1])
x[:, 2] = labelencoder_x_2.fit_transform(x[:, 2])
x[:, 3] = labelencoder_x_3.fit_transform(x[:, 3])
onehotencoder_1 = OneHotEncoder(categorical_features = [1])
x = onehotencoder_1.fit_transform(x).toarray()
onehotencoder_2 = OneHotEncoder(categorical_features = [4])
x = onehotencoder_2.fit_transform(x).toarray()
onehotencoder_3 = OneHotEncoder(categorical_features = [70])
x = onehotencoder_3.fit_transform(x).toarray()
labelencoder_y = LabelEncoder()
y = labelencoder_y.fit_transform(y)

#splitting the dataset into the training set and test set
from sklearn.model_selection import train_test_split
x_train, x_test, y_train, y_test = train_test_split(x, y, test_size = 0.3, random_state = 0)

#feature scaling
from sklearn.preprocessing import StandardScaler
sc_x = StandardScaler()
x_train = sc_x.fit_transform(x_train)
x_test = sc_x.transform(x_test)

# Fitting Decision Tree to the Training set
from sklearn import tree
# fit model no training data
classifier = tree.DecisionTreeClassifier()
classifier.fit(x_train, y_train)

# Predicting the Test set results
y_pred = classifier.predict(x_test)

# Making the Confusion Matrix
from sklearn.metrics import confusion_matrix
cm = confusion_matrix(y_test, y_pred)
# Applying k-Fold Cross Validation
from sklearn.model_selection import cross_val_score
accuracies = cross_val_score(estimator = classifier, X = x_train, y = y_train, cv = 10)

#%%
print("CV Accuracy: %0.4f (+/- %0.4f)" % (accuracies.mean(), accuracies.std() * 2))
#the performance of the classification model
print("Test Accuracy is : %0.4f" % ((cm[0,0] + cm[1,1])/
(cm[0,0] + cm[0,1] + cm[1,0] + cm[1,1])))
```

```
recall = cm[1,1]/(cm[0,1] + cm[1,1])
print("Recall is : %0.4f" % (recall))
print("False Positive rate: %0.4f" %(cm[1,0]/(cm[0,0] + cm[1,0])))
precision = cm[1,1]/(cm[1,0] + cm[1,1])
print("Precision is: %0.4f" %(precision))
print("F-measure is: %0.4f" % (2*((precision*recall)/
(precision + recall))))
from math import log
print("Entropy is: %0.4f" % (-precision*log(precision)))
print("\nConfusion Matrix:\n", cm)
```

5.2 钓鱼网站检测

全球通信在快速发展的网络技术中起着重要的作用，特别是在电子商务、电子银行、社交网络等方面。这些活动现在正在向网络空间转移。网络空间的威胁主要集中在不安全的互联网基础设施上。消费者安全并不是一件容易的事，因为它还包含钓鱼诈骗。钓鱼诈骗消失才是安全的。因此必须考虑到互联网的特点，并且必须把重点放在那些相对有经验的用户身上（Curtis, Rajivan, Jones, & Gonzalez, 2018）。大量的个人信息、金钱等关键数据和信息已经被网络攻击所破坏。近年来，钓鱼攻击的课题一直在研究。这已经成为黑客生成诈骗网站的流行工具。如果对这些网站进行进一步的审查分析，特别是在分析统一资源定位器（URL）时，可以看到它们含有恶意元素。攻击者的目的是利用URL尽可能多地获取受害者的个人信息、敏感信息、财务数据、密码、用户名等（Gupta, Arachchilage, & Psannis, 2018）。侦测钓鱼攻击在网上银行和交易中起着至关重要的作用，许多用户认为他们是安全的且不会受到这种攻击。这种攻击首先是发送一封携带所有合格凭证的初始电子邮件，试图通过电子邮件中共享的链接获取受害者的数据和信息。检测钓鱼者的方法不多。有两种已知的方法可用于检测钓鱼网站。一种是验证URL，包括检查该URL是否在黑名单上（Gastellier-Prevost, Granadillo, & Laurent, 2011）。另一种方法叫作元启发式方法。该方法会收集几个特征。通过使用识别程序收集这些信息并识别出该URL是合法网站还是钓鱼网站（Xiang, Hong, Rose, & Cranor, 2611）。机器学习方法被用于搜索和识别模式，并发现它们之间的联系（Han, Pei, & Kamber, 2011）。机器学习对决策很重要，因为决策基于学习者所遵循的规则（Mohammad, Thabtah, & McClus- key, 2014; Subasi & Kremic, 2019; Subasi, Molah, Almkallawi, & Chaudhery, 2017）。

例 5.2 以下的Python代码采用随机森林分类器对钓鱼网站数据进行分类，使用了单独的训练集和测试集，以及10折交叉验证，计算出分类准确率、精确度、召回率、F1得分、Cohen kappa得分和Matthews相关系数，同时给出分类报告和混淆矩阵。

数据集信息：该例子中，公开可用的钓鱼网站数据集下载自UCI机器学习库（UCI,

2017）。该数据由Mohammad等人收集并捐赠（Mohammad, Thabtah, & McCluskey, 2012; Mohammad et al., 2014; Mohammad, Thabtah, & McCluskey, 2015）。数据集中采用的特征由Mohammad等人（2015）定义。这些特征是：IP地址、带有"@"符号的URL、使用"//"符号重定向、"TinyURL"等短URL服务、隐藏可疑元素的长URL、子域和多子域、在域名上添加前缀或后缀(-)分隔、域名注册长度、网站图标、HTTPS（安全套接字层的超文本传输协议）、使用非标准端口、请求的URL、URL的域名部分是否有"HTTPS"标记、锚的URL、网站流量、服务器表单处理（SFH）、异常URL、状态栏定制、向电子邮件提交的信息、网站转发、页面排名、指向页面的链接数量、禁用右键、域名年龄、IFrame重定向、弹出窗口的使用情况、DNS记录、Google索引和基于报告的统计特征（Subasi & Kremic, 2019）。

```
#===================================================================
# Phishing website example
#===================================================================
"""
Created on Tue Jun 18 18:14:15 2019
@author: asubasi
"""
import pandas as pd
import warnings
#To prevent warnings
warnings.filterwarnings("ignore")
# load data
dataset = pd.read_csv("PhishingWebsiteDataset.csv")
#To prevent the "ValueError Input contains NaN, infinity or a value too large
for dtype('float32')"
dataset[:] = np.nan_to_num(dataset)
# split data into X and y
X = dataset.iloc[:,0:30]
y = dataset.iloc[:, 30]
class_names = dataset.iloc[:, 30]
#%%
#===================================================================
# Random forest example with training and test set
#===================================================================
#splitting the dataset into the training set and test set
from sklearn.model_selection import train_test_split
X_train, X_test, y_train, y_test = train_test_split(X, y, test_size = 0.3,
random_state = 0)
#%%
#feature scaling
from sklearn.preprocessing import StandardScaler
sc_X = StandardScaler()
X_train = sc_X.fit_transform(X_train)
X_test = sc_X.transform(X_test)
```

```python
#%%
# Fitting Random Forest to the Training set
from sklearn.ensemble import RandomForestClassifier
#In order to change to accuracy increase n_estimators
"""RandomForestClassifier(n_estimators = 'warn', criterion = 'gini', max_
depth = None,
min_samples_split = 2, min_samples_leaf = 1, min_weight_fraction_leaf = 0.0,
max_features = 'auto', max_leaf_nodes = None, min_impurity_decrease = 0.0,
min_impurity_split = None, bootstrap = True, oob_score = False, n_
jobs = None,
random_state = None, verbose = 0, warm_start = False, class_weight = None)"""
# fit model no training data
model = RandomForestClassifier(n_estimators = 100)
model.fit(X_train, y_train)

# Predicting the Test set results
y_pred = model.predict(X_test)
#%%
from sklearn import metrics
print('Test Accuracy:', np.round(metrics.accuracy_score(y_test,y_
pred),4))
print('Precision:', np.round(metrics.precision_score(y_test,
                        y_pred,average = 'weighted'),4))
print('Recall:', np.round(metrics.recall_score(y_test,y_pred,
                                    average = 'weighted'),4))
print('F1 Score:', np.round(metrics.f1_score(y_test,y_pred,
                                    average = 'weighted'),4))
print('Cohen Kappa Score:', np.round(metrics.cohen_kappa_score(y_test,y_
pred),4))
print('Matthews Corrcoef:', np.round(metrics.matthews_corrcoef(y_test,y_
pred),4))
print('\t\tClassification Report:\n', metrics.classification_report(y_
test,y_pred))

from sklearn.metrics import confusion_matrix
print("Confusion Matrix:\n",confusion_matrix(y_test,y_pred))
#%%
from sklearn.metrics import confusion_matrix
import seaborn as sns
# Making the Confusion Matrix
cm = confusion_matrix(y_test, y_pred)
#Print the Confusion Matrix
target_names = ['Legitimite','Phishing']
plt.figure(figsize = (6, 4))
sns.heatmap(cm,cmap = 'coolwarm',linecolor = 'white',linewidths = 1,
            xticklabels = target_names,
            yticklabels = target_names,
            annot = True,
            fmt = 'd')
```

```
plt.title('Confusion Matrix')
plt.ylabel('True Label')
plt.xlabel('Predicted Label')
plt.show()
#%%
#===================================================================
# Random forest example with cross-validation
#===================================================================
from sklearn.ensemble import RandomForestClassifier
#In order to change to accuracy increase n_estimators
"""RandomForestClassifier(n_estimators = 'warn', criterion = 'gini', max_depth = None,
min_samples_split = 2, min_samples_leaf = 1, min_weight_fraction_leaf = 0.0,
max_features = 'auto', max_leaf_nodes = None, min_impurity_decrease = 0.0,
min_impurity_split = None, bootstrap = True, oob_score = False, n_jobs = None,
random_state = None, verbose = 0, warm_start = False, class_weight = None)"""
# fit model no training data
model = RandomForestClassifier(n_estimators = 100)

CV = 10 #10-Fold Cross Validation
#Evaluate Model Using 10-Fold Cross Validation and Print Performance Metrics
Acc_scores = cross_val_score(model, X, y, cv = CV)
print("Accuracy: %0.3f (+/- %0.3f)" % (Acc_scores.mean(), Acc_scores.std() * 2))
f1_scores = cross_val_score(model, X, y, cv = CV, scoring = 'f1_macro')
print("F1 score: %0.3f (+/- %0.3f)" % (f1_scores.mean(), f1_scores.std() * 2))
Precision_scores = cross_val_score(model, X, y,
cv = CV, scoring = 'precision_macro')
print("Precision score: %0.3f (+/- %0.3f)" % (Precision_scores.mean(),
Precision_scores.std() * 2))
Recall_scores = cross_val_score(model, X, y, cv = CV, scoring = 'recall_macro')
print("Recall score: %0.3f (+/- %0.3f)" % (Recall_scores.mean(), Recall_scores.std() * 2))
from sklearn.metrics import cohen_kappa_score, make_scorer
kappa_scorer = make_scorer(cohen_kappa_score)
Kappa_scores = cross_val_score(model, X, y, cv = CV, scoring = kappa_scorer)
print("Kappa score: %0.3f (+/- %0.3f)" % (Kappa_scores.mean(), Kappa_scores.std() * 2))
```

5.3 垃圾邮件检测

电子邮件是最便宜、最快速、最准确、最流行的通信方式之一。它是我们日常生活的一部分，改变了我们的工作和生活方式。E-mail 被用作任务管理器、谈话的组织者、档案馆和文件的传递系统。这一成果的缺点是电子邮件的体积常常越来越大。大多数客户将电子邮件整理并归档成文件以避免这个问题，并在适当的时候可以自动检索。也不知道是谁想出了不管同时向多少人发送附带交易的广告邮件，至少有一个人会对电子邮件作出反应的主意。E-mail 提供了一种有效的方式可以免费发送数以百万计的广告，而且有几家公司已经在广泛地使用这种工具。因此，数以百万计的用户的收件箱中充满了这些所谓的不受欢迎的信息，常被称为"垃圾邮件"。可以随意地发送垃圾邮件引起了互联网社区的担忧，大量的垃圾流量导致重要邮件的投递延迟，使用拨号上网的人不得不将带宽用于接收垃圾邮件。在对不需要的邮件进行分类时，总会有误删重要邮件的风险。最后，还有大量的垃圾邮件不应该被儿童看到（Subasi, Alzahrani, Aljuhani, & Aljedani, 2018）。一些打击垃圾邮件的方法已经被提供出来。"私人"的解决办法有简单的人工介入和诉诸法律行动。一些技术手段包括限制垃圾邮件发送者的 IP 地址和过滤电子邮件。然而目前仍然没有一个完整的、最佳的方法来清除垃圾邮件，因此垃圾邮件的数量还在不断增加。随着过去几年电子邮件使用量的增长，由于大量来路不明的电子邮件信息所带来的问题逐渐浮出水面并进一步加剧。垃圾邮件的过滤集中在邮件的文字解释和附加信息，并试图对垃圾邮件进行分类。一旦识别出垃圾邮件，所采取的行动一般取决于过滤器的使用设置。当单个用户将其作为客户端过滤器使用时，垃圾邮件通常会被发送到一个只包含有垃圾邮件标签的文件夹中，从而更容易识别出这些邮件（Guzella & Caminhas, 2009; Subasi et al., 2018）。垃圾邮件过滤方法目前是基于自动分词规则或评估电子邮件特征的关键字实现的电子邮件分类（Androutsopoulos, Koutsias, Chandrinos, Paliouras, & Spyropoulos, 2000）。

> **例 5.3**　以下的 Python 代码采用随机森林分类器对垃圾邮件数据集进行分类，其使用了单独的训练集和测试集，以及 10 折交叉验证，计算出分类准确率、精确度、召回率、F1 得分、Cohen kappa 得分和 Matthews 相关系数。同时给出了分类报告和混淆矩阵。
>
> 　　**数据集信息**：电子邮件数据集下载自 UCI 机器学习数据集（Bache & Lichman, 2013）。垃圾邮件来自邮政局和保存垃圾邮件的个人。非垃圾邮件来自档案和个人邮件。这些对形成个性化的垃圾邮件过滤器时很有帮助。要想建立一个通用的垃圾邮件过滤器，人们要么必须忽视非垃圾邮件指标，要么收集大量的非垃圾邮件。该数据库有 4601 封邮件，其中 1813 封被归类为垃圾邮件。每封邮件信息由 57 个特征向量组成。
>
> ```
> #=====================================================
> # Spam e-mail filtering example
> #=====================================================
> """
> Created on Tue Jun 18 18:14:15 2019
> @author: asubasi
> """
> ```

```python
from sklearn.model_selection import train_test_split
import numpy as np
import pandas as pd
# load data
dataset = pd.read_csv("SpamEmail.csv")

# split data into X and Y
X = dataset.iloc[:,0:57]
y = dataset.iloc[:, 57]
class_names = dataset.iloc[:, 57]
#%%
#==================================================================
# Random forest example with training and test set
#==================================================================
#splitting the dataset into the training set and test set
from sklearn.model_selection import train_test_split
X_train, X_test, y_train, y_test = train_test_split(X, y, test_size = 0.3, random_state = 0)

#%%
#feature scaling
from sklearn.preprocessing import StandardScaler
sc_X = StandardScaler()
X_train = sc_X.fit_transform(X_train)
X_test = sc_X.transform(X_test)

#%%
# Fitting Random Forest to the Training set
from sklearn.ensemble import RandomForestClassifier
#In order to change to accuracy increase n_estimators
"""RandomForestClassifier(n_estimators = 'warn', criterion = 'gini', max_depth = None, 
min_samples_split = 2, min_samples_leaf = 1, min_weight_fraction_leaf = 0.0, 
max_features = 'auto', max_leaf_nodes = None, min_impurity_decrease = 0.0, 
min_impurity_split = None, bootstrap = True, oob_score = False, n_jobs = None, 
random_state = None, verbose = 0, warm_start = False, class_weight = None)"""
# fit model no training data
model = RandomForestClassifier(n_estimators = 100)
model.fit(X_train, y_train)

# Predicting the Test set results
y_pred = model.predict(X_test)
#%%
from sklearn import metrics
```

```python
print('Test Accuracy:', np.round(metrics.accuracy_score(y_test,y_pred),4))
print('Precision:', np.round(metrics.precision_score(y_test,
                            y_pred,average = 'weighted'),4))
print('Recall:', np.round(metrics.recall_score(y_test,y_pred,
                                        average = 'weighted'),4))
print('F1 Score:', np.round(metrics.f1_score(y_test,y_pred,
                                        average = 'weighted'),4))
print('Cohen Kappa Score:', np.round(metrics.cohen_kappa_score(y_test,y_
pred),4))
print('Matthews Corrcoef:', np.round(metrics.matthews_corrcoef(y_test,y_
pred),4))
print('\t\tClassification Report:\n', metrics.classification_report(y_
test,y_pred))

from sklearn.metrics import confusion_matrix
print("Confusion Matrix:\n",confusion_matrix(y_test,y_pred))
#%%
from sklearn.metrics import confusion_matrix
import seaborn as sns
import matplotlib.pyplot as plt
# Making the Confusion Matrix
cm = confusion_matrix(y_test, y_pred)
#Print the Confusion Matrix
target_names = ['NotSPAM','SPAM' ]
plt.figure(figsize = (6, 4))
sns.heatmap(cm,cmap = 'coolwarm',linecolor = 'white',linewidths = 1,
            xticklabels = target_names,
            yticklabels = target_names,
            annot = True,
            fmt = 'd')
plt.title('Confusion Matrix')
plt.ylabel('True Label')
plt.xlabel('Predicted Label')
plt.show()
#%%
#==================================================================
# Random forest example with cross-validation
#==================================================================
from sklearn.ensemble import RandomForestClassifier
#In order to change to accuracy increase n_estimators
"""RandomForestClassifier(n_estimators = 'warn', criterion = 'gini', max_
depth = None,
min_samples_split = 2, min_samples_leaf = 1, min_weight_fraction_
leaf = 0.0,
max_features = 'auto', max_leaf_nodes = None, min_impurity_decrease = 0.0,
min_impurity_split = None, bootstrap = True, oob_score = False, n_
jobs = None,
random_state = None, verbose = 0, warm_start = False, class_
weight = None)"""
```

```
# fit model no training data
model = RandomForestClassifier(n_estimators = 100)

# Applying k-Fold Cross Validation
from sklearn.model_selection import cross_val_score
CV = 10 #10-Fold Cross Validation
#Evaluate Model Using 10-Fold Cross Validation and Print Performance Metrics
Acc_scores = cross_val_score(model, X, y, cv = CV)
print("Accuracy: %0.3f (+/- %0.3f)" % (Acc_scores.mean(), Acc_scores.
std() * 2))
f1_scores = cross_val_score(model, X, y, cv = CV,scoring = 'f1_macro')
print("F1 score: %0.3f (+/- %0.3f)" % (f1_scores.mean(), f1_scores.std()
* 2))
Precision_scores = cross_val_score(model, X, y, cv = CV,scoring = 'precision_
macro')
print("Precision score: %0.3f (+/- %0.3f)" % (Precision_scores.mean(),
Precision_scores.std() * 2))
Recall_scores = cross_val_score(model, X, y, cv = CV,scoring = 'recall_
macro')
print("Recall score: %0.3f (+/- %0.3f)" % (Recall_scores.mean(), Recall_
scores.std() * 2))
from sklearn.metrics import cohen_kappa_score, make_scorer
kappa_scorer = make_scorer(cohen_kappa_score)
Kappa_scores = cross_val_score(model, X, y, cv = CV,scoring = kappa_scorer)
print("Kappa score: %0.3f (+/- %0.3f)" % (Kappa_scores.mean(), Kappa_
scores.std() * 2))
```

5.4 信用评分

信用评分是银行和其他金融机构信用风险管理系统的重要组成部分，因为它可以消除风险、提高可预测性和加快贷款决策的速度（Chuang & Huang, 2011; Serrano-Cinca & Gutiérrez-Nieto, 2016）。客户的信用管理是金融组织和商业银行的关键工作，它们必须小心处理客户的贷款以免因任何错误的决策导致资损。此外，错误的客户资信评估会对金融组织的稳定性产生重大影响。劳动密集型的客户资信评估耗费时间和资源。但是劳动密集型的评估工作极其依靠银行员工，因此基于计算机的数字信用评估系统被设计并实现出来消除该阶段的"人为因素"。这种基于计算机的自动信用评价模型可以为银行和金融组织提供是否应该发放贷款和贷款能否偿还的建议。目前人们已经实现了许多信贷模型，但它们中没有完美的分类效果，因为有些模型能生成可靠结果，而有些模型则不能（Ala'raj & Abbod, 2016）。信用评分是金融行业和研究人员经常研究的领域（Kumar & Ravi, 2007; Lin, Hu, & Tsai, 2012），许多模型被研究并通过各种算法实现，比如人工神经网络、决策树（Hung & Chen, 2009; Makowski, 1985）、支持向量机（Baesens et al., 2003; Huang, Chen, & Wang, 2007; Schebesch & Stecking, 2005）以及案例推断（Shin & Han, 2001; Wheeler & Aitken, 2000）。受金融危机的影响，巴塞尔银行监管委员会要求所有银行和金融机构在向公司或个人提供贷款之前建立健全的信用风险分析框架（Ala'raj & Abbod, 2016; Gicic´ & Subasi, 2019）。

例 5.4 以下的 Python 代码是采用 XGBoost 分类器对信用评分数据集进行分类，其使用了单独的训练集和测试集，以及 10 折交叉验证，计算出准确率、精确度、召回率、F1 得分、Cohen kappa 得分和 Matthew 相关系数。同时给出了分类报告和混淆矩阵。

数据集信息：该数据集下载自 UCI 机器学习资源库（UCI，2019）。原始数据集由 Hofmann 教授提供，包含分类和符号属性。共有两个类（Good 和 Bad），以及包含 20 属性的 1000 个实例，它们是：

- 现有支票账户的状态
- 持续月数
- 信用记录
- 目的
- 信贷额
- 储蓄账户 / 债券
- 当前职务
- 分期付款率（占可支配收入的百分比）
- 个人身份与性别
- 其他债务人 / 担保人
- 现居住地
- 财产
- 年龄
- 其他分期付款计划
- 住房
- 该行现有授信额度
- 工作
- 负责人数
- 电话号码
- 外国工作人员

```
#==============================================================
# Credit scoring example
#==============================================================
"""
Created on Tue Jun 18 18:14:15 2019
@author: asubasi
"""
```

```python
import numpy as np
import pandas as pd

# load data
dataset = pd.read_csv("german_credit.csv")
# split data into X and Y
X = dataset.iloc[:,1:20]
y = dataset.iloc[:, 0]
class_names = dataset.iloc[:,0]
#=====================================================================
# XGBoost example with training and test set
#=====================================================================
from sklearn.model_selection import train_test_split
# split data into train and test sets
Xtrain, Xtest, ytrain, ytest = train_test_split(X, y, test_size = 0.33, random_state = 7)

from xgboost import XGBClassifier
# fit model no training data
model = XGBClassifier()
model.fit(Xtrain, ytrain)
# make predictions for test data
ypred = model.predict(Xtest)
from sklearn import metrics
print('Accuracy:', np.round(metrics.accuracy_score(ytest,ypred),4))
print('Precision:', np.round(metrics.precision_score(ytest,
                            ypred,average = 'weighted'),4))
print('Recall:', np.round(metrics.recall_score(ytest,ypred,
                                        average = 'weighted'),4))
print('F1 Score:', np.round(metrics.f1_score(ytest,ypred,
                                        average = 'weighted'),4))
print('Cohen Kappa Score:', np.round(metrics.cohen_kappa_score(ytest,
ypred),4))
print('Matthews Corrcoef:', np.round(metrics.matthews_corrcoef(ytest,
ypred),4))
print('\t\tClassification Report:\n', metrics.classification_report(ypred,
ytest))

from sklearn.metrics import confusion_matrix
from io import BytesIO #neded for plot
import seaborn as sns; sns.set()
import matplotlib.pyplot as plt

mat = confusion_matrix(ytest, ypred)
sns.heatmap(mat.T, square = True, annot = True, fmt = 'd', cbar = False)
plt.xlabel('true label')
plt.ylabel('predicted label');

plt.savefig("SVM_Confusion.jpg")
# Save SVG in a fake file object.
```

```
f = BytesIO()
plt.savefig(f, format = "svg")

#%%
#================================================================
# XGBoost example with cross-validation
#================================================================
from xgboost import XGBClassifier
# fit model no training data
model = XGBClassifier()

# Applying k-Fold Cross Validation
from sklearn.model_selection import cross_val_score
CV = 10 #10-Fold Cross Validation
#Evaluate Model Using 10-Fold Cross Validation and Print Performance Metrics
Acc_scores = cross_val_score(model, X, y, cv = CV)
print("Accuracy: %0.3f (+/- %0.3f)" % (Acc_scores.mean(), Acc_scores.
std() * 2))
f1_scores = cross_val_score(model, X, y, cv = CV,scoring = 'f1_macro')
print("F1 score: %0.3f (+/- %0.3f)" % (f1_scores.mean(), f1_scores.std()
* 2))
Precision_scores = cross_val_score(model, X, y, cv = CV,scoring = 'precision_
macro')
print("Precision score: %0.3f (+/- %0.3f)" % (Precision_scores.mean(),
Precision_scores.std() * 2))
Recall_scores = cross_val_score(model, X, y, cv = CV,scoring = 'recall_
macro')
print("Recall score: %0.3f (+/- %0.3f)" % (Recall_scores.mean(), Recall_
scores.std() * 2))
from sklearn.metrics import cohen_kappa_score, make_scorer
kappa_scorer = make_scorer(cohen_kappa_score)
Kappa_scores = cross_val_score(model, X, y, cv = CV,scoring = kappa_scorer)
print("Kappa score: %0.3f (+/- %0.3f)" % (Kappa_scores.mean(), Kappa_
scores.std() * 2))
```

5.5 信用卡欺诈检测

据计算 2017 年全球银行卡欺诈损失为 228 亿美元，预计将继续增长（Jurgovsky et al., 2018）。最近机器学习开始成为检测大量交易的重要元素（Dal Pozzolo, Caelen, Le Borgne, Waterschoot, & Bontempi, 2014）。然而，目前的研究表明检测方法需要考虑到欺诈现象的特殊性（Dal Pozzolo, Boracchi, Caelen, Alippi, & Bontempi, 2018）。因此，开发一个有效的欺诈检测系统（FDS）不仅是实现一些传统的现成软件库，还需要对欺诈概念有深刻的理解。这意味着 FDS 系统不是立刻就有的，也不是简单地重新设置已有的 FDS 系统，如新的市场或新的支付方式（Lebichot, Le Borgne, He-Guelton, Oblé, & Bontempi, 2019）。也许人工智能算法的最佳测试平台之一是信用卡欺诈分析。尽管如此，这个问题还是涉及一些相关的挑战——客户的偏好会随着时间的推移而扩大，欺诈者也会调整策略，实际交易量远远超出欺诈的量，研究人员只能及时测验一小部分交易。但绝大多数针对欺诈检测提出的学习算法所基于的假设几乎不适用于现实世界中的欺诈检测系统。缺乏现实性主要有两个方面：交付监督信息的方法和时间与用于判断欺诈检测质量的指标。信用卡欺诈检测是机器学习和计算机智能领域关注的一个相关问题，已提出大量的自动化解决方案（Bhattacharyya, Jha, Tharakunnel, & Westland, 2011; Dal Pozzolo et al., 2014, 2018; Jha, Guillen, & Westland, 2012; Mahmoudi & Duman, 2015; Whitrow, Hand, Juszczak, Weston, & Adams, 2009）。

现实中的 FDS 系统中，大量的支付请求流被自动化工具迅速扫描以确定哪些交易需要授权。通常使用分类器分析所有授权交易，并对最可疑的交易发出警报。然后由专业的检查人员对警报进行审查，并通知持卡人确认每笔付款的真实性质（合法或欺诈）。这样研究人员会收到通过标记交易的形式出现的系统反馈，这些反馈可用于训练分类器来维持（或有可能提高）欺诈检测效率。由于明显的时间和成本限制，绝大多数交易无法由研究人员进行验证。直到客户发现并报告欺诈行为时，交易才会被标记，又或者在经过足够长的时间后无争议的交易被认为是真实的而被标记（Dal Pozzolo et al., 2018）。如今，企业和公共机构不得不面对持续出现的主动欺诈行为，他们亟须能检测欺诈行为的自动化系统（Delamaire, Abdou, & Pointon, 2009）。由于通常的交易数据样本量大、测量值多且实时报警，对人类分析人员来说要想识别交易数据中的欺诈行为通常不可行也不方便，因此自动化系统至关重要。检测问题通常采用两种不同的方法解决。在静态学习方法中，检测模型不断地从头开始重新学习。在在线学习方法中，只要有新数据产生，那么检测模型就会被更新。信用卡欺诈检测中另一个挑战性的问题是由于保密原则缺少可访问的数据，这造成研究社区很少有机会分享真实的数据集和评估当前的方法（Dal Pozzolo et al., 2014）。

信用卡欺诈检测旨在判断一笔交易是否存在欺诈或者是否基于历史数据。然而鉴定欺诈行为非常困难，例如在节假日期间消费者购买模式会变化，欺诈者本身也会制定策略，尤其是用来应对检测的手段。机器学习方法为解决这样的问题提供了一种积极的方式（Dal Pozzolo, Caelen, & Bontem- pi, 2015）。传统的欺诈监测系统具有分层管控能力，每一层都能自动化或者人为控制（Carcillo et al., 2018; Dal Pozzolo et al., 2018）。自动化层的一部分涉及机器学习技术，基于被注释的交易建立预测模型。在过去十年中，丰富的机器学习方法促进了针对信用卡欺诈检测的有监督、无监督和半监督算法的发展（Carcillo et al., 2019; Sethi & Gera, 2014）。

> **例 5.5** 以下的 Python 代码采用 Keras 深度神经网络分类器对信用评分数据集进行分类，使用了单独的训练集和测试集，以及 10 折交叉验证，计算出分类准确率、精确度、召回率、F1 得分、Cohen kappa 得分和 Matthews 相关系数。同时给出分类报告和混淆矩阵。本例改编自 *Beginning Anomaly Detection Using Python-Based Deep Learning*

一书（Alla & Adari, 2019）。

数据集信息：数据集下载自 Kaggle 网站（https://www.kaggle.com/mlg-ulb/creditcardfraud/version/3）。信用卡欺诈数据集由比利时的一家支付服务提供商提供。它包括 2012 年 2 月 1 日至 2013 年 5 月 20 日的交易子集的日志。该数据集按日划分成块，并且包含电子商务欺诈交易。原始变量包括交易金额、销售点、货币、交易国家、商家类型等诸多内容。但原始变量并不能描述持卡人的行为。因此在原有变量的基础上加入组合变量来描述用户行为，比如利用交易金额和卡的 ID 来计算一张卡每周和每月的平均消费额，前后两笔交易的差额等等。针对每笔交易和每张卡，都会记录前 3 个月（H=90 天）的交易数据来计算统计变量。因此，一张卡的周平均消费是过去 3 个月的周平均消费。这个数据集包括分类变量和连续变量。数据分块各包含一组每日交易数据，每块的平均交易量为 5218 笔（Dal Pozzolo et al., 2014）。该数据集包含 2 天内发生的 284 807 笔交易，其中有 492 笔欺诈交易。它只包括由 PCA 转换得出的数值输入变量。由于保密问题，我们没有原始特征和更多关于数据的背景信息。特征 V1，V2，…，V28 是用 PCA 得到的主成分，未进行 PCA 变换的特征只有"Time"和"Amount"。特征"Time"包含每笔交易与数据集中第一笔交易之间的秒数。特征"Amount"是交易金额，这个特征可以用于依赖样本、对成本敏感的学习。特征"Class"是响应变量，在欺诈的情况下取值为 1，否则取值为 0。该数据集来自比利时布鲁塞尔自由大学的研究小组 Worldline and the Machine Learning Group（http://mlg.ulb.ac.be），是在关于大数据挖掘和欺诈检测的研究合作中收集并分析的。https://www.researchgate.net/project/Fraud-detection-5 中 DefeatFraud 项目的主页有更多关于相关主题当前和过去项目的细节。

```
"""
Created on Thu Dec 12 07:56:20 2019
@author: absubasi
Adapted from "Beginning Anomaly Detection Using Python-Based Deep
Learning"
"""
#=====================================================================
# Credit card fraud detection using Keras
#=====================================================================
from keras.models import Sequential
from keras.layers import Dense
from keras.wrappers.scikit_learn import KerasClassifier
from keras.utils import np_utils
from sklearn.preprocessing import LabelEncoder
import pandas as pd
import numpy as np
import matplotlib.pyplot as plt
from sklearn.metrics import classification_report
from sklearn.metrics import confusion_matrix
import seaborn as sns
```

```python
from sklearn import metrics
from io import BytesIO #needed for plot
import seaborn as sns; sns.set()
#========================================================================
# Define utility functions
#========================================================================
def plot_model_accuracy_loss():
    plt.figure(figsize = (6, 4))
    plt.plot(history.history['accuracy'], 'r', label = 'Accuracy of training data')
    plt.plot(history.history['val_accuracy'], 'b', label = 'Accuracy of validation data')
    plt.plot(history.history['loss'], 'r--', label = 'Loss of training data')
    plt.plot(history.history['val_loss'], 'b--', label = 'Loss of validation data')
    plt.title('Model Accuracy and Loss')
    plt.ylabel('Accuracy and Loss')
    plt.xlabel('Training Epoch')
    plt.ylim(0)
    plt.legend()
    plt.show()

def print_confusion_matrix():
    matrix = confusion_matrix(y_test.argmax(axis = 1), max_y_pred_test.argmax(axis = 1))
    plt.figure(figsize = (10, 8))
    sns.heatmap(matrix, cmap = 'coolwarm', linecolor = 'white', linewidths = 1,
        annot = True,
        fmt = 'd')
    plt.title('Confusion Matrix')
    plt.ylabel('True Label')
    plt.xlabel('Predicted Label')
    plt.show()

def print_performance_metrics():
    print('Accuracy:', np.round(metrics.accuracy_score(y_test, max_y_pred_test),4))
    print('Precision:', np.round(metrics.precision_score(y_test,
                        max_y_pred_test, average = 'weighted'),4))
    print('Recall:', np.round(metrics.recall_score(y_test, max_y_pred_test,
                                average = 'weighted'),4))
    print('F1 Score:', np.round(metrics.f1_score(y_test, max_y_pred_test,
                                average = 'weighted'),4))
    print('Cohen Kappa Score:', np.round(metrics.cohen_kappa_score(y_test.argmax(axis = 1), max_y_pred_test.argmax(axis = 1)),4))
    print('Matthews Corrcoef:', np.round(metrics.matthews_corrcoef(y_test.argmax(axis = 1), max_y_pred_test.argmax(axis = 1)),4))
```

```python
    print('\t\tClassification Report:\n', metrics.classification_report(y_test, max_y_pred_test))
def print_confusion_matrix_and_save():
    mat = confusion_matrix(y_test.argmax(axis=1), max_y_pred_test.argmax(axis=1))
    sns.heatmap(mat.T, square=True, annot=True, fmt='d', cbar=False)
    plt.title('Confusion Matrix')
    plt.ylabel('True Label')
    plt.xlabel('Predicted Label')
plt.show()

    plt.savefig("Confusion.jpg")
    # Save SVG in a fake file object.
    f = BytesIO()
    plt.savefig(f, format="svg")
#%%
df = pd.read_csv("creditcard.csv", sep=",", index_col=None)
df['Amount'] = StandardScaler().fit_transform(df['Amount'].values.reshape(-1, 1))
df['Time'] = StandardScaler().fit_transform(df['Time'].values.reshape(-1, 1))

#%%
anomalies = df[df["Class"] == 1]
normal = df[df["Class"] == 0]
anomalies.shape, normal.shape
#%%
for f in range(0, 20):
    normal = normal.iloc[np.random.permutation(len(normal))]
data_set = pd.concat([normal[:5000], anomalies])
X = data_set.iloc[:,0:30]
y = data_set.iloc[:,30]

#%%
# encode class values as integers
encoder = LabelEncoder()
encoder.fit(y)
encoded_Y = encoder.transform(y)
# convert integers to dummy variables (i.e. one hot encoded)
dummy_y = np_utils.to_categorical(encoded_Y)
lenOfCoded = dummy_y.shape[1] # Dimension of binary coded output data
InputDataDimension = 30
#%%
#========================================================================
# Keras DNN example with cross-validation
#========================================================================
# define a Keras model
def my_model():
    # create model
    model = Sequential()
    model.add(Dense(128, input_dim=InputDataDimension,
```

```
        activation = 'relu'))
# model.add(Dense(32, activation = 'relu'))
# model.add(Dense(16, activation = 'relu'))
        model.add(Dense(lenOfCoded, activation = 'softmax'))
        # Compile model
        model.compile(loss = 'categorical_crossentropy', optimizer = 'adam',
        metrics = ['accuracy'])
        return model

estimator = KerasClassifier(build_fn = my_model, epochs = 25, batch_size = 5,
verbose = 1)
kfold = KFold(n_splits = 5, shuffle = True)
results = cross_val_score(estimator, X, dummy_y, cv = kfold)
print("Accuracy: %.2f%% (%.2f%%)" % (results.mean()*100, results.
std()*100))

#%%
#=====================================================================
# Keras DNN example with training and test set
#=====================================================================
# load data
from sklearn.model_selection import train_test_split
# split data into train and test sets
X_train, X_test, y_train, y_test = train_test_split(X, y, test_size = 0.33,
random_state = 7)

# One hot encode targets
y_train = np_utils.to_categorical(y_train)
y_test = np_utils.to_categorical(y_test)
num_classes = y_test.shape[1]
InputDataDimension = 30

#%%
#=====================================================================
# Create a model
#=====================================================================
model = Sequential()
model.add(Dense(128, input_dim = InputDataDimension, init = 'uniform',
activation = 'relu'))
model.add(Dense(32, activation = 'relu'))
model.add(Dense(num_classes, activation = 'softmax'))
#%%
#=====================================================================
# Compile the model
#=====================================================================
model.compile(loss = 'categorical_crossentropy', optimizer = 'adam',
metrics = ['accuracy'])
#=====================================================================
# Train and validate the model
#=====================================================================
```

```python
history = model.fit(X_train, y_train, validation_split = 0.33, epochs = 50, batch_size = 25,verbose = 0)
#%%
#=====================================================================
# Evaluate the model
#=====================================================================
test_loss, test_acc = model.evaluate(X_test, y_test, verbose = 0)
print('\nTest accuracy:', test_acc)

#%%
#=====================================================================
# Plot the history
#=====================================================================
#Plot the Model Accuracy and Loss for Training and Validation dataset
plot_model_accuracy_loss()

#%%
#Test the Model with testing data
y_pred_test = model.predict(X_test,)
# Round the test predictions
max_y_pred_test = np.round(y_pred_test)
#%%
#Print the Confusion Matrix
print_confusion_matrix()
#%%
#Evaluate the Model and Print Performance Metrics
print_performance_metrics()
#%%
#Print and Save the Confusion Matrix
print_confusion_matrix_and_save()
#%%
# ====================================================================
# Keras deep learning model
# ====================================================================
# load data
from sklearn.model_selection import train_test_split
# split data into train and test sets
X_train, X_test, y_train, y_test = train_test_split(X, y, test_size = 0.33, random_state = 7)

# One hot encode targets
y_train = np_utils.to_categorical(y_train)
y_test = np_utils.to_categorical(y_test)
num_classes = y_test.shape[1]
InputDataDimension = 30
#%%
#=====================================================================
# Build a deep model
#=====================================================================
```

```
model = Sequential()
model.add(Dense(128, input_dim = InputDataDimension, init = 'uniform',
activation = 'relu'))
model.add(Dense(64, activation = 'relu'))
model.add(Dense(32, activation = 'relu'))
model.add(Dense(num_classes, activation = 'softmax'))
#%%
#===============================================================
# Compile the model
#===============================================================
model.compile(loss = 'categorical_crossentropy', optimizer = 'adam',
metrics = ['accuracy'])
#===============================================================
# Train and validate the model
#===============================================================
history = model.fit(X_train, y_train, validation_split = 0.33, nb_epoch = 50,
batch_size = 25,verbose = 0)
#%%
#===============================================================
# Evaluate the model
#===============================================================
test_loss, test_acc = model.evaluate(X_test, y_test, verbose = 0)
print('\nTest accuracy:', test_acc)
#%%
#===============================================================
# Plot the history
#===============================================================
#Plot the Model Accuracy and Loss for Training and Validation dataset
plot_model_accuracy_loss()
#%%
#Test the Model with testing data
y_pred_test = model.predict(X_test,)
# Round the test predictions
max_y_pred_test = np.round(y_pred_test)
#%%
#Print the Confusion Matrix
print_confusion_matrix()
#%%
#Evaluate the Model and Print Performance Metrics
print_performance_metrics()
#%%
#Print and Save the Confusion Matrix
print_confusion_matrix_and_save()
```

5.6 使用CNN进行手写数字识别

手写数字识别是一个普遍存在的多类分类问题，它通常内置在手机银行应用程序和更传统的自动柜员机中以便用户能够自动存入纸质支票。每一类数据由数字0到9的几个手写版本（图像）组成，总共是 10 类数据。由于手写记录工具和功能强大的移动计算机等技术的发展，手写字符识别已经成为一个常见的研究领域（Elleuch, Maalej, & Kherallah, 2016）。然而，由于手写体对书写者的依赖性很强，因此开发一个高可靠且能够识别输入到应用程序中的每个手写体字符的识别系统是具有挑战的。光学字符识别（OCR）是字符识别和人工智能的研究领域之一（Pramanik & Bag, 2018）。10 多年来，在许多应用和识别算法中，数字识别在 OCR 手写领域得到了有效的研究。例如，这些算法包括支持向量机（SVM）、卷积神经网络（CNN）和随机森林（RF）等。不过实验的准确率大概为 95%。由于很多分类器不能很好地处理原始图像或细节，因此使用提取特征这一预处理方法来降低数据维度及整合有效信息（Lauer, Suen, & Bloch, 2007）。传统的特征选择是一个复杂且耗时的工作，并且它无法处理原始图像，相反，CNN 的自动提取方法可以直接从原始数据中识别特征（Bernard, Adam, & Heutte, 2007）。CNN 是一个前馈网络，它从图像中提取拓扑特征。该方法收集第一层原始图像的特征，并利用其最后一层来识别对象。尽管如此，大多数分类器（如 SVM 和 RF）不能有效地处理原始图像或数据，因为从复杂的模式中分出正确的特征是一个巨大的挑战（Pramanik & Bag, 2018）。另一方面，CNN 的自动特征提取方法将直接从原始图像中检索元素（Bernard et al., 2007; Zhao, 2018）。

CNN 是人工神经网络的一种改进，它重点模仿视觉皮层的行为。隐藏单元的目的是学习原始输入的非线性变化，这就是所谓的特征提取。接着这些隐藏的特征作为输入转移到最终的 GLM（广义线性模型）上。这种方法特别适用于原始输入特征不能独立提供信息的问题。CNN 是 MLP（多层感知器）的一种，特别适用于一维信号，如语音、生物医学信号或文本，或者二维信号，如图像（Murphy, 2012）。

> **例 5.6** 以下的 Python 代码用于使用 DCNN 分类器对 MNIST 手写数字数据集进行分类，该例子使用单独的训练集、验证集和测试集，计算出分类准确率、精确度、召回率、F1 得分、Cohen kappa 得分和 Matthews 相关系数，并给出分类报告和混淆矩阵。本例改编自机器学习权威网站（https://machinelearningmastery.com/handwritten-digit-recognition-using-convolutional-neural-networks-python-keras/）。
>
> 数据集信息：MNIST 的手写数字数据库可以从 Yann LeCun 的网页上获得（http://yann.lecun.com/exdb/mnist/）。该数据库的训练集有 60 000 个样本，测试集有 10 000 个样本。它是 NIST 提供的一个更大数据集的子类。这些数字经过大小归一化处理，并且以固定尺寸的图像为中心。对于那些想要在真实世界的数据上尝试机器学习技术的研究人员来说，这是一个不错的数据库，让他们可以不必在预处理和格式化上花费更多精力。来自 NIST 的原始黑白（bilevel）图像经过大小归一化后，在保持长宽比的同时适应 20×20 像素的方框。结果图像包括灰度等级，这是归一化算法所利用的抗锯齿方法的结果。通过计算像素的质量中心并将图像平移到 28×28 区域的中心位置使图像居中。

MNIST 中的数字图像最初是由 Chris Burges 和 Corinna Cortes 利用边界框归一化和居中选择出来的。Yann LeCun 通过在一个更大的窗口内根据质量中心定位来改进数据集。

```
"""
Created on Wed Oct 9 15:54:40 2019
@author: absubasi
https://machinelearningmastery.com/handwritten-digit-recognition-using-convolutional-neural-networks-python-keras/
"""
# ============================================================
# MNIST handwritten digit recognition
# ============================================================
from keras.datasets import mnist
from keras.models import Sequential
from keras.layers import Dense
from keras.layers import Dropout
from keras.layers import Flatten
from keras.layers.convolutional import Conv2D
from keras.layers.convolutional import MaxPooling2D
from keras.layers import BatchNormalization
from keras.utils import np_utils
import matplotlib.pyplot as plt
import matplotlib.pyplot as plt
from sklearn.metrics import classification_report
from sklearn.metrics import confusion_matrix
import seaborn as sns
import numpy as np
from sklearn import metrics
from io import BytesIO #needed for plot
import seaborn as sns; sns.set()

#============================================================
# Define utility functions
#============================================================
def plot_model_accuracy_loss():
    plt.figure(figsize = (6, 4))
    plt.plot(history.history['accuracy'], 'r', label = 'Accuracy of training data')
    plt.plot(history.history['val_accuracy'], 'b', label = 'Accuracy of validation data')
    plt.plot(history.history['loss'], 'r--', label = 'Loss of training data')
    plt.plot(history.history['val_loss'], 'b--', label = 'Loss of validation data')
    plt.title('Model Accuracy and Loss')
    plt.ylabel('Accuracy and Loss')
    plt.xlabel('Training Epoch')
    plt.ylim(0)
    plt.legend()
```

```python
    plt.show()
def print_confusion_matrix():
    matrix = confusion_matrix(y_test.argmax(axis = 1), max_y_pred_test.argmax(axis = 1))
    plt.figure(figsize = (10, 8))
    sns.heatmap(matrix,cmap = 'coolwarm',linecolor = 'white',linewidths = 1,
                annot = True,
                fmt = 'd')
    plt.title('Confusion Matrix')
    plt.ylabel('True Label')
    plt.xlabel('Predicted Label')
    plt.show()
def print_performance_metrics():
      print('Accuracy:', np.round(metrics.accuracy_score(y_test, max_y_pred_test),4))
    print('Precision:', np.round(metrics.precision_score(y_test,
                          max_y_pred_test,average = 'weighted'),4))
      print('Recall:', np.round(metrics.recall_score(y_test, max_y_pred_test,
                          average = 'weighted'),4))
    print('F1 Score:', np.round(metrics.f1_score(y_test, max_y_pred_test,
                          average = 'weighted'),4))
print('Cohen Kappa Score:', np.round(metrics.cohen_kappa_score(y_test.argmax(axis = 1), max_y_pred_test.argmax(axis = 1)),4))
print('Matthews Corrcoef:', np.round(metrics.matthews_corrcoef(y_test.argmax(axis = 1), max_y_pred_test.argmax(axis = 1)),4))
print('\t\tClassification Report:\n', metrics.classification_report(y_test, max_y_pred_test))

def print_confusion_matrix_and_save():
    mat = confusion_matrix(y_test.argmax(axis = 1), max_y_pred_test.argmax(axis = 1))
    sns.heatmap(mat.T, square = True, annot = True, fmt = 'd', cbar = False)
    plt.title('Confusion Matrix')
    plt.ylabel('True Label')
    plt.xlabel('Predicted Label')
    plt.show()

    plt.savefig("Confusion.jpg")
    # Save SVG in a fake file object.
    f = BytesIO()
    plt.savefig(f, format = "svg")

#%%
#================================================================
# Multilayer perceptrons model
#================================================================
```

```python
# load data
(X_train, y_train), (X_test, y_test) = mnist.load_data()
# Flatten 28*28 images to a 784 vector for each image
num_pixels = X_train.shape[1] * X_train.shape[2]
X_train = X_train.reshape((X_train.shape[0], num_pixels)).astype('float32')
X_test = X_test.reshape((X_test.shape[0], num_pixels)).astype('float32')
# Normalize inputs from 0-255 to 0-1
X_train = X_train / 255
X_test = X_test / 255
# One hot encode targets
y_train = np_utils.to_categorical(y_train)
y_test = np_utils.to_categorical(y_test)
num_classes = y_test.shape[1]
#================================================================
# Build a Keras model
#================================================================
model = Sequential()
model.add(Dense(num_pixels, input_dim = num_pixels,
    kernel_initializer = 'normal', activation = 'relu'))
model.add(Dense(num_classes, kernel_initializer = 'normal',
activation = 'softmax'))
#================================================================
# Compile the model
#================================================================
model.compile(loss = 'categorical_crossentropy', optimizer = 'adam',
metrics = ['accuracy'])
#================================================================
# Train and validate the model
#================================================================
history = model.fit(X_train, y_train, validation_split = 0.3, epochs = 10,
batch_size = 200, verbose = 2)
#%%
#================================================================
# Evaluate the model
#================================================================
test_loss, test_acc = model.evaluate(X_test, y_test, verbose = 0)
print('\nTest accuracy:', test_acc)
#%%
#================================================================
# Plot the history
#================================================================
#Plot the Model Accuracy and Loss for Training and Validation dataset
plot_model_accuracy_loss()

#%%
#Test the Model with testing data
y_pred_test = model.predict(X_test,)
```

```python
# Round the test predictions
max_y_pred_test = np.round(y_pred_test)
#Print the Confusion Matrix
print_confusion_matrix()
#%%
#Evaluate the Model and Print Performance Metrics
print_performance_metrics()
#%%
#Print and Save the Confusion Matrix
print_confusion_matrix_and_save()
#%%
#===================================================================
# 1 Convolutional neural network for MNIST digit recognition
#===================================================================
# load data
(X_train, y_train), (X_test, y_test) = mnist.load_data()
# reshape to be [samples][width][height][channels]
X_train = X_train.reshape((X_train.shape[0], 28, 28, 1)).astype('float32')
X_test = X_test.reshape((X_test.shape[0], 28, 28, 1)).astype('float32')
# normalize inputs from 0-255 to 0-1
X_train = X_train/255
X_test = X_test/255
# one hot encode outputs
y_train = np_utils.to_categorical(y_train)
y_test = np_utils.to_categorical(y_test)
num_classes = y_test.shape[1]
#%%
#===================================================================
# Build a model
#===================================================================
model = Sequential()
model.add(Conv2D(32, (5, 5), input_shape = (28, 28, 1), activation = 'relu'))
model.add(MaxPooling2D())
model.add(Dropout(0.2))
model.add(Flatten())
model.add(Dense(128, activation = 'relu'))
model.add(Dense(num_classes, activation = 'softmax'))
#===================================================================
# Compile model
#===================================================================
model.compile(loss = 'categorical_crossentropy', optimizer = 'adam', metrics = ['accuracy'])
#===================================================================
# Train the model
#===================================================================
history = model.fit(X_train, y_train, validation_split = 0.3, epochs = 10, batch_size = 200, verbose = 2)
#%%
```

```
#===============================================================
# Evaluate the model
#===============================================================
test_loss, test_acc = model.evaluate(X_test, y_test, verbose = 0)
print('\nTest accuracy:', test_acc)

#%%
#Plot the Model Accuracy and Loss for Training and Validation dataset
plot_model_accuracy_loss()
#%%
#Test the Model with testing data
y_pred_test = model.predict(X_test,)
# Round the test predictions
max_y_pred_test = np.round(y_pred_test)
#Print the Confusion Matrix
print_confusion_matrix()

#%%
#Evaluate the Model and Print Performance Metrics
print_performance_metrics()
#%%
#Print and Save the Confusion Matrix
print_confusion_matrix_and_save()

#%%
#===============================================================
# 3 Convolutional neural network for MNIST digit recognition
#===============================================================
# load data
(X_train, y_train), (X_test, y_test) = mnist.load_data()
# reshape to be [samples][width][height][channels]
X_train = X_train.reshape((X_train.shape[0], 28, 28, 1)).astype('float32')
X_test = X_test.reshape((X_test.shape[0], 28, 28, 1)).astype('float32')
# normalize inputs from 0-255 to 0-1
X_train = X_train / 255
X_test = X_test / 255
# one hot encode outputs
y_train = np_utils.to_categorical(y_train)
y_test = np_utils.to_categorical(y_test)
num_classes = y_test.shape[1]
#%%
#===============================================================
# Build model
#===============================================================
model = Sequential()
model.add(Conv2D(32, (5, 5), input_shape = (28, 28, 1), activation = 'relu'))
model.add(MaxPooling2D())
model.add(Conv2D(64, (3, 3), activation = 'relu'))
```

```python
model.add(MaxPooling2D((2, 2)))
model.add(Conv2D(64, (3, 3), activation = 'relu'))
model.add(Flatten())
model.add(Dense(64, activation = 'relu'))
model.add(Dense(num_classes, activation = 'softmax'))
#================================================================
# Compile model
#================================================================
model.compile(loss = 'categorical_crossentropy', optimizer = 'adam',
metrics = ['accuracy'])
#================================================================
# Train the model
#================================================================
history = model.fit(X_train, y_train, validation_split = 0.3, epochs = 10,
batch_size = 200, verbose = 2)
#%%
#================================================================
# Evaluate the model
#================================================================
test_loss, test_acc = model.evaluate(X_test, y_test, verbose = 0)
print('\nTest accuracy:', test_acc)
#%%
#Plot the Model Accuracy and Loss for Training and Validation dataset
plot_model_accuracy_loss()
#%%
#Test the Model with testing data
y_pred_test = model.predict(X_test,)
# Round the test predictions
max_y_pred_test = np.round(y_pred_test)
#Print the Confusion Matrix
print_confusion_matrix()
#%%
#Evaluate the Model and Print Performance Metrics
print_performance_metrics()
#%%
#Print and Save the Confusion Matrix
print_confusion_matrix_and_save()
#%%
#================================================================
# 4 Convolutional neural network for MNIST digit recognition
#================================================================
# load data
(X_train, y_train), (X_test, y_test) = mnist.load_data()
# reshape to be [samples][width][height][channels]
X_train = X_train.reshape((X_train.shape[0], 28, 28, 1)).
astype('float32')
X_test = X_test.reshape((X_test.shape[0], 28, 28, 1)).astype('float32')
# normalize inputs from 0-255 to 0-1
X_train = X_train / 255
X_test = X_test / 255
```

```python
# one hot encode outputs
y_train = np_utils.to_categorical(y_train)
y_test = np_utils.to_categorical(y_test)
num_classes = y_test.shape[1]
#%%
#======================================================================
# Build model
#======================================================================
model = Sequential()
model.add(Conv2D(32, (5, 5), input_shape = (28, 28, 1), activation = 'relu'))
model.add(BatchNormalization())
model.add(Conv2D(32, kernel_size = (3, 3), activation = 'relu'))
model.add(BatchNormalization())
model.add(MaxPooling2D((2, 2)))
model.add(Dropout(0.25))
model.add(Conv2D(64, (3, 3), activation = 'relu'))
model.add(BatchNormalization())
model.add(Dropout(0.25))
model.add(Conv2D(128, kernel_size = (3, 3), activation = 'relu'))
model.add(BatchNormalization())
model.add(MaxPooling2D(pool_size = (2, 2)))
model.add(Dropout(0.25))
model.add(Flatten())
model.add(Dense(512, activation = 'relu'))
model.add(BatchNormalization())
model.add(Dropout(0.5))
model.add(Dense(128, activation = 'relu'))
model.add(BatchNormalization())
model.add(Dropout(0.5))
model.add(Dense(num_classes, activation = 'softmax'))
#======================================================================
# Compile model
#======================================================================
model.compile(loss = 'categorical_crossentropy', optimizer = 'adam', metrics = ['accuracy'])
#======================================================================
# Train the model
#======================================================================
history = model.fit(X_train, y_train, validation_split = 0.3, epochs = 10, batch_size = 200, verbose = 2)
#======================================================================
#%%
#======================================================================
# Evaluate the model
#======================================================================
test_loss, test_acc = model.evaluate(X_test, y_test, verbose = 0)
print('\nTest accuracy:', test_acc)
#%%
#Plot the Model Accuracy and Loss for Training and Validation dataset
```

```
plot_model_accuracy_loss()
#%%
#Test the Model with testing data
y_pred_test = model.predict(X_test,)
# Round the test predictions
max_y_pred_test = np.round(y_pred_test)
#Print the Confusion Matrix
print_confusion_matrix()
#%%
#Evaluate the Model and Print Performance Metrics
print_performance_metrics()
#%%
#Print and Save the Confusion Matrix
print_confusion_matrix_and_save()
```

5.7 使用 CNN 进行 Fashion-MNIST 图像分类

　　CNN 不仅是一个具有许多隐藏层的深度神经网络，而且它与大脑的视觉皮层处理类似，还是一个模拟和理解刺激的大型网络。CNN 的输出层通常使用神经网络进行多类分类。CNN 在训练过程中使用了特征提取器，而不是手动实现它。CNN 的特征提取器由特殊类型的神经网络组成，该网络通过训练过程决定权重。当神经网络特征提取更深时（包含更多的层数），CNN 能达到更好的图像识别效果，其代价是某些时候因为学习方法复杂导致 CNN 效率低下和被忽视。CNN 不仅是提取输入图像特征的神经网络，也是对图像特征进行分类的神经网络。输入图像由特征提取网络使用。提取的特征信号被分类神经网络使用。接着神经网络分类在图像特征的基础上进行工作并产生输出。用于特征提取的神经网络包括卷积层堆和池化层集。顾名思义，卷积层利用卷积的过程对图像进行变换。这可以看作一系列的数字滤波器。池化层将相邻像素转化为单个像素。池化层降低了图像的维度。由于 CNN 主要关注的是图像，所以卷积层和池化层的程序直观上是在一个二维平面上。这也是 CNN 与其他神经网络的区别之一（Kim，2017）。

　　近日，Zalando 研究公布了一个新的图像数据库，它类似于著名的 MNIST 手写数字数据库。该数据库设计的初衷是用于机器学习的分类任务，它包含 60 000 个训练和 10 000 个测试图像，这些图像是 28×28 的灰度图像。每个训练和测试用例都与十个标签（0～9）中的一个相关。Zalando 的新数据集基本与原来的手写数字数据相同。但 Zalando 的数据没有 0～9 数字的图像而是 10 种不同时尚产品的图像。因此该数据集命名为 fashion-MNIST 数据集，可以从 GitHub（https://github.com/zalandoresearch/fashion-mnist）、Kaggle 网站（https://www.kaggle.com/zalando-research/fashionmnist）或者直接从 Keras 下载它（Zhang，2019）。

例 5.7 以下的 Python 代码使用深度 CNN 分类器对 fashion-MNIST 图像数据集进行分类，该例子使用单独的训练集、验证集和测试集。同时计算出分类准确率、精确度、召回率、F1 得分、Cohen kappa 得分和 Matthews 相关系数，并给出分类报告和混淆矩阵。本例改编自 TensorFlow 网页（https://www.tensorflow.org/tutorials/keras/classification）。

数据集信息：fashion-MNIST 是一个 Zalando 的物品图像数据集，该数据集由 60 000 个训练样本和 10 000 个测试样本组成。每个样本是一张 28×28 的灰度图像，同时每张图像关联某一种标签，标签的类型共 10 种。Zalando 打算在机器学习算法的基准测试中用 fashion-MNIST 直接替换原始的 MNIST 数据集。它的训练集和测试集具有相同的图像大小和结构。每个训练和测试样本被分配以下标签之一：

- t-shirt/top
- trouser
- pullover
- dress
- coat
- sandal
- shirt
- sneaker
- bag
- ankle boot

事实上，fashion-MNIST 数据设想是直接替代旧的 MNIST 手写数字数据，因为手写数字存在很多问题。举个例子，许多数字只需看几个像素就能正确区分。即使使用线性分类器也可以达到很高的分类精确度。fashion-MNIST 数据有望更加多样化，这样机器学习算法可以学习更高级的特征来实现可靠地分类。在本例子中，将创建几个基于 CNN 的深度学习模型来评估 fashion-MNIST 数据集的性能。将使用 Keras 框架来构建这些模型。原始训练数据（60 000 张图像）将被分为 80% 的训练（48 000 张图像）和 20% 的验证（12 000 张图像）用于优化分类器，同时保留测试数据（10 000 张图像）来最终评估模型在未知数据上的准确性。该方法有助于防止训练数据的过拟合，并判断若验证精确度高于训练精确度，是否应该降低学习率并训练更多次数，或者如果训练精确度高于验证精确度，是否停止过度训练（Zhang, 2019）。

```
#=====================================================================
# Classify FASHION-MNIST images of clothing with CNN
#=====================================================================
"""
```

```python
Created on Tue Dec 3 15:34:19 2019
@author: absubasi
"""
# TensorFlow and tf.keras
import tensorflow as tf
from tensorflow import keras
from keras.models import Sequential
from keras.layers import Dense
from keras.layers import Dropout
from keras.layers import Flatten
from keras.layers.convolutional import Conv2D
from keras.layers.convolutional import MaxPooling2D
from keras.layers import BatchNormalization
# Helper libraries
import numpy as np
import matplotlib.pyplot as plt
from sklearn.metrics import confusion_matrix
import seaborn as sns
from sklearn import metrics
from io import BytesIO #needed for plot
import seaborn as sns; sns.set()
#=====================================================================
# Define utility functions
#=====================================================================
def plot_model_accuracy_loss():
    plt.figure(figsize=(6, 4))
    plt.plot(history.history['accuracy'], 'r', label='Accuracy of training data')
    plt.plot(history.history['val_accuracy'], 'b', label='Accuracy of validation data')
    plt.plot(history.history['loss'], 'r--', label='Loss of training data')
    plt.plot(history.history['val_loss'], 'b--', label='Loss of validation data')
    plt.title('Model Accuracy and Loss')
    plt.ylabel('Accuracy and Loss')
    plt.xlabel('Training Epoch')
    plt.ylim(0)
    plt.legend()
    plt.show()

def print_confusion_matrix(y_test,y_pred):
# matrix=confusion_matrix(y_test.argmax(axis=1), max_y_pred_test.argmax(axis=1))
    matrix=confusion_matrix(y_test, y_pred)
    plt.figure(figsize=(10, 8))
    sns.heatmap(matrix,cmap='coolwarm',linecolor='white',linewidths=1,
                annot=True,
                fmt='d')
```

```python
        plt.title('Confusion Matrix')
        plt.ylabel('True Label')
        plt.xlabel('Predicted Label')
        plt.show()

def print_performance_metrics(y_test,y_pred):
    print('Accuracy:', np.round(metrics.accuracy_score(y_test, y_pred),4))
    print('Precision:', np.round(metrics.precision_score(y_test, y_pred,
                                                  average = 'weighted'),4))
    print('Recall:', np.round(metrics.recall_score(y_test, y_pred,
                                                  average = 'weighted'),4))
    print('F1 Score:', np.round(metrics.f1_score(y_test, y_pred,
                                                  average = 'weighted'),4))
    print('Cohen Kappa Score:', np.round(metrics.cohen_kappa_score(y_test,
y_pred),4))
    print('Matthews Corrcoef:', np.round(metrics.matthews_corrcoef(y_test,
y_pred),4))
    print('\t\tClassification Report:\n', metrics.classification_report(y_test, y_pred))

def print_confusion_matrix_and_save(y_test, y_pred):
    mat = confusion_matrix(y_test, y_pred)
    sns.heatmap(mat.T, square = True, annot = True, fmt = 'd', cbar = False)
    plt.title('Confusion Matrix')
    plt.ylabel('True Label')
    plt.xlabel('Predicted Label')
    plt.show()

    plt.savefig("Confusion.jpg")
    # Save SVG in a fake file object.
    f = BytesIO()
    plt.savefig(f, format = "svg")
#%%
#from keras.datasets import mnist
fashion_mnist = keras.datasets.fashion_mnist
(X_train, y_train), (X_test, y_test) = fashion_mnist.load_data()

class_names = ['T-shirt/top', 'Trouser', 'Pullover', 'Dress', 'Coat',
               'Sandal', 'Shirt', 'Sneaker', 'Bag', 'Ankle boot']
#%%
# Each image's dimension is 28 x 28
img_rows, img_cols = 28, 28
input_shape = (img_rows, img_cols, 1)

# Prepare the training images
X_train = X_train.reshape(X_train.shape[0], img_rows, img_cols, 1)
X_train = X_train.astype('float32')
```

```python
# Prepare the test images
X_test = X_test.reshape(X_test.shape[0], img_rows, img_cols, 1)
X_test = X_test.astype('float32')
"""
Scale these values to a range of 0 to 1 before feeding them to the neural
network model.
To do so, divide the values by 255.
It's important that the training set and the testing set be preprocessed in
the same way:"""
X_train = X_train / 255.0
X_test = X_test / 255.0
#%%
#====================================================================
# Build the model with 1 CNN
#====================================================================
num_classes = 10
model = Sequential()
model.add(Conv2D(32, (3, 3), activation = 'relu', input_shape = input_
shape))
model.add(MaxPooling2D(pool_size = (2, 2)))
model.add(Dropout(0.2))
model.add(Flatten())
model.add(Dense(128, activation = 'relu'))
model.add(Dense(num_classes, activation = 'softmax'))
#%%
#====================================================================
# Compile the model for numerical labels
#====================================================================
model.compile(optimizer = 'adam',
              loss = 'sparse_categorical_crossentropy',
              metrics = ['accuracy'])
#%%
#====================================================================
# Train and validate the model
#====================================================================
history = model.fit(X_train, y_train, validation_split = 0.3, epochs = 10,
batch_size = 200, verbose = 2)
#%%
#====================================================================
# Evaluate the model
#====================================================================
test_loss, test_acc = model.evaluate(X_test, y_test, verbose = 0)
print(" Test Accuracy is : %0.4f" % test_acc)
#%%
#Plot the Model Accuracy and Loss for Training and Validation dataset
plot_model_accuracy_loss()
#%%
#Test the Model with testing data
y_pred_test = model.predict(X_test,)
```

```python
# Round the test predictions
max_y_pred_test = np.round(y_pred_test)
#Convert binary labels into categorical
y_pred = max_y_pred_test.argmax(axis = 1)
#Print the Confusion Matrix
print_confusion_matrix(y_test, y_pred)
#%%
#Evaluate the Model and Print Performance Metrics
print_performance_metrics(y_test, y_pred)
#%%
#Print and Save the Confusion Matrix
print_confusion_matrix_and_save(y_test, y_pred)
#%%
#=====================================================================
# Build the model with 3 CNN
#=====================================================================
num_classes = 10
model = Sequential()
model.add(Conv2D(32, (3, 3), activation = 'relu', input_shape = input_shape))
model.add(MaxPooling2D((2, 2)))
model.add(Conv2D(64, (3, 3), activation = 'relu'))
model.add(MaxPooling2D((2, 2)))
model.add(Conv2D(64, (3, 3), activation = 'relu'))
model.add(Flatten())
model.add(Dense(64, activation = 'relu'))
model.add(Dense(num_classes, activation = 'softmax'))
#%%
#=====================================================================
# Compile the model for numerical labels
#=====================================================================
model.compile(optimizer = 'adam',
              loss = 'sparse_categorical_crossentropy',
              metrics = ['accuracy'])
#%%
#=====================================================================
# Train and validate the model
#=====================================================================
history = model.fit(X_train, y_train, validation_split = 0.3, epochs = 10,
batch_size = 200, verbose = 2)
#%%
#=====================================================================
# Evaluate the model
#=====================================================================
test_loss, test_acc = model.evaluate(X_test, y_test, verbose = 0)
print("Test Accuracy is : %0.4f" % test_acc)
#%%
#Plot the Model Accuracy and Loss for Training and Validation dataset
plot_model_accuracy_loss()
```

```python
#%%
#Test the Model with testing data
y_pred_test = model.predict(X_test,)
# Round the test predictions
max_y_pred_test = np.round(y_pred_test)
#Convert binary labels into categorical
y_pred = max_y_pred_test.argmax(axis = 1)
#Print the Confusion Matrix
print_confusion_matrix(y_test, y_pred)
#%%
#Evaluate the Model and Print Performance Metrics
print_performance_metrics(y_test, y_pred)
#%%
#Print and Save the Confusion Matrix
print_confusion_matrix_and_save(y_test, y_pred)
#%%
#================================================================
# Build the model with 4 CNN
#================================================================
num_classes = 10
model = Sequential()
model.add(Conv2D(32, (3, 3), activation = 'relu', input_shape = input_
shape))
model.add(BatchNormalization())

model.add(Conv2D(32, kernel_size = (3, 3), activation = 'relu'))
model.add(BatchNormalization())
model.add(MaxPooling2D((2, 2)))
model.add(Dropout(0.25))

model.add(Conv2D(64, (3, 3), activation = 'relu'))
model.add(BatchNormalization())
model.add(Dropout(0.25))
model.add(Conv2D(128, kernel_size = (3, 3), activation = 'relu'))
model.add(BatchNormalization())
model.add(MaxPooling2D(pool_size = (2, 2)))
model.add(Dropout(0.25))
model.add(Flatten())
model.add(Dense(512, activation = 'relu'))
model.add(BatchNormalization())
model.add(Dropout(0.5))

model.add(Dense(128, activation = 'relu'))
model.add(BatchNormalization())
model.add(Dropout(0.5))
model.add(Dense(num_classes, activation = 'softmax'))
#%%
#================================================================
# Compile the model for numerical labels
#================================================================
```

```python
model.compile(optimizer = 'adam',
              loss = 'sparse_categorical_crossentropy',
              metrics = ['accuracy'])
#%%
#===============================================================
# Train and validate the model
#===============================================================
history = model.fit(X_train, y_train, validation_split = 0.3, epochs = 10,
batch_size = 200, verbose = 2)
#%%
#===============================================================
# Evaluate the model
#===============================================================
test_loss, test_acc = model.evaluate(X_test, y_test, verbose = 0)
print("Test Accuracy is : %0.4f" % test_acc)
#%%
#Plot the Model Accuracy and Loss for Training and Validation dataset
plot_model_accuracy_loss()
#%%
#Test the Model with testing data
y_pred_test = model.predict(X_test,)
# Round the test predictions
max_y_pred_test = np.round(y_pred_test)
#Convert binary labels into categorical
y_pred = max_y_pred_test.argmax(axis = 1)
#Print the Confusion Matrix
print_confusion_matrix(y_test, y_pred)
#%%
#Evaluate the Model and Print Performance Metrics
print_performance_metrics(y_test, y_pred)
#%%
#Print and Save the Confusion Matrix
print_confusion_matrix_and_save(y_test, y_pred)
```

5.8 使用 CNN 进行 CIFAR 图像分类

目前，在几个主要任务上开展的广泛实验工作推动了机器学习的进步。然而，因为这些被选出来好几年的模型使用相同的测试集，所以性能最好的模型表现出的惊人准确性是值得怀疑的。CIFAR-10 分类器的可靠性评估可以通过构建一个新的真正未知对象测试集来了解过拟合的危险。虽然人们发现新的测试集已经很接近原始数据的分布，但是许多深度学习模型的准确率有 4% 到 10% 的大幅下降。然而，性能较高的新模型则表现出较小的降幅和较好的整体性能，这意味着这种降幅不可能是基于适应性的过拟合造成的。在过去的十年中，机器学习已经成为一个极具革命性的领域（Recht, Roelofs, Schmidt, Shankar, 2018）。最近，CNN 在许多计算机视觉问题上取得了较好的成绩。CNN 部分受到神经科学的启发，因而展现出大脑视觉系统的许多特性。尽管循环连接在视觉系统中很常见，但是显著的区别是 CNN 通常是一个前馈架构（Liang & Hu, 2015）。受深度学习研究兴起的推动，大多数已发表的论文都采用同一种模式去评判新的学习方法，即该方法在几个关键基准上的性能改进，与此同时却很少说明新方法相比过去的研究更好的原因。相反，我们对进步的感觉很大程度上取决于少量的标准基准，如 CIFAR-10、ImageNet 或 MuJoCo（Recht et al., 2018）。

例 5.8 以下的 Python 代码用于使用 DCNN 分类器对钓鱼网站数据集进行分类，代码采用 10 折交叉验证以及单独的训练和测试数据集。同时计算出分类准确率、精确度、召回率、F1 得分、Cohen kappa 得分和 Matthews 相关系数，并给出分类报告和混淆矩阵。本例改编自 TensorFlow 网页（https://www.tensorflow.org/tutorials/images/cnn）。

数据集信息：CIFAR-10 和 CIFAR-100 是被标记过的 8 000 万张小图像数据集的子集。它们由 Alex Krizhevsky、Vinod Nair 和 Geoffrey Hinton 收集。CIFAR-10 数据集包括 6 万张 32×32 的彩色图像，图像共分成 10 类，每类有 6000 张。其中有 50 000 张训练图像和 10 000 张测试图像。数据集分为 5 个训练批次和 1 个测试批次，每个训练批次有 1 万张图像。测试批次恰好包含从每个类中随机选取的 1 000 张图像。剩余的图像随机地划分到各训练批次中，但一些训练批次可能包括某个类的图像比另一个类更多。CIFAR-10 数据集可以从 CIFAR 网站（https://www.cs.toronto.edu/~kriz/cifar.html）或者直接从 Keras 下载。该数据集的分类如下

- airplane
- automobile
- bird
- cat
- deer
- dog
- frog
- horse
- ship
- truck

```
"""
Created on Tue Dec 3 15:45:50 2019
@author: absubasi
Convolutional Neural Network (CNN)
https://www.tensorflow.org/tutorials/images/cnn
"""
#=========================================================================
# Classification of CIFAR images using CNN
#=========================================================================
import tensorflow as tf
from keras.models import Sequential
from keras.layers import Dense
from keras.layers import Dropout
```

```python
from keras.layers import Flatten
from keras.layers.convolutional import Conv2D
from keras.layers.convolutional import MaxPooling2D
from keras.layers import BatchNormalization
from keras.utils import np_utils
from tensorflow.keras import datasets, layers, models
import matplotlib.pyplot as plt
import numpy as np
from sklearn.metrics import confusion_matrix
from sklearn import metrics
from io import BytesIO #needed for plot
import seaborn as sns; sns.set()
#=====================================================================
# Define utility functions
#=====================================================================
def plot_model_accuracy_loss():
    plt.figure(figsize = (6, 4))
    plt.plot(history.history['accuracy'], 'r', label = 'Accuracy of training data')
    plt.plot(history.history['val_accuracy'], 'b', label = 'Accuracy of validation data')
    plt.plot(history.history['loss'], 'r--', label = 'Loss of training data')
    plt.plot(history.history['val_loss'], 'b--', label = 'Loss of validation data')
    plt.title('Model Accuracy and Loss')
    plt.ylabel('Accuracy and Loss')
    plt.xlabel('Training Epoch')
    plt.ylim(0)
    plt.legend()
    plt.show()

def print_confusion_matrix():
    matrix = confusion_matrix(y_test, y_pred)
    plt.figure(figsize = (10, 8))
    sns.heatmap(matrix,cmap = 'coolwarm',linecolor = 'white',linewidths = 1,
                annot = True,
                fmt = 'd')
    plt.title('Confusion Matrix')
    plt.ylabel('True Label')
    plt.xlabel('Predicted Label')
    plt.show()

def print_performance_metrics():
    print('Accuracy:', np.round(metrics.accuracy_score(y_test, y_pred),4))
    print('Precision:', np.round(metrics.precision_score(y_test,
                               y_pred,average = 'weighted'),4))
    print('Recall:', np.round(metrics.recall_score(y_test, y_pred,
                            average = 'weighted'),4))
```

```python
        print('F1 Score:', np.round(metrics.f1_score(y_test, y_pred,
                                    average = 'weighted'),4))
        print('Cohen Kappa Score:', np.round(metrics.cohen_kappa_score(y_test,
y_pred),4))
        print('Matthews Corrcoef:', np.round(metrics.matthews_corrcoef(y_test,
y_pred),4))
        print('\t\tClassification Report:\n', metrics.classification_report(y_
test, y_pred))

def print_confusion_matrix_and_save():
    mat = confusion_matrix(y_test, y_pred)
    sns.heatmap(mat.T, square = True, annot = True, fmt = 'd', cbar = False)
    plt.title('Confusion Matrix')
    plt.ylabel('True Label')
    plt.xlabel('Predicted Label')
    plt.show()

    plt.savefig("Confusion.jpg")
    # Save SVG in a fake file object.
    f = BytesIO()
    plt.savefig(f, format = "svg")
#%%
# load data
(X_train, y_train), (X_test, y_test) = datasets.cifar10.load_data()
# Normalize pixel values to be between 0 and 1
X_train, X_test = X_train / 255.0, X_test / 255.0
#%%
class_names = ['airplane', 'automobile', 'bird', 'cat', 'deer',
               'dog', 'frog', 'horse', 'ship', 'truck']
plt.figure(figsize = (10,10))
for i in range(25):
    plt.subplot(5,5,i + 1)
    plt.xticks([])
    plt.yticks([])
    plt.grid(False)
    plt.imshow(X_train[i], cmap = plt.cm.binary)
    # The CIFAR labels happen to be arrays,
    # which is why you need the extra index
    plt.xlabel(class_names[y_train[i][0]])
plt.show()
#%%
#========================================================================
# 1 Convolutional neural network for CIFAR image classification
#========================================================================
# load data
(X_train, y_train), (X_test, y_test) = datasets.cifar10.load_data()
# Normalize pixel values to be between 0 and 1
X_train, X_test = X_train / 255.0, X_test / 255.0
# Each image's dimension is 32 x 32
img_rows, img_cols = 32, 32
```

```python
input_shape = (img_rows, img_cols, 3)
#%%
#================================================================
# Build model
#================================================================
num_classes = 10
model = Sequential()
model.add(Conv2D(32, (3, 3), activation = 'relu', input_shape = input_shape))
model.add(MaxPooling2D())
model.add(Dropout(0.2))
model.add(Flatten())
model.add(Dense(128, activation = 'relu'))
model.add(Dense(num_classes, activation = 'softmax'))
#%%
#================================================================
# Compile the model for numerical labels
#================================================================
model.compile(optimizer = 'adam',
    loss = 'sparse_categorical_crossentropy',
    metrics = ['accuracy'])
#%%
#================================================================
# Compile the model for binary labels
#================================================================
#model.compile(loss = 'categorical_crossentropy', optimizer = 'adam',
metrics = ['accuracy'])
#%%
#================================================================
# Train and validate the model
#================================================================
history = model.fit(X_train, y_train, validation_split = 0.3, epochs = 10,
batch_size = 200, verbose = 2)
#%%
#================================================================
# Evaluate the model
#================================================================
test_loss, test_acc = model.evaluate(X_test, y_test, verbose = 0)
print("Test Accuracy is : %0.4f" % test_acc)
#%%
#Plot the Model Accuracy and Loss for Training and Validation dataset
plot_model_accuracy_loss()
#%%
#Test the Model with testing data
y_pred_test = model.predict(X_test,)
# Round the test predictions
max_y_pred_test = np.round(y_pred_test)
#Print the Confusion Matrix
print_confusion_matrix()
#%%
```

```python
#Evaluate the Model and Print Performance Metrics
print_performance_metrics()
#%%
#Print and Save the Confusion Matrix
print_confusion_matrix_and_save()
#%%
#==================================================================
# 3 Convolutional neural network for CIFAR image classification
#==================================================================
# load data
(X_train, y_train), (X_test, y_test) = datasets.cifar10.load_data()
# Normalize pixel values to be between 0 and 1
X_train, X_test = X_train / 255.0, X_test / 255.0
# Each image's dimension is 32 x 32
img_rows, img_cols = 32, 32
input_shape = (img_rows, img_cols, 3)
#%%
#==================================================================
# Build model
#==================================================================
num_classes = 10
model = Sequential()
model.add(Conv2D(32, (3, 3), activation = 'relu', input_shape = input_
shape))
model.add(MaxPooling2D((2, 2)))
model.add(Conv2D(64, (3, 3), activation = 'relu'))
model.add(MaxPooling2D((2, 2)))
model.add(Conv2D(64, (3, 3), activation = 'relu'))
model.add(Flatten())
model.add(Dense(64, activation = 'relu'))
model.add(Dense(num_classes, activation = 'softmax'))
#%%
#==================================================================
# Compile the model for numerical labels
#==================================================================
model.compile(optimizer = 'adam',
              loss = 'sparse_categorical_crossentropy',
              metrics = ['accuracy'])
#%%
#==================================================================
# Compile the model for binary labels
#==================================================================
#model.compile(loss = 'categorical_crossentropy', optimizer = 'adam',
metrics = ['accuracy'])
#%%
#==================================================================
# Train and validate the model
#==================================================================
history = model.fit(X_train, y_train, validation_split = 0.3, epochs = 10,
batch_size = 200, verbose = 2)
```

```python
#%%
#==================================================================
# Evaluate the model
#==================================================================
test_loss, test_acc = model.evaluate(X_test, y_test, verbose = 0)
print("Test Accuracy is : %0.4f" % test_acc)
#%%
#Plot the Model Accuracy and Loss for Training and Validation dataset
plot_model_accuracy_loss()
#%%
#Test the Model with testing data
y_pred_test = model.predict(X_test,)
# Round the test predictions
max_y_pred_test = np.round(y_pred_test)
#Print the Confusion Matrix
print_confusion_matrix()
#%%
#Evaluate the Model and Print Performance Metrics
print_performance_metrics()
#%%
#Print and Save the Confusion Matrix
print_confusion_matrix_and_save()
#%%
#==================================================================
# 4 Convolutional neural network for CIFAR image classification
#==================================================================
# load data
(X_train, y_train), (X_test, y_test) = datasets.cifar10.load_data()
# Normalize pixel values to be between 0 and 1
X_train, X_test = X_train / 255.0, X_test / 255.0
#==================================================================
# Build model
#==================================================================
num_classes = 10
model = Sequential()
model.add(Conv2D(32, (3, 3), activation = 'relu', input_shape = (32, 32, 3)))
model.add(BatchNormalization())
model.add(Conv2D(32, kernel_size = (3, 3), activation = 'relu'))
model.add(BatchNormalization())
model.add(MaxPooling2D((2, 2)))
model.add(Dropout(0.25))
model.add(Conv2D(64, (3, 3), activation = 'relu'))
model.add(BatchNormalization())
model.add(Dropout(0.25))
model.add(Conv2D(128, kernel_size = (3, 3), activation = 'relu'))
model.add(BatchNormalization())
model.add(MaxPooling2D(pool_size = (2, 2)))
model.add(Dropout(0.25))
model.add(Flatten())
model.add(Dense(512, activation = 'relu'))
```

```python
model.add(BatchNormalization())
model.add(Dropout(0.5))
model.add(Dense(128, activation = 'relu'))
model.add(BatchNormalization())
model.add(Dropout(0.5))
model.add(Dense(num_classes, activation = 'softmax'))
#%%
#================================================================
# Compile the model for numerical labels
#================================================================
model.compile(optimizer = 'adam',
              loss = 'sparse_categorical_crossentropy',
              metrics = ['accuracy'])
#%%
#================================================================
# Compile the model for binary labels
#================================================================
#model.compile(loss = 'categorical_crossentropy', optimizer = 'adam',
metrics = ['accuracy'])
#%%
#================================================================
# Train and validate the model
#================================================================
history = model.fit(X_train, y_train, validation_split = 0.3, epochs = 10,
batch_size = 200, verbose = 2)
#%%
#================================================================
# Evaluate the model
#================================================================
test_loss, test_acc = model.evaluate(X_test, y_test, verbose = 0)
print("Test Accuracy is : %0.4f" % test_acc)
#%%
#Plot the Model Accuracy and Loss for Training and Validation dataset
plot_model_accuracy_loss()
#%%
#Test the Model with testing data
y_pred_test = model.predict(X_test,)
# Round the test predictions
max_y_pred_test = np.round(y_pred_test)
#Convert binary labels back to Numerical
y_pred = max_y_pred_test.argmax(axis = 1)
#Print the Confusion Matrix
print_confusion_matrix()
#%%
#Evaluate the Model and Print Performance Metrics
print_performance_metrics()
#%%
#Print and Save the Confusion Matrix
print_confusion_matrix_and_save()
```

5.9 文本分类

自动文本分类通常使用机器学习算法将文档划分为预定的类型。一般来说，最重要的技术之一是组织和利用大量存在于非结构化文本格式中的信息。文本分类是语言处理和文本挖掘研究中一个被广泛研究的领域。在传统的文本分类中，文档被表示为词袋，其中的词语是从其确切的上下文（即它们在句子或文档中的位置）中切出来的。由于在向量空间中只有文档上下文和某些类型的词频信息被广泛地利用。因此，词的语义——可以从词在句子中的位置及其与相邻词间的关系得出的含义——通常被忽略。但是，理解词和文档间的语义关系是获取语义的必要途径，该方法通常都能在分类中取得更好性能（Altinel & Ganiz, 2018）。

由于海量数据的来源广泛，如社交网络、博客/论坛、网站、电子邮件以及发表研究论文的数字图书馆等，因此近年来文本挖掘研究变得越来越重要。随着新技术的进步，如语音文本转换引擎、数字助理和个人智能助理，电子文本信息无疑会不断增加。一个基本的问题是如何自动加工、组织和处理这些文本数据。文本挖掘有几个重要的应用，比如文档的分类和过滤、摘要、情感分析/观点分类。机器学习可以和自然语言处理（NLP）技术协同检测并自动地对来自不同类型的文档模式进行分类（Altinel & Ganiz, 2018; Sebastiani, 2005）。

语义文本分类方法中考虑了语义词关系，其一般用于衡量文档之间的相似性。语义方法通过关注词义以及词与词之间隐藏的语言关系从而聚焦文档之间的关系（Altinel & Ganiz, 2018）。与传统的文本分类相比，语义文本分类的优点包括：

- 能够找到词语间显式的或者隐式的关联关系。
- 能够提取和使用词和文档间的潜在关联关系。
- 能够为现有的类型生成具有代表性的关键字。
- 对文本语义的理解能够提高分类的性能。
- 由于使用的是词之间的语义关系，与传统的文本分类算法相比，能够管理同义词和多义词。

不同的基于语义的技术都提出在文本分类中结合词与词之间的语义关系。这些技术可以分为五种类型，即基于领域知识（本体）的方法、基于语料库的方法、基于深度学习的方法、词/字符增强的方法和语言丰富的方法（Altinel & Ganiz, 2018）。

- 基于领域知识（语言依赖）的方法：基于领域知识的系统使用本体论或词典对文件中的概念进行分类。知识库的例子有字典、词典和百科全书资源。常见的知识库有 WordNet、Wiktionary 和 Wikipedia。其中 WordNet 是目前使用最多的知识库。
- 基于语料库（语言独立）的方法：这些方案利用数学计算识别学习语料库中单词之间的潜在相似性（Zhang, Gentile, & Ciravegna, 2012）。其中一个有名的基于语料库的算法是潜在语义分析（LSA）(Deerwester, Dumais, Furnas, Landauer, & Harshman, 1990)。
- 基于深度学习的方法：最近，深度学习在语义文本分析中得到了更多的关注。
- 词/字符序列增强的方法：词/字符序列增强方法将单词或字符视为字符串序列，并通过传统的字符串匹配方法从文档中提取出来。

- **语言丰富方法**：这些方法利用词法和句法规则从文档中提取名词短语、实体和术语以创建文档的表示（Altinel & Ganiz, 2018）。

> **例 5.9** 以下的 Python 代码使用不同的分类器对文本数据集进行分类，代码采用 10 折交叉验证并使用单独训练和测试数据集。同时计算出分类准确率、精确度、召回率、F1 得分、Cohen kappa 得分和 Matthews 相关系数，并给出分类报告和混淆矩阵。本例改编自 Kaggle 网页（https://www.kaggle.com/sanikamal/text-classification-with-python-and-keras）。
>
> 数据集信息：该实验使用三个真实世界的数据集。其中亚马逊数据集包括在 amazon.com 上销售的手机和配件类产品的分数和评论，该数据集是 McAuley 和 Leskovec 收集的一部分数据（McAuley & Leskovec, 2013）。分数是从 1 至 5 的整数，其中 4 分和 5 分被认为是正面的，而 1 分和 2 分为负面的。IMDb 是一个电影评论的情感数据库（https://www.kaggle.com/ lakshmi25npathi/imdb-dataset-of-50k-movie-reviews），它最初由 Mass 等人（Maas et al., 2011）创建用来作为情感分析的基准。该数据库总共包含 100 000 条发表在 imdb.com 上的电影评论。其中 50 000 条是未被标记的评论，剩余 50 000 条评论被分成包含 25 000 条评论训练集和包含 25 000 条评论的测试集。每条被标记的评论具有一个二元语义标签，positive 或者 negative。Yelp 是 Yelp 数据集挑战赛的一个数据集（https://www.yelp.com/dataset/challenge），它包含餐馆评价。分数是从 1 到 5 的整数，其中 4 分和 5 分被认为是正面的，而 1 分和 2 分为负面的（Kotzias, Denil, De Freitas, & Smyth, 2015）。
>
> ```
> #================================================================
> # Text classification example
> #================================================================
> """
> Created on Thu Dec 12 12:02:11 2019
> @author: absubasi
> Text Classification With Python and Keras
> """
> import pandas as pd # data processing, CSV file I/O (e.g. pd.read_csv)
> import matplotlib.pyplot as plt
> plt.style.use('ggplot')
> #importing the libraries
> import numpy as np
> from sklearn.feature_extraction.text import CountVectorizer
> from sklearn.model_selection import train_test_split
> from sklearn.linear_model import LogisticRegression
> from sklearn.preprocessing import LabelEncoder
> from sklearn.preprocessing import OneHotEncoder
> from sklearn.model_selection import RandomizedSearchCV
> from keras.models import Sequential
> from keras import layers
> from keras.preprocessing.text import Tokenizer
> from keras.preprocessing.sequence import pad_sequences
> from keras.wrappers.scikit_learn import KerasClassifier
> ```

```python
from sklearn.model_selection import cross_val_score
from sklearn import metrics
from sklearn.metrics import confusion_matrix
import seaborn as sns
import matplotlib.pyplot as plt
from io import BytesIO #needed for plot
import seaborn as sns; sns.set()
#====================================================================
# Define utility functions
#====================================================================
def kFold_Cross_Validation_Metrics(model,CV):
    Acc_scores = cross_val_score(model, X, y, cv = CV)
    print("Accuracy: %0.3f (+/- %0.3f)" % (Acc_scores.mean(), Acc_scores.std() * 2))
    f1_scores = cross_val_score(model, X, y, cv = CV,scoring = 'f1_macro')
    print("F1 score: %0.3f (+/- %0.3f)" % (f1_scores.mean(), f1_scores.std() * 2))
    Precision_scores = cross_val_score(model, X, y, cv = CV,scoring = 'precision_macro')
    print("Precision score: %0.3f (+/- %0.3f)" % (Precision_scores.mean(), Precision_scores.std() * 2))
    Recall_scores = cross_val_score(model, X, y, cv = CV,scoring = 'recall_macro')
    print("Recall score: %0.3f (+/- %0.3f)" % (Recall_scores.mean(), Recall_scores.std() * 2))
    from sklearn.metrics import cohen_kappa_score, make_scorer
    kappa_scorer = make_scorer(cohen_kappa_score)
    Kappa_scores = cross_val_score(model, X, y, cv = CV,scoring = kappa_scorer)
    print("Kappa score: %0.3f (+/- %0.3f)" % (Kappa_scores.mean(), Kappa_scores.std() * 2))

def Performance_Metrics(model,y_test,y_pred):
    print('Test Accuracy:', np.round(metrics.accuracy_score(y_test,y_pred),4))
    print('Precision:', np.round(metrics.precision_score(y_test,
                            y_pred,average = 'weighted'),4))
    print('Recall:', np.round(metrics.recall_score(y_test,y_pred,
                                average = 'weighted'),4))
    print('F1 Score:', np.round(metrics.f1_score(y_test,y_pred,
                                average = 'weighted'),4))
    print('Cohen Kappa Score:', np.round(metrics.cohen_kappa_score(y_test,y_pred),4))
    print('Matthews Corrcoef:', np.round(metrics.matthews_corrcoef(y_test,y_pred),4))
    print('\t\tClassification Report:\n', metrics.classification_report(y_test,y_pred))
    print("Confusion Matrix:\n",confusion_matrix(y_test,y_pred))
```

```python
def print_confusion_matrix(y_test, y_pred, target_names):
    matrix = confusion_matrix(y_test, y_pred)
    plt.figure(figsize = (10, 8))
    sns.heatmap(matrix,cmap = 'coolwarm',linecolor = 'white',linewidths = 1,
                xticklabels = target_names,
                yticklabels = target_names,
                annot = True,
                fmt = 'd')
    plt.title('Confusion Matrix')
    plt.ylabel('True Label')
    plt.xlabel('Predicted Label')
    plt.show()

def print_confusion_matrix_and_save(y_test, y_pred):
    #Print the Confusion Matrix
    matrix = confusion_matrix(y_test, y_pred)
    plt.figure(figsize = (6, 4))
    sns.heatmap(matrix, square = True, annot = True, fmt = 'd', cbar = False)
    plt.title('Confusion Matrix')
    plt.ylabel('True Label')
    plt.xlabel('Predicted Label')
    plt.show()
    #Save The Confusion Matrix
    plt.savefig("Confusion.jpg")
    # Save SVG in a fake file object.
    f = BytesIO()
    plt.savefig(f, format = "svg")

def print_metrics(X_test, y_test):
    score = classifier.score(X_test, y_test)
    print('Accuracy for {} data: {:.4f}'.format(source, score))
    # Predicting the Test set results
    y_pred = classifier.predict(X_test)
    precision = np.round(metrics.precision_score(y_test,
                         y_pred,average = 'weighted'),4)
    print('Precision for {} data: {:.4f}'.format(source, precision))
    recall = np.round(metrics.recall_score(y_test,y_pred,
                      average = 'weighted'),4)
    print('Recall for {} data: {:.4f}'.format(source, recall))
    f1score = np.round(metrics.f1_score(y_test,y_pred,
                       average = 'weighted'),4)
    print('F1 Score for {} data: {:.4f}'.format(source, f1score))
    kappa = np.round(metrics.cohen_kappa_score(y_test,y_pred),4)
    print('Cohen Kappa Score for {} data: {:.4f}'.format(source, kappa))
    matthews = np.round(metrics.matthews_corrcoef(y_test,y_pred),4)
    print('Matthews Corrcoef: for {} data: {:.4f}'.format(source, matthews))
    print('\t\tClassification Report:\n', metrics.classification_report(y_test,y_pred))
```

```python
def plot_history(history):
    accuracy = history.history['accuracy']
    val_accuracy = history.history['val_accuracy']
    loss = history.history['loss']
    val_loss = history.history['val_loss']
    x = range(1, len(accuracy) + 1)
    
    plt.figure(figsize = (12, 5))
    plt.subplot(1, 2, 1)
    plt.plot(x, accuracy, 'b', label = 'Training acc')
    plt.plot(x, val_accuracy, 'r', label = 'Validation acc')
    plt.title('Training and validation accuracy')
    plt.legend()
    plt.subplot(1, 2, 2)
    plt.plot(x, loss, 'b', label = 'Training loss')
    plt.plot(x, val_loss, 'r', label = 'Validation loss')
    plt.title('Training and validation loss')
    plt.legend()

def plot_model_accuracy_loss():
    plt.figure(figsize = (6, 4))
    plt.plot(history.history['accuracy'], 'r', label = 'Accuracy of training data')
    plt.plot(history.history['val_accuracy'], 'b', label = 'Accuracy of validation data')
    plt.plot(history.history['loss'], 'r--', label = 'Loss of training data')
    plt.plot(history.history['val_loss'], 'b--', label = 'Loss of validation data')
    plt.title('Model Accuracy and Loss')
    plt.ylabel('Accuracy and Loss')
    plt.xlabel('Training Epoch')
    plt.ylim(0)
    plt.legend()
    plt.show()
#%%
#===================================================================
# Load and prepare data
#===================================================================

filepath_dict = {'yelp': 'yelp_labelled.txt',
                 'amazon': 'amazon_cells_labelled.txt',
                 'imdb': 'imdb_labelled.txt'}
df_list = []
for source, filepath in filepath_dict.items():
    df = pd.read_csv(filepath, names = ['sentence', 'label'], sep = '\t')
    df['source'] = source # Add another column filled with the source name
    df_list.append(df)
```

```python
#%%
df = pd.concat(df_list)
df.iloc[0]
#%%
df_yelp = df[df['source'] == 'yelp']
sentences = df_yelp['sentence'].values
y = df_yelp['label'].values

sentences_train, sentences_test, y_train, y_test = train_test_
split(sentences, y, test_size = 0.33, random_state = 1000)
vectorizer = CountVectorizer()
vectorizer.fit(sentences_train)
X = vectorizer.transform(sentences)
X_train = vectorizer.transform(sentences_train)
X_test = vectorizer.transform(sentences_test)
#%%
#================================================================
# XGBoost example with cross-validation
#================================================================
from xgboost import XGBClassifier
# Create a Model
model = XGBClassifier()
#Evaluate Model Using 10-Fold Cross Validation and Print Performance
Metrics
CV = 10 #10-Fold Cross Validation
kFold_Cross_Validation_Metrics(model,CV)
#%%
#================================================================
# XGBoost example with training and testing set
#================================================================
model.fit(X_train, y_train)
# Predicting the Test set results
y_pred = model.predict(X_test)
target_names = ['NO','YES' ]
#Print the Confusion Matrix
print_confusion_matrix(y_test, y_pred, target_names)
# Print The Performance Metrics
Performance_Metrics(model,y_test,y_pred)
#Print and Save the Confusion Matrix
print_confusion_matrix_and_save(y_test, y_pred)
#%%
from xgboost import XGBClassifier
for source in df['source'].unique():
    df_source = df[df['source'] == source]
    sentences = df_source['sentence'].values
    y = df_source['label'].values
    sentences_train, sentences_test, y_train, y_test = train_test_split(
    sentences, y, test_size = 0.25, random_state = 1000)
    vectorizer = CountVectorizer()
    vectorizer.fit(sentences_train)
```

```python
    X_train = vectorizer.transform(sentences_train)
    X_test = vectorizer.transform(sentences_test)
    classifier = XGBClassifier()
    classifier.fit(X_train, y_train)
    score = classifier.score(X_test, y_test)
    print('Accuracy for {} data: {:.4f}'.format(source, score))
    print_metrics(X_test, y_test)
#%%
from sklearn.neural_network import MLPClassifier
"""mlp = MLPClassifier(hidden_layer_sizes = (100,), activation = 'relu', solver = 'adam',
            alpha = 0.0001, batch_size = 'auto', learning_rate =
            'constant',
            learning_rate_init = 0.001, power_t = 0.5, max_iter = 200,
            shuffle = True, random_state = None, tol = 0.0001, verbose =
            False,
            warm_start = False, momentum = 0.9, nesterovs_momentum =
            True,
            early_stopping = False, validation_fraction = 0.1, beta_1
            = 0.9,
            beta_2 = 0.999, epsilon = 1e-08, n_iter_no_change = 10)"""
for source in df['source'].unique():
    df_source = df[df['source'] == source]
    sentences = df_source['sentence'].values
    y = df_source['label'].values
    sentences_train, sentences_test, y_train, y_test = train_test_split(
            sentences, y, test_size = 0.25, random_state = 1000)
    vectorizer = CountVectorizer()
    vectorizer.fit(sentences_train)
    X_train = vectorizer.transform(sentences_train)
    X_test = vectorizer.transform(sentences_test)
    classifier = MLPClassifier(hidden_layer_sizes = (50,25), learning_rate_
init = 0.001,
                    alpha = 1, momentum = 0.9, max_iter = 1000)
    classifier.fit(X_train, y_train)
    print_metrics(X_test, y_test)
#%%
from sklearn.neighbors import KNeighborsClassifier
for source in df['source'].unique():
    df_source = df[df['source'] == source]
    sentences = df_source['sentence'].values
    y = df_source['label'].values
    sentences_train, sentences_test, y_train, y_test = train_test_split(
            sentences, y, test_size = 0.25, random_state = 1000)
    vectorizer = CountVectorizer()
    vectorizer.fit(sentences_train)
    X_train = vectorizer.transform(sentences_train)
    X_test = vectorizer.transform(sentences_test)
    classifier = KNeighborsClassifier(n_neighbors = 5)
    classifier.fit(X_train, y_train)
```

```python
        print_metrics(X_test, y_test)
#%%
from sklearn import svm
""" The parameters and kernels of SVM classifier can be changed as
follows
C = 10.0 # SVM regularization parameter
svm.SVC(kernel = 'linear', C = C)
svm.LinearSVC(C = C, max_iter = 10000)
svm.SVC(kernel = 'rbf', gamma = 0.7, C = C)
svm.SVC(kernel = 'poly', degree = 3, gamma = 'auto', C = C))
"""
C = 100.0 # SVM regularization parameter
for source in df['source'].unique():
    df_source = df[df['source'] == source]
    sentences = df_source['sentence'].values
    y = df_source['label'].values
    sentences_train, sentences_test, y_train, y_test = train_test_split(
        sentences, y, test_size = 0.25, random_state = 1000)
    vectorizer = CountVectorizer()
    vectorizer.fit(sentences_train)
    X_train = vectorizer.transform(sentences_train)
    X_test = vectorizer.transform(sentences_test)
    classifier =svm.LinearSVC(C = C, max_iter = 10000)
    classifier.fit(X_train, y_train)
    print_metrics(X_test, y_test)
#%%
from sklearn import metrics
# Fitting Random Forest to the Training set
from sklearn.ensemble import RandomForestClassifier
#In order to change to accuracy increase n_estimators
"""RandomForestClassifier(n_estimators = 'warn', criterion = 'gini', max_
depth = None,
min_samples_split = 2, min_samples_leaf = 1, min_weight_fraction_leaf = 0.0,
max_features = 'auto', max_leaf_nodes = None, min_impurity_decrease = 0.0,
min_impurity_split = None, bootstrap = True, oob_score = False, n_jobs =
None,
random_state = None, verbose = 0, warm_start = False, class_weight =
None)"""
for source in df['source'].unique():
    df_source = df[df['source'] == source]
    sentences = df_source['sentence'].values
    y = df_source['label'].values
    sentences_train, sentences_test, y_train, y_test = train_test_split(
        sentences, y, test_size = 0.25, random_state = 1000)
    vectorizer = CountVectorizer()
    vectorizer.fit(sentences_train)
    X_train = vectorizer.transform(sentences_train)
    X_test = vectorizer.transform(sentences_test)
    classifier =RandomForestClassifier(n_estimators = 100)
    classifier.fit(X_train, y_train)
```

```python
    print_metrics(X_test, y_test)
#%%
#================================================================
# Use Keras deep models
#================================================================
from sklearn.metrics import confusion_matrix
from sklearn import metrics
from io import BytesIO #needed for plot
import seaborn as sns; sns.set()
import matplotlib.pyplot as plt
import numpy as np
import seaborn as sns
#================================================================
# Create a Keras model
#================================================================
input_dim = X_train.shape[1] # Number of features

model = Sequential()
model.add(layers.Dense(100, input_dim = input_dim, activation = 'relu'))
model.add(layers.Dense(50, activation = 'relu'))
model.add(layers.Dense(1, activation = 'sigmoid'))
#%%
#================================================================
# Compile the model
#================================================================
model.compile(loss = 'binary_crossentropy',
              optimizer = 'adam',
              metrics = ['accuracy'])
#================================================================
# Train and validate the model
#================================================================
history = model.fit(X_train, y_train,
                    epochs = 20,
                    verbose = True,
                    validation_split = 0.33,
                    batch_size = 10)
#%%
#================================================================
# Evaluate the model
#================================================================
loss, accuracy = model.evaluate(X_train, y_train, verbose = False)
print("Training Accuracy: {:.4f}".format(accuracy))
loss, accuracy = model.evaluate(X_test, y_test, verbose = False)
print("Testing Accuracy: {:.4f}".format(accuracy))
#%%
#================================================================
# Plot the history
#================================================================
#Plot the Model Accuracy and Loss for Training and Validation dataset
plot_history(history)
```

```
plot_model_accuracy_loss()
#%%
#Test the Model with testing data
y_pred_test = model.predict(X_test,)
# Round the test predictions
y_pred = np.round(y_pred_test)
#%%
target_names = ['NO','YES' ]
#Print the Confusion Matrix
print_confusion_matrix(y_test, y_pred, target_names)
# Print The Perfromance Metrics
Performance_Metrics(model,y_test,y_pred)
#Print and Save the Confusion Matrix
print_confusion_matrix_and_save(y_test, y_pred)

#%%
tokenizer = Tokenizer(num_words = 5000)
tokenizer.fit_on_texts(sentences_train)

X_train = tokenizer.texts_to_sequences(sentences_train)
X_test = tokenizer.texts_to_sequences(sentences_test)

vocab_size = len(tokenizer.word_index) + 1 # Adding 1 because of reserved 0 index
print(sentences_train[2])
print(X_train[2])

#%%
for word in ['the', 'all','fan']:
    print('{}: {}'.format(word, tokenizer.word_index[word]))

#%%
maxlen = 100

X_train = pad_sequences(X_train, padding = 'post', maxlen = maxlen)
X_test = pad_sequences(X_test, padding = 'post', maxlen = maxlen)

print(X_train[0, :])

#%%
embedding_dim = 50
#================================================================
# Create a deep Keras model with embedding
#================================================================
model = Sequential()
model.add(layers.Embedding(input_dim = vocab_size,
                           output_dim = embedding_dim,
                           input_length = maxlen))
model.add(layers.Flatten())
#model.add(layers.Dense(300, activation = 'relu'))
```

```
model.add(layers.Dense(100, activation = 'relu'))
model.add(layers.Dense(50, activation = 'relu'))
model.add(layers.Dense(1, activation = 'sigmoid'))
#%%
#======================================================================
# Compile the model
#======================================================================
model.compile(loss = 'binary_crossentropy',
              optimizer = 'adam',
              metrics = ['accuracy'])
#======================================================================
# Train and validate the model
#======================================================================
#history = model.fit(X_train, y_train, validation_split = 0.33, nb_epoch
= 50, batch_size = 25,verbose = 0)
history = model.fit(X_train, y_train,
                    epochs = 20,
                    verbose = True,
                    validation_split = 0.33,
                    batch_size = 10)
loss, accuracy = model.evaluate(X_train, y_train, verbose = False)
print("Training Accuracy: {:.4f}".format(accuracy))
loss, accuracy = model.evaluate(X_test, y_test, verbose = False)
print("Testing Accuracy: {:.4f}".format(accuracy))
#%%
#======================================================================
# Plot the history
#======================================================================
#Plot the Model Accuracy and Loss for Training and Validation dataset
plot_history(history)
plot_model_accuracy_loss()
#%%
#Test the Model with testing data
y_pred_test = model.predict(X_test,)
# Round the test predictions
y_pred = np.round(y_pred_test)
#%%
#Test the Model with testing data
y_pred_test = model.predict(X_test,)
# Round the test predictions
y_pred = np.round(y_pred_test)
#%%
target_names = ['NO','YES' ]
#Print the Confusion Matrix
print_confusion_matrix(y_test, y_pred, target_names)
# Print The Perfromance Metrics
Performance_Metrics(model,y_test,y_pred)
#Print and Save the Confusion Matrix
print_confusion_matrix_and_save(y_test, y_pred)
#%%
```

```python
embedding_dim = 50
#================================================================
# Create a deep Keras model with embedding and pooling
#================================================================
model = Sequential()
model.add(layers.Embedding(input_dim = vocab_size,
                            output_dim = embedding_dim,
                            input_length = maxlen))
model.add(layers.GlobalMaxPool1D())
model.add(layers.Dense(10, activation = 'relu'))
model.add(layers.Dense(1, activation = 'sigmoid'))
model.compile(optimizer = 'adam',
              loss = 'binary_crossentropy',
              metrics = ['accuracy'])
model.summary()
#%%
history = model.fit(X_train, y_train,
                    epochs = 20,
                    verbose = True,
#                    validation_data = (X_test, y_test),
                    validation_split = 0.33,
                    batch_size = 10)
loss, accuracy = model.evaluate(X_train, y_train, verbose = 1)
print("Training Accuracy: {:.4f}".format(accuracy))
loss, accuracy = model.evaluate(X_test, y_test, verbose = 1)
print("Testing Accuracy: {:.4f}".format(accuracy))
#%%
#================================================================
# Plot the history
#================================================================
#Plot the Model Accuracy and Loss for Training and Validation dataset
plot_history(history)
plot_model_accuracy_loss()
#%%
#Test the Model with testing data
y_pred_test = model.predict(X_test,)
# Round the test predictions
y_pred = np.round(y_pred_test)
#%%
target_names = ['NO','YES' ]
#Print the Confusion Matrix
print_confusion_matrix(y_test, y_pred, target_names)
# Print The Perfromance Metrics
Performance_Metrics(model,y_test,y_pred)
#Print and Save the Confusion Matrix
print_confusion_matrix_and_save(y_test, y_pred)
#%%
embedding_dim = 100
#================================================================
```

```python
# Use Keras convolutional neural networks (CNN) model with word embedding
#=====================================================================
model = Sequential()
model.add(layers.Embedding(vocab_size, embedding_dim, input_length = maxlen))
model.add(layers.Conv1D(128, 5, activation = 'relu'))
#model.add(layers.Conv1D(64, 5, activation = 'relu'))
model.add(layers.GlobalMaxPooling1D())
model.add(layers.Dense(32, activation = 'relu'))
#model.add(layers.Dense(32, activation = 'relu'))
model.add(layers.Dense(1, activation = 'sigmoid'))
model.compile(optimizer = 'adam',
              loss = 'binary_crossentropy',
              metrics = ['accuracy'])
model.summary()
#%%
history = model.fit(X_train, y_train,
                    epochs = 20, verbose = True,
#                    validation_data = (X_test, y_test),
                    validation_split = 0.33, batch_size = 20)
loss, accuracy = model.evaluate(X_train, y_train, verbose = 1)
print("Training Accuracy: {:.4f}".format(accuracy))
loss, accuracy = model.evaluate(X_test, y_test, verbose = 1)
print("Testing Accuracy: {:.4f}".format(accuracy))

#%%
#=====================================================================
# Plot the history
#=====================================================================
#Plot the Model Accuracy and Loss for Training and Validation dataset
plot_history(history)
plot_model_accuracy_loss()
#%%
#Test the Model with testing data
y_pred_test = model.predict(X_test,)
# Round the test predictions
y_pred = np.round(y_pred_test)
#%%
target_names = ['NO','YES' ]
#Print the Confusion Matrix
print_confusion_matrix(y_test, y_pred, target_names)
# Print The Perfromance Metrics
Performance_Metrics(model, y_test, y_pred)
#Print and Save the Confusion Matrix
print_confusion_matrix_and_save(y_test, y_pred)
```

5.10 本章小结

机器学习已经开始在我们的日常生活中得到应用。对于在数据驱动行业工作的研究人员来说，应对每日增长的数据量将成为其关键技能。机器学习中的数据实例包括特征的数量。分类是机器学习的关键任务之一，用于将未知数据划分到已知的分组中。人们利用训练集训练或构造分类器。我们可以通过创造一个类似于人类专家可靠性的智能计算机来大大提高生活质量。假如创造一个软件来支持医生的决策，那么医生就会更容易治疗好病人。更好的天气预报将减少水资源短缺，增加粮食供应（Harrington, 2012）。本章介绍了多种机器学习算法和技术用于各种数据的分类及其不同实现。每一节的最后都有单独例子，其中包括完整的解决方案，这将有助于读者更加熟悉机器学习技术和原理。其中一些例子还可以作为学术工作和研究新问题的切入点。书中讨论并实现了众多的机器学习算法及其变体，可以直接使用这些算法快速解决现实世界的学习问题。对算法的详细解释和评价有助于其在其他学习场景实现和使用。

5.11 参考文献

Ala'raj, M., & Abbod, M. F. (2016). A new hybrid ensemble credit scoring model based on classifiers consensus system approach. *Expert Systems with Applications, 64*, 36–55 https://doi.org/10.1016/j.eswa.2016.07.017.

Alla, S., & Adari, S. K. (2019). In *Beginning anomaly detection using Python-based deep learning* (1st ed.). New York: Apress, Springer.

Altinel, B., & Ganiz, M. C. (2018). Semantic text classification: A survey of past and recent advances. *Information Processing & Management, 54*(6), 1129–1153.

Androutsopoulos, I., Koutsias, J., Chandrinos, K.V., Paliouras, G., & Spyropoulos, C.D. (2000). An evaluation of naive Bayesian anti-spam filtering, Proceedings of the 11th European Conference on Machine Learning, Barcelona

Bace, R. (1999). An introduction to intrusion detection and assessment for system and network security management, ICSA Intrusion Detection Systems Consortium Technical Report

Bache, K., & Lichman, M. (2013). UCI Machine Learning Repository [http://archive.Ics.Uci.Edu/ml]. University of California, School of Information and Computer Science. Irvine, CA.

Baesens, B., Van Gestel, T., Viaene, S., Stepanova, M., Suykens, J., & Vanthienen, J. (2003). Benchmarking state-of-the-art classification algorithms for credit scoring. *Journal of the Operational Research Society, 54*(6), 627–635.

Bernard, S., Adam, S., & Heutte, L. (2007). In *Using random forests for handwritten digit recognition* (pp. 1043–1047). Parana: Ninth International Conference on Document Analysis and Recognition (ICDAR 2007).

Bhattacharyya, S., Jha, S., Tharakunnel, K., & Westland, J. C. (2011). Data mining for credit card fraud: A comparative study. *On Quantitative Methods for Detection of Financial Fraud, 50*(3), 602–613 https://doi.org/10.1016/j.dss.2010.08.008.

Carcillo, F., Dal Pozzolo, A., Le Borgne, Y. -A., Caelen, O., Mazzer, Y., & Bontempi, G. (2018). Scarff: A scalable framework for streaming credit card fraud detection with spark. *Information Fusion, 41*, 182–194.

Carcillo, F., Le Borgne, Y. -A., Caelen, O., Kessaci, Y., Oblé, F., & Bontempi, G. (2019). Combining unsupervised and supervised learning in credit card fraud detection. *Information Sciences*.

Chuang, C., & Huang, S. (2011). A hybrid neural network approach for credit scoring. *Expert Systems, 28*(2), 185–196.

Curtis, S. R., Rajivan, P., Jones, D. N., & Gonzalez, C. (2018). Phishing attempts among the dark triad: Patterns of attack and vulnerability. *Computers in Human Behavior, 87*, 174–182.

Dal Pozzolo, A., Boracchi, G., Caelen, O., Alippi, C., & Bontempi, G. (2018). Credit card fraud detection: A realistic modeling and a novel learning strategy. *IEEE Transactions on Neural Networks and Learning Systems, 29*(8), 3784–3797.

Dal Pozzolo, A., Caelen, O., & Bontempi, G. (2015). In *When is undersampling effective in unbalanced classification tasks?* (pp. 200–215). Berlin: Springer.

Dal Pozzolo, A., Caelen, O., Le Borgne, Y. -A., Waterschoot, S., & Bontempi, G. (2014). Learned lessons in credit card fraud detection from a practitioner perspective. *Expert Systems with Applications, 41*(10), 4915–4928.

Deerwester, S., Dumais, S. T., Furnas, G. W., Landauer, T. K., & Harshman, R. (1990). Indexing by latent semantic analysis. *Journal of the American Society for Information Science, 41*(6), 391–407.

Delamaire, L., Abdou, H., & Pointon, J. (2009). Credit card fraud and detection techniques: A review. *Banks and Bank Systems, 4*(2), 57–68.

Elleuch, M., Maalej, R., & Kherallah, M. (2016). A new design based-SVM of the CNN classifier architecture with dropout for offline Arabic handwritten recognition. *Procedia Computer Science, 80*, 1712–1723.

Gastellier-Prevost, S., Granadillo, G. G., & Laurent, M. (2011). In *Decisive heuristics to differentiate legitimate from phishing sites* (pp. 1–9). La Rochelle, France: 2011 Conference on Network and Information Systems Security.

Gicic´, A., & Subasi, A. (2019). Credit scoring for a microcredit data set using the synthetic minority oversampling technique and ensemble classifiers. *Expert Systems, 36*(2), e12363.

Goel, S., & Hong, Y. (2015). Security Challenges in Smart Grid Implementation. In Smart Grid Security (pp. 1–39). Springer.

Govindarajan, M., & Chandrasekaran, R. (2011). Intrusion detection using neural based hybrid classification methods. *Computer Networks, 55*(8), 1662–1671.

Gupta, B. B., Arachchilage, N. A., & Psannis, K. E. (2018). Defending against phishing attacks: Taxonomy of methods, current issues and future directions. *Telecommunication Systems, 67*(2), 247–267.

Guzella, T. S., & Caminhas, W. M. (2009). A review of machine learning approaches to spam filtering. *Expert Systems with Applications, 36*(7), 10206–10222.

Han, J., Pei, J., & Kamber, M. (2011). *Data mining: Concepts and techniques*. Amsterdam: Elsevier.

Harrington, P. (2012). *Machine learning in action*. Shelter Island, NY: Manning Publications Co.

Huang, C. -L., Chen, M. -C., & Wang, C. -J. (2007). Credit scoring with a data mining approach based on support vector machines. *Expert Systems with Applications, 33*(4), 847–856.

Hung, C., & Chen, J. -H. (2009). A selective ensemble based on expected probabilities for bankruptcy prediction. *Expert Systems with Applications, 36*(3), 5297–5303.

Innella, P., & McMillan, O. (2017, February 17). An Introduction to IDS Symantec Connect. Retrieved February 17, 2017, from https://www.symantec.com/connect/articles/introduction-ids.

Jha, S., Guillen, M., & Westland, J. C. (2012). Employing transaction aggregation strategy to detect credit card fraud. *Expert Systems with Applications, 39*(16), 12650–12657.

Jurgovsky, J., Granitzer, M., Ziegler, K., Calabretto, S., Portier, P. -E., He-Guelton, L., & Caelen, O. (2018). Sequence classification for credit-card fraud detection. *Expert Systems with Applications, 100*, 234–245.

Kim, P. (2017). Convolutional neural network. In MATLAB deep learning (pp. 121–147). Springer.

Kotzias, D., Denil, M., De Freitas, N., & Smyth, P. (2015). In *From group to individual labels using deep features* (pp. 597–606). New York: ACM.

Kumar, P. R., & Ravi, V. (2007). Bankruptcy prediction in banks and firms via statistical and intelligent techniques–A review. *European Journal of Operational Research, 180*(1), 1–28.

Lauer, F., Suen, C. Y., & Bloch, G. (2007). A trainable feature extractor for handwritten digit recognition. *Pattern Recognition, 40*(6), 1816–1824.

Lebichot, B., Le Borgne, Y. -A., He-Guelton, L., Oblé, F., & Bontempi, G. (2019). In *Deep-learning domain adaptation techniques for credit cards fraud detection* (pp. 78–88). Berlin: Springer.

Liang, M., & Hu, X. (2015). In *Recurrent convolutional neural network for object recognition* (pp. 3367–3375). Boston: 2015 IEEE Conference on Computer Vision and Pattern Recognition (CVPR).

Lin, W., Hu, Y., & Tsai, C. (2012). Machine learning in financial crisis prediction. *IEEE Transactions on Systems, Man, and Cybernetics, Part C (Applications and Reviews), 42*(4), 421–436.

Maas, A. L., Daly, R. E., Pham, P. T., Huang, D., Ng, A. Y., & Potts, C. (2011). In *Learning word vectors for sentiment analysis* (pp. 142–150). Portland, OR: Proceedings of the 49th Annual Meeting of the Association for Computational Linguistics: Human Language Technologies.

Mahmoudi, N., & Duman, E. (2015). Detecting credit card fraud by modified Fisher discriminant analysis. *Expert Systems with Applications, 42*(5), 2510–2516.

Makowski, P. (1985). Credit scoring branches out. *Credit World, 75*(1), 30–37.

McAuley, J., & Leskovec, J. (2013). In *Hidden factors and hidden topics: Understanding rating dimensions with review text* (pp. 165–172). New York: ACM.

Mohammad, R. M., Thabtah, F., & McCluskey, L. (2012). In *An assessment of features related to phishing websites using an automated technique* (pp. 492–497). London: 2012 International Conference for Internet Technology and Secured Transactions.

Mohammad, R. M., Thabtah, F., & McCluskey, L. (2014). Intelligent rule-based phishing websites classification. *IET Information Security, 8*(3), 153–160.

Mohammad, R.M., Thabtah, F., & McCluskey, L. (2015). Phishing websites features. Unpublished. Available via: http://eprints.hud.ac.uk/id/eprint/24330/6/MohammadPhishing14July2015.pdf.

Murphy, K. P. (2012). *Machine learning: A probabilistic perspective*. Cambridge, MA: MIT press.

Pramanik, R., & Bag, S. (2018). Shape decomposition-based handwritten compound character recognition for Bangla OCR. *Journal of Visual Communication and Image Representation, 50*, 123–134.

Recht, B., Roelofs, R., Schmidt, L., & Shankar, V. (2018). Do CIFAR-10 classifiers generalize to CIFAR-10? ArXiv Preprint ArXiv:1806.00451.

Sahu, M., Mishra, B. K., Das, S. K., & Mishra, A. (2014). Intrusion detection system using data mining. *I-Manager's Journal on Computer Science, 2*(1), 19.

Schebesch, K. B., & Stecking, R. (2005). Support vector machines for classifying and describing credit applicants: Detecting typical and critical regions. *Journal of the Operational Research Society, 56*(9), 1082–1088.

Sebastiani, F. (2005). Text categorization. In Encyclopedia of Database Technologies and Applications (pp. 683–687). IGI Global.

Serrano-Cinca, C., & Gutiérrez-Nieto, B. (2016). The use of profit scoring as an alternative to credit scoring systems in peer-to-peer (P2P) lending. *Decision Support Systems, 89*, 113–122.

Sethi, N., & Gera, A. (2014). A revived survey of various credit card fraud detection techniques. *International Journal of Computer Science and Mobile Computing, 3*(4), 780–791.

Shin, K., & Han, I. (2001). A case-based approach using inductive indexing for corporate bond rating. *Decision Support Systems, 32*(1), 41–52.

Subasi, A., Alzahrani, S., Aljuhani, A., & Aljedani, M. (2018). In *Comparison of decision tree algorithms for spam e-mail filtering* (pp. 1–5). Riyadh, Saudi Arabia: 2018 1st International Conference on Computer Applications & Information Security (ICCAIS).

Subasi, A., & Kremic, E. (November 13, 2019). In *Comparison of Adaboost with multiboosting for phishing website detection*Malvern, PA: Presented at the Complex Adaptive Systems Conference.

Subasi, A., Molah, E., Almkallawi, F., & Chaudhery, T. J. (2017). In *Intelligent phishing website detection using random forest classifier* (pp. 1–5). Ras Al Khaimah, UAE: 2017 International Conference on Electrical and Computing Technologies and Applications (ICECTA).

Tiwari, M., Kumar, R., Bharti, A., & Kishan, J. (2017). Intrusion detection system. *International Journal of Technical Research and Applications, 5*, 2320–8163.

UCI Machine Learning Repository: Phishing Websites Data Set. (n.d.). Retrieved January 29, 2017, from https://archive.ics.uci.edu/ml/datasets/Phishing+Websites#.

UCI Machine Learning Repository: Statlog (German Credit Data) Data Set. (n.d.). Retrieved November 29, 2019, from https://archive.ics.uci.edu/ml/datasets/Statlog+(German+Credit+Data).

Wheeler, R., & Aitken, S. (2000). Multiple algorithms for fraud detection. *Knowledge-Based Systems, 13*(2–3), 93–99.

Whitrow, C., Hand, D. J., Juszczak, P., Weston, D., & Adams, N. M. (2009). Transaction aggregation as a strategy for credit card fraud detection. *Data Mining and Knowledge Discovery, 18*(1), 30–55.

Xiang, G., Hong, J., Rose, C. P., & Cranor, L. (2011). Cantina+: A feature-rich machine learning framework for detecting phishing web sites. *ACM Transactions on Information and System Security (TISSEC), 14*(2), 21.

Zhang, Y. (2019). Evaluation of CNN Models with Fashion MNIST Data (Iowa State University). Retrieved from https://lib.dr.iastate.edu/creativecomponents/364.

Zhang, Z., Gentile, A., & Ciravegna, F. (2012). *Recent advances in methods of lexical semantic relatedness*. Cabridge, UK: Cambridge University.

Zhao, K. (2018). Handwritten Digit Recognition and Classification Using Machine Learning. M. Sc. in Computing (Data Analytics),Technological University Dublin.

CHAPTER 6

第 6 章

回归示例

6.1 简介

在分类任务中，标签空间是一系列离散的类别。而在本章中，我们将考虑实值目标变量的情况。函数估计器，即回归器，具有这样的一种映射关系：$\tilde{f}: X \to \mathrm{R}$。而回归任务则是从样例 $(x_i, f(x_i))$ 中学习函数估计器。例如，我们想要基于一些经济指标，学习得到道琼斯指数或富时 100 指数的估计器。该过程似乎是对离散分类方法自然而然地泛化，但也不是毫无弊端。首先，我们需要从相对低分辨率的目标变量转换为一个无限分辨率的变量。其次，为了使该函数估计器达到一定的准确率，通常会导致过拟合。此外，由于一些可能的变化，实例中的一些目标值有可能是模型无法捕捉的。因此，较为合理的方法是假设实例是有噪声的，而且估计器只能捕捉一般的趋势或函数形状（Flach, 2012）。

在回归过程中，分组模型和分级模型之间存在着差异。分组模型理论是将实例空间分成若干段，在每段中尽可能简单地学习局部模型。例如，局部模型为决策树中的一个主分类器。同样，我们可能会为每个叶子节点估计一个常量值，从而得到一棵回归树。在单变量问题中，这将得到一个分段的常数曲线。这样的分组模型能够完全匹配给定的点，就如一个高阶多项式，然而，同样也应该警惕过拟合问题。一个经验法则是，根据数据估计的参数数量必须略低于数据点的数量，从而防止过拟合（Flach, 2012）。

我们看到，分类模型的评估可以通过对边际应用损失函数来完成：惩罚负边际（错误分类），奖励正边际（正确分类）。回归模型的评估是将残差（$f(x) - \tilde{f}(x)$）作为损失函数。与分类损失函数不同，回归损失函数通常在 0 附近对称，即使正残差和负残差的权重可能不同。因此，通常选择残差的平方作为损失函数。这样做的好处是具有统计学上的一致性，并且可以断定观测到的函数值是否受到累加影响或是正态分布噪声的影响。如果低估了模型中参数的数量，无论有多少训练数据，都无法将损失降低到零。另一方面，在参数较多的情况下，模型对训练样本的依赖性会更强，所以训练样本的微小变化会导致模型的明显不同。这就是所谓的偏差－方差困境：对于低复杂度模型，训练数据随机变化带来的差异影响较小，但可能导致系统性偏差；相反，高复杂度模型可以消除这种偏差，但可能会因为方差较大而产生非系统性误差（Flach, 2012）。

6.2 股票市场价格指数收益预测

有效市场假说认为，当前的股价反映了市场上所有可获得的信息。但是，根据这些信息，公众却无法进行成功的交易。另一些人则认为市场是无效的，部分原因是市场参与者的心理因素不同，再加上市场无法对新发布的信息作出即时反应。因此股票价格、股票市场指数、金融衍生品价格等金融变量是可预测的。通过分析提供给公众的信息，获得高于市场平均水平的收益，其结果优于随机发生的交易（Zhong & Enke, 2017）。

股价预测在金融界有着举足轻重的地位。在大数据时代，准确可靠的股市预测变得越来越重要，因为有太多的因素会造成股市波动。机器学习善于处理海量数据，并从错误中学习，能够为股市预测带来极大帮助。金融界最有趣的问题之一是对市场指数价格变动的估计。尤其是，统计方法不能提供自动的解决方案，因为要使每一步的目标值具有规律性和稳定性，涉及许多变化和变动。我们需要更有效的方法来降低传统统计方法的局限性，以便跟随市场的动态和非稳定特征。金融专业人士已经提出了一些不同的方法来做出合理的交易决策，比如机器学习方法（Şenyurt & Subaşı, 2015）。

股票市场会受到一些极其相关因素的影响，这些因素包括：1）经济变量，如汇率、商品价格、利率、货币增长率和总体经济状况；2）行业特定变量，如工业生产和消费价格的增长率；3）企业特定变量，如企业的政策变化、损益表、股息率等；4）投资者心理变量，如投资者的预期和机构投资者的选择；5）政治变量，如一些重要政治事件的发生和发布。这些因素都以非常复杂的方式相互影响。而且，股市是一个非线性的、动态的、非参数的、非稳定的、有噪声的、混乱的系统。由于潜在的市场效率低下，投资者希望通过在对金融变量进行越来越准确预测的基础上制定交易策略，以获得可观的收益。因此，分析股票市场的走势对投资者和研究者来说具有吸引力和挑战性。例如，金融时间序列的预测可以分为输入变量的单变量分析和多变量分析。在单变量分析中，只需要将金融时间序列本身作为输入变量，而在多变量分析中，输入变量可以是滞后的时间序列，也可以是其他形式的信息，例如技术指标（如表 6.1 所示）、基本面指标或市场间指标。因此，统计学和机器学习技术被用于分析股票市场（Zhong & Enke, 2017）。

表 6.1 股市技术指标（Şenyurt & Subaşı, 2015）

指标名称		公式
累积/派发摆动指标		$\dfrac{((收盘价-最低价)-(最高价-收盘价))}{(最高价-最低价)} \times 该周期内成交量$
蔡金摆动指标		累积/派发线（ADL）的 3 日指数平滑移动平均值（EMA）– ADL 的 10 日 EMA
资金流量乘数		$\dfrac{((收盘价-最低价)-(最高价-收盘价))}{(最高价-最低价)}$
资金流量		资金流量乘数 × 该周期内成交量
ADL		前期 ADL + 当前资金流量
异同移动平均线（MACD）		MACD = 12 日 EMA – 26 日 EMA MACD 反转信号 = MACD 的 9 日 EMA
随机摆动指标		$\%K = 100 \times \left[\dfrac{(C-L14)}{(H14-L14)}\right]$ C = 最近收盘价 $L14$ = 前 14 个交易日内的最低价 $H14$ = 相同的 14 个交易日内的最高价 $\%D$ = 3 个周期的 $\%K$ 移动平均值
加速震荡指标		间隔几个周期的两个动量之间的差
动量		$\left[\left(\dfrac{收盘价(J)}{收盘价(J\text{-}N)}\right) \times 100\right]$ J = 当前周期 N = 用于比较的周期之间的间隔
蔡金波动性指标		$\left[\dfrac{\text{EMA}(最高价-最低价)-\text{EMA}(最高价-10\ 日前最低价)}{\text{EMA}(最高价-10\ 日前最低价)}\right] \times 100$
快速随机指标	快速 %K	$\left[\dfrac{(收盘价-最低价)}{(最高价-最低价)}\right] \times 100$
	快速 %D	快速 $\%K$ 的简单移动平均值（3 个周期）
慢速随机指标	慢速 %K	等于快速 $\%D$，快速 $\%K$ 的移动平均值（3 个周期）
	慢速 %D	慢速 $\%K$ 的移动平均值（3 个周期）
威廉指标 %R		$\left[\dfrac{(n\ 日内最高价-第\ n\ 日收盘价)}{(n\ 日内最高价-n\ 日内最低价)}\right] \times 100$
负量指标（NVI）		$\left[\dfrac{(收盘价[今日]-收盘价[昨日])}{(收盘价[昨日])}\right] \times \text{NVI}[昨日]$
正量指标（PVI）		$\left[\dfrac{(收盘价[今日]-收盘价[昨日])}{(收盘价[昨日])}\right] \times \text{PVI}[昨日]$
相对强弱指标（RSI）		$100-\left[\dfrac{100}{1+RS^*}\right],\ RS^* = \dfrac{x\ 天内的平均收盘涨数}{x\ 天内的平均收盘跌数}$

(续)

指标名称		公式
布林线指标	中轨线	20 日的简单移动平均线（SMA）
	上轨线	中轨线 +20 日的标准差 ×2
	下轨线	中轨线 −20 日的标准差 ×2
n 日内最高价		过去 14 日内的最高价格向量
n 日内最低价		过去 14 日内的最低价格向量
中位价（MP）		MP =（最高价 + 最低价）/2
平衡交易量（OBV）		如果当期收盘价高于前期收盘价： OBV = OBV [前期] + 成交量 [当期] 如果当期收盘价低于前期收盘价： OBV= OBV [前期] − 成交量 [当期] 如果当期收盘价等于前期收盘价： OBV= OBV[前期]
价格变动率指标		（收盘价（今日）− 收盘价（前 n 日））/ 收盘价（前 n 日）
收盘价变动率		（收盘价（今日）− 收盘价（前 n 日））/ 收盘价（前 n 日）
价量趋势指标		变动率 × 交易量（今日）+ PVT（昨日）
典型价格		（最高价 + 最低价 + 收盘价）/3
成交量变动率		（成交量（今日）− 成交量（前 n 日））/ 成交量（前 n 日）×100
加权收盘价		（（收盘价 ×2）+ 最高价 + 最低价）/4
最高价真实波动幅度（TRH）		昨日收盘价或今日最高价，以较高者为准
最低价真实波动幅度（TRL）		昨日收盘价或今日最高价，以较低者为准
当日累积/派发（A/D）指标		今日收盘价 −TRL（如果今日收盘价高于昨日） 今日收盘价 −TRH（如果今日收盘价低于昨日） 0（如果今日收盘价等于昨日）
威廉 A/D 指标		当日 A/D 指标 + 昨日威廉 A/D 指标

在当今智能金融领域，股市价格预测专业及相关决策支持框架发挥着至关重要的作用。长期以来，科学家们进行了详细的研究，以了解股市的基本机制，并准确地预测市场价格。因为资金生意就是为了赚取高额的财务收益，所以市场上的每个参与者都在追求一些智能计算模型，从而能够持续地选择赢家并淘汰输家。股市的每日价格是一个数字序列，这些数据随时间变化，是计算模型需要处理的对象。虽然人工智能（AI）在金融领域的应用并不新颖，但能够解决问题才是有史以来对 AI 最大的推动力——包括使用新的思想和方法发展更好的技术，比如关注问题本身而不是算法方程式的深度学习。大量研究表明，人工智能技术在预测问题上比传统模型（如 ARMA、ARIMA 和线性回归）更具有效性，基于人工智能的方案具有更好的预测性能（Şenyurt & Subaşı, 2015）。

例6.1 以下 Python 代码使用不同的分类器和单独的训练和测试数据集来预测股票市场价格指数回报。统计质量测量，包括平均绝对误差（MAE）、均方误差（MSE）和相关系数（R）的计算。散点图、真实值和预测值也被绘制出来。

数据集信息：纳斯达克（NASDAQ）、道琼斯（DOWJONES）、标普（S&P）、罗素（RUSSEL）和纽约证券交易所（NYSE）的综合指数代表着全球不同的股票市场。数据集可以用以下命令从雅虎财经网站下载：

```
yahoo_data = pdr.data.get_data_yahoo('^IXIC', start_date, stop_date)
""" 
Created on Fri Dec 13 14:00:48 2019
@author: subasi
This example uses regression model to predict stock prices
For this example, Technical Indicators are also calculated on stock data.
Some part of this example is taken from
https://www.kaggle.com/kratisaxena/stock-market-technical-indicators-visualization
Please refer the above kernel for visualization part."""
#=====================================================================
# Stock market analysis
#=====================================================================
#%%
# Import Modules
import pandas as pd
import numpy as np
from datetime import datetime
import statsmodels.api as sm
import copy
import matplotlib.pyplot as plt
from IPython.display import Image
from matplotlib.pylab import rcParams
from statsmodels.tsa.stattools import adfuller
from statsmodels.tsa.stattools import acf, pacf
from sklearn.linear_model import LinearRegression
from statsmodels.tsa.arima_model import ARMA, ARIMA
from sklearn.metrics import explained_variance_score
from numpy import array
from pandas import read_csv
from sklearn.metrics import mean_squared_error, r2_score
from sklearn.metrics import mean_absolute_error
import matplotlib.pyplot as plt
#You should install mpl_finance module before using
#pip install mpl_finance
from mpl_finance import candlestick_ohlc #parse_yahoo_historical_ochl
rcParams['figure.figsize'] = 15, 5
#You should install pandas_datareader module before using
#!pip install pandas_datareader
import pandas_datareader as pdr
```

```
#===================================================================
# Define utility functions
#===================================================================
def plot_model_loss(history):
  plt.figure(figsize = (6, 4))
  plt.plot(history.history['loss'], 'r--', label = 'Loss of training
  data')
  plt.plot(history.history['val_loss'], 'b--', label = 'Loss of
  validation data')
  plt.title('Model Error')
  plt.ylabel('Loss')
  plt.xlabel('Training Epoch')
  plt.ylim(0)
  plt.legend()
  plt.show()

def print_performance_metrics(ytest, ypred):
# The mean absolute error
print("MAE = %5.3f" % mean_absolute_error(ytest, ypred))
# Explained variance score: 1 is perfect prediction
print("R^2 = %0.5f" % r2_score(ytest, ypred))
# The mean squared error
print("MSE = %5.3f" % mean_squared_error(ytest, ypred))

def plot_scatter(ytest, ypred):
fig, ax = plt.subplots()
ax.scatter(ytest, ypred, edgecolors = (0, 0, 0))
ax.text(10, ytest.max()-10, r'$R^2$ = %.2f, MAE = %.2f' % (
  r2_score(ytest, ypred), mean_absolute_error(ytest, ypred)))
ax.plot([ytest.min(), ytest.max()], [ytest.min(), ytest.max()], 'k--',
lw = 4)
ax.set_xlabel('Measured')
ax.set_ylabel('Predicted')
plt.show()

def plot_real_predicted(ytest, ypred):
  plt.plot(ytest[1:200], color = 'red', label = 'Real data')
  plt.plot(ypred[1:200], color = 'blue', label = 'Predicted data')
  plt.title('Prediction')
  plt.legend()
  plt.show()
#===================================================================
# Download data from Yahoo Finance and save it as a csv file
#===================================================================
start_date = pd.to_datetime('2009-12-01')
mid_date = pd.to_datetime('2014-12-01')
stop_date = pd.to_datetime('2019-12-01')

#NASDAQ
yahoo_data = pdr.data.get_data_yahoo('^IXIC', start_date, stop_date)
```

```python
yahoo_data.to_csv('NASDAQ.csv')
#DOW JONES
yahoo_data = pdr.data.get_data_yahoo('^DJI', start_date, stop_date)
yahoo_data.to_csv('DOWJONES.csv')
#S&P 500 (^GSPC)
yahoo_data = pdr.data.get_data_yahoo('^GSPC', start_date, stop_date)
yahoo_data.to_csv('S&P.csv')
#Russell 2000 (^RUT)
yahoo_data = pdr.data.get_data_yahoo('^RUT', start_date, stop_date)
yahoo_data.to_csv('RUSSEL.csv')
#NYSE
yahoo_data = pdr.data.get_data_yahoo('^NYA', start_date, stop_date)
yahoo_data.to_csv('NYSE.csv')
#=====================================================================
# Add various technical indicators in the data frame
#=====================================================================
"""There are four types of technical indicators. Let us take 4 sets
of indicators and test which performs better in prediction of stock
markets. These 4 sets of technical indicators are:
RSI, Volume (plain), Bollinger Bands, Aroon, Price Volume Trend,
acceleration bands
Stochastic, Chaikin Money Flow, Parabolic SAR, Rate of Change, Volume
weighted average Price, momentum
Commodity Channel Index, On Balance Volume, Keltner Channels, Triple
Exponential Moving Average, Normalized Averager True Range, directional
movement indicators
MACD, Money flowindex, Ichimoku, William %R, Volume MINMAX, adaptive
moving average"""
# Create copy of data to add columns of different sets of technical
indicators
dff = pd.read_csv("DOWJONES.csv")
techind = copy.deepcopy(dff)
#%%
#=====================================================================
# Calculate the relative strength index (RSI)
#=====================================================================
# Relative Strength Index
# Avg(PriceUp)/(Avg(PriceUP) + Avg(PriceDown)*100
# Where: PriceUp(t) = 1*(Price(t)-Price(t-1)){Price(t)- Price(t-1) > 0};
# PriceDown(t) = -1*(Price(t)-Price(t-1)){Price(t)- Price(t-1) < 0};
def rsi(values):
    up = values[values > 0].mean()
    down = -1*values[values < 0].mean()
    return 100 * up / (up + down)
# Add Momentum_1D column for all 15 stocks.
# Momentum_1D = P(t) - P(t-1)
techind['Momentum_1D'] = (techind['Close']-techind['Close'].shift(1)).fillna(0)
techind['RSI_14D'] = techind['Momentum_1D'].rolling(center = False,
window = 14).apply(rsi).fillna(0)
```

```python
#================================================================
# Calculate the volume (plain)
#================================================================
techind['Volume_plain'] = techind['Volume'].fillna(0)
#================================================================
# Calculate the Bollinger bands
#================================================================
def bbands(price, length=30, numsd=2):
    """ returns average, upper band, and lower band"""
    #ave = pd.stats.moments.rolling_mean(price,length)
    ave = price.rolling(window=length, center=False).mean()
    #sd = pd.stats.moments.rolling_std(price,length)
    sd = price.rolling(window=length, center=False).std()
    upband = ave + (sd*numsd)
    dnband = ave - (sd*numsd)
    return np.round(ave,3), np.round(upband,3), np.round(dnband,3)
techind['BB_Middle_Band'], techind['BB_Upper_Band'], techind['BB_Lower_Band'] = bbands(techind['Close'], length=20, numsd=1)
techind['BB_Middle_Band'] = techind['BB_Middle_Band'].fillna(0)
techind['BB_Upper_Band'] = techind['BB_Upper_Band'].fillna(0)
techind['BB_Lower_Band'] = techind['BB_Lower_Band'].fillna(0)
#================================================================
# Calculate the Aroon oscillator
#================================================================
def aroon(df, tf=25):
    aroonup = []
    aroondown = []
    x = tf
    while x < len(df['Date']):
        aroon_up = ((df['High'][x-tf:x].tolist().index(max(df['High'][x-tf:x])))/float(tf))*100
        aroon_down = ((df['Low'][x-tf:x].tolist().index(min(df['Low'][x-tf:x])))/float(tf))*100
        aroonup.append(aroon_up)
        aroondown.append(aroon_down)
        x += 1
    return aroonup, aroondown
listofzeros = [0] * 25
up, down = aroon(techind)
aroon_list = [x - y for x, y in zip(up,down)]
if len(aroon_list) ==0:
    aroon_list = [0] * techind.shape[0]
    techind['Aroon_Oscillator'] = aroon_list
else:
    techind['Aroon_Oscillator'] = listofzeros + aroon_list
#================================================================
# Calculate the price volume trend
#================================================================
#PVT = [((CurrentClose - PreviousClose) / PreviousClose) x Volume] + PreviousPVT
```

```python
techind["PVT"] = (techind['Momentum_1D']/ techind['Close'].
shift(1))*techind['Volume']
techind["PVT"] = techind["PVT"]-techind["PVT"].shift(1)
techind["PVT"] = techind["PVT"].fillna(0)
#==================================================================
# Calculate the acceleration bands
#==================================================================
def abands(df):
    #df['AB_Middle_Band'] = pd.rolling_mean(df['Close'], 20)
    df['AB_Middle_Band'] = df['Close'].rolling(window = 20, center = False).
    mean()
    # High * ( 1+4 * (High - Low) / (High + Low))
    df['aupband'] = df['High'] * (1 + 4 * (df['High']-df['Low'])/
    (df['High'] + df['Low']))
    df['AB_Upper_Band'] = df['aupband'].rolling(window = 20, center = False).
    mean()
    # Low *(1 - 4 * (High - Low)/ (High + Low))
    df['adownband'] = df['Low'] * (1 - 4 * (df['High']-df['Low'])/
    (df['High'] + df['Low']))
    df['AB_Lower_Band'] = df['adownband'].rolling(window = 20,
    center = False).mean()
abands(techind)
techind = techind.fillna(0)
#==================================================================
# Drop unwanted columns
#==================================================================
columns2Drop = ['Momentum_1D', 'aupband', 'adownband']
techind = techind.drop(labels = columns2Drop, axis = 1)
#==================================================================
# Calculate the stochastic oscillator (%K and %D)
#==================================================================
def STOK(df, n):
    df['STOK'] = ((df['Close'] - df['Low'].rolling(window = n,
    center = False).mean()) / (df['High'].rolling(window = n,
    center = False).max() - df['Low'].rolling(window = n, center = False).
    min())) * 100
    df['STOD'] = df['STOK'].rolling(window = 3, center = False).mean()

STOK(techind, 4)
techind = techind.fillna(0)
#==================================================================
# Calculate the Chaikin money flow
#==================================================================
def CMFlow(df, tf):
    CHMF = []
    MFMs = []
    MFVs = []
    x = tf

    while x < len(df['Date']):
```

```python
    PeriodVolume = 0
    volRange = df['Volume'][x-tf:x]
    for eachVol in volRange:
    PeriodVolume+=eachVol

    MFM = ((df['Close'][x] - df['Low'][x]) - (df['High'][x] - df['Close']
    [x])) / (df['High'][x] - df['Low'][x])
    MFV = MFM*PeriodVolume

    MFMs.append(MFM)
    MFVs.append(MFV)
    x += 1
y = tf
while y < len(MFVs):
PeriodVolume = 0
volRange = df['Volume'][x-tf:x]
for eachVol in volRange:
PeriodVolume+=eachVol
consider = MFVs[y-tf:y]
    tfsMFV = 0

    for eachMFV in consider:
    tfsMFV+=eachMFV

    tfsCMF = tfsMFV/PeriodVolume
    CHMF.append(tfsCMF)
    y += 1
    return CHMF
listofzeros = [0] * 40
CHMF = CMFlow(techind, 20)
if len(CHMF) ==0:
    CHMF = [0] * techind.shape[0]
    techind['Chaikin_MF'] = CHMF
 else:
    techind['Chaikin_MF'] = listofzeros + CHMF
#==================================================================
# Calculate the parabolic SAR
#==================================================================
def psar(df, iaf = 0.02, maxaf = 0.2):
    length = len(df)
    dates = (df['Date'])
    high = (df['High'])
    low = (df['Low'])
    close = (df['Close'])
    psar = df['Close'][0:len(df['Close'])]
    psarbull = [None] * length
    psarbear = [None] * length
    bull = True
    af = iaf
```

```
ep = df['Low'][0]
hp = df['High'][0]
lp = df['Low'][0]
for i in range(2,length):
    if bull:
        psar[i] = psar[i - 1] + af * (hp - psar[i - 1])
    else:
        psar[i] = psar[i - 1] + af * (lp - psar[i - 1])
    reverse = False
    if bull:
        if df['Low'][i] < psar[i]:
            bull = False
            reverse = True
            psar[i] = hp
            lp = df['Low'][i]
            af = iaf
    else:
        if df['High'][i] > psar[i]:
            bull = True
            reverse = True
            psar[i] = lp
            hp = df['High'][i]
            af = iaf
    if not reverse:
        if bull:
            if df['High'][i] > hp:
                hp = df['High'][i]
                af = min(af + iaf, maxaf)
            if df['Low'][i - 1] < psar[i]:
                psar[i] = df['Low'][i - 1]
            if df['Low'][i - 2] < psar[i]:
                psar[i] = df['Low'][i - 2]
        else:
            if df['Low'][i] < lp:
                lp = df['Low'][i]
                af = min(af + iaf, maxaf)
            if df['High'][i - 1] > psar[i]:
                psar[i] = df['High'][i - 1]
            if df['High'][i - 2] > psar[i]:
                psar[i] = df['High'][i - 2]
    if bull:
        psarbull[i] = psar[i]
    else:
        psarbear[i] = psar[i]
#return {"dates":dates, "high":high, "low":low, "close":close,
        "psar":psar, "psarbear":psarbear, "psarbull":psarbull}
#return psar, psarbear, psarbull
df['psar'] = psar
#df['psarbear'] = psarbear
#df['psarbull'] = psarbull
```

```
psar(techind)
#=================================================================
# Calculate the price rate of change
#=================================================================
# ROC = [(Close - Close n periods ago) / (Close n periods ago)] * 100
techind['ROC'] = ((techind['Close'] - techind['Close'].shift(12))/
(techind['Close'].shift(12)))*100
techind = techind.fillna(0)
#=================================================================
# Calculate the volume weighted average price
#=================================================================
techind['VWAP'] = np.cumsum(techind['Volume'] * (techind['High'] + techind
['Low'])/2) / np.cumsum(techind['Volume'])
techind = techind.fillna(0)
techind.tail()
#=================================================================
# Calculate the momentum
#=================================================================
techind['Momentum'] = techind['Close'] - techind['Close'].shift(4)
techind = techind.fillna(0)
techind.tail()
#=================================================================
# Calculate the commodity channel index
#=================================================================
def CCI(df, n, constant):
    TP = (df['High'] + df['Low'] + df['Close']) / 3
    CCI = pd.Series((TP - TP.rolling(window=n, center=False).mean())
    / (constant * TP.rolling(window=n, center=False).std())) #,
    name = 'CCI_' + str(n))
    return CCI
techind['CCI'] = CCI(techind, 20, 0.015)
techind = techind.fillna(0)
#=================================================================
# Calculate on balance volume
#=================================================================
""" "If the closing price is above the prior close price then: Current
OBV = Previous OBV + Current Volume
If the closing price is below the prior close price then: Current
OBV = Previous OBV - Current Volume
If the closing prices equals the prior close price then: Current
OBV = Previous OBV (no change)" " "
new = (techind['Volume'] * (~techind['Close'].diff().le(0) * 2 -1)).cum-
sum()
techind['OBV'] = new
#%%
#=================================================================
# Calcualte the Keltner channels
#=================================================================
def KELCH(df, n):
```

```python
    KelChM = pd.Series(((df['High'] + df['Low'] + df['Close']) /
        3).rolling(window=n, center = False).mean(), name = 'KelChM_' + str(n))
    KelChU = pd.Series(((4 * df['High'] - 2 * df['Low'] + df['Close']) /
        3).rolling(window=n, center = False).mean(), name = 'KelChU_' + str(n))
    KelChD = pd.Series((((-2 * df['High'] + 4 * df['Low'] + df['Close']) /
        3).rolling(window=n, center = False).mean(), name = 'KelChD_' + str(n))
    return KelChM, KelChD, KelChU
KelchM, KelchD, KelchU = KELCH(techind, 14)
techind['Kelch_Upper'] = KelchU
techind['Kelch_Middle'] = KelchM
techind['Kelch_Down'] = KelchD
techind = techind.fillna(0)
#================================================================
# Calculate the triple exponential moving average
#================================================================
""" "Triple Exponential MA Formula:
T-EMA = (3EMA - 3EMA(EMA)) + EMA(EMA(EMA))
Where:
EMA = EMA(1) + α * (Close - EMA(1))
α = 2 / (N + 1)
N = The smoothing period." " "
techind['EMA'] = techind['Close'].ewm(span = 3,min_periods = 0,adjust = True,ignore_na = False).mean()
techind = techind.fillna(0)
techind['TEMA'] = (3 * techind['EMA'] - 3 * techind['EMA'] *
    techind['EMA']) + (techind['EMA']*techind['EMA']*techind['EMA'])
#================================================================
# Calculation of normalized average true range¶
#================================================================
""" "True Range = Highest of (High - low, abs(High - previous close),
abs(low - previous close))
Average True Range = 14 day MA of True Range
Normalized Average True Range = ATR / Close * 100" " "
techind['HL'] = techind['High'] - techind['Low']
techind['absHC'] = abs(techind['High'] - techind['Close'].shift(1))
techind['absLC'] = abs(techind['Low'] - techind['Close'].shift(1))
techind['TR'] = techind[['HL','absHC','absLC']].max(axis = 1)
techind['ATR'] = techind['TR'].rolling(window = 14).mean()
techind['NATR'] = (techind['ATR'] / techind['Close']) *100
techind = techind.fillna(0)
#================================================================
# Calculate the average directional movement index (ADX)
#================================================================
""" "Calculating the DMI can actually be broken down into two parts.
First, calculating the+DI and -DI, and second, calculating the ADX.
To calculate the+DI and -DI you need to find the+DM and -DM (Directional
Movement).+DM and -DM are calculated using the High, Low, and Close for
each period. You can then calculate the following:
Current High - Previous High = UpMove Previous Low - Current
Low = DownMove
```

If UpMove > DownMove and UpMove > 0, then+DM = UpMove, else+DM = 0 If DownMove > Upmove and Downmove > 0, then -DM = DownMove, else -DM = 0
Once you have the current+DM and -DM calculated, the+DM and -DM lines can be calculated and plotted based on the number of user defined periods.
+DI = 100 times Exponential Moving Average of (+DM / Average True Range)
-DI = 100 times Exponential Moving Average of (-DM / Average True Range)
Now that - +DX and -DX have been calculated, the last step is calculating the ADX.
ADX = 100 times the Exponential Moving Average of the Absolute Value of (+DI - -DI) / (+DI + -DI)" " "

```
def DMI(df, period):
    df['UpMove'] = df['High'] - df['High'].shift(1)
    df['DownMove'] = df['Low'].shift(1) - df['Low']
    df['Zero'] = 0

    df['PlusDM'] = np.where((df['UpMove'] > df['DownMove']) &
    (df['UpMove'] > df['Zero']), df['UpMove'], 0)
    df['MinusDM'] = np.where((df['UpMove'] < df['DownMove']) &
    (df['DownMove'] > df['Zero']), df['DownMove'], 0)

    df['plusDI'] = 100 * (df['PlusDM']/df['ATR']).ewm(span = period,min_
    periods = 0,adjust = True,ignore_na = False).mean()
    df['minusDI'] = 100 * (df['MinusDM']/df['ATR']).ewm(span = period,min_
    periods = 0,adjust = True,ignore_na = False).mean()

    df['ADX'] = 100 * (abs((df['plusDI'] - df['minusDI'])/
    (df['plusDI'] + df['minusDI']))).ewm(span = period,min_
    periods = 0,adjust = True,ignore_na = False).mean()
DMI(techind, 14)
techind = techind.fillna(0)
#================================================================
# Drop unwanted columns
#================================================================
columns2Drop = ['UpMove', 'DownMove', 'ATR', 'PlusDM', 'MinusDM', 'Zero',
'EMA', 'HL', 'absHC', 'absLC', 'TR']
techind = techind.drop(labels = columns2Drop, axis = 1)
#================================================================
# Calculate the MACD
#================================================================
#MACD: (12-day EMA - 26-day EMA)
techind['26_ema'] = techind['Close'].ewm(span = 26,min_periods = 0,ad-
just = True,ignore_na = False).mean()
techind['12_ema'] = techind['Close'].ewm(span = 12,min_periods = 0,ad-
just = True,ignore_na = False).mean()
techind['MACD'] = techind['12_ema'] - techind['26_ema']
techind = techind.fillna(0)
#================================================================
# Calculate the money flow index
#================================================================
```

```python
""" "Typical Price = (High + Low + Close)/3
Raw Money Flow = Typical Price x Volume
The money flow is divided into positive and negative money flow.
Positive money flow is calculated by adding the money flow of all the days
where the typical price is higher than the previous day's typical price.
Negative money flow is calculated by adding the money flow of all the days
where the typical price is lower than the previous day's typical price.
If typical price is unchanged then that day is discarded.
Money Flow Ratio = (14-period Positive Money Flow)/(14-period Negative
Money Flow)
Money Flow Index = 100 - 100/(1 + Money Flow Ratio)" " "
def MFI(df):
    # typical price
    df['tp'] = (df['High'] + df['Low'] + df['Close'])/3
    #raw money flow
    df['rmf'] = df['tp'] * df['Volume']
    # positive and negative money flow
    df['pmf'] = np.where(df['tp'] > df['tp'].shift(1), df['tp'], 0)
    df['nmf'] = np.where(df['tp'] < df['tp'].shift(1), df['tp'], 0)
    # money flow ratio
    df['mfr'] = df['pmf'].rolling(window = 14, center = False).sum()/ \
        df['nmf'].rolling(window = 14, center = False).sum()
    df['Money_Flow_Index'] = 100 - 100 / (1 + df['mfr'])
MFI(techind)
techind = techind.fillna(0)
#==================================================================
# Calculate the Ichimoku cloud
#==================================================================
""" "Turning Line = ( Highest High + Lowest Low ) / 2, for the past 9 days
Standard Line = ( Highest High + Lowest Low ) / 2, for the past 26 days
Leading Span 1 = ( Standard Line + Turning Line ) / 2, plotted 26 days
ahead of today
Leading Span 2 = ( Highest High + Lowest Low ) / 2, for the past 52 days,
plotted 26 days ahead of today
Cloud = Shaded Area between Span 1 and Span 2" " "
def ichimoku(df):
    # Turning Line
    period9_high = df['High'].rolling(window = 9, center = False).max()
    period9_low = df['Low'].rolling(window = 9, center = False).min()
    df['turning_line'] = (period9_high + period9_low) / 2

    # Standard Line
    period26_high = df['High'].rolling(window = 26, center = False).max()
    period26_low = df['Low'].rolling(window = 26, center = False).min()
    df['standard_line'] = (period26_high + period26_low) / 2

    # Leading Span 1
    df['ichimoku_span1'] = ((df['turning_line'] + df['standard_line']) /
        2).shift(26)
```

```python
    # Leading Span 2
    period52_high = df['High'].rolling(window = 52, center = False).max()
    period52_low = df['Low'].rolling(window = 52, center = False).min()
    df['ichimoku_span2'] = ((period52_high + period52_low) / 2).shift(26)

    # The most current closing price plotted 22 time periods behind
    (optional)
    df['chikou_span'] = df['Close'].shift(-22) # 22 according to
    investopedia
ichimoku(techind)
techind = techind.fillna(0)
#================================================================
# Calculate the William %R
#================================================================
""" "%R = -100 * ( ( Highest High - Close) / ( Highest High - Lowest Low
) )" ""
def WillR(df):
    highest_high = df['High'].rolling(window = 14, center = False).max()
    lowest_low = df['Low'].rolling(window = 14, center = False).min()
    df['WillR'] = (-100) * ((highest_high - df['Close']) / (highest_high -
    lowest_low))
WillR(techind)
techind = techind.fillna(0)
#================================================================
# Calculate the MINMAX
#================================================================
def MINMAX(df):
    df['MIN_Volume'] = df['Volume'].rolling(window = 14, center = False).min()
    df['MAX_Volume'] = df['Volume'].rolling(window = 14, center = False).max()
MINMAX(techind)
techind = techind.fillna(0)
#================================================================
# Calculate the adaptive moving average
#================================================================
def KAMA(price, n = 10, pow1 = 2, pow2 = 30):
    """ kama indicator """
    """ accepts pandas dataframe of prices """
    absDiffx = abs(price - price.shift(1) )
    ER_num = abs( price - price.shift(n) )
    ER_den = absDiffx.rolling(window = n, center = False).sum()
    ER = ER_num / ER_den
    sc = ( ER*(2.0/(pow1 + 1)-2.0/(pow2 + 1.0)) + 2/(pow2 + 1.0) ) ** 2.0
    answer = np.zeros(sc.size)
    N = len(answer)
    first_value = True

    for i in range(N):
    if sc[i] !=sc[i]:
    answer[i] = np.nan
    else:
```

```python
    if first_value:
        answer[i] = price[i]
        first_value = False
    else:
        answer[i] = answer[i-1] + sc[i] * (price[i] - answer[i-1])
    return answer
techind['KAMA'] = KAMA(techind['Close'])
techind = techind.fillna(0)
#===================================================================
# Drop unwanted columns
#===================================================================
columns2Drop = ['26_ema', '12_ema','tp','rmf','pmf','nmf','mfr']
techind = techind.drop(labels = columns2Drop, axis = 1)

techind.index = techind['Date']
techind = techind.drop(labels = ['Date'], axis = 1)
#save Save dataset as a file
techind.to_csv('TechIndout.csv')
#%%
# = ====================================================
# Regression model
# = ====================================================
# importing necessary libraries
import numpy as np
import matplotlib
import matplotlib.pyplot as plt
from distutils.version import LooseVersion
from sklearn.model_selection import train_test_split
from sklearn.metrics import median_absolute_error, r2_score
from sklearn.metrics import mean_absolute_error, mean_squared_error

# Create the dataset
a = np.array(techind)
#Skip first 50 rows as they contain zeros
X = a[50:2400,8:43 ]
#y = a[50:2400,4 ]
y = a[51:2401,4 ] #To forecast one day ahead
X_train, X_test, y_train, y_test = train_test_split(X, y, random_state = 1)
######################################################################
#%%
from sklearn.datasets import make_regression
from sklearn.model_selection import train_test_split
from sklearn.compose import TransformedTargetRegressor
from sklearn.metrics import median_absolute_error, r2_score
from sklearn.ensemble.forest import RandomForestRegressor
from sklearn.preprocessing import QuantileTransformer, quantile_transform
#===================================================================
# Create random forest regressor model
#===================================================================
```

```python
regr = RandomForestRegressor(n_estimators = 100)
#================================================================
# Train the model
#================================================================
regr.fit(X_train, y_train)
#================================================================
# Predict unseen data with the model
#================================================================
y_pred = regr.predict(X_test)
#================================================================
# Evaluate the model and print performance metrics
#================================================================
print_performance_metrics(y_test,y_pred)
#%%
#================================================================
# Plot scatter
#================================================================
plot_scatter(y_test,y_pred)
#%%
#================================================================
# Plot real and predicted outputs
#================================================================
plot_real_predicted(y_test,y_pred)
#%%
#================================================================
# Create linear regression model
#================================================================
from sklearn import linear_model
# Create linear regression object
regr = linear_model.LinearRegression()
# Train the model using the training sets
regr.fit(X_train, y_train)
# Make predictions using the testing set
y_pred = regr.predict(X_test)
#================================================================
# Evaluate the model and print performance metrics
#================================================================
print_performance_metrics(y_test,y_pred)
#%%
#================================================================
# Plot scatter
#================================================================
plot_scatter(y_test,y_pred)
#%%
#================================================================
# Plot real and predicted outputs
#================================================================
plot_real_predicted(y_test,y_pred)
#%%
#================================================================
```

```python
# Create MLP regressor model
#================================================================
from sklearn.neural_network import MLPClassifier
# Create MLP regressor object
""" "mlp = MLPClassifier(hidden_layer_sizes = (100, ), activation = 'relu',
solver = 'adam',
    alpha = 0.0001, batch_size = 'auto', learning_rate = 'constant',
    learning_rate_init = 0.001, power_t = 0.5, max_iter = 200,
    shuffle = True, random_state = None, tol = 0.0001, verbose = False,
    warm_start = False, momentum = 0.9, nesterovs_momentum = True,
    early_stopping = False, validation_fraction = 0.1, beta_1 = 0.9,
    beta_2 = 0.999, epsilon = 1e-08, n_iter_no_change = 10)" " "
mlp = MLPClassifier(hidden_layer_sizes = (100, ), learning_rate_init = 0.001,
    alpha = 1, momentum = 0.9,max_iter = 1000)
# Train the model using the training sets
regr.fit(X_train, y_train)
# Make predictions using the testing set
y_pred = regr.predict(X_test)
#================================================================
# Evaluate the model and print performance metrics
#================================================================
print_performance_metrics(y_test,y_pred)
#%%
#================================================================
# Plot scatter
#================================================================
plot_scatter(y_test,y_pred)
#%%
#================================================================
# Plot real and predicted outputs
#================================================================
plot_real_predicted(y_test,y_pred)
#%%
#================================================================
# Create k-NN regressor model
#================================================================
from sklearn.neighbors import KNeighborsRegressor
# Create a Regression object
print("Training Regressor...")
regr = KNeighborsRegressor(n_neighbors = 2)
# Train the model using the training sets
regr.fit(X_train, y_train)
# Make predictions using the testing set
y_pred = regr.predict(X_test)
#================================================================
# Evaluate the model and print performance metrics
#================================================================
print_performance_metrics(y_test,y_pred)
#%%
#================================================================
```

```
# Plot scatter
#===============================================================
plot_scatter(y_test,y_pred)
#%%
#===============================================================
# Plot real and predicted outputs
#===============================================================
plot_real_predicted(y_test,y_pred)
#%%
#===============================================================
# Create SVR regressor model
#===============================================================
from sklearn.svm import SVR
# Create a Regression object
print("Training Regressor...")
regr = SVR(gamma = 'scale', C = 200.0, epsilon = 0.2)
# Train the model using the training sets
regr.fit(X_train, y_train)
# Make predictions using the testing set
y_pred = regr.predict(X_test)
#===============================================================
# Evaluate the model and print performance metrics
#===============================================================
print_performance_metrics(y_test,y_pred)
#%%
#===============================================================
# Plot scatter
#===============================================================
plot_scatter(y_test,y_pred)
#%%
#===============================================================
# Plot real and predicted outputs
#===============================================================
plot_real_predicted(y_test,y_pred)
```

6.3 通货膨胀预测

通货膨胀是指平均物价水平变化的百分比，是衡量经济活动最重要的指标之一。它会对家庭、投资者、政府以及政策制定者产生影响。高通胀率会使经济增长水平恶化，降低实际报酬，增加生产成本。同样，低通胀环境也被认为是负面的经济指标，它与较低的经济需求水平相关。因此，对不同时间段的通货膨胀进行预测是至关重要的。机器学习模型用于通货膨胀预测以及其他宏观经济变量的预测（Ülke, Sahin, & Subasi, 2018）。

例 6.2 以下 Python 代码用于通货膨胀预测，其中采用了不同的分类方法，使用了独立的训练数据集和测试数据集。本例对统计质量进行了度量，包括计算平均绝对误差

（MAE）、均方误差（MSE）和相关系数（R），同时绘制了散点图、实际值以及预测值。

数据集信息：通货膨胀的预测可以使用月度物价指数的四个度量标准。它们分别是消费者价格指数（CPI）、不包含食物和能源的消费者价格指数（核心CPI）、个人消费支出物价指数（PCE），以及不包含食物和能源的个人消费支出物价指数（核心PCE）。我们将考虑以下六种经济活动，作为通货膨胀预测的模型：即失业率（UNEM）、工业生产指数（IP）、实际个人消费支出（INC）、非农就业人数（WORK）、房屋开工率（HS）和期限利差（SPREAD）——即长期债券利率（5年期国债）减去短期债券（3月期国债）利率的差额。这些数据由美国圣路易斯联邦储备银行数据库FRED进行收集，时间跨度从1984年1月到2014年12月。此外，这些数据序列的稳定条件、变化形势、波动（噪声），以及分布情况是预测的关键点。对于通货膨胀的预测，可以分别考虑CPI、核心CPI、PCE和核心PCE（Ülke et al., 2018）。

```python
""" 
Created on Tue May 21 19:39:42 2019
@author: asubasi" """
#======================================================================
# Inflation forecasting
#======================================================================
# importing necessary libraries
import numpy as np
import pandas as pd
import matplotlib.pyplot as plt
from sklearn.model_selection import train_test_split
from sklearn.metrics import mean_absolute_error,mean_squared_error, r2_score

# Load Inflation data set
Dataset = pd.read_csv("Inflation.csv")
# Create the dataset
rng = np.random.RandomState(1)
a = np.array(Dataset)
X = a[0:370,5:10 ]
y = a[0:370,10 ]
X_train, X_test, y_train, y_test = train_test_split(X, y, test_size = 0.3, random_state = 1)
#%%
# Create a regression model
from sklearn.ensemble.forest import RandomForestRegressor
regr = RandomForestRegressor(n_estimators = 100)
# Train the model using the training sets
regr.fit(X_train, y_train)
# Make predictions using the testing set
y_pred = regr.predict(X_test)
#======================================================================
# Evaluate the model and print performance metrics
```

```
#================================================================
# The mean absolute error
print("MAE = %5.3f" % mean_absolute_error(y_test, y_pred))
# Explained variance score: 1 is perfect prediction
print("R^2 = %0.5f" % r2_score(y_test, y_pred))
# The mean squared error
print("MSE = %5.3f" % mean_squared_error(y_test, y_pred))
#%%
#================================================================
# Plot scatter
#================================================================
fig, ax = plt.subplots()
ax.scatter(y_test, y_pred, edgecolors = (0, 0, 0))
ax.text(-10, 15, r'$R^2$ = %.2f, MAE = %.2f' % (
    r2_score(y_test, y_pred), mean_absolute_error(y_test, y_pred)))
ax.plot([y_test.min(), y_test.max()], [y_test.min(), y_test.max()], 'k--', lw = 4)
ax.set_xlabel('Measured')
ax.set_ylabel('Predicted')
#ax.set_xlim([0, 10])
#ax.set_ylim([0, 10])
plt.show()
#%%
#================================================================
# Plot real and predicted outputs
#================================================================
plt.plot(y_test[1:200], color = 'red', label = 'Real data')
plt.plot(y_pred[1:200], color = 'blue', label = 'Predicted data')
plt.title('Prediction')
plt.legend()
plt.show()
```

6.4 电力负荷预测

在电力生产过程中，负荷对于设备方和交易方都是非常关键的信息，特别是在生产计划、日常运行、机组组合和经济调度方面。根据需求，负荷预测被划分为不同的时间间隔：长期负荷预测包括一年到几年，用于工厂和基础设施投资决策；中期负荷预测包括几天到几个月，用于维护调度和远期合同谈判；短期负荷预测（STLF）包括1小时到几天，用于实时发电控制和能源交易规划（Bozkurt, Biricik, & Tayşi, 2017）。从时间跨度来看，短期、中期、长期的规划或预测是文献中常见的概念。短期规划通常是指定期的以小时或周为单位的电力系统运行规划，在该阶段，最关键的是可靠、准确的电力需求预测。在电力市场上，可靠的短期预测是必要的，因为这可能影响向消费者提供的电价。另外，中期电力系统运营规划通常以一个月或多个月为时间跨度，而长期规划通常以年为单位。长期电力需求预测通常以几年为单位提出，并且作为建设新发电机组规划过程中的输入指标（Sikiric,

Avdakovic, & Subasi, 2013）。

智能电网旨在建立一体化的高效能源供应网络，提高供电的可靠性和质量，以及网络安全、能源效率和需求管理。现代化的配电组织结构由先进的监控基础设施来实现，它们产生了大量的数据，可以进行细粒度的分析，提高预测性能。在能源领域，电力负荷预测是一项非常重要的任务，它有助于决策制定、促进最佳定价策略、易于整合可再生能源以及降低维护成本。像智能电网这样的重要基础设施的管理和成功运营，需要利用准确预测电力需求带来的巨大优势，但由于其非线性的性质，该任务仍然具有挑战性。近年来，深度学习的出现，在从图像分类到机器翻译等众多不同任务中取得了令人印象深刻的表现。因此，深度学习模型在电力负荷预测问题上的应用越来越受到研究人员以及业界的关注。本例使用了一个真实的数据集，通过深度学习模型进行短期的（以小时为单位）电力负荷预测。特别地，我们应用了循环神经网络——序列到序列的模型和时间卷积神经网络以及在信号处理和负载预测领域中广为人知的变体架构（Gasparin, Lukovic, & Alippi, 2019）。

众所周知的供需平衡法则也适用于能源市场：在需求量较大的时段，价格会上涨；而在需求量较小的时段，如晚上、周末和节假日，价格会下降。在一个庞大的发电厂中，需求是按小时进行开发的，不能即时启动或停止输出，因此，开发计划大多按天进行。因此，短期负荷预测（STLF）在电力市场的运营管理中发挥着至关重要的作用。电力负荷是一个典型的时间序列，它涉及每小时的连续测量值。时间序列分析的一个重要组成部分是预测，它主要是根据时间序列来预测未来的事件（Bozkurt et al., 2017）。由于电力分配管制的放松和可再生能源的广泛使用对正常的市场价格有很大的影响，STLF 对有效的电力供应具有根本性的意义（Gasparin et al., 2019）。工程师们在预测过程中尽可能地找出影响电力消费的关键因素，如 GDP 增长、人口和气候变化、消费者标准和偏好等。气温是中短期预测的重要因素之一。在预测过程中，环境温度、人类社会行为等变量也会提供关键信息，使预报结果更加可靠（Sikiric et al., 2013）。

例 6.3 以下 Python 代码用于电力负荷预测，其中采用了 LSTM 模型，使用了独立的训练数据集和测试数据集。本例对统计质量进行了度量，包括计算平均绝对误差（MAE）、均方误差（MSE）和相关系数（R），同时绘制了散点图、实际值以及预测值。本例改编自 machinelearningmastery 网站（https://machinelearningmastery.com/how-to-develop-lstm-models-for-time-series-forecasting/）。

数据集信息：预测中最复杂的时间序列类型之一是短期负荷数据，它们是非线性的，而且有很多季节内的因素，这使得问题的解决更加困难。由于短期负荷预测是一个复杂的预测问题，我们选择了深度学习模型。通过使用马来西亚 Johor 市供电公司 2009 年和 2010 年的以小时为单位的负荷数据，实现深度学习 Keras 模型进行负荷预测。同时，将温度的时间序列与每小时的负荷数据相结合，以提高模型的性能（Sadaei, e Silva, de,

Guimarães, & Lee, 2019）。该数据集可以从 Mendeley 网站下载（https://data.mendeley.com/datasets/f4fcrh4tn9/1）。

```
" " "
Created on Thu Oct 10 11:07:13 2019
@author: asubasi
" " "
#================================================================
# Load forecasting with univariate LSTM
#================================================================
from numpy import array
from keras.models import Sequential
from keras.layers import LSTM
from keras.layers import Dense
from pandas import read_csv
from sklearn.metrics import mean_squared_error, r2_score
from sklearn.metrics import mean_absolute_error
import matplotlib.pyplot as plt
#================================================================
# Define utility functions
#================================================================
def plot_model_loss(history):
plt.figure(figsize = (6, 4))
plt.plot(history.history['loss'], 'r--', label = 'Loss of training data')
plt.plot(history.history['val_loss'], 'b--', label = 'Loss of validation data')
plt.title('Model Error')
plt.ylabel('Loss')
plt.xlabel('Training Epoch')
    plt.ylim(0)
    plt.legend()
    plt.show()

def plot_real_predicted(ytest, ypred):
    plt.plot(ytest[1:200], color = 'red', label = 'Real data')
    plt.plot(ypred[1:200], color = 'blue', label = 'Predicted data')
    plt.title('Prediction')
    plt.legend()
    plt.show()

def print_performance_metrics(ytest, ypred):
    # The mean absolute error
    print("MAE = %5.3f" % mean_absolute_error(ytest, ypred))
    # Explained variance score: 1 is perfect prediction
    print("R^2 = %0.5f" % r2_score(ytest, ypred))
    # The mean squared error
    print("MSE = %5.3f" % mean_squared_error(ytest, ypred))

def plot_scatter(ytest, ypred):
```

```python
    fig, ax = plt.subplots()
    ax.scatter(ytest, ypred, edgecolors = (0, 0, 0))
    ax.text(10, ytest.max()-10, r'$R^2$ = %.2f, MAE = %.2f' % (
    r2_score(ytest, ypred), mean_absolute_error(ytest, ypred)))
    ax.plot([ytest.min(), ytest.max()], [ytest.min(), ytest.max()], 'k--',
    lw = 4)
    ax.set_xlabel('Measured')
    ax.set_ylabel('Predicted')
    plt.show()

# split a univariate sequence into samples
def split_sequence(sequence, n_steps):
    X, y = list(), list()
    for i in range(len(sequence)):
        # find the end of this pattern
        end_ix = i + n_steps
        # check if we are beyond the sequence
        if end_ix > len(sequence)-1:
            break
        # gather input and output parts of the pattern
        seq_x, seq_y = sequence[i:end_ix], sequence[end_ix]
        X.append(seq_x)
        y.append(seq_y)
    return array(X), array(y)
#=====================================================================
# Prepare dataset
#=====================================================================
# Load all data
dataset = read_csv('malaysia_all_data_for_paper.csv')
# define input sequence
raw_seq = dataset.values[:,2]
# choose a number of time steps
X_train=dataset.values[0:10000,2]
X_test = dataset.values[10000:17000,2]
n_steps = 5
# Split data into training and testing samples
Xtrain, ytrain = split_sequence(X_train, n_steps)
Xtest, ytest = split_sequence(X_test, n_steps)
#%%
n_features = 1
Xtrain = Xtrain.reshape((Xtrain.shape[0], Xtrain.shape[1], n_features))
#=====================================================================
# Build and compile the model
#=====================================================================
model = Sequential()
model.add(LSTM(50, activation = 'relu', input_shape = (n_steps,
n_features)))
model.add(Dense(1))
model.compile(optimizer = 'adam', loss = 'mse')
#=====================================================================
```

```
# Train the model
#==============================================================
#model.fit(Xtrain, ytrain, epochs = 50, verbose = 1)
history = model.fit(Xtrain, ytrain, validation_split = 0.3,
        epochs = 50, batch_size = 20, verbose = 2)
#==============================================================
# Plot the model loss for training and validation dataset
#==============================================================
plot_model_loss(history)
#%%
#==============================================================
# Predict unseen data with the model
#==============================================================
Xtest = Xtest.reshape((Xtest.shape[0], Xtest.shape[1], n_features))
ypred = model.predict(Xtest, verbose = 2)
#==============================================================
# Evaluate the model and print performance metrics
#==============================================================
print_performance_metrics(ytest, ypred)
#%%
#==============================================================
# Plot scatter
#==============================================================
plot_scatter(ytest, ypred)
#%%
#==============================================================
# Plot real and predicted outputs
#==============================================================
```

例6.4 以下Python代码用于电力负荷预测，其中采用了卷积神经网络模型，使用了独立的训练数据集和测试数据集。本例对统计质量进行了度量，包括计算平均绝对误差（MAE）、均方误差（MSE）和相关系数（R），同时绘制了散点图、实际值以及预测值。本例改编自machinelearningmastery网站（https://machinelearningmastery.com/how-to-develop-convolutional-network-models-for-time-series-forecasting/）。

数据集信息：预测中最复杂的时间序列类型之一是短期负荷数据，它们是非线性的，而且有很多季节内的因素，这使得问题的解决更加困难。由于短期负荷预测是一个复杂的预测问题，我们选择了深度学习模型。通过使用马来西亚Johor市供电公司2009年和2010年的以小时为单位的负荷数据，实现深度学习Keras模型进行负荷预测。同时，将温度的时间序列与每小时的负荷数据相结合，以提高模型的性能（Sadaei et al., 2019）。该数据集可以从Mendeley网站下载（https://data.mendeley.com/datasets/f4fcrh4tn9/1）。

```
"""
Created on Mon Oct 14 10:32:03 2019
@author: asubasi """ """
#==================================================================
# Load forecasting with convolutional neural network model
#==================================================================
from numpy import array
from keras.models import Sequential
from keras.layers import Dense
from keras.layers import Flatten
from keras.layers.convolutional import Conv1D
from keras.layers.convolutional import MaxPooling1D
from pandas import read_csv
from sklearn.metrics import mean_squared_error, r2_score
from sklearn.metrics import mean_absolute_error
import matplotlib.pyplot as plt
#==================================================================
# Define utility functions
#==================================================================
def plot_model_loss(history):
    plt.figure(figsize = (6, 4))
    plt.plot(history.history['loss'], 'r--', label = 'Loss of training
    data')
    plt.plot(history.history['val_loss'], 'b--', label = 'Loss of
    validation data')
    plt.title('Model Error')
    plt.ylabel('Loss')
    plt.xlabel('Training Epoch')
    plt.ylim(0)
    plt.legend()
    plt.show()

def plot_real_predicted(ytest, ypred):
    plt.plot(ytest[1:200], color = 'red', label = 'Real data')
    plt.plot(ypred[1:200], color = 'blue', label = 'Predicted data')
    plt.title('Prediction')
    plt.legend()
    plt.show()

def print_performance_metrics(ytest, ypred):
    # The mean absolute error
    print("MAE = %5.3f" % mean_absolute_error(ytest, ypred))
    # Explained variance score: 1 is perfect prediction
    print("R^2 = %0.5f" % r2_score(ytest, ypred))
    # The mean squared error
    print("MSE = %5.3f" % mean_squared_error(ytest, ypred))

def plot_scatter(ytest, ypred):
    fig, ax = plt.subplots()
    ax.scatter(ytest, ypred, edgecolors = (0, 0, 0))
```

```python
    ax.text(10, ytest.max()-10, r'$R^2$ = %.2f, MAE = %.2f' % (
    r2_score(ytest, ypred), mean_absolute_error(ytest, ypred)))
    ax.plot([ytest.min(), ytest.max()], [ytest.min(), ytest.max()], 'k--',
    lw = 4)
    ax.set_xlabel('Measured')
    ax.set_ylabel('Predicted')
    plt.show()

# split a univariate sequence into samples
def split_sequence(sequence, n_steps):
    X, y = list(), list()
    for i in range(len(sequence)):
    # find the end of this pattern
    end_ix = i + n_steps
    # check if we are beyond the sequence
    if end_ix > len(sequence)-1:
    break
    # gather input and output parts of the pattern
    seq_x, seq_y = sequence[i:end_ix], sequence[end_ix]
    X.append(seq_x)
    y.append(seq_y)
    return array(X), array(y)
#%%
#=====================================================================
# Univariate convolutional neural network models for load forecasting
#=====================================================================
# load all data
dataset = read_csv('malaysia_all_data_for_paper.csv')

X_train=dataset.values[0:10000,2]
X_test = dataset.values[10000:17000,2]

# split a univariate sequence into samples
def split_sequence(sequence, n_steps):
    X, y = list(), list()
    for i in range(len(sequence)):
    # find the end of this pattern
    end_ix = i + n_steps
    # check if we are beyond the sequence
    if end_ix > len(sequence)-1:
    break
    # gather input and output parts of the pattern
    seq_x, seq_y = sequence[i:end_ix], sequence[end_ix]
    X.append(seq_x)
    y.append(seq_y)
    return array(X), array(y)
n_steps = 5
# split into samples
Xtrain, ytrain = split_sequence(X_train, n_steps)
Xtest, ytest = split_sequence(X_test, n_steps)
```

```python
# reshape from [samples, timesteps] into [samples, timesteps, features]
n_features = 1
Xtrain = Xtrain.reshape((Xtrain.shape[0], Xtrain.shape[1], n_features))
#%%
#==================================================================
# Build and compile the model
#==================================================================
model = Sequential()
model.add(Conv1D(filters = 64, kernel_size = 2, activation = 'relu', input_shape = (n_steps, n_features)))
model.add(MaxPooling1D(pool_size = 2))
model.add(Flatten())
model.add(Dense(50, activation = 'relu'))
model.add(Dense(1))
model.compile(optimizer = 'adam', loss = 'mse')
#==================================================================
# Train the model
#==================================================================
history = model.fit(Xtrain, ytrain, validation_split = 0.3,
        epochs = 50, batch_size = 20, verbose = 2)
#==================================================================
# Plot the model loss for training and validation dataset
#==================================================================
plot_model_loss(history)
#%%
#==================================================================
# Predict unseen data with the model
#==================================================================
Xtest = Xtest.reshape((Xtest.shape[0], n_steps, n_features))
ypred = model.predict(Xtest, verbose = 0)
#==================================================================
# Evaluate the model and print performance metrics
#==================================================================
print_performance_metrics(ytest, ypred)
#%%
#==================================================================
# Plot scatter
#==================================================================
plot_scatter(ytest, ypred)
#%%
#==================================================================
# Plot real and predicted outputs
#==================================================================
plot_real_predicted(ytest, ypred)
```

6.5 风速预测

目前，各国政府、监管机构和能源公司使用了诸多不同的方式以促进各种可再生能源技术的实施，包括对项目的奖励、资助、现金和税收抵免。近来，越来越多的人认识到风力发电在众多方面的优势，包括减少排放和对石油的依赖、能源供应多样化以及提供低成本电力等。许多国家制定了支持探索可再生能源技术的政策，特别是太阳能和风能。风速的大小和非线性波动是空气动力负荷预测和风力发电机效率的主要影响因子。确定风速的特性对于测量发电量、转子叶片上的负荷和压力，以及结构组件的疲劳程度至关重要。与传统的发电厂不同，风力发电机的发电量主要取决于天气状况，特别是风速的大小（Shen, Zhou, Li, Fu, & Lie, 2018）。风力发电机的容量与风速的立方成正比。因此，我们强烈建议准确预测风速分布，以便有效地收集风力发电机组产生的电量，特别是当我们知道风速是波动的。调研给定地的风能发电量，需要深入研究风能在可用性、方向、小时分布、变化和频率等方面的分布。风速预报可以分为极短期（30 分钟以内）、短期（30 分钟至 6 小时）、中期（6 小时至 24 小时）和长期（7 天以内）（Okumus & Dinler, 2016）。近些年来，有许多技术被用来测量风速。它们基于：1）物理建模，其中大量数据来自天气预报；2）统计建模，利用数据的输入和输出寻找关系模式；3）混合建模，利用混合数据进行建模（Lei, Shiyan, Chuanwen, Hongling, & Yan 2009）。物理建模需要大量不必要的计算力，且性能较差。统计建模使用基于时间序列的研究，应用机器学习算法试图寻找现在和未来之间的关系（Sideratos & Hatziargyriou, 2007）。因此，预测风速的常用方法是利用人工智能和机器学习工具，采用大量的历史输入数据作为训练，学习输入/输出数据之间的关系和依赖。

> **例 6.5** 以下 Python 代码用于风速预测，其中采用了卷积神经网络（CNN），使用了独立的训练数据集和测试数据集。本例对统计质量进行了度量，包括计算平均绝对误差（MAE）、均方误差（MSE）和相关系数（R），同时绘制了散点图、实际值以及预测值。本例改编自 machinelearningmastery 网站（https://machinelearningmastery.com/how-to-develop-convolutional-network-models-for-time-series-forecasting/）。
>
> 数据集信息：调研各地风力发电机的风能生产情况，需要对风能分布的可用性、方向、小时分布、昼夜变化和频率进行严格研究。本例所使用的数据由阿卜杜拉国王经济城（K.A.CARE）（https://rratlas.kacare.gov.sa）收集，作为沙特阿拉伯可再生能源资源环境测量和监控（RRMM）项目的一部分。该数据集提供了 2013 年 5 月至 2016 年 7 月的每小时数据，包括气温、风向和风速、总水平辐射（GHI）、相对湿度和气压等不同属性。数据的收集使用旋转的影带辐射计测量散射水平辐射（DHI）、法向直接辐射（DNI）和总水平辐射（GHI）。使用气温探测器测量空气温度，气压计和相对湿度探测器分别测量压力和相对湿度。最后，使用风速仪和风向标分别测量风速和风向。
>
> ```
> """
> Created on Mon Oct 14 10:32:03 2019
> @author: absubasi
> """
> ```

```python
#================================================================
# Wind speed forecasting
#================================================================
from numpy import array
from keras.models import Sequential
from keras.layers import Dense
from keras.layers import Flatten
from keras.layers.convolutional import Conv1D
from keras.layers.convolutional import MaxPooling1D
from pandas import read_csv
from sklearn.metrics import mean_squared_error, r2_score
from sklearn.metrics import mean_absolute_error
import matplotlib.pyplot as plt
#================================================================
# Define utility functions
#================================================================
def plot_model_loss(history):
    plt.figure(figsize = (6, 4))
    plt.plot(history.history['loss'], 'r--', label = 'Loss of training data')
    plt.plot(history.history['val_loss'], 'b--', label = 'Loss of validation data')
    plt.title('Model Error')
    plt.ylabel('Loss')
    plt.xlabel('Training Epoch')
    plt.ylim(0)
    plt.legend()
    plt.show()
def plot_real_predicted(ytest, ypred):
    plt.plot(ytest[1:200], color = 'red', label = 'Real data')
    plt.plot(ypred[1:200], color = 'blue', label = 'Predicted data')
    plt.title('Prediction')
    plt.legend()
    plt.show()

def print_performance_metrics(ytest, ypred):
    # The mean absolute error
    print("MAE = %5.3f" % mean_absolute_error(ytest, ypred))
    # Explained variance score: 1 is perfect prediction
    print("R^2 = %0.5f" % r2_score(ytest, ypred))
    # The mean squared error
    print("MSE = %5.3f" % mean_squared_error(ytest, ypred))

def plot_scatter(ytest, ypred):
    fig, ax = plt.subplots()
    ax.scatter(ytest, ypred, edgecolors = (0, 0, 0))
    ax.text(10, ytest.max()-10, r'$R^2$ = %.2f, MAE = %.2f' % (
        r2_score(ytest, ypred), mean_absolute_error(ytest, ypred)))
    ax.plot([ytest.min(), ytest.max()], [ytest.min(), ytest.max()], 'k--',
        lw = 4)
```

```python
    ax.set_xlabel('Measured')
    ax.set_ylabel('Predicted')
    plt.show()
#%%#===============================================================
# Convolutional neural network models for wind forecasting
#===================================================================
from numpy import array
from keras.models import Sequential
from keras.layers import Dense
from keras.layers import Flatten
from keras.layers.convolutional import Conv1D
from keras.layers.convolutional import MaxPooling1D
from pandas import read_csv
# load all data
dataset = read_csv('Wind.csv')

Xtrain=dataset.values[0:17000,0:12]
ytrain=dataset.values[1:17001,12] #Forecast One hour ahead
Xtest = dataset.values[17000:27000,0:12]
ytest = dataset.values[17001:27001,12] #Forecast One hour ahead
# reshape from [samples, timesteps] into [samples, timesteps, features]
n_features = 1
Xtrain = Xtrain.reshape((Xtrain.shape[0], Xtrain.shape[1], n_features))
Xtest = Xtest.reshape((Xtest.shape[0], Xtrain.shape[1], n_features))
#%%
#===================================================================
# Build and compile the model
#===================================================================
model = Sequential()
model.add(Conv1D(filters = 128, kernel_size = 2, activation = 'relu', input_
shape = (Xtrain.shape[1], n_features)))
model.add(MaxPooling1D(pool_size = 2))
model.add(Flatten())
model.add(Dense(64, activation = 'relu'))
model.add(Dense(1))
model.compile(optimizer = 'adam', loss = 'mse')
#===================================================================
# Train the model
#===================================================================
#history = model.fit(Xtrain, ytrain, epochs = 100, verbose = 1)
history = model.fit(Xtrain, ytrain, validation_split = 0.3, epochs = 50,
batch_size = 20, verbose = 2)
#===================================================================
# Plot the model loss for training and validation dataset
#===================================================================
plot_model_loss(history)
#%%
#===================================================================
# Predict unseen data with the model
#===================================================================
```

```python
ypred = model.predict(Xtest, verbose = 0)
#================================================================
# Evaluate the model and print performance metrics
#================================================================
print_performance_metrics(ytest, ypred)
#%%
#================================================================
# Plot scatter
#================================================================
plot_scatter(ytest, ypred)
#%%
#================================================================
# Plot real and predicted outputs
#================================================================
plot_real_predicted(ytest, ypred)
#%%
#================================================================
# Multivariate CNN model for wind speed forecasting
#================================================================
# multivariate cnn example
from numpy import array
from numpy import hstack
from keras.models import Sequential
from keras.layers import Dense
from keras.layers import Flatten
from keras.layers.convolutional import Conv1D
from keras.layers.convolutional import MaxPooling1D
from pandas import read_csv
# load all data
dataset = read_csv('Wind.csv')

Xtrain=dataset.values[0:17000,0:12]
ytrain=dataset.values[1:17001,12] #Forecast One hour ahead
Xtest = dataset.values[17000:27000,0:12]
ytest = dataset.values[17001:27001,12] #Forecast One hour ahead
n_steps = 12
# reshape from [samples, timesteps] into [samples, timesteps, features]
n_features = 1
Xtrain = Xtrain.reshape((Xtrain.shape[0], Xtrain.shape[1], n_features))
Xtest = Xtest.reshape((Xtest.shape[0], Xtrain.shape[1], n_features))
#%%
#================================================================
# Build and compile the model
#================================================================
model = Sequential()
model.add(Conv1D(filters = 128, kernel_size = 2, activation = 'relu',
    input_shape = (n_steps, n_features)))
model.add(MaxPooling1D(pool_size = 2))
model.add(Flatten())
model.add(Dense(64, activation = 'relu'))
```

```
model.add(Dense(1))
model.compile(optimizer = 'adam', loss = 'mse')
#================================================================
# Train the model
#================================================================
#history = model.fit(Xtrain, ytrain, epochs = 100, verbose = 1)
history = model.fit(Xtrain, ytrain, validation_split = 0.3, epochs = 50,
batch_size = 20, verbose = 2)
#================================================================
# Plot the model loss for training and validation dataset
#================================================================
plot_model_loss(history)
#%%
#================================================================
# Predict unseen data with the model
#================================================================
Xtest = Xtest.reshape((Xtest.shape[0], n_steps, n_features))
ypred = model.predict(Xtest, verbose = 0)
#================================================================
# Evaluate the model and print performance metrics
#================================================================
print_performance_metrics(ytest, ypred)
#%%
#================================================================
# Plot scatter
#================================================================
plot_scatter(ytest, ypred)
#%%
#================================================================
# Plot real and predicted outputs
#================================================================
plot_real_predicted(ytest, ypred)
```

6.6 旅游需求预测

总的来说，旅游需求呈现稳定增长。然而，由于决定性因素和外部措施的不确定性，旅游行业经历了数次波动。科学家、政策制定者和从业人员重点关注了旅游增长和需求波动的周期，以试图预测未来的游客流量。旅游需求的决定因素包括游客数量、旅游费用和停留时间。这类统计与多种类型的变化相关，如原始地和目的地区域的季节性变化、与汇率和收入水平相关的商业周期，或是与气候变化或特殊事件相关的各种环境影响。由于社会、文化、政治、技术等方面的变化，旅游业正从发达国家向新兴工业化国家扩展，国际旅游业迅速发展。随着更多的旅游目的地国家/地区对稀缺资源的争夺，这种增长会导致成本和收益的混合效应。因此，对旅游目的地的准确预测至关重要，决策者会试图对旅游业因势利导，以及平衡当地的生态和社会承载能力。在这种情况下，国际旅游需求的预测者会考虑可能影响其客流量的因素，包括旅游市场、目的地,甚至是邻国或竞争国家/地区的总体情况(Song, Li, & Cao, 2018; Song, Qiu, & Park, 2019)。

时间序列模型根据历史趋势预测旅游需求。这种模型通过利用连续时期内产生的测量序列来确定时间序列数据之间的模式、斜率和周期。与随机抽样方法不同的是,时间序列预测模型是基于连续的数值,以每月、每季度或每年等有规律的间隔时间进行连续测量。时间序列模型的趋势一旦形成,就可以对即将到来的时间序列进行预测(Song et al., 2019)。

空置的酒店房间、未购买的活动门票,以及未使用的食品反映了不必要的成本和未实现的销售,它们对财务稳定构成了潜在威胁。简而言之,大多数旅游用品和酒店用品是无法为未来的需求进行保留的,因此对旅游需求的准确预测至关重要。精确的旅游需求预测能够为政治、战略以及运营决策提供重要支持。例如,政府需要具体的旅游标准,以便就容纳能力建设和住宿地点规划等问题作出知情决策;组织需要预测,以便作出与旅游宣传册有关的战略决策;旅游接待专业人员需要准确的预测,以便作出人员配置和时间安排等组织决策。因此,准确的旅游需求预测是旅游业的一个重要组成部分,为与旅游相关的决策提供了关键信息。许多对于旅游需求预测的研究都属于由来已久的定量方法,它们将过去的游客数量和特定的旅游需求预测因子作为训练数据,建立模型。因此,在数据集中找出所有潜在的良好关系,能够更加灵活地创建准确的预测模型。时间序列、计量经济学和人工智能模型提供了出色的预测效率,打破了基于目的市场领域知识的功能工程屏障(Law, Li, Fong, & Han, 2019)。

对于公共或者私有从业人员来说,旅游业的需求预测都具有重要的经济意义。通常我们没有能力储存大部分的旅游用品,因此需求预测的准确性对于提高旅游业的决策水平、管理的生产力、竞争力,以及经济的可持续增长具有重要作用。传统的统计方法是时间序列分析预测中最常用的模型,但这种线性模型具有一定的局限性。对于大多数现实问题而言,变量之间的关系并不是线性的,因此使用线性模型是无效的。常规的统计方法如多元线性回归,只适用于具有特定模式的数据,如趋势性、季节性和周期性(Cankurt & Subasi, 2016)。人工智能,尤其是深度学习方面的最新进展,为规避上述障碍提供了解决方案,使旅游需求预测更加准确(Pouyanfar et al., 2019)。深度网络架构通过使用两层以上的非线性计算扩展了人工神经网络模型,并且在多种应用中被证明是高效的。它们的成功通常归功于其内置的工程能力,促使我们在机器学习领域能够同时突破这两个障碍。就时间序列的背景信息研究而言,深度网络架构在健壮性和层次非线性关系方面也具有一定优势。尤其是循环神经网络(RNN)、长短期记忆(LSTM)以及注意力机制尤其能够处理和学习长期的依赖关系。这些资源使得深度学习成为旅游需求预测的替代方案(Law et al., 2019)。

例 6.6 以下 Python 代码用于旅游需求预测，其中采用了不同的回归模型，使用了独立的训练数据集和测试数据集。本例对统计质量进行了度量，包括计算平均绝对误差（MAE）、均方误差（MSE）和相关系数（R），同时绘制了散点图、实际值以及预测值。目标转换也被用于观测目标变量的变换效果，从本例中可以看出，目标转换对某些回归模型有效，但对其他的无效。

数据集信息：本例使用了土耳其的旅游需求预测数据。旅游业是世界上最大的产业，而土耳其又是旅游市场上最大的参与者之一，因此，对于土耳其旅游业的每一项工作都在产生巨大的经济价值，并为旅游部门做出重大贡献。在欧洲和地中海地区，土耳其除了具有现代化和完善的旅游基础设施，稳定的价格和高质量的旅游服务外，还拥有悠久丰富的文化和历史遗产、美丽的自然风光、分明的四季气候以及热情好客的人民，具有极大的潜力。本数据集收集了从 1992 年至 2010 年，在土耳其的旅游排行榜前 24 个国家的月度游客量数据，用于衡量其旅游需求。模型包括土耳其政府许可酒店床位数、消费价格指数（CPI）和土耳其游客汇率，作为可能会影响土耳其境外旅游需求的环境和经济方面的时间序列数据。

输入变量包括批发物价指数清单、美元抛售、1 盎司黄金的伦敦美元售价、土耳其酒店床位数、土耳其主要客户国家的 CPI（即奥地利、比利时、加拿大、丹麦、法国、德国、希腊、意大利、荷兰、挪威、波兰、西班牙、瑞典、瑞士、土耳其、英国、美国、俄罗斯联邦）、土耳其主要客户国家的游客数量（即德国、俄罗斯、法国、伊朗、保加利亚、格鲁吉亚、希腊、乌克兰、阿塞拜疆、奥地利、比利时、丹麦、荷兰、英国、西班牙、瑞典、瑞士、意大利、挪威、波兰、罗马尼亚、美国、伊拉克、叙利亚）、土耳其主要客户国家的汇率（加拿大元、丹麦克朗、挪威克朗、波兰兹罗提、瑞典克朗、瑞士法郎、土耳其里拉、英镑、俄罗斯卢布）、年、月、季，以及以前游客的数量。月度的时间序列数据收集自土耳其共和国文化旅游部（www.turizm.gov.tr）、土耳其国家统计局（www.die.gov.tr）、土耳其共和国中央银行数据库（http://evds.tcmb.gov.tr）、土耳其旅行社协会（TÜRSAB）（www.tursab.org.tr）和世界银行数据库（http://databank.worldbank.org）(Cankurt & Subasi, 2016)。

```
"""
Created on Fri Aug 2 01:23:04 2019
@author: asubasi
"""
#================================================================
# Tourism demand forecasting example
#================================================================
import matplotlib.pyplot as plt
import numpy as np
```

```python
from sklearn.metrics import mean_squared_error, r2_score
from sklearn.metrics import mean_absolute_error
from sklearn.model_selection import train_test_split
#===============================================================
# Define utility functions
#===============================================================
def print_performance_metrics(ytest, ypred):
    # The mean absolute error
    print("MAE = %5.3f" % mean_absolute_error(ytest, ypred))
    # Explained variance score: 1 is perfect prediction
    print("R^2 = %0.5f" % r2_score(ytest, ypred))
    # The mean squared error
    print("MSE = %5.3f" % mean_squared_error(ytest, ypred))

def plot_scatter(ytest, ypred):
    fig, ax = plt.subplots()
    ax.scatter(ytest, ypred, edgecolors = (0, 0, 0))
    ax.text(ypred.max()-4.5, ytest.max()-0.1, r'$R^2$ = %.2f, MAE = %.2f' % (
        r2_score(ytest, ypred), mean_absolute_error(ytest, ypred)))
    ax.plot([ytest.min(), ytest.max()], [ytest.min(), ytest.max()], 'k--',
        lw = 4)
    ax.set_xlabel('Measured')
    ax.set_ylabel('Predicted')
    plt.show()

def plot_real_predicted(ytest, ypred):
    plt.plot(ytest[1:200], color = 'red', label = 'Real data')
    plt.plot(ypred[1:200], color = 'blue', label = 'Predicted data')
    plt.title('Prediction')
    plt.legend()
    plt.show()
# ==============================================================
# Load the tourism dataset
# ==============================================================
import pandas as pd
dataset = pd.read_csv('TurkeyVisitors.csv')
AllData = dataset.iloc[:, :].values
X = AllData[:,1:60]
y = AllData[:,60]
Xtrain, Xtest, ytrain, ytest = train_test_split(X, y, test_size = 0.33,
    random_state = 0)
#%%
#===============================================================
# Linear regression example
#===============================================================
import matplotlib.pyplot as plt
from sklearn import linear_model
#===============================================================
# Forecasting without target transformation
#===============================================================
```

```python
# Create linear regression object
regr = linear_model.LinearRegression()
# Train the model using the training sets
regr.fit(Xtrain, ytrain)
# Make predictions using the testing set
ypred = regr.predict(Xtest)
#================================================================
# Evaluate the model and print performance metrics
#================================================================
print_performance_metrics(ytest, ypred)
#%%
#================================================================
# Plot scatter
#================================================================
plot_scatter(ytest, ypred)
#%%
#================================================================
# Plot real and predicted outputs
#================================================================
plot_real_predicted(ytest, ypred)
#%%
#================================================================
# Forecasting with target transformation
#================================================================
from sklearn.preprocessing import QuantileTransformer
from sklearn.compose import TransformedTargetRegressor
transformer = QuantileTransformer(output_distribution = 'normal')
# Create linear regression object
regr = linear_model.LinearRegression()
regr = TransformedTargetRegressor(regressor = regr,
      transformer = transformer)
X_train, X_test, y_train, y_test = train_test_split(X, y, test_
size = 0.33, random_state = 0)
regr.fit(X_train, y_train)
y_pred = regr.predict(X_test)
#================================================================
# Evaluate the model and print performance metrics
#================================================================
print_performance_metrics(y_test, y_pred)
#%%
#================================================================
# Plot scatter
#================================================================
plot_scatter(y_test, y_pred)
#%%
#================================================================
# Plot real and predicted outputs
#================================================================
plot_real_predicted(y_test, y_pred)
#%%
```

```python
from sklearn.neural_network import MLPRegressor
#================================================================
# ANN regression example
#================================================================
#================================================================
# Forecasting without target transformation
#================================================================
# Create MLP regression object
print("Training MLPRegressor...")
regr = MLPRegressor(activation = 'logistic')
# Train the model using the training sets
regr.fit(Xtrain, ytrain)
# Make predictions using the testing set
ypred = regr.predict(Xtest)
#================================================================
# Evaluate the model and print performance metrics
#================================================================
print_performance_metrics(ytest, ypred)
#%%
#================================================================
# Plot scatter
#================================================================
plot_scatter(ytest, ypred)
#%%
#================================================================
# Plot real and predicted outputs
#================================================================
plot_real_predicted(ytest, ypred)
#%%
#================================================================
# Forecasting with target transformation
#================================================================
from sklearn.preprocessing import QuantileTransformer
from sklearn.compose import TransformedTargetRegressor
transformer = QuantileTransformer(output_distribution = 'normal')
# Create MLP regression object
print("Training MLPRegressor...")
regr = MLPRegressor(activation = 'logistic')
regr = TransformedTargetRegressor(regressor = regr,
    transformer = transformer)
regr.fit(X_train, y_train)
y_pred = regr.predict(X_test)
#================================================================
# Evaluate the model and print performance metrics
#================================================================
print_performance_metrics(y_test,y_pred)
#%%
#================================================================
# Plot scatter
#================================================================
```

```python
plot_scatter(y_test, y_pred)
#%%
#================================================================
# Plot real and predicted outputs
#================================================================
plot_real_predicted(y_test, y_pred)
#%%
#================================================================
# k-NN regression example
#================================================================
from sklearn.neighbors import KNeighborsRegressor
#================================================================
# Forecasting without target transformation
#================================================================
# Create a Regression object
print("Training Regressor...")
regr = KNeighborsRegressor(n_neighbors = 2)
# Train the model using the training sets
regr.fit(Xtrain, ytrain)
# Make predictions using the testing set
ypred = regr.predict(Xtest)
#================================================================
# Evaluate the model and print performance metrics
#================================================================
print_performance_metrics(ytest, ypred)
#%%
#================================================================
# Plot scatter
#================================================================
plot_scatter(ytest, ypred)
#%%
#================================================================
# Plot real and predicted outputs
#================================================================
plot_real_predicted(ytest, ypred)

#%%
#================================================================
# Forecasting with target transformation
#================================================================
from sklearn.preprocessing import QuantileTransformer
from sklearn.compose import TransformedTargetRegressor
transformer = QuantileTransformer(output_distribution = 'normal')
# Create a Regression object
print("Training Regressor...")
regr = KNeighborsRegressor(n_neighbors = 2)
regr = TransformedTargetRegressor(regressor = regr,
    transformer = transformer)
regr.fit(X_train, y_train)
y_pred = regr.predict(X_test)
```

```python
#===============================================================
# Evaluate the model and print performance metrics
#===============================================================
print_performance_metrics(y_test,y_pred)
#%%
#===============================================================
# Plot scatter
#===============================================================
plot_scatter(y_test, y_pred)
#%%
#===============================================================
# Plot real and predicted outputs
#===============================================================
plot_real_predicted(y_test, y_pred)
#%%
#===============================================================
# Random forest regressor example
#===============================================================
from sklearn.ensemble.forest import RandomForestRegressor
#===============================================================
# Forecasting without target transformation
#===============================================================
# Create Random Forest Regressor Model
regr = RandomForestRegressor(n_estimators = 100)
# Train the model using the training sets
regr.fit(Xtrain, ytrain)
# Make predictions using the testing set
ypred = regr.predict(Xtest)
#===============================================================
# Evaluate the model and print performance metrics
#===============================================================
print_performance_metrics(ytest, ypred)
#%%
#===============================================================
# Plot scatter
#===============================================================
plot_scatter(ytest, ypred)
#%%
#===============================================================
# Plot real and predicted outputs
#===============================================================
plot_real_predicted(ytest, ypred)
#%%
#===============================================================
# Forecasting with target transformation
#===============================================================
from sklearn.compose import TransformedTargetRegressor
from sklearn.preprocessing import QuantileTransformer
transformer = QuantileTransformer(output_distribution = 'normal')
regr = RandomForestRegressor(n_estimators = 100)
```

```python
regr = TransformedTargetRegressor(regressor = regr,
    transformer = transformer)
regr.fit(X_train, y_train)
y_pred = regr.predict(X_test)
#===================================================================
# Evaluate the model and print performance metrics
#===================================================================
print_performance_metrics(y_test,y_pred)
#%%
#===================================================================
# Plot scatter
#===================================================================
plot_scatter(y_test, y_pred)
#%%
#===================================================================
# Plot real and predicted outputs
#===================================================================
plot_real_predicted(y_test, y_pred)
#%%
#===================================================================
# SVR regression example
#===================================================================
from sklearn.svm import SVR
#===================================================================
# Forecasting without target transformation
#===================================================================
#Create Regression object
regr = SVR(gamma = 'scale', C = 200.0, epsilon = 0.2)
# Train the model using the training sets
# Train the model using the training sets
regr.fit(Xtrain, ytrain)
# Make predictions using the testing set
ypred = regr.predict(Xtest)
#===================================================================
# Evaluate the model and print performance metrics
#===================================================================
print_performance_metrics(ytest, ypred)
#%%
#===================================================================
# Plot scatter
#===================================================================
plot_scatter(ytest, ypred)
#%%
#===================================================================
# Plot real and predicted outputs
#===================================================================
plot_real_predicted(ytest, ypred)
#%%
#===================================================================
# Forecasting with target transformation
```

```python
#==================================================================
from sklearn.compose import TransformedTargetRegressor
from sklearn.preprocessing import QuantileTransformer
transformer = QuantileTransformer(output_distribution = 'normal')
regr = TransformedTargetRegressor(regressor = regr,
    transformer = transformer)
regr.fit(X_train, y_train)
y_pred = regr.predict(X_test)
#==================================================================
# Evaluate the model and print performance metrics
#==================================================================
print_performance_metrics(y_test,y_pred)
#%%
#==================================================================
# Plot scatter
#==================================================================
plot_scatter(y_test,y_pred)
#%%
#==================================================================
# Plot real and predicted outputs
#==================================================================
plot_real_predicted(y_test,y_pred)

#%%
#==================================================================
# Gradient boosting regression example
#==================================================================
from sklearn.ensemble import GradientBoostingRegressor
#==================================================================
# Create a regressor model
#==================================================================
regr = GradientBoostingRegressor(n_estimators = 100, learning_rate = 0.1,
    max_depth = 1, random_state = 0, loss = 'ls').fit(X_train, y_train)
#==================================================================
# Forecasting without target transformation
#==================================================================
# Train the model using the training sets
regr.fit(Xtrain, ytrain)
# Make predictions using the testing set
ypred = regr.predict(Xtest)
#==================================================================
# Evaluate the model and print performance metrics
#==================================================================
print_performance_metrics(ytest, ypred)
#%%
#==================================================================
# Plot scatter
#==================================================================
plot_scatter(ytest, ypred)
#%%
```

```python
#================================================================
# Plot real and predicted outputs
#================================================================
plot_real_predicted(ytest, ypred)
#%%
#================================================================
# Forecasting with target transformation
#================================================================
from sklearn.compose import TransformedTargetRegressor
from sklearn.preprocessing import QuantileTransformer
transformer = QuantileTransformer(output_distribution = 'normal')
regr = GradientBoostingRegressor(n_estimators = 100, learning_rate = 0.1,
    max_depth = 1, random_state = 0, loss = 'ls').fit(X_train, y_train)
regr = TransformedTargetRegressor(regressor = regr,
        transformer = transformer)
X_train, X_test, y_train, y_test = train_test_split(X, y, random_
state = 0)
regr.fit(X_train, y_train)
y_pred = regr.predict(X_test)
#================================================================
# Evaluate the model and print performance metrics
#================================================================
print_performance_metrics(y_test,y_pred)
#%%
#================================================================
# Plot scatter
#================================================================
plot_scatter(y_test,y_pred)
#%%
#================================================================
# Plot real and predicted outputs
#================================================================
plot_real_predicted(y_test,y_pred)
```

6.7 房价预测

　　二级抵押贷款市场的不断增长，需要使用更加复杂的分析方法、计量经济学方法和机器学习方法对抵押贷款信用风险进行定价评估。无论是抵押贷款的定价和核销，还是一般的住房市场研究，都需要进行房价预测。由于难以收集到及时的、一致的区域经济数据，单变量的时间序列方法会优于结构性的住房市场模型，尤其是对于短期预测。时间序列模型被广泛应用于金融建模中。近年来，对于不断变化的经济条件，不对称性和非线性的作用受到越来越多的关注。直观地讲，状态转换范式对于房地产的应用是非常有意义的，因为房地产市场历来容易受到繁荣和衰退的影响。该模型的基本思想是，在不同的经济条件下，住房市场表现不同，导致房价指数的特征在时间序列中发生明显的变化。例如，在经济衰退时期，对住房市场的冲击可能会更加持久，或者房价波动可能会加大，而在经济繁荣时期则可能出现相反的情况。由于区域经济和人口结构的差异，各种制度的特征可能因地理区域的不同而不同（Crawford & Fratantoni, 2003）。例如，美国各地的实际收入增长率和增长水平会有很大差异。这种差异性也应该表现在房价上面。无论是可观测的具体冲击，如利率和油价的上涨，或者是不可观测的因素，如科技的变革，都会对不同地区的房价造成不同的影响（Holly, Pesaran, & Yamagata, 2010）。

近年来，欧元区住宅房价走势呈现出明显的波动性。欧元区房价在这十年的中期阶段一直稳步增长，只是近年来急剧下降。房价上涨以及随后的下跌，显示了实际房价与其供求关系的基本决定因素之间的时间差异。最近的这些发展似乎符合长期的趋势特征，即大范围常规住房需求和供应基本面长期演化带来的房价短暂波动。再加上收入等需求决定因素的波动性较低，供给反应较慢，许多被广泛用于计算房价均衡值的指标都标志着房价相对于某些基本面有不同程度的高估或低估。因此，建立良好的房价预测模型，比如使用机器学习方法，是非常有必要的。具体而言，该方法被用于接收特定冲击的脉冲响应函数——临时住房需求的冲击，长期融资成本、经济技术和住房工艺的冲击，以及基于这些冲击解释的方差分解和长短期贡献。机器学习方法不仅对于朴素时间序列模型具有较好的预测性能，而且这种协整关系有助于预测近期房价的发展，其准确率高于没有长期均衡条件的模型（Gattini & Hiebert, 2010）。

例 6.7 以下 Python 代码用于加利福尼亚房价预测，其中采用了不同的回归模型，使用了独立的训练数据集和测试数据集。本例对统计质量进行了度量，包括计算平均绝对误差（MAE）、均方误差（MSE）和相关系数（R），同时绘制了散点图、实际值以及预测值。

数据集信息：本数据集由 1990 年美国加州各街区的人口普查收集得到。其中包括 20 640 个观测实例，每个实例包括 9 个属性，如下所示（Pace & Barry, 1997）：

经度：衡量房屋靠西的距离，值越高越靠西

纬度：衡量房屋靠北的距离，值越高越靠北

房屋年限中位数：街区内房屋年限的中位数，数值越低说明建筑物越新

房间总数：街区内的房间总数

卧室总数：街区内的卧室总数

该地区人口：街区内居住的总人数

住户数：住户总数，指街区内以家庭为单位的居住人群

收入中位数：街区内住户家庭的收入中位数（以万美元计算）

房价中位数：街区内住户家庭的房价中位数（以美元计算）

```
#================================================================
# House prices prediction using California housing data set
#================================================================
import matplotlib.pyplot as plt
import numpy as np
from sklearn.neighbors import KNeighborsRegressor
from sklearn.datasets.california_housing import fetch_california_housing
```

```python
from sklearn.metrics import mean_squared_error, r2_score
from sklearn.metrics import mean_absolute_error
from sklearn.model_selection import train_test_split
from sklearn.metrics import mean_squared_error, r2_score
from sklearn.metrics import mean_absolute_error
import matplotlib.pyplot as plt
#===================================================================
# Define utility functions
#===================================================================
def print_performance_metrics(ytest, ypred):
    # The mean absolute error
    print("MAE = %5.3f" % mean_absolute_error(ytest, ypred))
    # Explained variance score: 1 is perfect prediction
    print("R^2 = %0.5f" % r2_score(ytest, ypred))
    # The mean squared error
    print("MSE = %5.3f" % mean_squared_error(ytest, ypred))
def plot_scatter(ytest, ypred):
    fig, ax = plt.subplots()
    ax.scatter(ytest, ypred, edgecolors = (0, 0, 0))
    ax.text(ypred.max()-4.5, ytest.max()-0.1, r'$R^2$ = %.2f, MAE = %.2f' % (
            r2_score(ytest, ypred), mean_absolute_error(ytest, ypred)))
    ax.plot([ytest.min(), ytest.max()], [ytest.min(), ytest.max()], 'k--',
    lw = 4)
    ax.set_xlabel('Measured')
    ax.set_ylabel('Predicted')
    plt.show()

def plot_real_predicted(ytest, ypred):
    plt.plot(ytest[1:200], color = 'red', label = 'Real data')
    plt.plot(ypred[1:200], color = 'blue', label = 'Predicted data')
    plt.title('Prediction')
    plt.legend()
    plt.show()
# ===================================================================
# Load the California housing dataset
# ===================================================================
cal_housing = fetch_california_housing()
X, y = cal_housing.data, cal_housing.target
names = cal_housing.feature_names
Xtrain, Xtest, ytrain, ytest = train_test_split(X, y, test_size = 0.3,
random_state = 0)

#%%
#===================================================================
# Linear regression example
#===================================================================
import matplotlib.pyplot as plt
from sklearn import linear_model
# Create linear regression object
regr = linear_model.LinearRegression()
```

```python
# Train the model using the training sets
regr.fit(Xtrain, ytrain)
# Make predictions using the testing set
ypred = regr.predict(Xtest)
#================================================================
# Evaluate the model and print performance metrics
#================================================================
print_performance_metrics(ytest, ypred)
#%%
#================================================================
# Plot scatter
#================================================================
plot_scatter(ytest, ypred)
#%%
#================================================================
# Plot real and predicted outputs
#================================================================
plot_real_predicted(ytest, ypred)
#%%
from sklearn.neural_network import MLPRegressor
#================================================================
# ANN regression example
#================================================================
# Create MLP regression object
print("Training MLPRegressor...")
regr = MLPRegressor(activation = 'logistic')
# Train the model using the training sets
regr.fit(Xtrain, ytrain)
# Make predictions using the testing set
ypred = regr.predict(Xtest)
#================================================================
# Evaluate the model and print performance metrics
#================================================================
print_performance_metrics(ytest, ypred)
#%%
#================================================================
# Plot scatter
#================================================================
plot_scatter(ytest, ypred)
#%%
#================================================================
# Plot real and predicted outputs
#================================================================
plot_real_predicted(ytest, ypred)

#%%
#================================================================
# k-NN regression example
#================================================================
from sklearn.neighbors import KNeighborsRegressor
```

```python
# Create a Regression object
print("Training Regressor...")
regr = KNeighborsRegressor(n_neighbors = 2)
# Train the model using the training sets
regr.fit(Xtrain, ytrain)
# Make predictions using the testing set
ypred = regr.predict(Xtest)
#================================================================
# Evaluate the model and print performance metrics
#================================================================
print_performance_metrics(ytest, ypred)
#%%
#================================================================
# Plot scatter
#================================================================
plot_scatter(ytest, ypred)
#%%
#================================================================
# Plot real and predicted outputs
#================================================================
plot_real_predicted(ytest, ypred)

#%%
#================================================================
# Random forest regressor example
#================================================================
from sklearn.ensemble.forest import RandomForestRegressor
# Create Random Forest Regressor Model
regr = RandomForestRegressor(n_estimators = 100)
# Train the model using the training sets
regr.fit(Xtrain, ytrain)
# Make predictions using the testing set
ypred = regr.predict(Xtest)
#================================================================
# Evaluate the model and print performance metrics
#================================================================
print_performance_metrics(ytest, ypred)
#%%
#================================================================
# Plot scatter
#================================================================
plot_scatter(ytest, ypred)
#%%
#================================================================
# Plot real and predicted outputs
#================================================================
plot_real_predicted(ytest, ypred)
#%%
#================================================================
# SVR regression example
```

```python
#===================================================================
from sklearn.svm import SVR
#Create Regression object
regr = SVR(gamma = 'scale', C = 50.0, epsilon = 0.2)
# Train the model using the training sets
regr.fit(Xtrain, ytrain)
# Make predictions using the testing set
ypred = regr.predict(Xtest)
#===================================================================
# Evaluate the model and print performance metrics
#===================================================================
print_performance_metrics(ytest, ypred)
#%%
#===================================================================
# Plot scatter
#===================================================================
plot_scatter(ytest, ypred)
#%%
#===================================================================
# Plot real and predicted outputs
#===================================================================
plot_real_predicted(ytest, ypred)
#%%
#===================================================================
# Gradient boosting regressor example
#===================================================================
from sklearn.ensemble import GradientBoostingRegressor
#===================================================================
# Create a regressor model
#===================================================================
regr = GradientBoostingRegressor(n_estimators = 100, learning_rate = 0.1,
    max_depth = 1, random_state = 0, loss = 'ls')
# Train the model using the training sets
regr.fit(Xtrain, ytrain)
# Make predictions using the testing set
ypred = regr.predict(Xtest)
#===================================================================
# Evaluate the model and print performance metrics
#===================================================================
print_performance_metrics(ytest, ypred)
#%%
#===================================================================
# Plot scatter
#===================================================================
plot_scatter(ytest, ypred)
#%%
#===================================================================
# Plot real and predicted outputs
#===================================================================
plot_real_predicted(ytest, ypred)
```

```
#%%
#================================================================
# AdaBoost regressor example
#================================================================
from sklearn.ensemble import AdaBoostRegressor
#================================================================
# Create a regressor model
#================================================================
regr = AdaBoostRegressor(random_state = 0, n_estimators = 100)
# Train the model using the training sets
regr.fit(Xtrain, ytrain)
# Train the model using the training sets
regr.fit(Xtrain, ytrain)
# Make predictions using the testing set
ypred = regr.predict(Xtest)
#================================================================
# Evaluate the model and print performance metrics
#================================================================
print_performance_metrics(ytest, ypred)
#%%
#================================================================
# Plot scatter
#================================================================
plot_scatter(ytest, ypred)
#%%
#================================================================
# Plot real and predicted outputs
#================================================================
plot_real_predicted(ytest, ypred)
```

例 6.8 以下 Python 代码用于波士顿房价预测，其中采用了不同的回归模型，使用了独立的训练数据集和测试数据集。本例对统计质量进行了度量，包括计算平均绝对误差（MAE）、均方误差（MSE）和相关系数（R），同时绘制了散点图、实际值以及预测值。目标转换也用于观测目标变量的变换效果，从本例中可以看出，目标转换对某些回归模型有效，但对其他的无效。

数据集信息：本数据集包括美国人口普查局收集的有关马萨诸塞州波士顿地区的住房信息。该数据集来自 StatLib 归档（http://lib.stat.cmu.edu/datasets/boston），被广泛用于文献中测试基准算法。该数据集最初由 Harrison 和 Rubinfeld 发表（Harrison & Rubinfeld, 1978），其中包含 506 个实例，每个实例包括 14 个属性。属性信息如下：

CRIM：城镇人均犯罪率

ZN：住宅用地超过 25 000 平方英尺的比例

INDUS：城镇非零售商用土地的比例

CHAS：查尔斯河虚拟变量（如果边界是河流，则为1；否则为0）

NOX：一氧化氮浓度（千万分之一）

RM：住宅平均房间数

AGE：1940 年之前建成的自用房屋比例

DIS：到波士顿五个中心区域的加权距离

RAD：辐射式公路的可达性指数

TAX：每 10 000 美元的全值财产税率

PTRATIO：城镇师生比例

B：1000（Bk-0.63）^2，其中 Bk 指代城镇中黑人的比例

LSTAT：人口中地位低下者的比例

MEDV：自住房房价的中位数，以千美元计

```
"""
Created on Fri Aug 2 01:23:04 2019
@author: asubasi
"""
#================================================================
# House prices prediction example with Boston house prices dataset
#================================================================
import matplotlib.pyplot as plt
import numpy as np
from sklearn.metrics import mean_squared_error, r2_score
from sklearn.metrics import mean_absolute_error
from sklearn.model_selection import train_test_split
#================================================================
# Define utility functions
#================================================================
def print_performance_metrics(ytest, ypred):
    # The mean absolute error
    print("MAE = %5.3f" % mean_absolute_error(ytest, ypred))
    # Explained variance score: 1 is perfect prediction
    print("R^2 = %0.5f" % r2_score(ytest, ypred))
    # The mean squared error
    print("MSE = %5.3f" % mean_squared_error(ytest, ypred))

def plot_scatter(ytest, ypred):
    fig, ax = plt.subplots()
    ax.scatter(ytest, ypred, edgecolors = (0, 0, 0))
    ax.text(ypred.max()-4.5, ytest.max()-0.1, r'$R^2$ = %.2f, MAE = %.2f' % (
        r2_score(ytest, ypred), mean_absolute_error(ytest, ypred)))
    ax.plot([ytest.min(), ytest.max()], [ytest.min(), ytest.max()], 'k--',
        lw = 4)
    ax.set_xlabel('Measured')
    ax.set_ylabel('Predicted')
    plt.show()
```

```python
def plot_real_predicted(ytest, ypred):
    plt.plot(ytest[1:200], color = 'red', label = 'Real data')
    plt.plot(ypred[1:200], color = 'blue', label = 'Predicted data')
    plt.title('Prediction')
    plt.legend()
    plt.show()
#================================================================
# Load the Boston house prices dataset
#================================================================
from sklearn.datasets import load_boston
boston = load_boston()
X = boston.data
y = boston.target
Xtrain, Xtest, ytrain, ytest = train_test_split(X, y, test_size = 0.33,
    random_state = 0)
#%%
#================================================================
# Linear regression example
#================================================================
import matplotlib.pyplot as plt
from sklearn import linear_model
#================================================================
# Forecasting without target transformation
#================================================================
# Create linear regression object
regr = linear_model.LinearRegression()
# Train the model using the training sets
regr.fit(Xtrain, ytrain)
# Make predictions using the testing set
ypred = regr.predict(Xtest)
#================================================================
# Evaluate the model and print performance metrics
#================================================================
print_performance_metrics(ytest, ypred)
#%%
#================================================================
# Plot scatter
#================================================================
plot_scatter(ytest, ypred)
#%%
#================================================================
# Plot real and predicted outputs
#================================================================
plot_real_predicted(ytest, ypred)
#%%
#================================================================
# Forecasting with target transformation
#================================================================
from sklearn.preprocessing import QuantileTransformer
from sklearn.compose import TransformedTargetRegressor
```

```python
transformer = QuantileTransformer(output_distribution = 'normal')
# Create linear regression object
regr = linear_model.LinearRegression()
regr = TransformedTargetRegressor(regressor = regr,
    transformer = transformer)
X_train, X_test, y_train, y_test = train_test_split(X, y, test_size = 0.33, random_state = 0)
regr.fit(X_train, y_train)
y_pred = regr.predict(X_test)
#================================================================
# Evaluate the model and print performance metrics
#================================================================
print_performance_metrics(y_test,y_pred)
#%%
#================================================================
# Plot scatter
#================================================================
plot_scatter(y_test, y_pred)
#%%
#================================================================
# Plot real and predicted outputs
#================================================================
plot_real_predicted(y_test, y_pred)

#%%
#================================================================
# ANN regression example
#================================================================
from sklearn.neural_network import MLPRegressor
#================================================================
# Forecasting without target transformation
#================================================================
# Create MLP regression object
print("Training MLPRegressor...")
regr = MLPRegressor(activation = 'logistic')
# Train the model using the training sets
regr.fit(Xtrain, ytrain)
# Make predictions using the testing set
ypred = regr.predict(Xtest)
#================================================================
# Evaluate the model and print performance metrics
#================================================================
print_performance_metrics(ytest, ypred)
#%%
#================================================================
# Plot scatter
#================================================================
plot_scatter(ytest, ypred)
#%%
#================================================================
```

```python
# Plot real and predicted outputs
#================================================================
plot_real_predicted(ytest, ypred)

#%%
#================================================================
# Forecasting with target transformation
#================================================================
from sklearn.preprocessing import QuantileTransformer
from sklearn.compose import TransformedTargetRegressor
transformer = QuantileTransformer(output_distribution = 'normal')
# Create MLP regression object
print("Training MLPRegressor...")
regr = MLPRegressor(activation = 'logistic')
regr = TransformedTargetRegressor(regressor = regr,
    transformer = transformer)
regr.fit(X_train, y_train)
y_pred = regr.predict(X_test)
#================================================================
# Evaluate the model and print performance metrics
#================================================================
print_performance_metrics(y_test,y_pred)
#%%
#================================================================
# Plot scatter
#================================================================
plot_scatter(y_test, y_pred)
#%%
#================================================================
# Plot real and predicted outputs
#================================================================
plot_real_predicted(y_test, y_pred)
#%%
#================================================================
# k-NN regression example
#================================================================
from sklearn.neighbors import KNeighborsRegressor
#================================================================
# Forecasting without target transformation
#================================================================
# Create a Regression object
print("Training Regressor...")
regr = KNeighborsRegressor(n_neighbors = 2)
# Train the model using the training sets
regr.fit(Xtrain, ytrain)
# Make predictions using the testing set
ypred = regr.predict(Xtest)
#================================================================
# Evaluate the model and print performance metrics
#================================================================
```

```python
print_performance_metrics(ytest, ypred)
#%%
#===============================================================
# Plot scatter
#===============================================================
plot_scatter(ytest, ypred)
#%%
#===============================================================
# Plot real and predicted outputs
#===============================================================
plot_real_predicted(ytest, ypred)

#%%
#===============================================================
# Forecasting with target transformation
#===============================================================
from sklearn.preprocessing import QuantileTransformer
from sklearn.compose import TransformedTargetRegressor
transformer = QuantileTransformer(output_distribution = 'normal')
# Create a Regression object
print("Training Regressor...")
regr = KNeighborsRegressor(n_neighbors = 2)
regr = TransformedTargetRegressor(regressor = regr,
      transformer = transformer)
regr.fit(X_train, y_train)
y_pred = regr.predict(X_test)
#===============================================================
# Evaluate the model and print performance metrics
#===============================================================
print_performance_metrics(y_test, y_pred)
#%%
#===============================================================
# Plot scatter
#===============================================================
plot_scatter(y_test, y_pred)
#%%
#===============================================================
# Plot real and predicted outputs
#===============================================================
plot_real_predicted(y_test, y_pred)
#%%
#===============================================================
# Random forest regressor example
#===============================================================
from sklearn.ensemble.forest import RandomForestRegressor
#===============================================================
# Forecasting without target transformation
#===============================================================
# Create random forest regressor model
regr = RandomForestRegressor(n_estimators = 100)
```

```python
# Train the model using the training sets
regr.fit(Xtrain, ytrain)
# Make predictions using the testing set
ypred = regr.predict(Xtest)
#=================================================================
# Evaluate the model and print performance metrics
#=================================================================
print_performance_metrics(ytest, ypred)
#%%
#=================================================================
# Plot scatter
#=================================================================
plot_scatter(ytest, ypred)
#%%
#=================================================================
# Plot real and predicted outputs
#=================================================================
plot_real_predicted(ytest, ypred)
#%%
#=================================================================
# Forecasting with target transformation
#=================================================================
from sklearn.compose import TransformedTargetRegressor
from sklearn.preprocessing import QuantileTransformer
transformer = QuantileTransformer(output_distribution='normal')
regr = RandomForestRegressor(n_estimators = 100)
regr = TransformedTargetRegressor(regressor = regr,
      transformer = transformer)
regr.fit(X_train, y_train)
y_pred = regr.predict(X_test)
#=================================================================
# Evaluate the model and print performance metrics
#=================================================================
print_performance_metrics(y_test, y_pred)
#%%
#=================================================================
# Plot scatter
#=================================================================
plot_scatter(y_test, y_pred)
#%%
#=================================================================
# Plot real and predicted outputs
#=================================================================
plot_real_predicted(y_test, y_pred)
#%%
#=================================================================
# SVR regression example
#=================================================================
from sklearn.svm import SVR
#=================================================================
```

```python
# Forecasting without target transformation
#=====================================================================
#Create Regression object
regr = SVR(gamma = 'scale', C = 200.0, epsilon = 0.2)
# Train the model using the training sets
# Train the model using the training sets
regr.fit(Xtrain, ytrain)
# Make predictions using the testing set
ypred = regr.predict(Xtest)
#=====================================================================
# Evaluate the model and print performance metrics
#=====================================================================
print_performance_metrics(ytest, ypred)
#%%
#=====================================================================
# Plot scatter
#=====================================================================
plot_scatter(ytest, ypred)
#%%
#=====================================================================
# Plot real and predicted outputs
#=====================================================================
plot_real_predicted(ytest, ypred)
#%%
#=====================================================================
# Forecasting with target transformation
#=====================================================================
from sklearn.compose import TransformedTargetRegressor
from sklearn.preprocessing import QuantileTransformer
transformer = QuantileTransformer(output_distribution = 'normal')
regr = TransformedTargetRegressor(regressor = regr,
    transformer = transformer)
regr.fit(X_train, y_train)
y_pred = regr.predict(X_test)
#=====================================================================
# Evaluate the model and print performance metrics
#=====================================================================
print_performance_metrics(y_test, y_pred)
#%%
#=====================================================================
# Plot scatter
#=====================================================================
plot_scatter(y_test, y_pred)
#%%
#=====================================================================
# Plot real and predicted outputs
#=====================================================================
plot_real_predicted(y_test, y_pred)

#%%
```

```python
#================================================================
# Gradient boosting regression example
#================================================================
from sklearn.ensemble import GradientBoostingRegressor
#================================================================
# Create a regressor model
#================================================================
regr = GradientBoostingRegressor(n_estimators=100, learning_rate=0.1,
    max_depth=1, random_state=0, loss='ls').fit(X_train, y_train)
#================================================================
# Forecasting without target transformation
#================================================================
# Train the model using the training sets
regr.fit(Xtrain, ytrain)
# Make predictions using the testing set
ypred = regr.predict(Xtest)
#================================================================
# Evaluate the model and print performance metrics
#================================================================
print_performance_metrics(ytest, ypred)
#%%
#================================================================
# Plot scatter
#================================================================
plot_scatter(ytest, ypred)
#%%
#================================================================
# Plot real and predicted outputs
#================================================================
plot_real_predicted(ytest, ypred)
#%%
#================================================================
# Forecasting with target transformation
#================================================================
from sklearn.compose import TransformedTargetRegressor
from sklearn.preprocessing import QuantileTransformer
transformer = QuantileTransformer(output_distribution='normal')
regr = GradientBoostingRegressor(n_estimators=100, learning_rate=0.1,
    max_depth=1, random_state=0, loss='ls').fit(X_train, y_train)
regr = TransformedTargetRegressor(regressor=regr,
        transformer=transformer)
X_train, X_test, y_train, y_test = train_test_split(X, y, random_state=0)
regr.fit(X_train, y_train)
y_pred = regr.predict(X_test)
#================================================================
# Evaluate the model and print performance metrics
#================================================================
print_performance_metrics(y_test, y_pred)
#%%
#================================================================
```

```
# Plot scatter
#===============================================================
plot_scatter(y_test,y_pred)
#%%
#===============================================================
# Plot real and predicted outputs
#===============================================================
plot_real_predicted(y_test,y_pred)
```

6.8 单车使用情况预测

随着越来越多的人生活在城市中，人口的增长导致污染、噪音、拥堵和温室气体排放的增加。使用单车共享系统（BSS）是解决这些问题的一个可行方法。在许多城市中，BBS 是城市交通的一个重要部分，而且是可持续的、对环境友好的。随着城市密度及其相关问题的增加，未来可能会有更多的 BBS 系统出现。BBS 的资金和运营成本相对较低，安装方便，能够为无法长途跋涉或在复杂地形上行走的人提供脚踏支持，以及更好地追踪单车（Ashqar et al., 2017; DeMaio, 2009）。单车共享系统已经在众多城市部署，用于促进绿色出行和健康的生活方式。优化此类系统效用的关键点是将单车站放在最能满足骑行者出行需求的地方。一般来说，城市规划者需要依据专门的调查来了解当地的单车出行需求，这在时间和工作上都是相当昂贵的，尤其是当他们需要对比多个地点时。近年来，越来越多的城市发起了共享单车的倡议，鼓励环境的可持续发展，倡导健康的生活方式（Pucher, Dill, & Handy, 2010）。这些系统让人们能够在社区内的自助服务站取放公共单车，进行短途出行。鉴于支持共享单车计划需要大量的基础设施投资，如规划停车设施、使道路更适合单车，因此，最大化共享单车的价值对城市规划来说非常重要。鼓励市民参与单车共享系统的关键因素之一是将自行车站放在最能满足潜在用户出行需求的位置（Chen et al., 2015; García-Palomares, Gutiérrez, & Latorre, 2012）。

机器学习方法有助于构建一个精确的预测模型，用于估计两地之间的单车出行需求。因此，即使其中一个或两个地点目前没有设置单车站点，也可以利用机器学习模型进行预测，将其作为规划工具，确定如何扩展社区的单车共享网络。通过使用机器学习方法对始发 – 目的地之间的出行需求进行预测，不仅可以估计如果新增一个单车租借点将实现多少进 / 出站需求，而且还能提供进 / 出站需求的始发地和目的地，用于预测新增站点对现有网络的影响（Divya, Somya, & Peter, 2015）。

例 6.9 以下 Python 代码用于预测单车使用情况，采用了随机森林回归模型，使用了独立的训练数据集和测试数据集。本例对统计质量进行了度量，包括计算平均绝对误差（MAE）、均方误差（MSE）和相关系数（R），同时绘制了散点图、实际值以及预测值。

数据集信息：该单车共享数据集被 Fanaee-T 和 Gama 所采用（Fanaee-T & Gama,

2014），可以从 UCI 机器学习库下载（https://archive.ics.uci.edu/ml/datasets/bike+sharing+dataset）。该数据集包括两个独立的信息集合：hour.csv 和 day.csv，它们都包含以下字段，除了 hr 字段在 day.csv 中不可用。

- instant：记录索引
- dteday：日期
- season：季节（1：冬季、2：春季、3：夏季、4：秋季）
- yr：年（0：2011，1：2012）
- mnth：月（1 到 12）
- hr：小时（0 到 23）
- holiday：是否是节假日（提取自 [Web Link]）
- weekday：星期几
- workingday：如果不是节假日，值为 1；否则，值为 0
- weathersit：
- 1：晴、少云、部分多云
- 2：雾 + 多云、雾 + 碎云、雾 + 少云、雾
- 3：小雪、小雨 + 雷暴 + 散云、小雨 + 散云
- 4：大雨 + 冰雹 + 雷暴 + 雾、雪 + 雾
- temp：归一化温度（摄氏度），数值由以下公式计算得出：（t-t_min）/（t_max-t_min），t_min = −8, t_max = +39（仅以小时为单位）
- atemp：归一化体感温度（摄氏度），数值由以下公式计算得出：（t-t_min）/（t_max-t_min），t_min = −16, t_max = +50（仅以小时为单位）
- hum：归一化湿度，数值除以 100（最大值）
- windspeed：归一化风速，数值除以 67（最大值）
- casual：临时用户数
- registered：注册用户数
- cnt：租用单车的总数（包括临时用户数和注册用户数）

```
"""
Created on Fri Aug 2 01:23:04 2019
@author: asubasi
"""
#================================================================
# Bike-sharing example
#================================================================
import pandas as pd
import numpy as np
```

```python
import matplotlib.pyplot as plt
import zipfile
import requests, io, os
import warnings
from sklearn.model_selection import train_test_split
from sklearn.ensemble import RandomForestRegressor
from sklearn.metrics import mean_squared_error, r2_score, mean_absolute_error
from scipy import stats
#=====================================================================
# Define utility functions
#=====================================================================
def print_performance_metrics(ytest, ypred):
    # The mean absolute error
    print("MAE = %5.3f" % mean_absolute_error(ytest, ypred))
    # Explained variance score: 1 is perfect prediction
    print("R^2 = %0.5f" % r2_score(ytest, ypred))
    # The mean squared error
    print("MSE = %5.3f" % mean_squared_error(ytest, ypred))

def plot_scatter(ytest, ypred):
    fig, ax = plt.subplots()
    ax.scatter(ytest, ypred, edgecolors = (0, 0, 0))
    ax.text(ypred.max()-4.5, ytest.max()-0.1, r'$R^2$ = %.2f, MAE = %.2f' % (
        r2_score(ytest, ypred), mean_absolute_error(ytest, ypred)))
    ax.plot([ytest.min(), ytest.max()], [ytest.min(), ytest.max()], 'k--',
        lw = 4)
    ax.set_xlabel('Measured')
    ax.set_ylabel('Predicted')
    plt.show()
def plot_real_predicted(ytest, ypred):
    plt.plot(ytest[1:200], color = 'red', label = 'Real data')
    plt.plot(ypred[1:200], color = 'blue', label = 'Predicted data')
    plt.title('Prediction')
    plt.legend()
    plt.show()

#Unzip downloaded data
def unzip_from_UCI(UCI_url, dest = ''):
    #Downloads and unpacks datasets from UCI in zip format
    response = requests.get(UCI_url)
    compressed_file = io.BytesIO(response.content)
    z = zipfile.ZipFile(compressed_file)
    print ('Extracting in %s' % os.getcwd() + '\\' + dest)
    for name in z.namelist():
    if '.csv' in name:
    print ('\tunzipping %s' %name)
    z.extract(name, path = os.getcwd() + '\\' + dest)
    # data cleaning i.e. removing outliers
    def remove_outliers(data, type):
```

```python
        print("Shape of {} Data frame before removing Outliers: {}".
        format(type, data.shape))
        no_outliers = data[(np.abs(stats.zscore(data)) < 3).all(axis = 1)]
        print("Shape of {} Data frame before removing Outliers: {}".
        format(type, no_outliers.shape))
        return no_outliers

    #%%
    #Ignore Warnings
    warnings.filterwarnings("ignore")
    #Download data from UCI website
    UCI_url = 'https://archive.ics.uci.edu/ml/machine-learning-databas-
    es/00275/Bike-Sharing-Dataset.zip'
    unzip_from_UCI(UCI_url, dest = 'bikesharing')

    #Read and drop missing values if any
    hourly_data = pd.read_csv("bikesharing/hour.csv", na_values = '?').
    dropna()
    daily_data = pd.read_csv("bikesharing/day.csv", na_values = '?').drop-
    na()
    #%%
    #change date to int
    list_dh = []
    for i in hourly_data['dteday']:
       list1 = i.split('-')
  list_dh.append(int(list1[2]))
dfh = pd.DataFrame(list_dh, columns = ['dteday'])
hourly_data[['dteday']] = dfh[['dteday']]

list_dd = []
for i in daily_data['dteday']:
  list2 = i.split('-')
  list_dd.append(int(list2[2]))
dfd = pd.DataFrame(list_dd, columns = ['dteday'])
daily_data[['dteday']] = dfd[['dteday']]

no_outliers = remove_outliers(hourly_data,'Hourly')
y_hour = no_outliers.cnt
x_hour = no_outliers.drop(['cnt','instant','registered','casual'],ax
is = 1)

no_outliers = remove_outliers(daily_data,'Daily')
y_day = no_outliers.cnt
x_day = no_outliers.drop(['cnt','instant','registered','casual'],ax
is = 1)

#%%
# choosing alpha as 0.8 where coefficients of holiday, atemp and
windspeed are becoming zero.
x_hour = x_hour.drop(['holiday','atemp','windspeed'],axis = 1)
```

```
# choosing alpha as 0.3 where coefficients of holiday is becoming zero.
x_day = x_day.drop(['holiday','dteday','mnth'],axis = 1)

#Splitting data
X_hour_train, X_hour_test, Y_hour_train, Y_hour_test = train_test_
split(x_hour, y_hour, test_size = 0.3, random_state = 5)
X_day_train, X_day_test, Y_day_train, Y_day_test = train_test_split(x_
day, y_day, test_size = 0.3, random_state = 5)

#Random Forest Regressor
rgr = RandomForestRegressor(n_estimators = 200, n_jobs = -1, min_
samples_split = 4)
print('\nRandom Forest')

#%%
#General Model training and Testing for Hourly Data
print("\n\t Hour Dataset")
rgr.fit(X_hour_train, Y_hour_train)
predictions = rgr.predict(X_hour_test)
print_performance_metrics(Y_hour_test, predictions)
# Plot outputs
plot_scatter(Y_hour_test, predictions)
plot_real_predicted(np.array(Y_hour_test), predictions)
#%%
#General Model training and Testing for Daily Data
print("\n\t Day Dataset")
rgr.fit(X_day_train, Y_day_train)
predictions = rgr.predict(X_day_test)
print_performance_metrics(Y_day_test, predictions)
# Plot outputs
plot_scatter(Y_day_test, predictions)
plot_real_predicted(np.array(Y_day_test), predictions)
```

6.9 本章小结

在本章中，我们介绍了许多与回归学习问题相关的例子，通过使用已知数据，尽可能准确地预测或预测正确、真实地估计未知数据标签。回归是常见的机器学习问题，有着不同的应用场景，本章旨在对其进行解释和研究。回归是确定分组和分级模型之间差异的过程。分组模型理论是将实例空间划分为若干段，并在每段中尽可能简单地学习局部模型。在前面的章节中，学习任务主要集中在分类问题上。而在本章中，介绍了将不同的机器学习算法和深度学习算法用于回归，包括有限和无限的假设集。同时，详细讨论了这些算法在不同领域的应用，包括股票市场价格指数收益预测、通货膨胀预测、电力负荷预测、风速预测、旅游需求预测、房价预测和单车使用情况预测。

6.10 参考文献

Ashqar, H. I., Elhenawy, M., Almannaa, M. H., Ghanem, A., Rakha, H. A., & House, L. (2017). In *Modeling bike availability in a bike-sharing system using machine learning* (pp. 374–378). IEEE.

Bozkurt, Ö. Ö., Biricik, G., & Tayşi, Z. C. (2017). Artificial neural network and SARIMA based models for power load forecasting in Turkish electricity market. PloS One, 12(4), e0175915.

Cankurt, S., & Subasi, A. (2016). Tourism demand modelling and forecasting using data mining techniques in multivariate time series: A case study in Turkey. Turkish Journal of Electrical Engineering & Computer Sciences, 24(5), 3388–3404.

Chen, L., Zhang, D., Pan, G., Ma, X., Yang, D., Kushlev, K., & Li, S. (2015). In *Bike sharing station placement leveraging heterogeneous urban open data* (pp. 571–575). ACM.

Crawford, G. W., & Fratantoni, M. C. (2003). Assessing the forecasting performance of regime-switching, ARIMA and GARCH models of house prices. Real Estate Economics, 31(2), 223–243.

DeMaio, P. (2009). Bike-sharing: History, impacts, models of provision, and future. Journal of Public Transportation, 12(4), 3.

Divya, S., Somya, S., & Peter, I. (2015). Predicting bike usage for New York City's bike sharing system. Association for the Advancement of Artificial Intelligence, 110–114.

Fanaee-T, H., & Gama, J. (2014). Event labeling combining ensemble detectors and background knowledge. Progress in Artificial Intelligence, 2(2-3), 113–127.

Flach, P. (2012). Machine learning: The art and science of algorithms that make sense of data. Cambridge, UK: Cambridge University Press.

García-Palomares, J. C., Gutiérrez, J., & Latorre, M. (2012). Optimizing the location of stations in bike-sharing programs: A GIS approach. Applied Geography, 35(1-2), 235–246.

Gasparin, A., Lukovic, S., & Alippi, C. (2019). Deep learning for time series forecasting: The electric load case. *ArXiv Preprint ArXiv:1907.09207*.

Gattini, L., & Hiebert, P. (2010). Forecasting and assessing Euro area house prices through the lens of key fundamentals.

Harrison, D., Jr., & Rubinfeld, D. L. (1978). Hedonic housing prices and the demand for clean air. Journal of Environmental Economics and Management, 5(1), 81–102.

Holly, S., Pesaran, M. H., & Yamagata, T. (2010). A spatio-temporal model of house prices in the USA. Journal of Econometrics, 158(1), 160–173.

Law, R., Li, G., Fong, D. K. C., & Han, X. (2019). Tourism demand forecasting: A deep learning approach. Annals of Tourism Research, 75, 410–423.

Lei, M., Shiyan, L., Chuanwen, J., Hongling, L., & Yan, Z. (2009). A review on the forecasting of wind speed and generated power. Renewable and Sustainable Energy Reviews, 13(4), 915–920.

Okumus, I., & Dinler, A. (2016). Current status of wind energy forecasting and a hybrid method for hourly predictions. Energy Conversion and Management, 123, 362–371.

Pace, R. K., & Barry, R. (1997). Sparse spatial autoregressions. Statistics & Probability Letters, 33(3), 291–297.

Pouyanfar, S., Sadiq, S., Yan, Y., Tian, H., Tao, Y., Reyes, M. P., & Iyengar, S. (2019). A survey on deep learning: Algorithms, techniques, and applications. ACM Computing Surveys (CSUR), 51(5), 92.

Pucher, J., Dill, J., & Handy, S. (2010). Infrastructure, programs, and policies to increase bicycling: An international review. Preventive Medicine, 50, S106–S125.

Sadaei, H. J., e Silva, P. C., de, L., Guimarães, F. G., & Lee, M. H. (2019). Short-term load forecasting by using a combined method of convolutional neural networks and fuzzy time series. Energy, 175, 365–377.

Şenyurt, G., & Subaşı, A. (2015). Effects of technical market indicators on stock market index direction forecasting. Modeling of Artificial Intelligence, 6(2), 137–149.

Shen, X., Zhou, C., Li, G., Fu, X., & Lie, T. (2018). Overview of wind parameters sensing methods and framework of a novel MCSPV recombination sensing method for wind turbines. Energies, 11(7), 1747.

Sideratos, G., & Hatziargyriou, N. D. (2007). An advanced statistical method for wind power forecasting. IEEE Transactions on Power Systems, 22(1), 258–265.

Sikiric, G., Avdakovic, S., & Subasi, A. (2013). Comparison of machine learning methods for electricity demand forecasting in Bosnia and Herzegovina. Southeast Europe Journal of Soft Computing, 2(2).

Song, H., Li, G., & Cao, Z. (2018). Tourism and economic globalization: An emerging research agenda. Journal of Travel Research, 57(8), 999–1011.

Song, H., Qiu, R. T., & Park, J. (2019). A review of research on tourism demand forecasting. Annals of Tourism Research, 75, 338–362.

Ülke, V., Sahin, A., & Subasi, A. (2018). A comparison of time series and machine learning models for inflation forecasting: Empirical evidence from the USA. Neural Computing and Applications, 30(5), 1519–1527.

Zhong, X., & Enke, D. (2017). A comprehensive cluster and classification mining procedure for daily stock market return forecasting. Neurocomputing, 267, 152–168.

CHAPTER 7

第 7 章

聚类示例

7.1 简介

聚类是最常用的实验数据分析方法之一。从社会科学到生物学,再到计算机科学的所有学科,人们通过在数据点之间定义有意义的类别,试图获得对其结果的初步理解。例如,零售商根据客户资料将客户聚类,以进行有针对性的营销;计算生物学家在不同的研究中根据基因表达的相似性对基因进行聚类;天文学家根据星星之间的接近情况将其聚类。当然,我们首先要回答的问题是:什么是聚类?聚类是对对象集合进行直观分组的过程,以使相同的对象最终属于同一类别,并将不相似的对象分为不同的组。虽然这个定义是不精确的,而且可能比较模糊。然而,要找到更准确的定义并不容易。有如下几个原因。首先,我们需要考虑的基本问题是,在许多情况下,前一陈述中所述的两个目标相互矛盾。从数学上讲,相似性(或接近性)不是传递关系,而聚类共享是等价关系,甚至是传递关系。更具体地说,存在一个序列对象 $x_1,\cdots,x_m$,其中 x_i 与其两个近邻 x_{i-1} 和 x_{i+1} 都很相似,但是 x_1 和 x_m 十分不同。若我们想要让两个相似元素共享同一个聚类,那么必须将所有序列元素都放在同一个聚类中。此时,我们将会因为 x_1 和 x_m 在同一聚类中而违反第二个条件。聚类算法强调不分离相邻的点,通过在两条线上水平划分来聚类该输入。另一方面,聚类方法强调距离远的点不能被放在同一聚类中,它通过垂直划分相同的输入来进行聚类操作(Shalev-Shwartz & Ben-David, 2014)。

聚类的另一个基本问题是没有"基本事实",这是无监督学习的一个常见问题。到目前为止,书中主要处理的是监督学习(例如,从标签数据中学习分类器的问题)。监督学习的目的很简单——训练一个分类器,以尽可能准确地预测未来样例的标签。另外,通过估计经验损失,有监督的学习器可以利用标记的训练数据来估计假设的成功或风险。而聚类是一种无监督学习问题。也就是说,我们并非试图预测任何标签,而是想要使用一些实用的方法来组织数据。因此，没有直接的聚类性能评估方法。此外，由于我们对数据的"正确"聚类是什么尚不清楚，甚至是在已经了解数据分布的情况下，也无法对于所建议的聚类方法进行评估（Shalev-Shwartz & Ben-David, 2014）。

7.2 聚类

聚类是一种将相关的对象绑定在一起的机制。通过不同的输入方式，有两种不同的聚类：基于相似性聚类和基于特征聚类。在基于相似性聚类的方法中，算法输入是不相似性矩阵或距离矩阵 D。基于特征聚类的方法的输入是一个特征矩阵，或是设计好的基于 X 个向量的 $N \times D$ 的矩阵。基于相似性聚类方法的优点是可以方便地包含领域特定的相似性或核函数。基于特征聚类的好处是它适用于"原始"数据，虽然该数据可能包含噪声。除了两种输入类型，还有两种潜在的输出类型：平面聚类（也称为分区聚类），将对象划分为不相交的集合；层次聚类，形成嵌套的分区树（Murphy, 2012）。

在不相似性矩阵 D 中，每个元素代表距离，其中 $d(i, i) = 0$ 且 $d(i, j) \geq 0$。从严格意义上讲，主观确定的度量不相似性的指标很少是距离，因为在 D 中三角不等式 $d(i, j) \leq d(i, k) + d(j, k)$ 往往不成立。有些算法声称 D 是一个真正的距离矩阵，但其他算法则不然。如果我们有一个相似性矩阵 S，通过应用任何单调递减的函数，例如，$D = \max(S) - S$，可以将其转换为衡量不相似的矩阵。描述对象相似性的最常见方式是用它们的属性相似性来描述。平方（欧几里得）距离、城市街区距离、相关系数和汉明距离是一些常见的衡量不相似性的函数。

k 均值聚类算法选择 k 个初始点代表初始聚类中心，然后将所有数据点分配给离其最接近的一个中心，计算每个聚类中点的平均值以形成其当前聚类核心，这个过程将会不断重复，直到没有聚类更改为止。仅当事先知道聚类的个数时，k 均值聚类才能有效。本节将会从不知道聚类个数的数据开始，介绍相应的方法。首先，我们研究"聚集"以构建分层聚类结构的策略，即从单个实例开始，然后将它们依次合并为聚类。因此，我们就好像是在看一个渐进工作的系统，即每一个新的数据实例都在出现时立刻被处理。最后，我们将研究一种基于混合模型的聚类统计方法，使用不同的概率分布，每个聚类一个分布。它不会像 k 均值一样将实例分成不相交的聚类，而是根据概率将实例分配给相应的类（Witten, Frank, Hall, & Pal, 2016）。

聚类是人类最基本的心理实践之一，我们使用它来处理每天获得的海量信息。人们通常将实体聚类（即对象、个体和事件）后再进行信息处理，因为处理单个实体的信息十分困难。每个聚类能够对其所包含的实体的特定属性进行特征化。在监督学习的情况下，我们必须假定所有模式都是根据形成一维特征向量的特征来定义的。通常，建立聚类要采取的基本步骤如下：

- 特征选择：我们应该选择适合的特征用以在值函数中最大化保留信息。再次，特征选择的主要目标是简约，因此需要最小化知识之间的重复。如同在监督分类中一样，

使用特征之前可能需要对特征进行预处理
- 近邻测度：该测度定义两个特征向量之间的相似性。在预处理过程中，需要确保所有选定的特征均对近邻测度做出相同的贡献，并且保证没有任何特征会主导其他特征
- 聚类标准：该标准可能依赖于领域专家根据数据集的聚类形式对该项的定义。聚类的标准可以通过代价函数或其他一些规则来表达
- 聚类算法：该阶段是指根据近邻测度和聚类标准的定义，选择一种特定的聚类算法，发现数据集中的聚类结构
- 结果验证：一旦获得了聚类结果，我们需要验证其正确性。通常需要根据适当的评估指标进行验证
- 结果解释：在很多情况下，为了能够获得正确的结论，应用专家需要结合聚类结果与其他的实验证据和解释

在很多情况下，一个被称为"聚类趋势"的阶段也需要考虑在内。该阶段主要包括使用各种不同的测试方法来确定数据中是否存在聚类模式。例如，若数据分布完全随机，那么在这种数据上使用聚类方法是没有意义的。不同的特征选择、近邻测度、聚类标准以及聚类算法都会导致不同的聚类结果（Theodoridis, Pikrakis, Koutroumbas, & Cavouras, 2010）。

7.2.1 评估聚类输出

聚类分析中最难也最令人沮丧的部分是验证聚类结构的正确性。如果没有花费正确的精力在聚类结果认证上，聚类分析仍旧是只被那些有丰富经验的专家和信徒所使用的方法。聚类是一种无监督学习技术，很难对任何给定技术的输出质量进行评估。如果我们使用概率模型，我们可以评估测试集的似然性，然而这种方法有两个不足：首先，该方法并非对算法所找到的聚类给出直接评价；其次，该方法无法应用于非概率方法。所以现在，我们讨论一些基于非概率方法的评估指标。理论上，聚类的目的在于将相似的点分配给同一个聚类中，同时将不相似的数据点放在不同的聚类中。该特质可以用不同的方式衡量。但是，通过这种特质进行聚类评估的好处有限。一种替代方法是使用任意外部数据类型来验证系统。例如，如果对每个对象都设置标签，则可以使用不同的度量标准（例如轮廓线）将聚类与标签进行比较（Murphy, 2012）。

7.2.2 聚类分析的应用

在一些应用中，聚类是一个主要工具。聚类分析可以有效地应用于以下领域（Theodoridis et al., 2010）：

- 数据归约：在某些情况下，数据的数量 N 可能是一个很大的值，从而使得数据处理变得极富挑战性。为了能够将数据组织成几个"重要"的聚类并且将每个聚类视作单个实体，我们可以使用聚类分析。例如，在数据传输中指定每个聚类的代表。相

较于直接发送样本，我们发送与每个特定样本所在的聚类代表相对应的代码号，这样就完成了数据压缩
- 假设生成：在这种情况下，我们通常将聚类分析应用于数据集，以得出有关数据性质的一些假设。此处将聚类作为工具来得到假设。注意，使用其他数据集来检验这些假设同样很重要
- 假设检验：使用聚类分析对需要验证的假设进行判断
- 基于分组的预测：在这种情况下，将聚类分析应用于现有数据集，然后根据形成聚类模式的特征来识别后续聚类。在后续步骤中，当出现一个模棱两可的模式时，我们可以评估它更可能属于的聚类，并根据各个聚类类别对其进行定义

7.2.3 可能的聚类数

不同的近邻测度对于相似和不相似有着不同的定义，从而会使得聚类过程中产生不同的聚类。正如之前所说，专家可以根据不同的近邻测度以及聚类模式解释不同的聚类结果。将一个集合 X 的特征向量 x_i, $i=1,2,..N$，指定为聚类的最好方法是描述所有可能的分区，并根据先前选择的标准选择最合理的分区。但是即使是对一个相对不是很大的 N 值，这也是极具挑战性的任务。

7.2.4 聚类算法种类

聚类算法可视为通过仅考虑包含所有可能的 X 分区集合中的一小部分来提供敏感聚类的方案。聚类的结果取决于所使用的算法和标准。因此，聚类算法是一个学习过程，它试图识别数据集中聚类的特定特征。聚类算法主要分为以下类别：
- 顺序算法：该算法创建一个单一聚类。这类算法十分直接且高效。在大多算法中，所有特征向量都会被赋给算法一次或几次。通常，最终结果取决于向量被赋予算法的顺序。根据所使用的近邻测度，这些技术倾向于生成紧凑的、超球形或超椭圆形的聚类。
- 层次聚类算法：这些算法可以被归为两类。
  - 凝聚算法，此算法在每个阶段都会减少聚类数量。新的聚类都是通过将前一阶段的聚类合并所得。单连接和全连接算法是凝聚算法的关键代表。这样的算法非常适合大型聚类和紧凑型聚类的恢复
  - 分裂算法，此算法的工作方向相反。也就是说，它们在每个阶段都生成一个 m 的聚类序列。新的聚类是通过对前一阶段的聚类进行划分生成的
- 基于代价函数优化的聚类算法：该类算法包含了一个代价函数 J，用以量化聚类的敏感度。通常，聚类的数量保持不变。这些算法中的大多数在尝试优化 J 时都会使用微分原理。当确定 J 的局部最优值时结束。同样，这类算法也称为迭代函数优化技术，包括以下子类：

- 硬聚类算法，应用于一个向量只属于一个聚类的情况。将向量分配给各个聚类是在公认的最佳性标准的基础上进行的。Isodata 或 Lloyd 算法是该组中最受欢迎的算法
- 概率分类算法，是一种特殊的硬聚类算法。它采用贝叶斯分类方法，将每个向量 x 分配给聚类 C_i，使得 $P(C_i|x)$（即后验概率）最大化。这个概率可以通过已经定义好的优化过程进行计算
- 模糊聚类算法，应用于向量只是某种程度上属于一个聚类的情况
- 概率聚类算法，应用于当一个向量 x 只部分属于一个聚类 C_i 的情况
- 边界检测算法，是迭代地更新聚类所在区域的边界，而不是通过特征向量来识别聚类本身。尽管这些算法是从成本函数优化理论发展而来的，但它们与之前描述的算法不同（Theodoridis et al., 2010）

除了这些聚类算法之外，分支定界聚类算法、遗传聚类算法、随机松弛方法、谷类寻求聚类算法、竞争性学习算法、基于形态转换技术的算法、基于密度的算法、子空间聚类算法和基于核的算法方法也是聚类算法的类型（Theodoridis et al., 2010）。

7.3 k 均值聚类算法

k 均值聚类算法从描述一组可能聚类的参数化的代价函数开始，并且聚类算法的目标是找到最小代价划分（聚类）。在该模型下，聚类变成了优化问题。目标函数是根据输入对 (X, d) 和建议的聚类解 $C=(C_1, ..., C_k)$ 得到正实数的函数。聚类算法的目的可以被描述为替给定的输入 (X, d) 寻找一个聚类 C，使得目标函数 $G((X, d), C)$ 最小化的过程。为了实现这个目标，必须使用合适的搜索算法。因此，k 均值聚类是一种特定的通用近似算法，而不是代价函数或最小化问题的相应精确解。大多数的目标函数参数包含聚类的个数 k。实际上，通常是由使用聚类算法的人选择适合该聚类问题的 k 值。我们随后会给出一些常见的目标函数定义。k 均值目标函数是常见的目标函数之一。目标函数 k 均值测量的是从 X 中每个点到聚类的质心的平方距离。例如，在数字通信任务中，X 个成员可以解释为一组要传输的信号。k 均值目标函数很重要。在实际的聚类应用中，k 均值的目标函数十分常见。然而，找到 k 均值的最优解在计算上是不可行的。因此，通常使用简单的迭代算法进行计算。所以，在多数情况下，术语"k 均值聚类"指的是该算法的结果，而不是使 k 均值的目标代价最小化的聚类（Shalev-Shwartz & Ben-David, 2014）。

> **例 7.1** 以下 Python 代码使用 scikit-learn 库的 API 来进行 k 均值聚类来寻找乳腺癌数据的聚类中心。在下面的例子中，我们使用了 sklearn.datasets 中的乳腺癌数据集。将聚类中心画在散点图上，用以表现聚类算法的有效性。该例改编自 scikit-learn。

```python
# ================================================================
# K-means clustering example
# ================================================================
from sklearn import datasets
import matplotlib.pyplot as plt
import numpy as np
from sklearn.cluster import KMeans
#%%
# ####################################################################
# Import some data to play with
Breast_Cancer = datasets.load_breast_cancer()
X = Breast_Cancer.data
y = Breast_Cancer.target

# Plot the original data points
plt.scatter(X[:, 0], X[:, 1], c = y, cmap = plt.cm.Set1,
        edgecolor = 'k')
plt.xlabel('Attribute I')
plt.ylabel('Attribute II')
plt.title('Original data Scatter')
plt.xticks(())
plt.yticks(())
#%%
"""

sklearn.cluster.KMeans(n_clusters = 8, init = 'k-means + +', n_
init = 10, max_iter = 300, tol = 0.0001,
precompute_distances = 'auto', verbose = 0, random_state = None,
copy_x = True, n_jobs = None,
algorithm = 'auto')
"""
#Find Cluster Centers
kmeans = KMeans(n_clusters = 2)
kmeans.fit(X)
y_kmeans = kmeans.predict(X)

#Plot the Cluster Centers
plt.scatter(X[:, 0], X[:, 1], c = y_kmeans, s = 50, cmap = 'viridis')
centers = kmeans.cluster_centers_
plt.scatter(centers[:, 0], centers[:, 1], c = 'black', s = 200,
alpha = 0.5);
plt.xlabel('Attribute I')
plt.ylabel('Attribute II')
plt.title('Cluster Centers')
plt.xticks(())
plt.yticks(())
```

7.4 k 中心点聚类算法

在 k 均值算法中,每个聚类由其向量的均值表示,但是在 k 中心点聚类方法中,该聚类由从 X 元素中选择的向量表示,我们将其称为中心点。除了中心点之外,每个聚类还包括所有 X 中的向量,没有被其他聚类选作中心点,并且这些向量相较于其他聚类的中心点而言,更靠近自己聚类的中心点。相比于 k 均值聚类算法,使用中心点表示聚类有两个优势。首先,它可用于源自连续域或离散域的数据集,而 k 均值仅适用于连续域,因为数据向量子集的均值本质上并不是离散域上下文中的点。第二,k 中心点算法对离群值的敏感性低于 k 均值算法。但是,应该记住,聚类的均值具有很强的几何和统计意义,而对于类中心点则不一定如此。并且,相较于 k 均值算法,计算聚类最好的中心点需要更多的计算力。PAM(围绕中心点划分)、CLARA(大型应用中的聚类方法)和 CLARANS(基于随机搜索的大规模应用聚类)都是著名的基于 k 中心点的算法。最后两个算法是基于 PAM 算法的,但是比 PAM 算法能更好地处理大型数据集(Theodoridis et al., 2010)。

例 7.2 以下 Python 代码在模拟数据集和 Mall_Customer(https://www.kaggle.com/akram24/mall-customers)的数据集上利用 KMedoids 聚类函数,使用 k 中心点算法进行聚类。我们将聚类中心画在散点图上,用以表现聚类算法的有效性。

```
# ==================================================================
# K-medoids clustering example
# ==================================================================
from k_medoids import KMedoids
import numpy as np
import matplotlib.pyplot as plt
#Define a distance utility function
def example_distance_func(data1, data2):
    "'example distance function"'
    return np.sqrt(np.sum((data1 - data2)**2))
#%%
# K-Medoids Clustering using synthetic data with 3 clusters
from sklearn.datasets import make_blobs
# ##################################################################
# Generate sample data
np.random.seed(0)
batch_size = 45
centers = [[1, 1], [-1, -1], [1, -1]]
n_clusters = len(centers)
X, labels_true = make_blobs(n_samples = 300, centers = centers, cluster_std = 0.7)

model = KMedoids(n_clusters = n_clusters, dist_func = example_distance_func)
model.fit(X, plotit = True, verbose = True)
```

```
plt.show()
#%%
# K-Medoids clustering using Mall_Customers data
import pandas as pd
#loading the dataset
# Importing the Mall_Customers dataset by pandas
dataset = pd.read_csv('Mall_Customers.csv')
X = dataset.iloc[:, [2,3,4]].values
model = KMedoids(n_clusters = 5, dist_func = example_distance_func)
model.fit(X, plotit = True, verbose = True)
plt.show()
```

7.5 层次聚类

层次聚类算法具有不同的原理。特别是，它们生成聚类的层次结构，而不是生成单个聚类。这种算法通常应用于在社会科学和生物分类学中。层次聚类算法可以构建嵌套聚类的层次结构。更具体地说，这些算法包括了 N 步，与数据向量的数量相同。在 t 步时，这些算法根据前 $t-1$ 步生成聚类的基础上进行聚类（Theodoridis et al., 2010）。层次结构聚类中存在两种主要方法：自底向上（聚集）和自顶向下（分散）。这些方法将对象之间的相异矩阵作为输入。在每个步骤中，将最相似的组以自下而上的方法组合。自顶向下的方法使用不同的标准将组划分。请记住，聚集聚类和分散聚类都只是启发式方法，不会优化任何定义明确的目标函数。因此，从任何形式上讲，很难评估它们创建聚类的质量。实际上，即使数据没有任何结构（如随机噪声），它们也可以根据输入数据生成聚类。

7.5.1 聚集聚类算法

聚集聚类从 N 个组开始，每个组最初包含一个实体，然后两个最相似的组在每个阶段合并，直到有一个包含所有数据的组。当 N 的值比较大的时候，一种典型启发式方法是先运行 k 均值算法，然后将分层聚类应用于已经估计的聚类中心。该合并过程可以通过被称为树状图的二叉树表示。初始组（对象）在叶子上（在图的底部），并且每当两个组合并时，我们就将它们加入树中。划分的高度是当前对象与要加入的组之间的差异。树根（位于顶部）是一个包含所有数据的类。如果在任何给定的高度切割树，我们都会得到给定大小的聚类。此外，根据定义对象类别之间的差异，聚集聚类有三种变体（Murphy, 2012）。

换言之可以假设，如果第 t 层的两个向量在单个聚类中组合在一起，则对于所有后续聚类，它们将保留在同一聚类中。这是查看嵌套属性的另一种方法。嵌套属性的缺点是无法从"较差"的聚类中恢复，这可能是在较早的层上出现的。阈值树状图，或简称为树状图，是描述聚集聚类算法生成的聚类序列的有效方法。一般聚集方案（GAS）的每个阶段都与树状图阶段有关。切割树状图可能会产生特定级别的聚类。邻近树状图是考虑了两个聚类首次合并的相似度的树状图。一旦确定了衡量相似度（非相似）的指标，距离树状图又被称为相似（非相似）树状图。该方法可在任何阶段表示当前聚类是自然或强迫生成的。类似地，为了得到合适的层级，必须计算出一个合适的切割层级。

例 7.3 以下 Python 代码使用了 Mall_Customers 数据（https://www.kaggle.com/akram24/mall-customers）以及标准 scikit-learn 方法库，应用了聚集聚类的方法，将客户分成了 Careful、Standard、Target、Careless 和 Sensible 类。基于欧氏距离，我们绘制了客户树状图。另外，我们还将五个不同的聚类使用散点图表现出来。该例源自于网页（https://www.kdnuggets.com/2019/09/hierarchical-clustering.html）。

```
# =====================================================================
# Agglomerative clustering example
# =====================================================================
import numpy as np
import matplotlib.pyplot as plt
import pandas as pd
dataset = pd.read_csv('Mall_Customers.csv')
#%%
" " "Out of all the features, CustomerID and Genre are irrelevant fields
and can be dropped and create a matrix of independent variables by select
only Age and Annual Income." " "
X = dataset.iloc[:, [3, 4]].values
import scipy.cluster.hierarchy as sch
dendrogrm = sch.dendrogram(sch.linkage(X, method = 'ward'))
plt.title('Dendrogram')
plt.xlabel('Customers')
plt.ylabel('Euclidean distance')
plt.show()

#%%
from sklearn.cluster import AgglomerativeClustering
hc = AgglomerativeClustering(n_clusters = 5, affinity = 'euclidean', linkage = 'ward')
y_hc = hc.fit_predict(X)
# Visualising the clusters
plt.scatter(X[y_hc == 0, 0], X[y_hc == 0, 1], s = 50, c = 'red', label = 'Careful')
plt.scatter(X[y_hc == 1, 0], X[y_hc == 1, 1], s = 50, c = 'blue', label = 'Standard')
plt.scatter(X[y_hc == 2, 0], X[y_hc == 2, 1], s = 50, c = 'green', label = 'Target')
plt.scatter(X[y_hc == 3, 0], X[y_hc == 3, 1], s = 50, c = 'cyan', label = 'Careless')
plt.scatter(X[y_hc == 4, 0], X[y_hc == 4, 1], s = 50, c = 'magenta', label = 'Sensible')
```

```
plt.title('Clusters of customers')
plt.xlabel('Annual Income (k$)')
plt.ylabel('Spending Score (1-100)')
plt.legend()
plt.show()
```

例 7.4 以下 Python 代码使用了 Mall_Customers 数据（https://www.kaggle.com/akram24/mall-customers）以及标准 scikit-learn 方法库，应用了聚集聚类的方法。该例中使用了散点图来表现算法的有效性。在此示例中，我们展示了使用连接图以捕获客户数据中的局部结构的效果。通过连接图，我们可以看到实现连接的两个暗示。首先，使用连接矩阵可以更快地进行聚类。其次，使用连接矩阵时，单个、平均和完整的连接不稳定，并且往往会创建一些增长非常快的聚类。但是，平均和完全连接通过合并两个聚类之间的所有距离来解决此过滤行为（而仅将聚类之间的最短距离视为夸大行为）。连接图消除了平均连接和总连接的过程，使其看起来像更脆弱的单个连接。图中的邻居非常少，其几何形状类似于单个连接的几何形状，众所周知，这种连接具有渗透的不稳定性。在此示例中对此进行了介绍。请注意，此示例改编自 scikit-learn。

```
# Authors: Gael Varoquaux, Nelle Varoquaux
# License: BSD 3 clause
import time
import matplotlib.pyplot as plt
import numpy as np

from sklearn.cluster import AgglomerativeClustering
from sklearn.neighbors import kneighbors_graph
import pandas as pd
#2 Importing the Mall_Customers dataset by pandas
dataset = pd.read_csv('Mall_Customers.csv')
X = dataset.iloc[:, [3,4]].values

# Create a graph capturing local connectivity. Larger number of neighbors
# will give more homogeneous clusters to the cost of computation
# time. A very large number of neighbors gives more evenly distributed
# cluster sizes, but may not impose the local manifold structure of
# the data
knn_graph = kneighbors_graph(X, 30, include_self = False)

for connectivity in (None, knn_graph):
    for n_clusters in (4, 5, 6):
        plt.figure(figsize = (10, 4))
        for index, linkage in enumerate(('average',
        'complete',
        'ward',
        'single')):
```

```
        plt.subplot(1, 4, index + 1)
        model = AgglomerativeClustering(linkage = linkage,
        connectivity = connectivity,
        n_clusters = n_clusters)
        t0 = time.time()
        model.fit(X)
        elapsed_time = time.time() - t0
        plt.scatter(X[:, 0], X[:, 1], c = model.labels_,
        cmap = plt.cm.nipy_spectral)
        plt.title('linkage = %s\n(time %.2fs)' % (linkage, elapsed_time),
        fontdict = dict(verticalalignment = 'top'))
        plt.axis('equal')
        plt.axis('off')
        plt.subplots_adjust(bottom = 0, top = .89, wspace = 0,
        left = 0, right = 1)
        plt.suptitle('n_cluster = %i, connectivity = %r' %
        (n_clusters, connectivity is not None), size = 17)
plt.show()
```

7.5.2 分裂聚类算法

分裂算法采用与聚集方案相反的策略。第一个聚类 X 中包含了单一的集合。我们需要寻找在第一步中将 X 划分为两个聚类的最佳方法。直接的方法是考虑两个集合中所有可能的 X 分区，并根据预定标准选择能够最大值。然后，在得到的新的聚类中，我们重复这一过程。最后的聚类的结果包括 N 个聚类，每个聚类中有一个 X 向量。根据 N 的不同，产生不同的算法。可以很容易地观察到，即使对于一般的 N 值，分裂聚类算法对于计算要求很高。与聚集聚类相比，这是它的主要缺点。因此，如果想要将这些方案应用于实际中，则需要一些进一步的计算简化。一种选择是相较于寻找所有可能的类的划分，有针对性地进行搜索。这可以通过在不合理的预设标准下排除几个不合理的分区来实现。聚类划分基于先前算法中特征向量的所有特征（坐标）。这些类型的算法也称为多元算法。另一方面，在每个阶段都有分裂算法，可基于单个功能实现聚类划分。这些是被称为单一算法的算法（Theodoridis et al., 2010）。

分裂聚类算法从包含所有数据点的单个聚类开始，使用自顶向下的方法，将每个聚类分裂成更小的聚类。由于有 $2^{N-1}-1$ 种将 N 个项目划分成两组的方法，计算最优的划分十分困难，因此算法通常会采用不同的启发式算法。有一个方法是选择距离最大的聚类，然后使用 k 均值或是 k 中心点的方法划分聚类（k=2）。这种方法也称为二分 k 均值算法（Steinbach, Karypis, & Kumar, 2000）。我们可以重复此过程，直到获得所需数量的聚类为止。该方法可用作标准 k 均值的替代方法，但是也可以视作层次聚类。另一种策略是从不同的图构造最小生成树，然后通过断开与最大差异有关的连接来建立新的聚类。分裂聚类不如集聚聚类常见，但有两个好处。首先，它可以更快，因为如果我们中断恒定数量的层次，则只需要 O（N）时间。其次，分裂算法会根据所有结果做出拆分决策，而自下而上的方法做出的合并决策是根据局部情况做出的（Murphy, 2012）。

例 7.5 以下 Python 代码利用分裂聚类绘制人类基因氨基酸序列的树状图。在这个例子中，利用了人类基因的氨基酸序列。树状图显示了该算法的有效性。注意，该例改编自 github（https://github.com/ronak-07/Divisive-Hierarchical-Clustering）。系统发育树或进化树是分支图或"树"，基于其物理或遗传特征的相似性和差异，显示了不同生物物种之间的隐含进化关系。该示例的目标是基于数据库中使用分裂（自顶向下）层次聚类的物种的 DNA / 蛋白质序列构建系统树。

```
# ====================================================================
# Divisive clustering
# ====================================================================
import numpy as np
import scipy
import matplotlib.pyplot as plt
from scipy.cluster import hierarchy
global g
import time
# ====================================================================
# Define utility functions
# ====================================================================
def subtract(indices,splinter):
l3 = [x for x in indices if x not in splinter]
return l3

def divisive(a,indices,splinter,sub):
if(len(indices) = =1):
return
avg = []
flag = 0
for i in indices:
if(i not in splinter):
sum = 0
for j in indices:
if(j not in splinter):
sum = sum + a[i][j]
if((len(indices)-len(splinter)-1) = =0):
avg.append(sum)
else:
avg.append(sum/(len(indices)-len(splinter)-1))
if(splinter):
k = 0
for i in sub:
total = 0
for j in splinter:
total = total + a[i][j]
```

```
avg[k] = avg[k] - (total/(len(splinter)))
k + = 1
positive = []
for i in range(0,len(avg)):
if(avg[i] > 0):
positive.append(avg[i])
flag = 1
if(flag = =1):
splinter.append(sub[avg.index(max(positive))])
sub.remove(sub[avg.index(max(positive))])
divisive(a,indices,splinter,sub)
else:
splinter.append(indices[avg.index(max(avg))])
sub[:] = subtract(indices,splinter)
divisive(a,indices,splinter,sub)

def original_subset(indices):
sp = np.zeros(shape = (len(indices),len(indices)))
for i in range(0,len(indices)):
for j in range(0,len(indices)):
sp[i][j] = a[indices[i]][indices[j]]
return sp

def original_max(x):
new = original_subset(x)
return new.max()

def diameter(l):
return original_max(l)

def recursive(a,indices,u,v,clusters,g):
clus_s.append(len(indices))
d.append(diameter(indices))
parents[g] = indices
g- = 1
divisive(a,indices,u,v)
clusters.append(u)
clusters.append(v)
new = []
for i in range(len(clusters)):
new.append(clusters[i])
final.append(new)
x = []
y = []
store_list = []
max = -1
f = 0
for list in clusters:
if(diameter(list) > max):
if(len(list)! = 1):
```

```
f = 1
max = diameter(list)
store_list = (list)
if(f = =0):
return
else:
clusters.remove(store_list)
recursive(a,store_list,x,y,clusters,g)

def augmented_dendrogram(*args, **kwargs):
data = scipy.cluster.hierarchy.dendrogram(*args, **kwargs)
if not kwargs.get('no_plot', False):
for i, d in zip(data['icoord'], data['dcoord']):
x = 0.5 * sum(i[1:3])
y = d[1]
plt.plot(x, y, 'ro')
plt.annotate("%.3g" % y, (x, y), xytext = (0,12),textcoords = 'offset
points',va = 'top', ha = 'center')
return data

# =====================================================================
# Main program
# =====================================================================
a = np.load('distance_matrix.npy')
size = len(a)
g = (size-1)*2
parents = {}
final = []
clusters = []
indices = []
clus_s = []
d = []
Z = np.zeros(shape = (size-1,4))
p = []
q = []
ans = []
for i in range(0,len(a)):
indices.append(i)

for i in range(0,size):
list = []
list.append(i)
parents[i] = list

start = time.time()
recursive(a,indices,p,q,clusters,g)
print("Clustering done\t" + str(time.time()-start))
for i in range(0,len(d)):
Z[size-i-2][2] = d[i]
Z[size-i-2][3] = clus_s[i]
```

```
for i in range(len(final)-1,0,-1):
    for j in range(0,len(final[i-1])):
        if final[i-1][j] not in final[i]:

ans.append(final[i-1][j])
ans.append(indices)
for i in range(0,len(ans)):
    if(len(ans[i]) < = 2):
        Z[i][0] = ans[i][0]
        Z[i][1] = ans[i][1]
    else:
        s = 0
        add = []
        common = []
        for j in range(len(ans)-1,-1,-1):
            if(set(ans[j]) < set(ans[i])):
                common = ans[j]
                break;
        x = (subtract(ans[i],common))
        for key in parents.keys():
            if(parents[key] = =common):
                Z[i][0] = key
                break;
        for key in parents.keys():
            if(set(parents[key]) = =set(x)):
                Z[i][1] = key
                s = 1
                break;
        if(s = =0):
            print(Z[i][0],Z[i][1],x)
names = [i for i in range(0,size)]
#%%
# ======================================================================
# Plot dendrogram of divisive clustering
# ======================================================================
plt.figure(figsize = (15, 15))
plt.title('Hierarchical Clustering Dendrogram (Divisive)')
plt.xlabel('Sequence No.')
plt.ylabel('Distance')
augmented_dendrogram(Z,labels = names,show_leaf_
counts = True,p = 25,truncate_mode = 'lastp')
plt.show()
```

7.6 模糊 c 均值聚类算法

与概率算法相关的挑战之一是概率密度函数（PDF）的存在，为此必须假定合适的模型。但是，当聚类不是紧凑的而是壳形的时候，处理实例并不容易。模糊聚类算法是一系列可以摆脱这种约束的聚类算法。在过去三十年中，这些算法吸引了大量研究者。区分这两种方法的要点是，向量在模糊聚类算法中可以同时属于一个以上的聚类，而每个向量在概率算法中仅属于一个聚类。聚类的数量及其形状被假定是已知先验的。聚类形状由所采用的参数集定义。大多数众所周知的模糊聚类算法都是通过最小化代价函数来进行聚类的（Theodoridis et al., 2010）。

广泛研究和实现的模糊 c 均值（FCM）聚类算法需要聚类数量作为先验知识。如果 FCM 预测到了所需的聚类数量，并且有可能猜到每个聚类中心的位置，那么输出规则就强烈依赖于初始值的选择。FCM 算法生成适当的聚类模式，以通过迭代最小化基于聚类位置的目标函数。也可以通过山峰聚类（mountain clustering）过程中可用的搜索技术自动确定聚类中心的数量和初始位置。通过在每个网格点上测量称为山形函数的搜索度量，这种方法将每个不同的网格点视为可能的聚类核心。这是一种通过增加计算量进行聚类的减法方法，其中数据点被视为聚类中心而非网格点的候选者。通过应用此方法，需要估计的值与数据点的数量成正比，而与问题的维度无关。在此过程中，作为距离测量函数的高势能数据点被称为聚类中心，靠近新聚类中心的数据点将受到惩罚以监测新聚类中心的出现。有时，两个聚类的逐渐隶属关系可以视为聚类中心之间的点。当然，这可以通过转变"低"和"高"的含义来补偿。模糊 c 均值算法通过定义隶属度允许某个数据点在某种程度上属于一个聚类，因此每个数据点可以属于几个聚类。模糊 c 均值算法将一组识别为 m 维向量的 K 个数据点划分为 c 个模糊聚类，并在每个聚类中找到一个聚类中心以最小化目标函数。模糊 c 均值与非模糊 c 均值不同，主要是因为它使用模糊分区，即一个点可以隶属于具有隶属度的多个聚类。隶属矩阵 M 中每个元素值都可以在 [0，1] 之间，以满足模糊划分的需求。但是，要保持 M 矩阵的属性，一个点的所有聚类的总隶属度必须始终等于单位（Sumathi & Paneerselvam, 2010）。

例 7.6 以下 Python 代码利用模糊 c 均值聚类算法查找 Iris 数据集聚类的中心。在此示例中，使用了 sklearn.datasets 中存在的 Iris 数据集。该示例中使用了散点图来表现算法的有效性。聚类中心以散点图的形式绘制。请注意，此示例改编自网页（https://github.com/omadson/fuzzy-c-means）。你可以通过"pip install fuzzy-c-means"或从网页（https://pypi.org/project/fuzzy-c-means/）下载算法库来运行本例。

```
# ========================================================================
# Fuzzy c-means clustering example
# ========================================================================
#pip install fuzzy-c-means
from fcmeans import FCM
from sklearn.datasets import make_blobs
from matplotlib import pyplot as plt
from seaborn import scatterplot as scatter
import numpy as np
import pandas as pd
```

```
from sklearn import datasets
from sklearn.datasets import load_iris
iris = load_iris()
X = iris['data']
y = iris['target']
# Plot the original data points
plt.scatter(X[y == 0, 0], X[y == 0, 1], s = 80, c = 'orange',
label = 'Iris-setosa')
plt.scatter(X[y == 1, 0], X[y == 1, 1], s = 80, c = 'yellow',
label = 'Iris-versicolour')
plt.scatter(X[y == 2, 0], X[y == 2, 1], s = 80, c = 'green',
label = 'Iris-virginica')
plt.xlabel('Sepal length')
plt.ylabel('Sepal width')
plt.title('Original data points')
plt.legend()
#%%
# fit the fuzzy-c-means
fcm = FCM(n_clusters = 3)
fcm.fit(X)
# outputs
fcm_centers = fcm.centers
fcm_labels = fcm.u.argmax(axis = 1)
# plot result
f, axes = plt.subplots(1, 2, figsize = (11,5))
scatter(X[:,0], X[:,1], ax = axes[0])
scatter(X[:,0], X[:,1], ax = axes[1], hue = fcm_labels)
scatter(fcm_centers[:,0], fcm_centers[:,1],
ax = axes[1],marker = "s",s = 200)
plt.show()
```

7.7 基于密度的聚类算法

基于密度的聚类通常被称为在 X 点"密集"的一维空间区域。许多基于密度的算法对最终的聚类形状没有任何要求。因此，这类算法可以生成任意形状的聚类并且能够有效处理异常点。此外，这类算法有较小的时间复杂度，可以处理大型的数据集。DBSCAN、DBCLASD、DENCLUE 和 OPTICS 都是流行的基于密度的算法。虽然这些算法在解决聚类问题上有着类似的基本思路，但是它们对于密度有不同的定义（Theodoridis et al., 2010）。

7.7.1 DBSCAN 算法

DBSCAN（基于密度的带噪声的空间聚类应用）中的"密度"是基于点 x 计算的，即 X 中落在 x 点周围的某个一维空间区域内的个数。该算法的结果受到参数 ε 和 q 的影响极大，在不同的参数值下，算法可能生成完全不同的聚类结果。参数值的选择需要保证算法能够检测出"密度"最小的聚类。在实践中，为了能够确定最佳的参数值组合，需要对 ε

和 q 的值进行测试。DBSCAN 算法不适合密度区别较大的聚类有很大优势，也无法处理好高维度数据。OPTICS（基于点序识别聚类结构）是 DBSCAN 算法的扩展，它克服了需要仔细选择参数 ε 和 q 的缺点。它产生基于密度的聚类顺序，以易于理解的方式描述数据集的固有层次聚类结构。实验表明，OPTICS 的计算复杂度约为 DBSCAN 的 1.6 倍。但是，在实际应用中，为了找到合适的参数 ε 和 q 值，需要多次运行 DBSCAN（Theodoridis et al., 2010）。

例 7.7 以下 Python 代码利用 DBSCAN 聚类算法数据集聚类。在此示例中，使用了 sklearn 的 API。本例使用模拟数据集，并且使用散点图来表现算法的有效性。同时计算不同的聚类评估指标。注意，本例改编自 scikit-learn。

```python
# =====================================================================
# DBSCAN clustering
# =====================================================================
import numpy as np
from sklearn.cluster import DBSCAN
from sklearn import metrics
from sklearn.datasets import make_blobs
from sklearn.preprocessing import StandardScaler

# #####################################################################
# Generate sample data
centers = [[1, 1], [-1, -1], [1, -1]]
X, labels_true = make_blobs(n_samples = 750, centers = centers, cluster_std = 0.4,
    random_state = 0)
X = StandardScaler().fit_transform(X)
# #####################################################################
# Compute DBSCAN
db = DBSCAN(eps = 0.3, min_samples = 10).fit(X)
core_samples_mask = np.zeros_like(db.labels_, dtype = bool)
core_samples_mask[db.core_sample_indices_] = True
labels = db.labels_

# Number of clusters in labels, ignoring noise if present.
n_clusters_ = len(set(labels)) - (1 if -1 in labels else 0)
n_noise_ = list(labels).count(-1)

print('Estimated number of clusters: %d' % n_clusters_)
print('Estimated number of noise points: %d' % n_noise_)
print("Homogeneity: %0.3f" % metrics.homogeneity_score(labels_true, labels))
print("Completeness: %0.3f" % metrics.completeness_score(labels_true, labels))
print("V-measure: %0.3f" % metrics.v_measure_score(labels_true, labels))
print("Adjusted Rand Index: %0.3f"
```

```
        % metrics.adjusted_rand_score(labels_true, labels))
print("Adjusted Mutual Information: %0.3f"
        % metrics.adjusted_mutual_info_score(labels_true, labels))
print("Silhouette Coefficient: %0.3f"
        % metrics.silhouette_score(X, labels))
#%%
# #####################################################################
# Plot result
import matplotlib.pyplot as plt

# Black removed and is used for noise instead.
unique_labels = set(labels)
colors = [plt.cm.Spectral(each)
          for each in np.linspace(0, 1, len(unique_labels))]
for k, col in zip(unique_labels, colors):
if k == -1:
            # Black used for noise.
            col = [0, 0, 0, 1]

class_member_mask = (labels == k)

xy = X[class_member_mask & core_samples_mask]
plt.plot(xy[:, 0], xy[:, 1], 'o', markerfacecolor = tuple(col),
markeredgecolor = 'k', markersize = 14)
xy = X[class_member_mask & ~core_samples_mask]
plt.plot(xy[:, 0], xy[:, 1], 'o', markerfacecolor = tuple(col),
markeredgecolor = 'k', markersize = 6)
plt.title('Estimated number of clusters: %d' % n_clusters_)
plt.show()
```

7.7.2 OPTICS 聚类算法

我们可以对 DBSCAN 算法进行扩展，使其能够同时处理多个距离参数，即针对不同的密度同时构建基于密度的聚类。但是，我们需要遵循特定的处理对象顺序，同时扩展聚类以实现一致的结果。我们总是必须选择一个相对于最低 ε 值可以达到密度的对象，以确保相对于较高的密度（即较小的 ε 值）首先完成聚类。从理论上讲，OPTICS 算法可作为扩展的 DBSCAN 算法，用于无数个小于"生成距离"ε 的距离参数。唯一的区别是，我们不将这些单位分配到聚类中，而是将对象处理的顺序存储下来，以便扩展的 DBSCAN 算法可以用此信息来将对象分配到合适的聚类。OPTICS 算法生成数据库顺序，额外存储了核心距离或每个对象以及对象的可达距离。我们将会看到，这些知识足以消除基于密度任意小于 ε 值生成的聚类。由于与 DBSCAN 算法概念相似，OPTICS 算法的运行时间与 DBSCAN 类似。实际上，OPTICS 算法的运行时间几乎是 DBSCAN 的 1.6 倍。抽象上来说，数据集上的聚类顺序包含了数据集内在的聚类结构。

例7.8 以下 Python 代码利用 OPTICS 聚类算法对数据集聚类。在此示例中，使用了 scikit-learn 的 API。本例使用了模拟数据集，并且保证模拟数据集中的聚类密度有很大差异。sklearn.cluster.OPTICS 类首先使用了 Xi 聚类检测方法，然后在与类 sklearn.cluster.DBSCAN 相关的可达性上设置阈值。我们可以看到，通过 DBSCAN 中不同的阈值选择，可以恢复 OPTICS Xi 方法的不同聚类。本例给出了可达性图和散点图，以证明该算法的有效性。请注意，此示例改编自 scikit-learn。

```
# ======================================================================
# Optics clustering example
# ======================================================================
# Authors: Shane Grigsby <refuge@rocktalus.com>
# Adrin Jalali <adrin.jalali@gmail.com>
# License: BSD 3 clause
from sklearn.cluster import OPTICS, cluster_optics_dbscan
import matplotlib.gridspec as gridspec
import matplotlib.pyplot as plt
import numpy as np
import pandas as pd
# Generate synthetic sample data
np.random.seed(0)
n_points_per_cluster = 250
C1 = [-5, -2] + .8 * np.random.randn(n_points_per_cluster, 2)
C2 = [4, -1] + .1 * np.random.randn(n_points_per_cluster, 2)
C3 = [1, -2] + .2 * np.random.randn(n_points_per_cluster, 2)
C4 = [-2, 3] + .3 * np.random.randn(n_points_per_cluster, 2)
C5 = [3, -2] + 1.6 * np.random.randn(n_points_per_cluster, 2)
C6 = [5, 6] + 2 * np.random.randn(n_points_per_cluster, 2)
X = np.vstack((C1, C2, C3, C4, C5, C6))
#Use OPTICS for Clustering
clust = OPTICS(min_samples = 50, xi = .05, min_cluster_size = .05)

# Run the fit
clust.fit(X)

labels_050 = cluster_optics_dbscan(reachability = clust.reachability_,
    core_distances = clust.core_distances_,
    ordering = clust.ordering_, eps = 0.5)
labels_200 = cluster_optics_dbscan(reachability = clust.reachability_,
    core_distances = clust.core_distances_,
    ordering = clust.ordering_, eps = 2)

space = np.arange(len(X))
reachability = clust.reachability_[clust.ordering_]
labels = clust.labels_[clust.ordering_]

plt.figure(figsize = (15, 10))
G = gridspec.GridSpec(2, 3)
```

```python
ax1 = plt.subplot(G[0, :])
ax2 = plt.subplot(G[1, 0])
ax3 = plt.subplot(G[1, 1])
ax4 = plt.subplot(G[1, 2])

# ====================================================================
# Reachability plot
# ====================================================================
colors = ['g.', 'r.', 'b.', 'y.', 'c.']
for klass, color in zip(range(0, 5), colors):
Xk = space[labels == klass]
Rk = reachability[labels == klass]
ax1.plot(Xk, Rk, color, alpha = 0.3)
ax1.plot(space[labels == -1], reachability[labels == -1], 'k.', alpha = 0.3)
ax1.plot(space, np.full_like(space, 2., dtype = float), 'k-', alpha = 0.5)
ax1.plot(space, np.full_like(space, 0.5, dtype = float), 'k-.', alpha = 0.5)
ax1.set_ylabel('Reachability (epsilon distance)')
ax1.set_title('Reachability Plot')

# ====================================================================
# Plot OPTICS clustering results
# ====================================================================
colors = ['g.', 'r.', 'b.', 'y.', 'm.']
for klass, color in zip(range(0, 5), colors):
Xk = X[clust.labels_ == klass]
ax2.plot(Xk[:, 0], Xk[:, 1], color, alpha = 0.3)
ax2.plot(X[clust.labels_ == -1, 0], X[clust.labels_ == -1, 1], 'k + ',
alpha = 0.1)
ax2.set_title('Automatic Clustering\nOPTICS')

# ====================================================================
# Plot DBSCAN at 0.5 clustering results
# ====================================================================
colors = ['r', 'greenyellow', 'olive', 'g', 'b', 'c']
for klass, color in zip(range(0, 6), colors):
Xk = X[labels_050 == klass]
ax3.plot(Xk[:, 0], Xk[:, 1], color, alpha = 0.3, marker = '.')
ax3.plot(X[labels_050 == -1, 0], X[labels_050 == -1, 1], 'k + ',
alpha = 0.1)
ax3.set_title('Clustering at 0.5 epsilon cut\nDBSCAN')

# ====================================================================
# Plot DBSCAN at 2. clustering results
# ====================================================================
colors = ['r.', 'm.', 'y.', 'c.']
for klass, color in zip(range(0, 4), colors):
Xk = X[labels_200 == klass]
ax4.plot(Xk[:, 0], Xk[:, 1], color, alpha = 0.3)
ax4.plot(X[labels_200 == -1, 0], X[labels_200 == -1, 1], 'k + ',
alpha = 0.1)
```

```
ax4.set_title('Clustering at 2.0 epsilon cut\nDBSCAN')
plt.tight_layout()
plt.show()
```

7.8 基于期望最大化的混合高斯模型聚类算法

使用混合高斯模型的问题主要来自以下未知的两点：我们不知道训练实例来自何种分布，以及混合模型的五个参数的取值。因此，我们使用 k 均值聚类的技术进行迭代。这五个参数从估计的初始值开始，利用它们来衡量每个实例的聚类概率，并且根据这些概率来重新估计参数值。一直重复该估计过程。这一参数估计的方法称为 EM 算法，用来最大化期望。第一步，计算聚类概率，即"期望"的类，被称为"期望"。第二步，计算参数分布，用来根据当前数据"最大化"概率分布（Witten et al., 2016）。

现在我们看到了两种分布的高斯混合模型，让我们考虑如何将其应用于更具体的条件。只要预先给出正态分布的数量 k，就很容易将算法从两类问题扩展至多类问题。只要假设各个属性是独立的，就可以轻松地将模型从每个实例的单个数值属性扩展为多个属性。将每个属性的概率相乘以获得实例的联合概率。当我们已经了解数据集中的数据包含相关属性时，独立性假设不再成立。相反，二元正态分布可以同时对两个属性建模。虽然每个属性都有自己的平均值，但是两个标准差可以被具有四个参数的"协方差矩阵"代替。在已知实例及其类别概率时，我们可以使用标准的概率统计技术估计实例的类别概率、均值以及协方差矩阵。多元分布可以容纳多个相关属性。参数的数量随共同变化的属性数量的平方的增加而增加。期望，即根据分布参数计算每个实例所属的聚类，就像计算未知实例的类。最大化，即从分类实例中估计参数，就像评估训练实例中属性值的概率一样，唯一的小差别在于 EM 算法是根据概率分配实例，而非类别（Witten et al., 2016）。

例 7.9 以下 Python 代码利用高斯混合模型（GMM）聚类算法寻找聚类，使用了 scikit-learn 库的 API 进行实现。该例使用了 Iris 数据集。虽然 GMM 通常应用于聚类，我们可以将找到的聚类与实际的类别进行对比。根据 Iris 数据集上的各种 GMM 协方差类型，我们在训练和保留的测试数据集上绘制了预测标签。该例比较了 GMM 的球面、对角线、完全以及约束协方差，根据递增顺序，比较了聚类效果。尽管我们可以预期全协方差一般情况下获得最佳性能，但它容易在小型数据集上过拟合，并且不能很好地推广到保留的测试数据上。在图中，我们用点表示训练数据，用十字表示测试数据。尽管 Iris 数据集是四维的，但此处仅显示前两个维度。注意，此示例改编自 scikit-learn。

```python
# ================================================================
# GMM clustering
# ================================================================
# Author: Ron Weiss <ronweiss@gmail.com > , Gael Varoquaux
# Modified by Thierry Guillemot <thierry.guillemot.work@gmail.com>
# License: BSD 3 clause

import matplotlib as mpl
import matplotlib.pyplot as plt

import numpy as np

from sklearn import datasets
from sklearn.mixture import GaussianMixture
from sklearn.model_selection import StratifiedKFold

colors = ['navy', 'red', 'green']

def make_ellipses(gmm, ax):
    for n, color in enumerate(colors):
        if gmm.covariance_type == 'full':
            covariances = gmm.covariances_[n][:2, :2]
        elif gmm.covariance_type == 'tied':
            covariances = gmm.covariances_[:2, :2]
        elif gmm.covariance_type == 'diag':
            covariances = np.diag(gmm.covariances_[n][:2])
        elif gmm.covariance_type == 'spherical':
            covariances = np.eye(gmm.means_.shape[1]) * gmm.covariances_[n]
        v, w = np.linalg.eigh(covariances)
        u = w[0] / np.linalg.norm(w[0])
        angle = np.arctan2(u[1], u[0])
        angle = 180 * angle / np.pi # convert to degrees
        v = 2. * np.sqrt(2.) * np.sqrt(v)
        ell = mpl.patches.Ellipse(gmm.means_[n, :2], v[0], v[1],
                                   180 + angle, color = color)
        ell.set_clip_box(ax.bbox)
        ell.set_alpha(0.5)
        ax.add_artist(ell)
        ax.set_aspect('equal', 'datalim')

iris = datasets.load_iris()
# Break up the dataset into non-overlapping training (75%) and testing
# (25%) sets.
skf = StratifiedKFold(n_splits = 4)
# Only take the first fold.
train_index, test_index = next(iter(skf.split(iris.data, iris.target)))

X_train = iris.data[train_index]
y_train = iris.target[train_index]
X_test = iris.data[test_index]
```

```python
    y_test = iris.target[test_index]

    n_classes = len(np.unique(y_train))

    # Try GMMs using different types of covariances.
    estimators = {cov_type: GaussianMixture(n_components = n_classes,
            covariance_type = cov_type, max_iter = 20, random_state = 0)
            for cov_type in ['spherical', 'diag', 'tied', 'full']}
    n_estimators = len(estimators)

    plt.figure(figsize = (3 * n_estimators // 2, 6))
    plt.subplots_adjust(bottom = .01, top = 0.95, hspace = .15,
    wspace = .05,
    left = .01, right = .99)
    for index, (name, estimator) in enumerate(estimators.items()):
            # Since we have class labels for the training data, we can
            # initialize the GMM parameters in a supervised manner.
            estimator.means_init = np.array([X_train[y_train ==
            i].mean(axis = 0)
            for i in range(n_classes)])

            # Train the other parameters using the EM algorithm.
            estimator.fit(X_train)

            h = plt.subplot(2, n_estimators // 2, index + 1)
            make_ellipses(estimator, h)

            for n, color in enumerate(colors):
            data = iris.data[iris.target == n]
            plt.scatter(data[:, 0], data[:, 1], s = 0.8, color = color,
            label = iris.target_names[n])
            # Plot the test data with crosses
            for n, color in enumerate(colors):
            data = X_test[y_test == n]
            plt.scatter(data[:, 0], data[:, 1], marker = 'x', color = color)
            y_train_pred = estimator.predict(X_train)
            train_accuracy = np.mean(y_train_pred.ravel() == y_train.ravel()) * 100
            plt.text(0.05, 0.9, 'Train accuracy: %.1f' % train_accuracy,
            transform = h.transAxes)
            y_test_pred = estimator.predict(X_test)
            test_accuracy = np.mean(y_test_pred.ravel() == y_test.ravel()) * 100
            plt.text(0.05, 0.8, 'Test accuracy: %.1f' % test_accuracy,
             transform = h.transAxes)
            plt.xticks(())
            plt.yticks(())
            plt.title(name)
plt.legend(scatterpoints = 1, loc = 'lower right', prop = dict(size = 12))
plt.show()
```

7.9 贝叶斯聚类

GMM算法有个缺点，即过拟合。如果不能确定哪些属性之间互相相关，那么，为什么不安全地将所有的属性都认为是协变的呢？原因是参数越多，得到的结构在训练集上过拟合概率越大。协方差会显著增加参数的数量。在所有的机器学习问题中，过拟合问题都存在，概率聚类也不例外。发生过拟合的情况有两种：定义了过多的聚类和为分布定义了过多的参数。当每个数据点都是一个独立聚类时，就会出现聚类过多的极端情况。显然，此时数据将会被过度拟合。此外，当任何正态分布变得如此之小以至于聚类仅基于一个数据点时，基于EM的GMM模型就会出现问题。在实现时，通常在聚类中至少有两个不同值。当参数过多时，会出现过度拟合的问题。如果不确定哪些属性是协变的，可以尝试不同的可能性，并选择一种使整体数据可能性最大化的方案。但是，参数越多，结果出现的平均可能性就越大。这种结果的出现未必是因为模型找到了更好的聚类，也可能是因为过拟合。引入更多的参数可以更容易地找到看似较好的聚类。然而，我们也需要考虑对于引入新参数模型的惩罚。最近提出的完全贝叶斯层次聚类技术会对所有可能表示该数据集的层次结构生成一个概率分布作为输出。做到这一点的主要方法之一是遵循贝叶斯方法，其中每个参数都具有先验的概率分布。因此，无论何时添加新参数，将其先验概率集成到总体概率图中都是很重要的。由于这包括将总似然度乘以小于1（先前的似然度）的数字，因此可以将其视为惩罚引入新参数的方法之一。为了提高总体概率，只有当产生超过成本的收益时才进行更新。AutoClass是一种详尽的贝叶斯聚类策略，它利用了具有先验分布的有限混合模型的所有参数。它允许使用数值和标称属性，并使用EM算法估计概率分布参数，以便能够更好地拟合数据。由于不能保证EM算法将收敛到全局最优，该过程将会在几个不同的初始值上进行重复迭代。AutoClass考虑各种聚类数，并且可以考虑各种协方差量和数字属性概率基础分布的不同类型（Witten et al., 2016）。

例7.10 以下代码比较两种高斯混合模型（GMM）聚类算法。它绘制了由期望最大化（"GaussianMixture"）和变分推断（具有Dirichlet先验过程的"BayesianGaussianMixture"）产生的置信椭圆。本例使用合成数据，生成的数据中，聚类具有不同的密度。两种模型都可以使用五个成分来拟合数据。注意，期望最大化模型需要采用所有五个成分。另一方面，变分推理模型仅有效地使用了能够进行良好的拟合所需的数量。我们还可以看到，由于期望最大化会尝试拟合过多的成分，而会对成分进行随机划分。但是Dirichlet过程模型会自动调整状态数。请注意，此示例改编自scikit-learn。

```
===========================================
Comparison of Gaussian mixture models with EM and Bayesian
===========================================
```

```python
import itertools
import numpy as np
from scipy import linalg
import matplotlib.pyplot as plt
import matplotlib as mpl
from sklearn import mixture

color_iter = itertools.cycle(['navy', 'c', 'cornflowerblue', 'gold',
'    darkorange'])

def plot_results(X, Y_, means, covariances, index, title):
    splot = plt.subplot(2, 1, 1 + index)
    for i, (mean, covar, color) in enumerate(zip(
    means, covariances, color_iter)):
        v, w = linalg.eigh(covar)
        v = 2. * np.sqrt(2.) * np.sqrt(v)
        u = w[0] / linalg.norm(w[0])
        # as the DP will not use every component it has access to
        # unless it needs it, we shouldn't plot the redundant
        # components.
        if not np.any(Y_ == i):
        continue
        plt.scatter(X[Y_ == i, 0], X[Y_ == i, 1], .8, color = color)

        # Plot an ellipse to show the Gaussian component
        angle = np.arctan(u[1] / u[0])
        angle = 180. * angle / np.pi # convert to degrees
        ell = mpl.patches.Ellipse(mean, v[0], v[1], 180. + angle,
        color = color)
        ell.set_clip_box(splot.bbox)
        ell.set_alpha(0.5)
        splot.add_artist(ell)

        plt.xlim(-9., 5.)
        plt.ylim(-3., 6.)
        plt.xticks(())
        plt.yticks(())
        plt.title(title)

# Number of samples per component
n_samples = 500

# Generate random sample, two components
np.random.seed(0)
C = np.array([[0., -0.1], [1.7, .4]])
X = np.r_[np.dot(np.random.randn(n_samples, 2), C),
        .7 * np.random.randn(n_samples, 2) + np.array([-6, 3])]
# Fit a Gaussian mixture with EM using five components
gmm = mixture.GaussianMixture(n_components = 5, covariance_
type = 'full').fit(X)
```

```
plot_results(X, gmm.predict(X), gmm.means_, gmm.covariances_, 0,
        'Gaussian Mixture')
# Fit a Dirichlet process Gaussian mixture using five components
dpgmm = mixture.BayesianGaussianMixture(n_components = 5,
        covariance_type = 'full').fit(X)
plot_results(X, dpgmm.predict(X), dpgmm.means_, dpgmm.covariances_, 1,
        'Bayesian Gaussian Mixture with a Dirichlet process prior')
plt.show()
```

7.10 轮廓分析

那么,如何判断聚类算法的好坏呢?轮廓分析是一个有用的技术。对于每个被归进聚类的点 x,轮廓分析将会对其排序并且画出 $s(x)$。在这种特定情况下,我们使用基于平方欧几里得距离的聚类进行轮廓构造,该方法可以扩展到其他的距离度量上。此时可以清楚地看到第一个聚类比第二个聚类强。除了图形表示之外,还可以估计每个聚类以及整个数据集的平均轮廓值(Flach, 2012)。轮廓分析可以用于检查聚类之间的分离程度。轮廓图显示了聚类中每个点与相邻聚类中点的距离,从而提供了一种直观地确定参数(如聚类数)的方法。此度量的范围为 [-1,1]。轮廓系数接近 +1 表明样本距离邻近的聚类很远。值为 0 表示样本在两个相邻聚类之间的决策边界上或非常接近,而负值表示样本可能分配到错误的聚类。

例 7.11 以下代码使用了 k 均值聚类算法来寻找营销数据的轮廓,基于 scikit-learn 库的 API 实现。本示例使用市场营销数据查看聚类中心的轮廓。绘制了不同情况下的轮廓和聚类中心。请注意,此示例取自 scikit-learn。

在此示例中,轮廓分析用于选择"n_clusters"的最佳值。轮廓图显示,对于给定的数据,"n_clusters"值为 2、3 和 6 时是不好的选择,这是由于轮廓线得分低于平均水平的聚类和轮廓图大小的波动很大。在确定 4 到 5 之间时,轮廓分析更具矛盾性。另外,从轮廓图的厚度可以看出聚类大小。当"n_clusters"等于 2 时,第 0 个聚类的轮廓图的尺寸较大,这是因为将 3 个聚类分组为一个大的聚类。然而,从右侧的散点图可以看出,当"n_clusters"等于 4 或 5 时,所有图的厚度大约类似,因此这些聚类具有相似的大小。

```
===========================================================================
Selecting the number of clusters with silhouette analysis on k-means
clustering
===========================================================================
from sklearn.cluster import KMeans
from sklearn.metrics import silhouette_samples, silhouette_score
```

```python
import matplotlib.pyplot as plt
import matplotlib.cm as cm
import numpy as np
import pandas as pd
# Import the Mall Customers dataset by pandas
dataset = pd.read_csv('Mall_Customers.csv')
X = dataset.iloc[:, [3,4]].values
range_n_clusters = [2, 3, 4, 5, 6]
for n_clusters in range_n_clusters:
    # Create a subplot with 1 row and 2 columns
    fig, (ax1, ax2) = plt.subplots(1, 2)
    fig.set_size_inches(18, 7)
    # The 1st subplot is the silhouette plot
    # The silhouette coefficient can range from -1, 1 but in this example all
    # lie within [-0.1, 1]
    ax1.set_xlim([-0.1, 1])
    # The (n_clusters + 1)*10 is for inserting blank space between silhouette
    # plots of individual clusters, to demarcate them clearly.
    ax1.set_ylim([0, len(X) + (n_clusters + 1) * 10])

    # Initialize the clusterer with n_clusters value and a random generator
    # seed of 10 for reproducibility.
    clusterer = KMeans(n_clusters = n_clusters, random_state = 10)
    cluster_labels = clusterer.fit_predict(X)

    # The silhouette_score gives the average value for all the samples.
    # This gives a perspective into the density and separation of the formed
    # clusters
    silhouette_avg = silhouette_score(X, cluster_labels)
    print("For n_clusters =", n_clusters,
    "The average silhouette_score is :", silhouette_avg)

    # Compute the silhouette scores for each sample
    sample_silhouette_values = silhouette_samples(X, cluster_labels)
    y_lower = 10
    for i in range(n_clusters):
    # Aggregate the silhouette scores for samples belonging to
    # cluster i, and sort them
        ith_cluster_silhouette_values = \
        sample_silhouette_values[cluster_labels == i]
        ith_cluster_silhouette_values.sort()
        size_cluster_i = ith_cluster_silhouette_values.shape[0]
        y_upper = y_lower + size_cluster_i
        color = cm.nipy_spectral(float(i) / n_clusters)
        ax1.fill_betweenx(np.arange(y_lower, y_upper),
```

```python
                0, ith_cluster_silhouette_values,
                facecolor = color, edgecolor = color, alpha = 0.7)

    # Label the silhouette plots with their cluster numbers at the middle
    ax1.text(-0.05, y_lower + 0.5 * size_cluster_i, str(i))
    # Compute the new y_lower for next plot
    y_lower = y_upper + 10 # 10 for the 0 samples
    ax1.set_title("The silhouette plot for the various clusters.")
    ax1.set_xlabel("The silhouette coefficient values")
    ax1.set_ylabel("Cluster label")

    # The vertical line for average silhouette score of all the values
    ax1.axvline(x = silhouette_avg, color = "red", linestyle = "--")

    ax1.set_yticks([]) # Clear the yaxis labels / ticks
    ax1.set_xticks([-0.1, 0, 0.2, 0.4, 0.6, 0.8, 1])

    # 2nd Plot showing the actual clusters formed
    colors = cm.nipy_spectral(cluster_labels.astype(float) / n_clusters)
    ax2.scatter(X[:, 0], X[:, 1], marker = '.', s = 30, lw = 0, alpha = 0.7,
    c = colors, edgecolor = 'k')
    # Labeling the clusters
    centers = clusterer.cluster_centers_
    # Draw white circles at cluster centers
    ax2.scatter(centers[:, 0], centers[:, 1], marker = 'o',
    c = "white", alpha = 1, s = 200, edgecolor = 'k')
    for i, c in enumerate(centers):
        ax2.scatter(c[0], c[1], marker = '$%d$' % i, alpha = 1,
    s = 50, edgecolor = 'k')

    ax2.set_title("The visualization of the clustered data.")
    ax2.set_xlabel("Feature space for the 1st feature")
    ax2.set_ylabel("Feature space for the 2nd feature")

     plt.suptitle(("Silhouette analysis for KMeans clustering on sample
       data "
    "with n_clusters = %d" % n_clusters),
    fontsize = 14, fontweight = 'bold')
plt.show()
```

7.11 基于聚类的图像分割

图像是已知的最有效的信息传递方式之一。机器学习的一个关键方面是理解图像并从中提取信息,以将其应用于不同的任务中。例如,将图像用于机器人导航。其他诸如从人体扫描中提取恶性组织等应用,也是医学诊断不可或缺的一部分。识别图像的第一步就是对它们进行分割并在其中找到各种对象。直方图和频域变换都是可以实现这一目的的特征（Tatiraju & Mehta, 2008）。

在图像识别和计算机视觉中，图像分割是重要的预处理过程。图像分割，即对具有相同属性的图像在多个不重叠的相关区域中进行分解。图像分割是数字图像处理中的一项关键技术，分割的准确性直接影响后续任务的有效性。考虑到图像分割问题的复杂性和难度，当前的分割技术已经在不同程度上取得了成功。然而，这一方面的研究仍然面临许多问题。聚类分析算法根据一定的标准将数据集中的数据进行分组，因此在图像分割中具有广泛的应用。图像分割作为主要的数字图像处理技术之一，与相关的专业领域知识相融合并应用于实际生活中，是一个值得深入探索的方向。目前，其通常用于机器视觉、面部识别、指纹识别、交通控制系统、卫星图像跟踪对象（道路、树林等）、行人检测、医学影像以及许多其他领域（Zheng, Lei, Yao, Gong, & Yin, 2018）。

由于图像分割在图像处理的许多应用中起着至关重要的作用，为了解决这个问题，最近几十年已经开发了几种图像分割算法。然而，由于图像分割问题极具挑战性并且对后续流程影响极大，为了能够进一步提高图像分割的质量，这些算法一直在被不断改进。虽然聚类算法最初并不是为图像处理而开发的，但是计算机视觉社区将其应用于图像分割问题。例如，在应用需要预先知道聚类数（k）的 k 均值算法时，图像的每个像素都会被重复迭代地分配给离质心最接近的聚类。基于分配给该聚类的像素，算法将确定每个聚类的质心。聚类中像素的成员隶属度基于计算距离进行选择。通常，因为易于计算，选取欧氏距离作为距离指标。但值得注意的是，选取欧几里得距离可能会导致图像的最终分割中出现误差（Gaura, Sojka, & Krumnikl, 2011）。

例 7.12 以下 Python 代码用于将希腊硬币的图像分割成多个区域，该实现基于 scikit-learn 库 API。在此示例中，使用了在 skimage.data 中的硬币数据集。本示例利用从图像上体素之间的差异创建的图形上的 "spectral_clustering" 将图像划分为多个部分均匀的区域。图像上的光谱聚类过程是找到归一化图割的有效近似解。有两个分配标签的选项：

- "K 均值"光谱聚类应用了 k 均值聚类算法，将采样在嵌入空间中进行聚类
- "Discrete"将会在嵌入空间中迭代搜索最近的分割空间

注意，该例改编自 scikit-learn。

```
# =====================================================================
# Image segmentation with clustering
# =====================================================================
# Author: Gael Varoquaux <gael.varoquaux@normalesup.org> , Brian Cheung
# License: BSD 3 clause
import time
import numpy as np
```

```python
from distutils.version import LooseVersion
from scipy.ndimage.filters import gaussian_filter
import matplotlib.pyplot as plt
import skimage
from skimage.data import coins
from skimage.transform import rescale
from sklearn.feature_extraction import image
from sklearn.cluster import spectral_clustering

# these were introduced in skimage-0.14
if LooseVersion(skimage.__version__) >= '0.14':
        rescale_params = {'anti_aliasing': False, 'multichannel': False}
else:
        rescale_params = {}
# load the coins as a numpy array
orig_coins = coins()

# Resize it to 20% of the original size to speed up the processing
# Applying a Gaussian filter for smoothing prior to down-scaling
# reduces aliasing artifacts.
smoothened_coins = gaussian_filter(orig_coins, sigma = 2)
rescaled_coins = rescale(smoothened_coins, 0.2, mode = "reflect",
        **rescale_params)

# Convert the image into a graph with the value of the gradient on the
# edges.
graph = image.img_to_graph(rescaled_coins)

# Take a decreasing function of the gradient: an exponential
# The smaller beta is, the more independent the segmentation is of the
# actual image. For beta = 1, the segmentation is close to a voronoi
beta = 10
eps = 1e-6
graph.data = np.exp(-beta * graph.data / graph.data.std()) + eps

# Apply spectral clustering (this step goes much faster if you have pyamg
# installed)
N_REGIONS = 25

#################################################################
# Visualize the resulting regions

for assign_labels in ('kmeans', 'discretize'):
    t0 = time.time()
    labels = spectral_clustering(graph, n_clusters = N_REGIONS,
        assign_labels = assign_labels, random_state = 42)
    t1 = time.time()
    labels = labels.reshape(rescaled_coins.shape)
```

```
        plt.figure(figsize = (5, 5))
        plt.imshow(rescaled_coins, cmap = plt.cm.gray)
        for l in range(N_REGIONS):
        plt.contour(labels == l,
         colors = [plt.cm.nipy_spectral(l / float(N_REGIONS))])
        plt.xticks(())
        plt.yticks(())
        title = 'Spectral clustering: %s, %.2fs' % (assign_labels, (t1 - t0))
        print(title)
        plt.title(title)
plt.show()
```

7.12 基于聚类的特征提取

k 均值聚类通过为每个聚类或每组中的数据点找到合适的代表或质心来降低数据维度。这样，每个聚类中的所有元素均可以该元素所在的聚类的质心为特征。因此，聚类的问题可以看作将数据划分为具有相似特征的聚类，并且在使用 k 均值算法时，该特征在特征空间中具有几何紧密性。当这些特征可以被清楚表示时，我们可以创建一个机器学习问题，来准确恢复聚类质心，从而忽略一些不适用的表示。假设我们使用 c_k 表示第 k 个聚类的质心，S_k 表示这 P 个数据的索引集，并且 $x_1 \cdots x_p$ 属于该聚类，那么对于所有 $k=1 \cdots K$ 而言，所有在第 k 个聚类中的数据点必须与其质心相近。我们可以通过逐列堆叠获得质心矩阵来更好地表示这些必要关系：

$$C = c_1 c_2 \cdot c_K \qquad (7.1)$$

若使用 e_k 表示第 k 个标准基向量（即在 $K \times 1$ 的向量，只有第 k 个元素为 1，其余元素为 0 的向量），可以得到 $C_{ek} = c_k$。因此对于每一个 k 值，式（7.1）可以表示为：

$$C_{ek} \approx x_p, p \in S_k \qquad (7.2)$$

接下来，为了更好地表达这些关系，我们将数据按列堆叠到数据矩阵 $X = x_1 x_2 \cdots x_p$ 中，并生成一个 $K \times P$ 的分配矩阵 W。该矩阵的第 p 列，即 w_p，是第 p 个点所属的聚类有关的标准基础向量，若 $p \in S_k$，则 $w_p = e_k$。当使用 w_p 进行表示时，对 $p \in S_k$，我们将式（7.2）中的每个方程表示为 $Cw_p \approx x_p$，或者同时使用矩阵符号表示所有 K，即

$$CW \approx X \qquad (7.3)$$

我们可以忘记已知质心矩阵 C 和赋值矩阵 W 的准确描述的假设，即知道聚类质心的位置，并且知道为它们分配了哪些点。我们想学习这两个矩阵的正确值。特别地，期望得到满足式（7.3）所描述的紧凑关系的 C 和 W 的理想值。即 $CW \approx X$，或换句话说在 $CW-X2F$ 取值小，而 W 是由与数据点与它们各自质心的相关标准基本向量所组成的。注意，该目标是非凸函数，由于不能同时最小化 C 和 W，我们需要使用交替最小化来解决。即保持另一个变量不变的情况下交替固定 C 或 W，对其中一个变量最小化目标函数的方法（Watt,

Borhani, & Katsaggelos, 2016）。

例 7.13 以下 Python 代码介绍了 k 均值和 GMM 聚类算法作为特征提取器的用法。我们将利用鸢尾花（Iris）数据集，该数据包括三种类型（类）的鸢尾花（Setosa、Versicolour 和 Virginica），具有四个属性：萼片长度、萼片宽度、花瓣长度和花瓣宽度。在此示例中，我们利用 sklearn.cluster.KMeans 和 sklearn.mixture.GaussianMixture 提取鸢尾花据集的特征。在 scikit-learn 中，k 均值和 GMM 被实现为 sklearn.cluster.KMeans 和 sklearn.mixture.GaussianMixture 聚类器对象，并用于特征提取。注意，本例改编自 Python–scikit-learn。

```python
# ====================================================================
# Feature extraction with k-means and GMM clustering
# ====================================================================
"""
Created on Mon Dec 23 11:35:28 2019
@author: absubasi
"""
from sklearn import metrics
from sklearn.metrics import confusion_matrix
import seaborn as sns
import matplotlib.pyplot as plt
from io import BytesIO #needed for plot
# ====================================================================
# Define utility functions
# ====================================================================
def print_confusion_matrix_and_save(y_test, y_pred):
    #Print the Confusion Matrix
    matrix = confusion_matrix(y_test, y_pred)
    plt.figure(figsize = (6, 4))
    sns.heatmap(matrix, square = True, annot = True, fmt = 'd',
    cbar = False)
    plt.title('Confusion Matrix')
    plt.ylabel('True Label')
    plt.xlabel('Predicted Label')
    plt.show()
    #Save The Confusion Matrix
    plt.savefig("Confusion.jpg")
    # Save SVG in a fake file object.
    f = BytesIO()
    plt.savefig(f, format = "svg")

def Performance_Metrics(y_test,y_pred):
    print('Test Accuracy:', np.round(metrics.accuracy_score(y_test,y_pred),4))
    print('Precision:', np.round(metrics.precision_score(y_test,y_pred,average = 'weighted'),4))
    print('Recall:', np.round(metrics.recall_score(y_test,y_pred,average = 'weighted'),4))
```

```python
print('F1 Score:', np.round(metrics.f1_score(y_test,y_pred,
average = 'weighted'),4))
print('Cohen Kappa Score:', np.round(metrics.cohen_kappa_score(y_
test,y_pred),4))
print('Matthews Corrcoef:', np.round(metrics.matthews_corrcoef(y_
test,y_ pred),4))
print('\t\tClassification Report:\n', metrics.classification_report(y_
test,y_pred))
print("Confusion Matrix:\n",confusion_matrix(y_test,y_pred))

# ======================================================================
# Random forest classifier with k-means for feature extraction
# ======================================================================
#load Data
from sklearn.datasets import load_iris
import numpy as np
iris = load_iris()
X = iris['data']
y = iris['target']
#Extract Features
from sklearn.cluster import KMeans
kmeans = KMeans(n_clusters = 6).fit(X)
distances = np.column_stack([np.sum((X - center)**2, axis = 1)**0.5 for
center in kmeans.cluster_centers_])

from sklearn.model_selection import train_test_split
from sklearn.ensemble import RandomForestClassifier
# Split dataset into training set and test set
# 70% training and 30% test
Xtrain, Xtest, ytrain, ytest = train_test_split(distances, y,test_
size = 0.3, random_state = 0)
#In order to change to accuracy increase n_estimators
#Classify Data
" " "RandomForestClassifier(n_estimators = 'warn', criterion = 'gini',
max_depth = None,
min_samples_split = 2, min_samples_leaf = 1, min_weight_fraction_
leaf = 0.0,
max_features = 'auto', max_leaf_nodes = None, min_impurity_
decrease = 0.0,
min_impurity_split = None, bootstrap = True, oob_score = False, n_
jobs = None,
random_state = None, verbose = 0, warm_start = False, class_
weight = None)" " "
clf = RandomForestClassifier(n_estimators = 200)
#Create the Model
#Train the model with Training Dataset
clf.fit(Xtrain,ytrain)
#Test the model with Testset
ypred = clf.predict(Xtest)
```

```python
#Evaluate the Model and Print Performance Metrics
Performance_Metrics(ytest,ypred)
print_confusion_matrix_and_save(ytest, ypred)
#%%
# =======================================================================
# Random forest classifier with GMM for feature extraction
# =======================================================================
#load Data
from sklearn.datasets import load_iris
import numpy as np
iris = load_iris()
X = iris['data']
y = iris['target']
#Extract Features
from sklearn.mixture import GaussianMixture
gmm = GaussianMixture(n_components = 8).fit(X)
proba = gmm.predict_proba(X)

from sklearn.model_selection import train_test_split
from sklearn.ensemble import RandomForestClassifier
# Split dataset into training set and test set
# 70% training and 30% test
Xtrain, Xtest, ytrain, ytest = train_test_split(proba, y,test_size = 0.3, random_state = 0)
#In order to change to accuracy increase n_estimators
" " "RandomForestClassifier(n_estimators = 'warn', criterion = 'gini', max_depth = None,
min_samples_split = 2, min_samples_leaf = 1, min_weight_fraction_leaf = 0.0,
max_features = 'auto', max_leaf_nodes = None, min_impurity_decrease = 0.0,
min_impurity_split = None, bootstrap = True, oob_score = False, n_jobs = None,
random_state = None, verbose = 0, warm_start = False, class_weight = None)" " "
clf = RandomForestClassifier(n_estimators = 200)
#Create the Model
#Train the model with Training Dataset
clf.fit(Xtrain,ytrain)
#Test the model with Testset
ypred = clf.predict(Xtest)

#Evaluate the Model and Print Performance Metrics
Performance_Metrics(ytest,ypred)
print_confusion_matrix_and_save(ytest, ypred)
#%%
# =======================================================================
# k-NN classifier with k-means for feature extraction
# =======================================================================
```

```python
#load Data
from sklearn.datasets import load_iris
import numpy as np
iris = load_iris()
X = iris['data']
y = iris['target']
#Extract Features
from sklearn.cluster import KMeans
kmeans = KMeans(n_clusters = 6).fit(X)
distances = np.column_stack([np.sum((X - center)**2, axis = 1)**0.5 for center in kmeans.cluster_centers_])

from sklearn.model_selection import train_test_split
# Split dataset into training set and test set
# 70% training and 30% test
Xtrain, Xtest, ytrain, ytest = train_test_split(distances, y,test_size = 0.3, random_state = 0)
from sklearn.neighbors import KNeighborsClassifier
clf = KNeighborsClassifier(n_neighbors = 1)
#Create the Model
#Train the model with Training Dataset
clf.fit(Xtrain,ytrain)
#Test the model with Testset
ypred = clf.predict(Xtest)

#Evaluate the Model and Print Performance Metrics
Performance_Metrics(ytest,ypred)
print_confusion_matrix_and_save(ytest, ypred)

 #%%
# =====================================================================
# k-NN classifier with GMM for feature extraction
# =====================================================================
#load Data
from sklearn.datasets import load_iris
import numpy as np
iris = load_iris()
X = iris['data']
y = iris['target']
#Extract Features
from sklearn.mixture import GaussianMixture
gmm = GaussianMixture(n_components = 4).fit(X)
proba = gmm.predict_proba(X)

from sklearn.model_selection import train_test_split
# Split dataset into training set and test set
# 70% training and 30% test
Xtrain, Xtest, ytrain, ytest = train_test_split(proba, y,test_size = 0.3, random_state = 0)
```

```python
from sklearn.neighbors import KNeighborsClassifier
clf = KNeighborsClassifier(n_neighbors = 1)
#Create the Model
#Train the model with Training Dataset
clf.fit(Xtrain,ytrain)
#Test the model with Testset
ypred = clf.predict(Xtest)

#Evaluate the Model and Print Performance Metrics
Performance_Metrics(ytest,ypred)
print_confusion_matrix_and_save(ytest, ypred)

#%%
# =======================================================================
# MLP classifier with k-means for feature extraction
# =======================================================================
#load Data
from sklearn.datasets import load_iris
import numpy as np
iris = load_iris()
X = iris['data']
y = iris['target']
#Extract Features
from sklearn.cluster import KMeans
kmeans = KMeans(n_clusters = 6).fit(X)
distances = np.column_stack([np.sum((X - center)**2, axis = 1)**0.5 for
center in kmeans.cluster_centers_])

from sklearn.model_selection import train_test_split
# Split dataset into training set and test set
# 70% training and 30% test
Xtrain, Xtest, ytrain, ytest = train_test_split(distances, y,test_
size = 0.3, random_state = 0)
from sklearn.neural_network import MLPClassifier
clf = MLPClassifier(hidden_layer_sizes = (100, ), learning_rate_
init = 0.001,
alpha = 1, momentum = 0.9,max_iter = 1000)
#Create the Model
#Train the model with Training Dataset
clf.fit(Xtrain,ytrain)
#Test the model with Testset
ypred = clf.predict(Xtest)
#Evaluate the Model and Print Performance Metrics
Performance_Metrics(ytest,ypred)
print_confusion_matrix_and_save(ytest, ypred)
#%%
# =======================================================================
# MLP classifier with GMM for feature extraction
# =======================================================================
#load Data
```

```
from sklearn.datasets import load_iris
import numpy as np
iris = load_iris()
X = iris['data']
y = iris['target']
#Extract Features
from sklearn.mixture import GaussianMixture
gmm = GaussianMixture(n_components = 8).fit(X)
proba = gmm.predict_proba(X)

from sklearn.model_selection import train_test_split
# Split dataset into training set and test set
# 70% training and 30% test
Xtrain, Xtest, ytrain, ytest = train_test_split(proba, y,test_
size = 0.3, random_state = 0)
from sklearn.neural_network import MLPClassifier
clf = MLPClassifier(hidden_layer_sizes = (100, ), learning_rate_
init = 0.001,
alpha = 1, momentum = 0.9,max_iter = 1000)
#Create the Model
#Train the model with Training Dataset
clf.fit(Xtrain,ytrain)
#Test the model with Testset
ypred = clf.predict(Xtest)

#Evaluate the Model and Print Performance Metrics
Performance_Metrics(ytest,ypred)
print_confusion_matrix_and_save(ytest, ypred)
```

7.13 基于聚类的分类

如何对于没有标签的数据集进行分类？可以利用朴素贝叶斯方法，使用EM迭代聚类算法从一个小的、有标签的数据集中学习类，然后将其扩展到一个大的、没有标签的数据集中。因此，第一步是使用有标签的数据来训练分类器。第二步将其应用于没有标签的数据以进行类别概率标记（"期望"步骤）。第三步使用所有数据标签来训练新的分类器（"最大化"步骤）。最后一步，迭代直到收敛。EM方法保证了在每次迭代中找到的模型参数都具有相同或是比前一次更高的似然度。我们可以凭经验回答一个关键问题，即这些较高概率的参数是否可以改善分类性能。凭直觉而言是可以的。EM方法使用它们来概括学习的模型，以便使用未出现在标记数据集中的数据。EM迭代地推广模型以正确分类数据。这可以与任何用于分类和迭代聚类的算法一起使用。但这本质上是一种自我引导技术，需要注意确保提供正反馈。相较于使用确定的决策边界，使用概率具有更大的优势，因为它可以帮助计算过程缓慢收敛，而不是直接得出错误的结论。标准概率EM技术和朴素贝叶斯可以互相代替，因为它们都具有相同的基本假设，即属性之间的独立性。更具体地说，类属性之间的条件独立性。然而，通过将朴素贝叶斯和 EM 结合起来的方法可以很好地用于文档分类。使用少于三分之一的带标签的训练实例和五倍的未标记训练实例，它可以达到传统学习器在特定分类任务中的表现。如果获取数据标签的成本很高，但得到无标签的实例基本上是免费的时候，结合 EM 和朴素贝叶斯是一个很好的权衡：使用少量带标签的文档，通过添加其他无标签的文档，可以显著提高分类的精确度（Witten et al., 2016）。

对于当前的方法，有两种可以提高性能的改进。首先是根据实验得到的启发表明，当具有标签的数据较多时，包含更多的无标签数据会降低性能，而非提高性能。从本质上说，人工标记的标签相较于自动标记的标签应该包含更少的噪声。补救措施是添加一个加权参数，用来减少未标记数据的贡献。此时，通过最大化标记和未标记实例的加权概率来实现最终的分类，我们可以将其集成到 EM 的最大化阶段。第二个改进是允许每个类包含多个聚类。EM 聚类算法假设每类的数据是由许多混合概率分布随机产生的。最初，每个带标签的文档组成部分都以概率的方式随机分配给一个类中的多个聚类。此时，EM 算法的最大化步骤保持不变，但是期望步骤进行了调整，使其不仅可以用类对每个实例进行概率标记，还可以将标签分配给类中的组成部分（Witten et al., 2016）。

例 7.14 以下 Python 代码基于 scikit-learn 库的 API 实现了 k 均值算法对手写数字进行分类。本例中的手写数字数据集来自 sklearn.datasets。我们对分类的准确率、精确度、召回率、F1 得分、Cohen kappa 得分和 Matthews 相关系数进行了计算。本例给出了分类报告以及混淆矩阵。注意，本例改编自 scikit-learn。

```
# ====================================================================
# Clustering as a classifier
# ====================================================================
#k-means on digits
import seaborn as sns
import matplotlib.pyplot as plt
from io import BytesIO #needed for plot
from sklearn.metrics import confusion_matrix
from sklearn import metrics
import numpy as np
# ====================================================================
# Define utility functions
# ====================================================================
def print_confusion_matrix_and_save(y_test, y_pred):
#Print the Confusion Matrix
matrix = confusion_matrix(y_test, y_pred)
plt.figure(figsize = (6, 4))
sns.heatmap(matrix, square = True, annot = True, fmt = 'd',
cbar = False)
plt.title('Confusion Matrix')
plt.ylabel('True Label')
plt.xlabel('Predicted Label')
```

```python
plt.show()
#Save The Confusion Matrix
plt.savefig("Confusion.jpg")
# Save SVG in a fake file object.
f = BytesIO()
plt.savefig(f, format = "svg")

def Performance_Metrics(y_test,y_pred):
 print('Test Accuracy:', np.round(metrics.accuracy_score(y_test,y_pred),4))
 print('Precision:', np.round(metrics.precision_score(y_test,y_pred,average = 'weighted'),4))
 print('Recall:', np.round(metrics.recall_score(y_test,y_pred, average = 'weighted'),4))
 print('F1 Score:', np.round(metrics.f1_score(y_test,y_pred, average = 'weighted'),4))
 print('Cohen Kappa Score:', np.round(metrics.cohen_kappa_score(y_test,y_pred),4))
 print('Matthews Corrcoef:', np.round(metrics.matthews_corrcoef(y_test,y_pred),4))
 print('\t\tClassification Report:\n', metrics.classification_report(y_test,y_pred))
 print("Confusion Matrix:\n",confusion_matrix(y_test,y_pred))
# ====================================================================
from sklearn.cluster import KMeans
from sklearn.datasets import load_digits
digits = load_digits()
digits.data.shape

kmeans = KMeans(n_clusters = 10, random_state = 0)
clusters = kmeans.fit_predict(digits.data)
kmeans.cluster_centers_.shape

fig, ax = plt.subplots(2, 5, figsize = (8, 3))
centers = kmeans.cluster_centers_.reshape(10, 8, 8)
for axi, center in zip(ax.flat, centers):
 axi.set(xticks = [], yticks = [])
 axi.imshow(center, interpolation = 'nearest', cmap = plt.cm.binary)

#%%
from scipy.stats import mode
labels = np.zeros_like(clusters)
for i in range(10):
 mask = (clusters == i)
 labels[mask] = mode(digits.target[mask])[0]

# Evaluate and Print the Performance Metrics
Performance_Metrics(digits.target, labels)
#Print and Save Confusion Matrix
print_confusion_matrix_and_save(digits.target, labels)
```

```
#%%
from sklearn.manifold import TSNE

# Project the data: this step will take several seconds
tsne = TSNE(n_components = 2, init = 'random', random_state = 0)
digits_proj = tsne.fit_transform(digits.data)

# Compute the clusters
kmeans = KMeans(n_clusters = 10, random_state = 0)
clusters = kmeans.fit_predict(digits_proj)

# Permute the labels
labels = np.zeros_like(clusters)
for i in range(10):
    mask = (clusters == i)
    labels[mask] = mode(digits.target[mask])[0]
# Evaluate and Print the Perfromance Metrics
Performance_Metrics(digits.target, labels)
#Print and Save Confusion Matrix
print_confusion_matrix_and_save(digits.target, labels)
```

7.14 本章小结

本章介绍了许多与聚类问题相关的例子，包括了无监督学习的技术。因为在机器学习领域中，聚类问题是一类重要的任务，并且在实际中有较多应用，所以本章专门学习和解释这个问题。聚类是自动将对象进行分组，使得相似的对象在一类，不同的对象在不同组的过程。例如，零售商根据顾客画像进行聚类，从而进行精准营销；计算生物学家根据基因表达的相似度进行聚类，从而进行基因多样性的研究；天文学家根据天体的距离进行聚类。本章之前主要集中讨论了监督学习问题。本章中，我们主要介绍了许多针对聚类问题的无监督学习算法。除了使用聚类方法对无标签数据进行分组外，它们还可以被用于图像分割、特征提取以及分类问题。本章中还详细地讨论了这些算法在不同领域中的应用。

7.15 参考文献

Ankerst, M., Breunig, M. M., Kriegel, H. -P., & Sander, J. (1999). OPTICS: ordering points to identify the clustering structure. *ACM, 28*, 49–60.

Flach, P. (2012). *Machine learning: The art and science of algorithms that make sense of data*. Cambridge, United Kingdom: Cambridge University Press.

Gaura, J., Sojka, E., & Krumnikl, M. (2011). *Image segmentation based on k-means clustering and energy-transfer proximity*. Berlin: Springer 567–577.

Murphy, K. P. (2012). *Machine learning: A probabilistic perspective*. Cambridge, MA: MIT press.

Shalev-Shwartz, S., & Ben-David, S. (2014). *Understanding machine learning: From theory to algorithms*. Cambridge, United Kingdom: Cambridge University Press.

Steinbach, M., Karypis, G., & Kumar, V. (2000). A comparison of document clustering techniques. *KDD Workshop on Text Mining, 400*, 525–526 Boston.

Sumathi, S., & Paneerselvam, S. (2010). *Computational intelligence paradigms: Theory & applications using MATLAB.* Boca Raton, FL: CRC Press.

Tatiraju, S., & Mehta, A. (2008). Image Segmentation using k-means clustering, EM and Normalized Cuts. *Department of EECS, 1,* 1–7.

Theodoridis, S., Pikrakis, A., Koutroumbas, K., & Cavouras, D. (2010). *Introduction to pattern recognition: A matlab approach.* Cambridge, MA: Academic Press.

Watt, J., Borhani, R., & Katsaggelos, A. K. (2016). *Machine learning refined: Foundations, algorithms, and applications.* Cambridge, United Kingdom: Cambridge University Press.

Witten, I. H., Frank, E., Hall, M. A., & Pal, C. J. (2016). *Data Mining: Practical machine learning tools and techniques.* Burlington, MA: Morgan Kaufmann.

Zheng, X., Lei, Q., Yao, R., Gong, Y., & Yin, Q. (2018). Image segmentation based on adaptive K-means algorithm. *EURASIP Journal on Image and Video Processing, 1,* 68.

推荐阅读

神经网络与深度学习

作者：邱锡鹏　ISBN：978-7-111-64968-7　定价：149.00元

深度学习进阶：卷积神经网络和对象检测

作者：Umberto Michelucci　ISBN：978-7-111-66092-7　定价：79.00元

TensorFlow 2.0神经网络实践

作者：Paolo Galeone　ISBN：978-7-111-65927-3　定价：89.00元

深度学习：基于案例理解深度神经网络

作者：Umberto Michelucci　ISBN：978-7-111-63710-3　定价：89.00元